11th International Seminar on Ocean, Coastal Engineering, Environmental and Natural Disaster Management 2023

IOP Conference Series: Earth and Environmental Science
Volume 1298

Surabaya, Indonesia
27 September 2023

Editors:

S. T. Suntoyo
S. T. Silvianita
Dendy Satrio
Nguyen T. Viet

ISBN: 978-1-7138-9123-9
ISSN: 1755-1307

Printed from e-media with permission by:

Curran Associates, Inc.
57 Morehouse Lane
Red Hook, NY 12571

Institute of Physics
Dirac House, Temple Back
Bristol BS1 6BE UK

Phone: 44 1 17 929 7481
Fax: 44 1 17 920 0979

techtracking@iop.org

Additional copies of this publication are available from:

Curran Associates, Inc.
57 Morehouse Lane
Red Hook, NY 12571 USA
Phone: 845-758-0400
Fax: 845-758-2633
Email: curran@proceedings.com
Web: www.proceedings.com

TABLE OF CONTENTS

IOP Conf. Series: Earth and Environmental Science 1298 (2024) 011001 doi:10.1088/1755-1315/1298/1/011001

ISOCEEN 2023

The 11th International Seminar on Ocean and Coastal Engineering, Environmental and Natural Disaster Management

CONFERENCE PROCEEDING OF ISOCEEN 2023

Development of Ocean Science and Technology for Sustainable Blue Economy

The 11th International Seminar on Ocean and Coastal Engineering, Environmental and Natural Disaster Management
Institut Teknologi Sepuluh Nopember
Surabaya, 27th September 2023

IOP Conf. Series: Earth and Environmental Science 1298 (2024) 011001 doi:10.1088/1755-1315/1298/1/011001

Preface

It is a great honor and pleasure to present the proceeding of the 11th International Seminar on Ocean and Coastal Engineering, Environmental and Natural Disaster Management (ISOCEEN) and the agreement of collaboration between Institut Teknologi Sepuluh Nopember and Thuyloi University Vietnam. The signing of Memorandum of Understanding represents a commitment to strengthen academic ties, research collaborations, and cultural exchanges between Institut Teknologi Sepuluh Nopember and Thuyloi University Vietnam.

Started by the second vice-chancellor of Institut Teknologi Sepuluh Nopember, Ir. Mas Agus Mardyanto, M.E., Ph.D., the event brings the theme Development of Ocean Science and Technology for Sustainable Blue Economy. Our coastal, ocean, and offshore activities and products must adopt more innovative thoughts and efforts to ensure that we achieve sustainable development goals. Through this seminar, engineers and scientists whose activities are related to research and solving problems in the coastal, marine, maritime, hydrodynamics, oil, and gas are expected to develop science and technology in Coastal, Ocean, and Offshore Engineering in an integrated manner. Different perspectives are addressed by keynote speakers:

- Prof. Dr.Eng. Hitoshi Tanaka from Tohoku University, Japan,
- Ir. Suharyanto, M.Sc., Director of Marine Spatial Planning – KKP
- Prof. Dr. Eng. Nguyen Trung Viet, Thuyloi University, Vietnam

With contributions from over 100 participants and over 35 accepted papers, the seminar was successfully conducted in Swissbel Hotel Manyar, Surabaya, Indonesia, on September 27th, 2023. Selected papers presented in ISOCEEN 2023 will be published in The IOP Conference Series: Earth and Environmental Science (ESS) indexed in SCOPUS.

On behalf of the organizing committee of ISOCEEN 2023, I would like to warmly thank all the authors who, with their presentations, generously contributed to the lively exchange of scientific information that is so vital to the endurance of scientific conferences of this kind. Special thanks to all committees for their efforts in preparing the manuscripts and managing the sessions, respectively. Also, all lecturers within ITS, Contributing Authors, Seminar delegates, experts, scholars, and all students participate in ISOCEEN 2023.

Best regards,

Prof. Suntoyo, S.T, M.Eng., Ph.D.

Chairman

IOP Conf. Series: Earth and Environmental Science 1298 (2024) 011001 doi:10.1088/1755-1315/1298/1/011001

Editors

1. Prof. Suntoyo S.T., M.Eng., Ph.D.
   Department of Ocean Engineering
   Institut Teknologi Sepuluh Nopember
   suntoyo@oe.its.ac.id

2. Silvianita, S.T., M.Sc., Ph.D.
   Department of Ocean Engineering
   Institut Teknologi Sepuluh Nopember
   silvianita@oe.its.ac.id

3. Dr. Dendy Satrio, S.ST.
   Department of Ocean Engineering
   Institut Teknologi Sepuluh Nopember
   dendy.satrio@its.ac.id

4. Prof. Nguyen Trung Viet (Guest Editor)
   Thuyloi University
   nguyentrungviet@tlu.edu.vn

IOP Conf. Series: Earth and Environmental Science 1298 (2024) 011001 doi:10.1088/1755-1315/1298/1/011001

Committee

1. Advisory Committee:
   - Prof. Dr. Ir. Mochamad Ashari, M.Eng.
     Rector of Institut Teknologi Sepuluh Nopember, Indonesia
   - Dr. Eng. Trika Pitana, S.T., M.Sc.
     Dean of Marine Technology Faculty – ITS, Indonesia
   - Herman Pratikno, S.T., M.T., Ph.D.
     Head of Ocean Engineering Department, FTK-ITS, Indonesia

2. Organizing Committee
   - **Prof. Suntoyo, Ph.D. (Chairman)**
   - Silvianita, Ph.D. (Secretary)
   - Dr. Dendy Satrio (Paper and Proceeding Chair)
   - Wimala L. Dhanista, M.T. (Finance Chair)
   - Dr. Wahyudi (ITS, Indonesia)
   - Yoyok Setyo Hadiwidodo, Ph.D. (ITS, Indonesia)
   - Dr. Hasan Ikhwani (ITS, Indonesia)
   - Dr. Eng. Rudi Walujo Prastianto (ITS, Indonesia)
   - Dr. Eng. Muhammad Zikra (ITS, Indonesia)
   - Dr. Eng. Shade Rahmawati (ITS, Indonesia)
   - Sujantoko, M.T. (ITS, Indonesia)

3. International Scientific Committee
   - Prof. Hitoshi Tanaka, Dr. Eng. (Tohoku University, Japan)
   - Dr. Ir. Irwandi Idris, MSc. (Chairman, Indonesian Coastal Management Expert Association)
   - Prof. Widi Agoes Pratikto, PhD (ITS, Indonesia)
   - Prof. Daniel M. Rosyid, PhD (ITS, Indonesia)
   - Prof. Eko Budi Djatmiko, PhD (ITS, Indonesia)
   - Prof. Mukhtasor, Ph.D. (ITS, Indonesia)
   - Prof. Suntoyo, Ph.D (ITS, Indonesia)
   - Assoc. Prof. Dr. Alvaro Semedo (IHE-Delft, Netherlands)
   - Makoto Umeda, Dr.Eng (Tohoku University, Japan)
   - Dr. Joost Stronkhorst (HZ University of Applied Sciences, Netherlands)
   - Ir. Liliane Geerling R.T.D. (HZ Universityof Applied Sciences, Netherlands)
   - Assoc. Prof. Ahmad Sana, Ph.D. (Sultan Qaboos University, Oman)
   - Priyantha Ranjan, PhD (Curtin University, Australia)
   - Prof. Dr.Eng. Nguyen Trung Viet (Thuyloi University, Vietnam)
   - Dr. Eng. Muhammad Zikra (ITS, Indonesia)
   - Dr. Eng. Shade Rahmawati (ITS, Indonesia)

- Agro Wisudawan, M.T. (University College Cork, Ireland)
- Dr.Eng Totok Suprijo (ITB, Indonesia)
- Prof. Dr. Muhammad Arsyad Thaha (Hasanuddin University, Indonesia)
- Dr. Taufiqur Rachman (Hasanuddin University, Indonesia)
- Dr.Eng Sitang Pilailar (Kasetsart University, Thailand)
- Assoc. Prof Ts Dr Mohd Faris Khamidi (Qatar University, Qatar)
- Assoc. Prof. Ir. Dr. Eng. Ahmad Fitriadhy (Universiti Malaysia Terengganu, Malaysia)
- Dr.Eng. Purwanto Bekti Santoso (Unsoed, Indonesia)

4. Technical Committee
   - Rizky Dwi Putra, S.E.
   - Ferdita Syalsabila, S.T.
   - Muhammad Rizky Syarifudin, S.T.

IOP Conf. Series: Earth and Environmental Science 1298 (2024) 011002 doi:10.1088/1755-1315/1298/1/011002

Peer Review Statement

All papers published in this volume have been reviewed through processes administered by the Editors. Reviews were conducted by expert referees to the professional and scientific standards expected of a proceedings journal published by IOP Publishing.

1. Type of peer review: Single anonymous
2. Conference submission management system: Morressier
3. Number of submissions received: 58
4. Number of submissions sent for review: 58
5. Number of submissions accepted: 37
6. Acceptance Rate (Submissions Accepted / Submissions Received × 100): 63.7 %
7. Average number of reviews per paper: 1.00
8. Total number of reviewers involved: 9
9. Contact person for queries:
   Name: Dr. Dendy Satrio, S.ST.
   Affiliation: Institut Teknologi Sepuluh Nopember (ITS)
   Email: dendy.satrio@its.ac.id

Effect of Interparticle Distance on Smoothed Particle Hydrodynamics on Wave Transformation Through Submerged Breakwater

Sujantoko[1*], I. F. Rasman[1], M. Mustain[1], W. A. Pratikto[1], S. Frestiqauli[1]

[1] Department of Ocean Engineering, Institut Teknlogi Sepuluh Nopember, Surabaya, Indonesia
[*]Email: sujantoko@oe.its.ac.id

Abstract. Beach abrasion and erosion are problems that many beaches in Indonesia experience. Waves are the most influential processes in this phenomenon. When the waves move towards the coast, they will transform, generating currents near the coast. Currents moving along the coast displace sediment on the beach, causing changes in the shoreline. So it is necessary to carry out coastal protection efforts so that abrasion does not occur, which results in the erosion of coastal areas so that the island's area is reduced and the coastline retreats. To overcome this problem, one of the supporting facilities is a submerged breakwater. In this paper, researchers will numerically study the effect of particle size on smoothed particle hydrodynamics (SPH) when waves pass through a submerged breakwater structure. The simulation is carried out on a Numerical Wave Tank (NWT), the piston is at the end of the tank, and the structure of the submerged breakwater is set according to the experimental scenario. The distance between the particles varied from 0.03, 0.025, 0.02, 0.015, and 0.01. The results show that the smaller the distance between the particles used, the wave elevation results are closer to the experimental model.

1. Introduction

Indonesia is an archipelagic country with a long coastline. As an area bordering the ocean, the coast also has the potential to experience damage, such as erosion, sedimentation, and damage to coastal areas which can cause changes in the existing coastline [1]-[7] and disrupt beach tourism [8] [9]. These coastal problems can be overcome by natural protection with mangroves or bamboo [10]-[12] or by building protective structures to protect the coastline from existing problems. Coastal protection structures can be either floating or fixed breakwaters.

Floating breakwaters have been extensively studied both experimentally and numerically by researchers [13]-[24]. This floating breakwater has advantages, including that it can be applied in seas with depths of more than 10 feet and soft seabed soil conditions [13] [14], causing minimal impact on water circulation, sediment transport, and fish migration [13], effective absorb waves of less than 2 meters [13], this structure can also be moved to another location with a different layout [13] [14] [16]. However, this structure is less effective at damping short waves (wave period 6 s and frequency 1.6 rad/s). If the mooring system fails, it will cause an overall structural failure [21], and the cost of maintaining a floating breakwater is higher than that of a conventional breakwater [16].

Coastal protection of fixed breakwater types can be in seawalls, seawalls, bulkheads, revetments, groins, and breakwaters [1]. This structure can work effectively in shallow water. *Breakwater* is a structure that effectively reduces wave energy before it reaches the coast and protects the beach. Waves that hit the breakwater, some of their energy will be reflected, transmitted, and partially destroyed (dissipated). However, breakwaters can also hinder the beauty of a beach. The best solution is with a submerged breakwater structure. A submerged breakwater is a structure whose top is below or equal to sea level. So that breakwaters will not hinder tourists who enjoy the beauty of the beach. In addition, another advantage of submerged breakwaters is that the waves that pass through this structure do not experience a change in wave direction, resulting in no negative impact on the surrounding coast. Several submerged breakwaters have been applied in Indonesia, including Tanah Lot Bali, Peburan Bali Beach, Pasir Putih Karawang Beach, etc.

The transformation of waves passing through submerged structures is related to the movement of waves on the surface. This phenomenon can be investigated experimentally and numerically and is a challenge for experts working in the coastal sector. This study used a numerical approach with the smoothed particle hydrodynamics (SPH) method, which is very good at producing wave simulations. Moreover, SPH can be visualized to get a clear simulation picture. Research on the application of the SPH model has been widely studied by researchers [12] [25]-[28]. This research was carried out in a numerical wave tank (NWT) with a piston at the end of the tank and a structure in the form of a submerged breakwater. The effect of particle size will be tested for its accuracy and ability to transmit waves using the SPH method when it passes through a submerged breakwater.

2. Methods

Smoothed Particle Hydrodynamics (SPH) is a non-mesh Lagrangian method often used for applications in Computational Fluid Dynamics (CFD) where particles represent flow, interact with structures, and can exhibit large deformations with moving boundaries [29]. The constituent equations of this model (equations 1-4) consist of the momentum equation, Lagrange, states, continuity in the delta-SPH form, which functions to reduce fluctuations, and the momentum equation with a fluid particle approach.

Experimental data that has been done before [30] is used for the simulation of the SPH model. The wave parameters used in the numerical model also refer to the time the experiment was carried out; namely, the wave height and period were 0.1 m and 1.47 s, with the water depth set to 0.6 m.

$$\frac{dv}{dt} = -\frac{1}{\rho}\nabla P + g + \Gamma \tag{1}$$

$$\frac{dv_a}{dt} = -\sum_b m_b \left(\frac{P_b}{\rho_b^2} + \frac{P_a}{\rho_a^2} + \Pi_{ab}\right)\nabla_a W_{ab} + g \tag{2}$$

$$P = B\left[\left(\frac{\rho}{\rho_0}\right)^{\gamma} - 1\right] \tag{3}$$

$$\frac{d\rho_a}{dt} = \sum_b m_b \mathbf{u}_{ab} \cdot \nabla_a W_{ab} + 2\delta_\varphi h c_0 \sum_b (\rho_b - \rho_a)\frac{\mathbf{r}_{ab} \cdot \nabla_a W_{ab}}{\mathbf{r}_{ab}^2}\frac{m_b}{\rho_b} \tag{4}$$

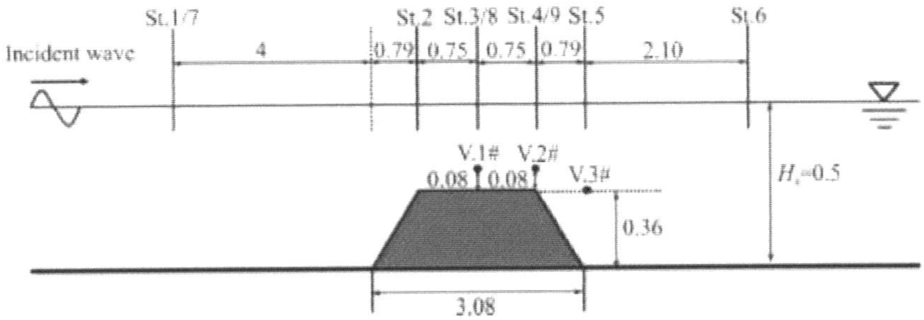

Figure 1. Submerged breakwater experimental layout model [30]

3. Analysis and Discussion

SPH numerical modeling was carried out by analyzing five different distances between particles. The most optimum distance between particles will be obtained for this study. The distance between the particles used is as follows: 0.03 m, 0.025 m, 0.02 m, 0.015 m, and 0.01 m. Validation is comparing the results of numerical modeling with experimental data. In the study referring to the experiment [30], model validation occurred when the wave transformation passed through the submerged breakwater. Changes in water level elevation are recorded on wave probes 1- 6, as shown in Fig. 1 above. The results of the running program at five variations of the distance between particles can be seen by visualizing the results in the Paraview software (Fig. 2). The model results show that the smaller the distance between the particles, the better and conversely, the larger the distance, the less noticeable the particles are. This condition will certainly affect the accuracy of the simulation results.

The weakness of the distance between particles that produces better and clearer images is that the running time becomes longer. Nevertheless, this also depends on the specifications of the PC used. This DualSPHysics software is intended to use the GPU (Graphics Processing Unit) but can still use the CPU (Central Processing Unit). The difference between the CPU and GPU on DualSPhysics is the running time. This software is intended for utility use of the GPU. Therefore, if it is run with the GPU, it can be faster than using the CPU. The better the CPU or GPU specifications, the faster it can run. This research was conducted on a PC with the exact specifications for a simulation time of 120 seconds with a distance between particles of 0.03m, 0.025, 0.02, 0.015, and 0.01, requiring a running time of 4-6 hours, 10-14 hours, 19-26 hours, 38 -54 hours, and 72-94 hours, respectively (Table 1). It can be seen that the time difference is quite significant. Therefore, the running time is included in the criteria for selecting the distance between particles to obtain maximum results, the simulation time is efficient, and the error rate is low between the numerical and experimental results.

After running the DualSPHysics software, the results of the water level elevation in the form of a wave time series will be the output of this study. These results will be compared with the experimental results on the breakwater model and the input wave parameters. The smaller the deviation of the running numerical results from the experimental results, the better. The mean absolute percentage error (MAPE) method is used [31] to calculate the deviations. This method measures the accuracy of a forecast model made from the actual model. The smaller the value, the smaller the deviation of the forecast model from the actual model or, in other words, the closer the forecast model is to the actual model.

After running on DualSPHysics, use one of the post-processing tools on DualSPHysics to get wave elevation data that occurs at specific predetermined points according to experimental data, namely "MeasureTool". This tool adjusts the location for recording wave elevation data to the wave probe point in the experiment (Fig. 1). Based on Table 1, the running time and the deviation between numerical and experimental means that at a distance between particles of 0.01 m, the average error is at least 2.95%, but the time needed to complete the process is 72-94 hours. At a distance between particles of 0.015m, the error result is not too far from 0.01m; it requires a running time of 38-54, which is about 1.5 times faster than 0.01m. For distances of 0.03 m, 0.025 m, and 0.02 m, the resulting error results are no better than those of 0.01 m and 0.015 m, even though the time required for running is faster among the five distances between particles. Therefore in this study, the distance between particles of 0.015 m is the most efficient, considering that the resulting error is not too far at 0.01 m, and the time needed is more efficient and faster than 0.01m.

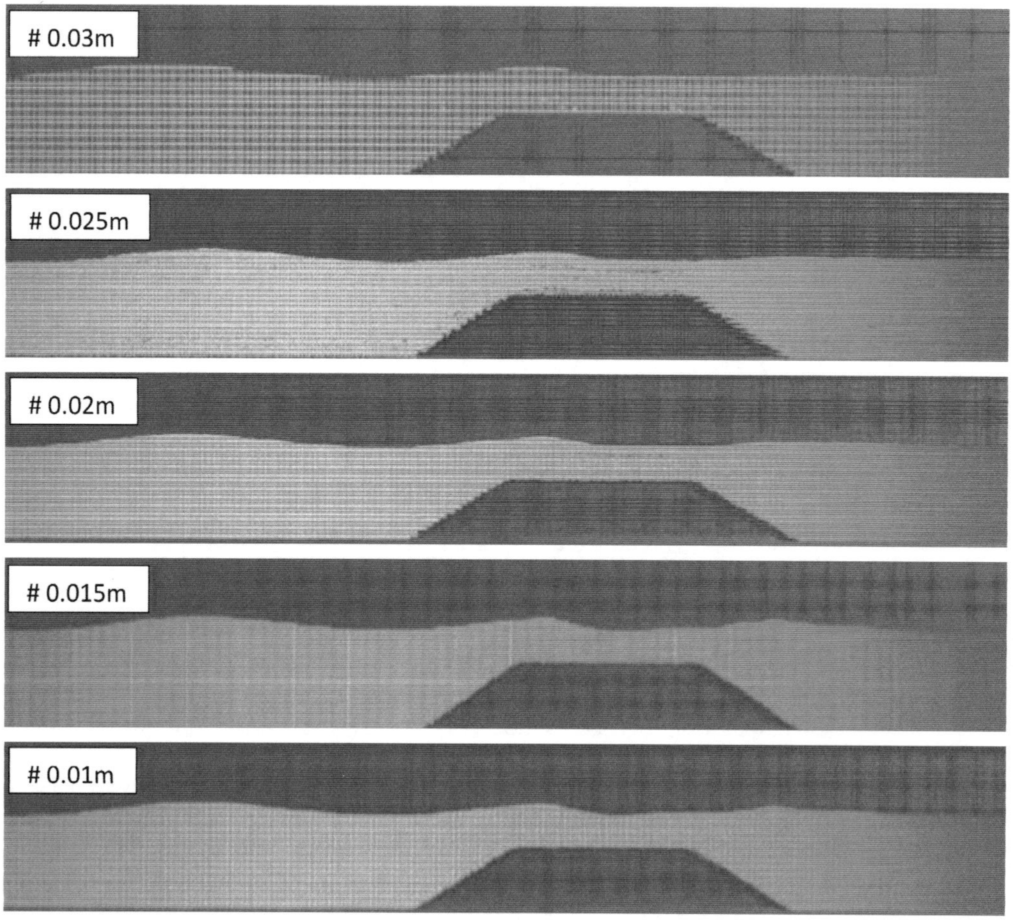

Figure 2. Visualization in Paraview software results from running models with various distance variations between particles.

Table 1. Error calculation between numerical and experimental model data

Distance between particles (m)	Estimated running time (hours)	MAPE calculation results (%)						
		wp1	wp2	wp3	wp4	wp5	wp6	avg
0.01	72-94	1.73	2.64	2.84	4.23	3.04	3.23	2.95
0.015	38-54	1.73	2.89	2.91	4.19	3.42	3.45	3.10
0.02	19-26	2.36	3.19	3.18	4.28	3.55	3.48	3.34
0.025	10-14	2.54	3.40	3.33	4.46	3.69	3.43	3.48
0.03	4-6	2.39	3.71	3.66	4.66	4.00	3.60	3.67

IOP Conf. Series: Earth and Environmental Science 1298 (2024) 012001 doi:10.1088/1755-1315/1298/1/012001

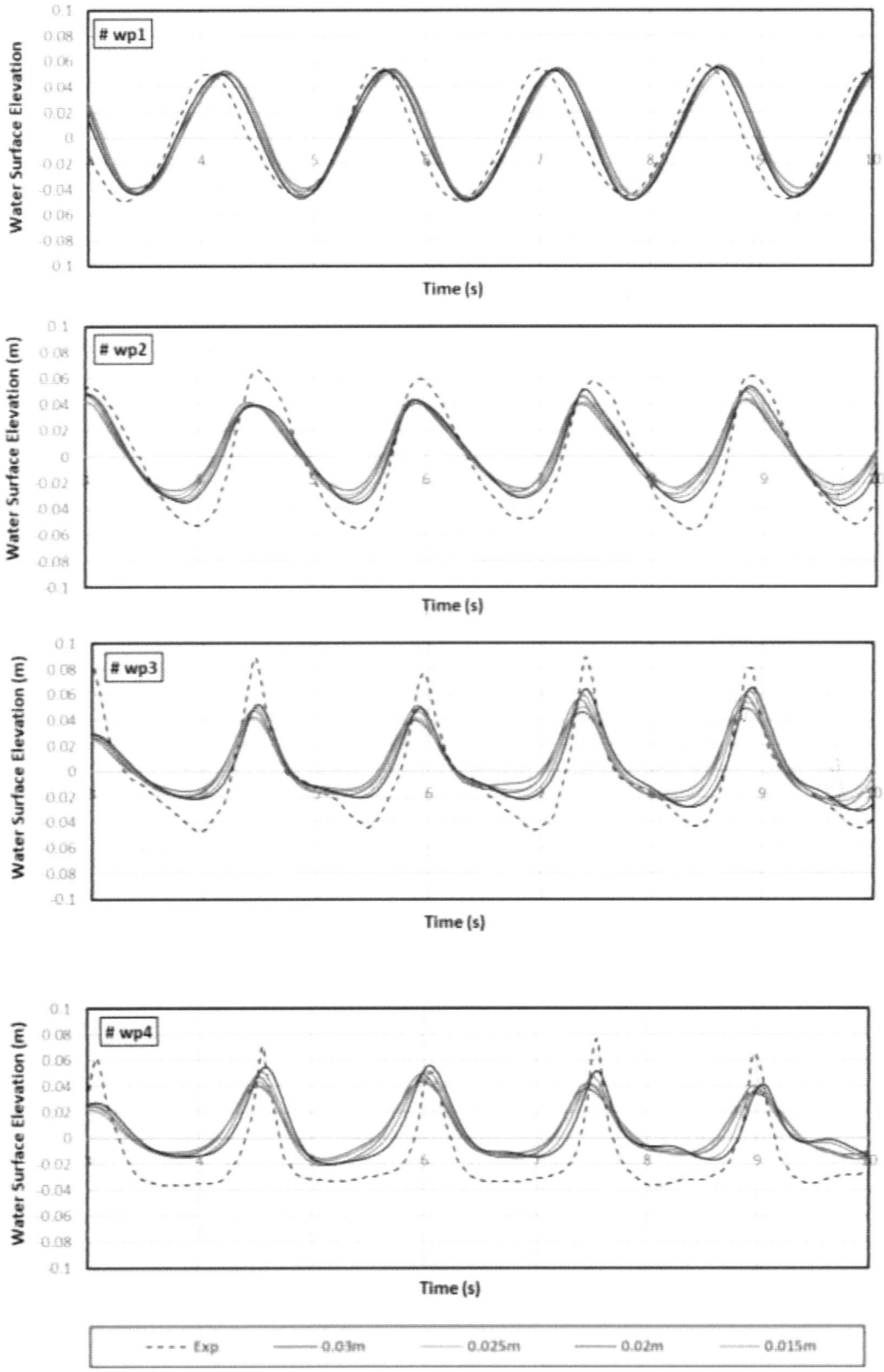

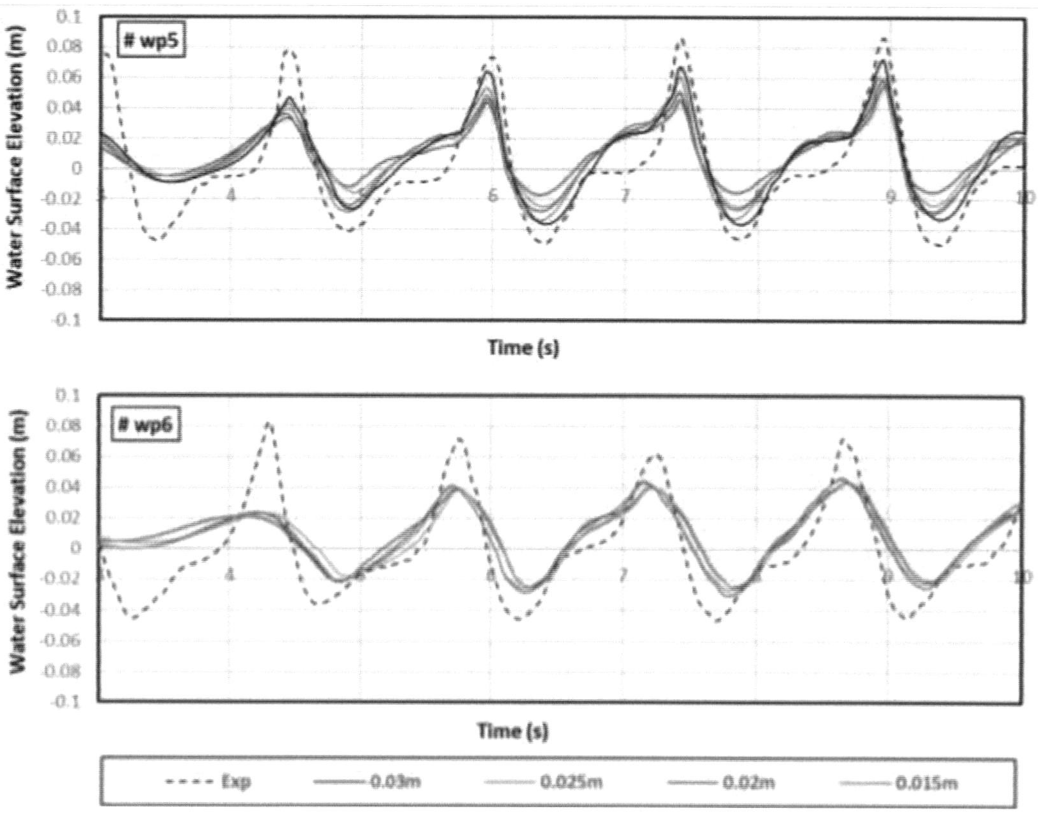

Figure 3. Simulation of the SPH model of water level fluctuations when passing through a submerged breakwater at various distances between particles for each wave probe

4. Conclusion

This study concluded that the SPH method could simulate wave transformations well, as evidenced by the relatively minor deviations in the water level elevation fluctuations from the model and experiment results. The smaller the distance between the particles, the smaller the deviation that occurs; in this study, 0.01 m is the best. However, 0.01 m requires the longest running time. Therefore for optimal results with more efficient time, 0.015 m is the best choice.

Acknowledgment

Researchers would like to thank the Department of Ocean Engineering, DRPM-ITS, and Institut Teknologi Sepuluh Nopember, Surabaya, Indonesia, for the financial support and facilities provided.

References

[1] CERC, 1984. Shore Protection Manual. Coastal Engineering Research Centre, Dept. of the Army Waterways Experiment Station, US Army Corps of Engineers, Washington.
[2] Sujantoko and Natakusuma, D., 2003. Model simulasi interaksi gelombang dan arus di perairan dangkal, *Jurnal Teknik Sipil*, vol. 10 (3), 99-108.

[3] Sujantoko, Pratikto, W.A., Prastianto, R. W. Maulana, M. I. and Vebriyanti, A., 2021. Study of changes in coastal morphology due to utilization of the Surabaya city coastal area. *International Journal of Marine Engineering Innovation and Research*, 7 (1), pp. 26-32.
[4] Sujantoko, Pangestu, P. A. Saputro, D., Ekianto, B. P. P., Hasanudin, Kurniati, N., Kusumawardhani, R., and Hartanto, D., 2023. Hydrodynamic model simulation at the port of Tanjung Rhu Belitung. *International Journal of Marine Engineering Innovation and Research*, 8 (2), pp.141-149.
[5] Sujantoko, Mustain, M., Wahyudi, and Ikhwani, H., 2023. Hydrodynamic model due to reclamation in Lamong Bay. IOP Conference Series: Earth and Environmental Science, Volume 1198, 012004.
[6] Mustain, M., and Sujantoko, 2023. Spiral analysis of vertical currents in Lamong Bay. IOP Conference Series: Earth and Environmental Science, Volume 1166, 012024.
[7] Mustain, M., and Sujantoko, 2021. Simple Analysis of Determination of Tidal Types of Sea Water: A Case Study in Lamong Bay Waters, East Java. *Proceeding - IEEE Ocean Engineering Technology and Innovation Conference: Ocean Observation, Technology and Innovation in Support of Ocean Decade of Science (OETIC)*, pp. 13–17.
[8] Mustain, M., Sujantoko, and Prasetyo, M. A., 2022. Empowerment and optimalization of Kenjeran beach tourism potential, Surabaya-East Java. *International Review of Civil Engineering*, 13 (4), pp. 246-254
[9] Mustain, M., Garang, I. J., Ikhwani, H., and Sujantoko, 2022. Analysis and maximize the ecotourism potential of the Wonorejo mangrove in East Surabaya. *International Journal on Engineering Applications*, 10 (2), pp. 158-166.
[10] Salem, M. E., and Mercer, D. E., 2012. The economic value of mangroves: a meta-analysis. *Sustainability*, 4, pp. 359–383.
[11] Winterwerp, J. C., Erftemeijer, P. L. A., Suryadiputra, N., Eijk, P.V., and Zhang, L., 2013. Defining eco-morphodynamic requirements for rehabilitating eroding mangrove-mud coasts. *Wetlands*, 33, pp. 515–526.
[12] Armono, H.D., Bromo, B.H., Sholihin, and Sujantoko, 2022, Numerical study of bamboo breakwater for wave reduction, *Fluids*, 7 (1), pp. 1-13.
[13] Hales, L. Z, 1981. *Floating Breakwater: State of the art Literature Review*. Technical Report No. 81-1. U.S. Army Coastal Engineering Research Center, Fort Belvoir.
[14] McCartney, B. L., 1985. Floating breakwater design. *Journal of Waterway, Port, Coastal and Ocean Engineering,* 111 (2), pp. 304-318.
[15] Kelly, H., 1999. *Discussion about floating breakwaters*, Naval Facilities Engineering Command, Norfolk, Virginia.
[16] Tsinker, G. P., 1995. *Marine Structures Engineering: Specialized Application*. New York, An International Thomson Publishing Company.
[17] Sujantoko, Wardhana, W., Djatmiko, E. B., Armono, H. D., Putro, W. S., and Almuzaki, R. H., 2021. Study of wave characteristics of the floating breakwater for piling and tethered type. *Jurnal Teknik Hidraulik (Indonesian)*, 12(1), pp. 39-51.
[18] Sujantoko, Djatmiko, E. B., Wardhana, W., Hidayatullah, A., 2021.Analysis of mooring tension on the saw-type floating breakwater with physical modeling. *Jurnal Teknik Sipil (Indonesian)*, 28(3), pp. 289-300.
[19] Sujantoko, Djatmiko, E. B., Wardhana, W., 2022. Experimental investigation of the performance of porous-slope floating breakwater. *International Journal on Engineering and Applications*, 10 (2), pp. 149-157.
[20] Sujantoko, Armono, H. D., Djatmiko, E. B., and Putra, R. D., 2022. Stability analysis of concrete block anchor on steep-slope floating breakwater. *Fluids*, 7 (8), 259.
[21] Sujantoko, Djatmiko, E. B., and Wardhana, W., 2022. Experimental study on mooring tension of porous-slope floating breakwater. *International Journal on Engineering and Applications*, 10 (5), pp. 345-352.

[22] Sujantoko, Djatmiko, E.B., Wardhana, W., and Firmansyah, I. G., 2023. Numerical modeling of hydrodynamic performance on porous slope type floating breakwater. *Nasemore*, 70(2), pp. 149-157.

[23] Sujantoko, Djatmiko, E. B., Wardhana, W., and Mustain, M., 2023. Analysis of wave spectrum change of porous-saw floating breakwater. *International Journal on Engineering and Applications*, 11 (5).

[24] Sujantoko, Djatmiko, E. B., Wardhana, W., Mustain, M., and Ahidah, D. R., 2023. Numerical Modeling of Wave Hydrodynamic Performance with Various Layouts of Single-Slope Floating Breakwater Arrays, *International Journal on Engineering and Applications*, vol.11

[25] Islam, M.R., Hayano, K., and Nadeesha, A.M.K.N., 2013. Simulating the effect of flow volume on scouring process around breakwater under tsunami condition by SPH method, *Int. Journal of Geomate*, 5 (2), pp. 723-728.

[26] Armono, H.A, Manurung, A., Sujantoko, and Suastika, I.K., 2022. Wave transmission model over hexagonal artificial reefs using DualSPHysics. *IOP Conference Series: Earth and Environmental Science*, vol. 1095, 012025.

[27] Meringolo, D. D., Aristodemo, F, and Veltri, p., 2015. SPH numerical modeling of wave–perforated breakwater interaction, *Coastal Engineering*, vol.101, pp 48-68.

[28] Rogers, B. D., 2009. Simulation of caisson breakwater movement using 2-D SPH, *Journal of Hydraulic Research*, 48(1), pp.135-141.

[29] Crespo, A.J.C., Domínguez, J.M., Rogers, B.D., Gómez-Gesteira, M., Longshaw, S., Canelas, R., Vacondio, R., Barreiro, A., and García-Feal, O., 2014. DualSPHysics: Open-source parallel CFD solver based on Smoothed Particle Hydrodynamics (SPH). Computer Physics Communications, 187, pp. 204-216.

[30] Liu, B. jin, Cheng, D., Sun, Zhao-Chen, Zhao, Xi-Zeng, Chen, Y., Lin, and Wei-dong, 2019. Experimental and numerical study of regular waves past a submerged breakwater. *Journal of Hydrodynamics*. 31, pp. 641–653.

[31] Lewis, C.D., 1982. *International and Business Forecasting Methods*. Butterworths, London.

Numerical simulation of water wave propagation using DualSPHysics

A Trimulyono[1,*], B R Dhanureski [1], S Samuel [1], S Jokosisworo [1]

[1] Department of Naval Architecture, Faculty of Engineering, Universitas Diponegoro

*Email: anditrimulyono@live.undip.ac.id

Abstract. In this study, the study of water waves was carried out using one of major the mesh-free CFD i.e. Smoothed Particle Hydrodynamics (SPH). DualSPHysics ver 5.0 which is an open-source SPH solver based on Weakly Compressible Smoothed Particle Hydrodynamics (WCSPH) was used to reproduce three dimension of water waves in large and medium numerical wave tanks (NWT). This study analyses water wave profiles and the accuracy of DualSPHysics for regular waves with period 1.15 second and two NWT were used, i.e., 35 and 15 m. In addition, advanced post-processing using VisualSPHysics was used to mimic water waves in real conditions. Single water depth was used, i.e. 0.4 meters for large and medium NWT. The results showed SPH has limitations for reproducing large NWT. However medium NWT successfully reproduce water wave with advanced post-processing using VisualSPHysics.

1. Introduction

Water Water wave propagation is a demanding problem in marine and coastal engineering field. One of the difficulties of free-surface flow problems is generating intended nonlinear water waves and predicting complicated interactions between water waves and fixed or floating objects, which is necessary for the utilization and development of further marine problems. Experimental and numerical methods have been used in several investigations. Because experiments are more expensive and costly, researchers typically utilize computer models instead. The finite element method (FEM), the finite difference method (FDM), the finite volume method (FVM), and smoothed particle hydrodynamics (SPH) are well-known numerical techniques to perform numerical models of water waves [1,2].

Most researchers use the SPH method in analyzing nonlinear free-surface flow problems and the interactions on solid structures. Such as research on breaking wave modelling and the interaction of coastal structures with water waves [3,4], sloshing in the prismatic tank [5–8], and the impact of water waves on offshore structures [9]. These studies showed the potential of using SPH method in marine and coastal engineering applications and application for actual problem in engineering field.

DualSPHysics is an SPH numerical simulation software based on the Smoothed Particle Hydrodynamics (SPH) method. It is used for modeling fluid flows, especially those involving free surfaces and complex interactions [10,11], and engineering problems in coastal areas [12]. The SPH method discretises the water domain into many small particles. Each particle has specific physical properties such as mass, position, velocity and density. The momentum and continuity equations govern the motion of each SPH particle. Reviews regarding the SPH method and its application can be seen in reference [13].

This study uses the DualSPHysics application to carry out further development of previous research conducted by Trimulyono and Hashimoto [4]. In previous research, three-dimensional wave simulation in large numerical wave tanks (NWT) experienced a decrease in accuracy compared to two-dimensional wave simulation and had yet to carry out advanced post-processing in the research. In this study, the authors conducted further parameter analysis to increase the level of accuracy, which aims to test the ability of the SPH method to carry out three-dimensional wave simulations. In addition, three-dimensional wave simulation in medium numerical wave tanks (NWT) was carried out with advanced post-processing using VisualSPHysics.

2. Methods

Smoothed Particle Hydrodynamics (SPH) is a computational method for solving fluid dynamics problems using the Lagrangian approach. SPH method was originally used by Gingold and Monaghan in the field of astrophysics field [14]. Then, free-surface simulation using the SPH method was introduced by Monaghan [15] and began to develop in problems in the field of fluid dynamics.

The basis of the SPH formulation is integral representation or what is commonly called the kernel approximation. The integral representation is estimated by adding up the values of the closest neighboring particles which produce a particle approximation of the function at a point or a discrete particle whose results can be seen in equation (1) where W is kernel function and r is vector position.

$$f(r) = \int_{\Omega} f(r')W(r - r', h)dr' \tag{1}$$

Equation (1) can be approximated into discrete form by replacing the integral as number of neighboring particles, where a is the particle and r is the position. So, the particle equation in equation (2) becomes equation (3), where m is mass, ρ is density and v is velocity.

$$f(r_\alpha) \approx \sum_b f(r_b)W(r_\alpha - r_b, h)\frac{m_b}{\rho_b} \tag{2}$$

$$fra \approx \sum_a m_b \frac{A_b}{\rho_b} w_{ab} \tag{3}$$

The momentum equation (Navier-Stokes) is used to set the motion of the particle due to the influence of the pressure generated by the neighboring particle. The form of the Lagrangian formula which forms the basis of the Navier-Stokes equation is as follows:

$$\frac{D_\rho}{D_t} = -\rho\nabla v, \tag{4}$$

$$\frac{D_v}{D_t} = -\frac{1}{\rho}\nabla P + g + \tau \tag{5}$$

$$\frac{D_r}{D_t} = v \tag{6}$$

The kernel function in this study uses the Wendland kernel function for computational efficiency, as stated by Cao et al [16]. This function can be seen in equation (7).

$$W(q) = \alpha_d \left(1 - \frac{q}{2}\right)^4 (2q + 1) 0 \leq q \leq 2 \tag{7}$$

The momentum equation in the water phase is described in equation (8)

$$\frac{dv_a}{dt} = -\sum_b mb\left(\frac{P_a+P_b}{\rho_a \cdot \rho_b} + \Pi_{ab}\right) + \nabla_a W_{ab} + g, \tag{8}$$

$$\Pi_{ab} = \begin{cases} \frac{-\propto \overline{c_{ab}} \mu_{ab}}{\overline{\rho_{ab}}} & \upsilon_{ab} \cdot r_{ab} < 0 \\ 0 & \upsilon_{ab} \cdot r_{ab} > 0 \end{cases} \tag{9}$$

DualSPHysics is based on weakly compressible smoothed particle hydrodynamics (WCSPH) by treating fluids as weakly compressed. Thus, the equation of state is used in equation (10) which is used to find the pressure field based on the density of the particles.

$$P = b\left[\left(\frac{\rho}{\rho_0}\right)^{\gamma} - 1\right] \tag{10}$$

Where $b = \frac{c_0^2 \rho_0}{\gamma}\gamma$, $c_0 = c_s(\rho_0) = \sqrt{\left|\frac{\partial P}{\partial \rho}\right|} \rho 0$, and $\gamma = 7$ with c_0, ρ_0 and γ are the speed of sound at the reference density, the reference density and polytropic constant, respectively. This study used two main models to be analyzed on reproducing the observed water waves in the experiment that conducted by Trimulyono and Hashimoto [4], i.e., large numerical wave tank (NWT) and medium numerical wave tank (NWT). The large NWT has dimensions of 35 meters in length, 0.5 meter in width, 1.5 meters in depth and a wave probe distance of 24.6 meters with a simulation time of 25 seconds and a piston-type wave generator as shown in Figure 1. Medium NWT has dimensions of 15 meters in length, 1 meter in width, 1.5 meters in depth and a wave probe distance of 6 meters with a simulation time of 30 seconds and a piston-type wave generator as show in Figure 2.

The parameters used in DualSPHysics were set as shown in Table 1. Coefh is defined as $h/dp\sqrt{2}$ where *dp* is particle spacing. Artificial viscosity affects on proper dissipation in water wave propagation problem. For this subject problem the optimal value of artificial viscosity is 0.001. Coefsound is defined as $Cs/\sqrt{g.d_{swl}}$ where Cs and d_{swl} are the speed of sound and the height of water, respectively.

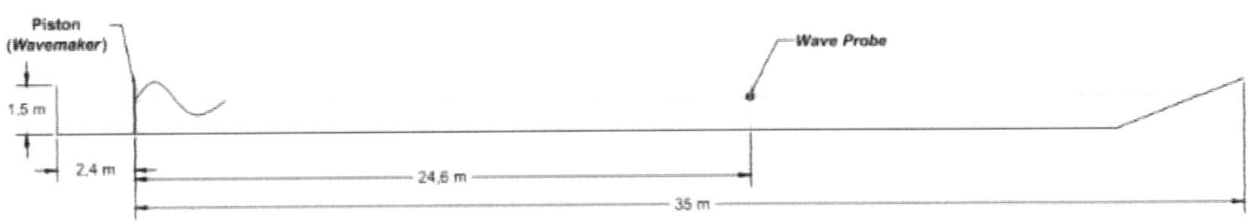

Figure 1. Schematic view of large numerical wave tank

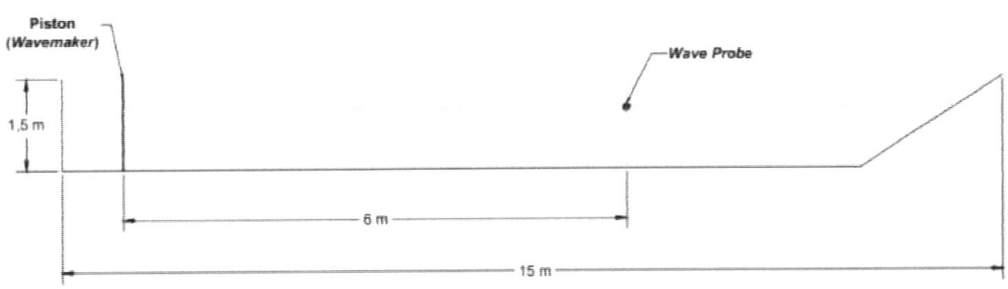

Figure 2. Schematic view of medium numerical wave tank

Table 1. Numerical setting in DualSPHysics

Parameter setup	
Kernel function	Wendland
Time step algorithm	Symplectic
Particle spacing/*dp* (mm)	0.006
α (artificial viscosity)	0.001
Coefsound	20
Coefh	1.2
Simulation time (*s*)	25 (large NWT) 30 (medium NWT)

3. Results and Discussion

3.1. Grid Study

A grid study was conducted to investigate the sensitivity of the particle spacing to the numerical accuracy. Figure 3 shows several particle spacing used for grid study on the medium NWT model with water depth d = 0.4 m. It shows that the value of particle spacing (*dp*) will affect the wave elevation and phase. Particle spacing 0.006 m was chosen over the others because it shows the best accordance in the wave elevation with the experiment. Decreasing particle spacing causes the number of particles to increase, but it also will increase computation times and the total number of particles.

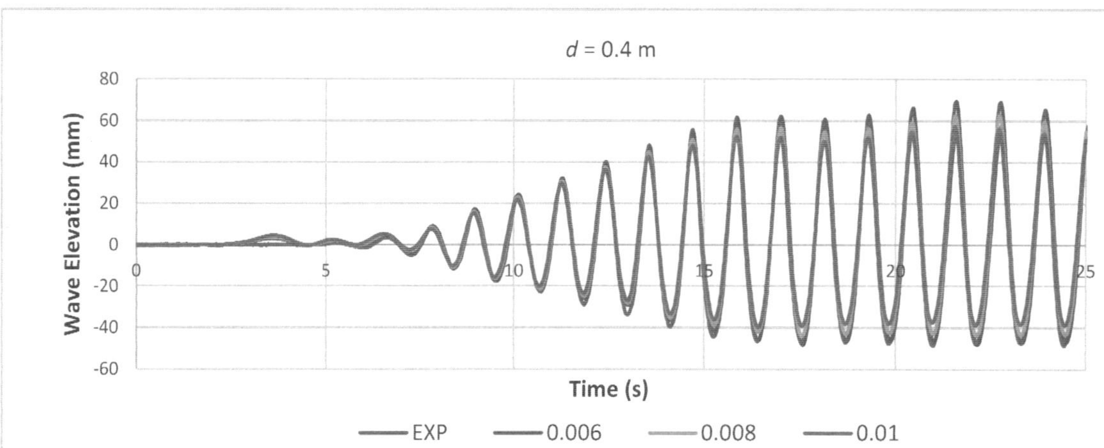

Figure 3. Grid study on medium NWT with $d = 0.4$ m

3.2. Wave Propagation on Large Numerical Wave Tank (NWT)

Wave propagation simulation on large NWT with water depth d = 0.4 m uses the parameters in Table 1. Figure 4 shows the comparison results of wave elevation at 24.6 meters distance from the wave generator between the experiment and SPH simulation of the large NWT, which shows that the phase difference is quite large. The SPH simulated wave elevation is not in accordance with the experimental wave elevation. It shows that the wave generated from the SPH simulation for large NWT has poor accuracy, and further parameter improvements are needed. Figure 5 shows the three-dimensional wave profile of the large numerical wave tank (NWT).

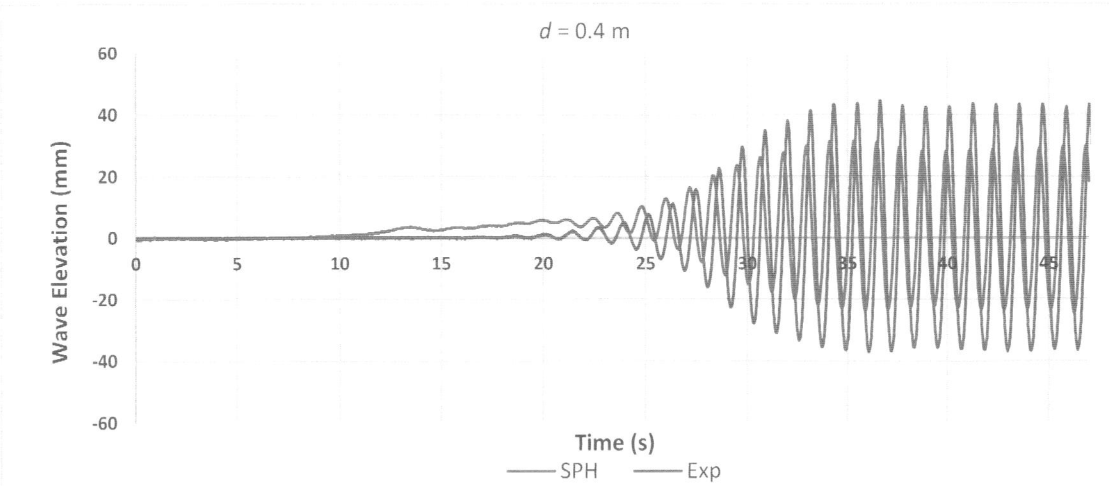

Figure 4. Comparison results of wave elevation between experiment and SPH simulation

Figure 5. Wave propagation simulation at 25 seconds

3.3. Wave Propagation on Medium Numerical Wave Tank (NWT)

Wave propagation simulation on medium NWT with water depth d = 0.4 m uses the parameters in Table 1. Figure 6 shows the comparison results of wave elevation at 6 meters distance from the wave generator between the experiment and SPH simulation of the medium NWT, which indicates that the SPH simulated wave has no phase difference compared to the experimental results. The wave elevation from the SPH simulation is following the experiment. It shows that the waves generated from the SPH simulation have good accuracy. Figure 7 shows the three-dimensional wave profile of the medium numerical wave tank (NWT).

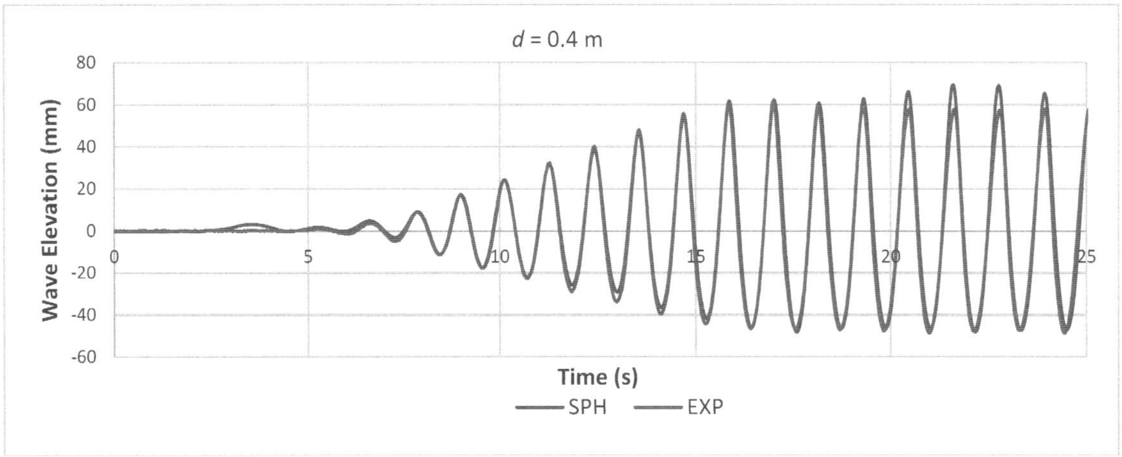

Figure 6. Comparison results of wave elevation between experiment and SPH simulation

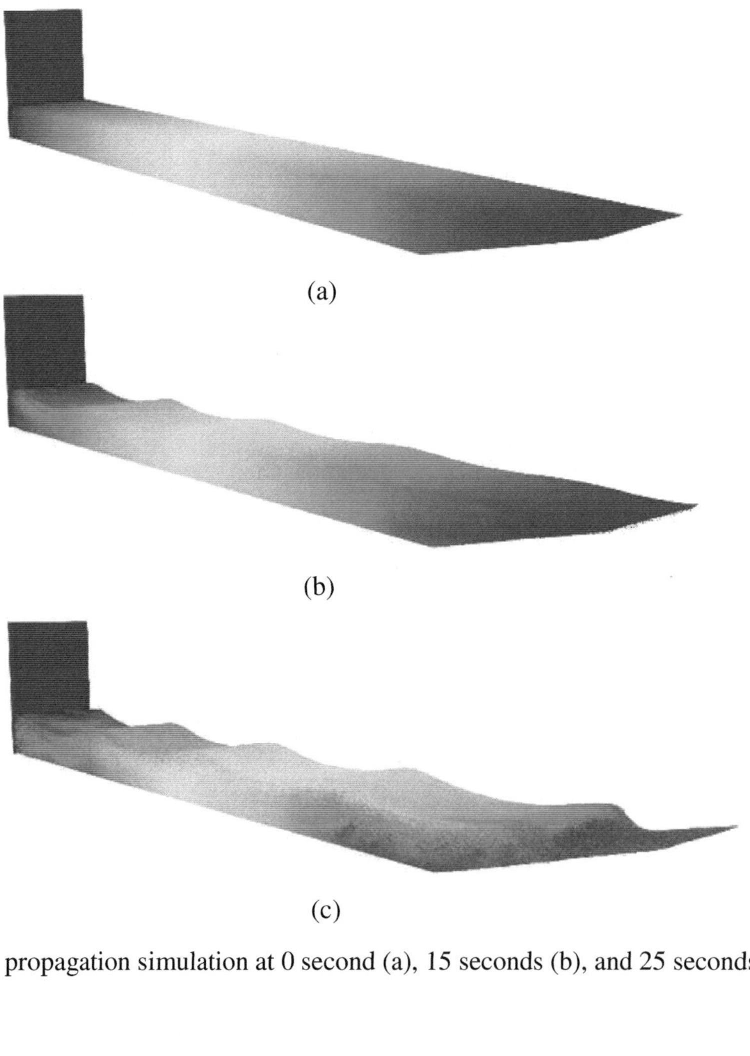

Figure 7. Wave propagation simulation at 0 second (a), 15 seconds (b), and 25 seconds (c)

IOP Conf. Series: Earth and Environmental Science 1298 (2024) 012002 doi:10.1088/1755-1315/1298/1/012002

3.4. Advanced Post-Processing

DualSPHysics DualSPHysics can generate VTK (Visualization ToolKit) files that can be used in post-processing. VTK files show particle positions and support many data types, such as scalar, vector, tensor, and texture. These VTK files can be used in post-processing software like Paraview and Blender. Paraview and Blender software were used in this study to visualize the wave propagation simulation.

Paraview can read the VTK files generated from DualSPHysics directly, but in Blender, an add-on called VisualSPHysics must be installed first. VisualSPHysics is an add-on for Blender that allows users to render realistic images and animations for DualSPHysics simulations. VisualSPHysics is integrated into the Blender user interface. So, objects imported from DualSPHysics simulations behave like regular Blender objects.

Figure 8 shows the result of the post-processing application in Paraview and Blender software that is visualized from VTK files generated from DualSPHysics. Paraview shows simple visualization from wave propagation. Blender offers complex and realistic visualization that can represent simulation results more realistically..

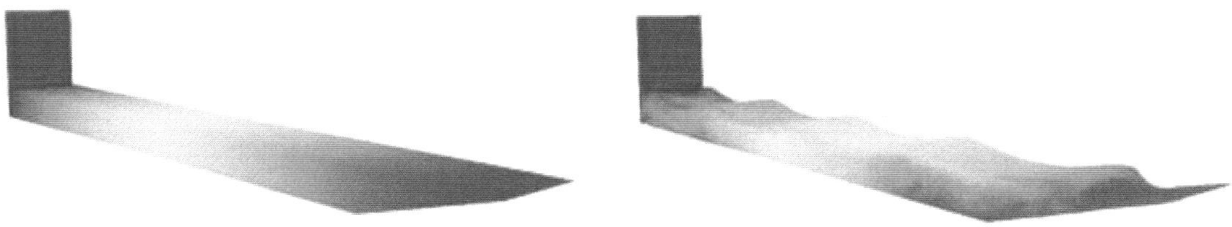

(a)

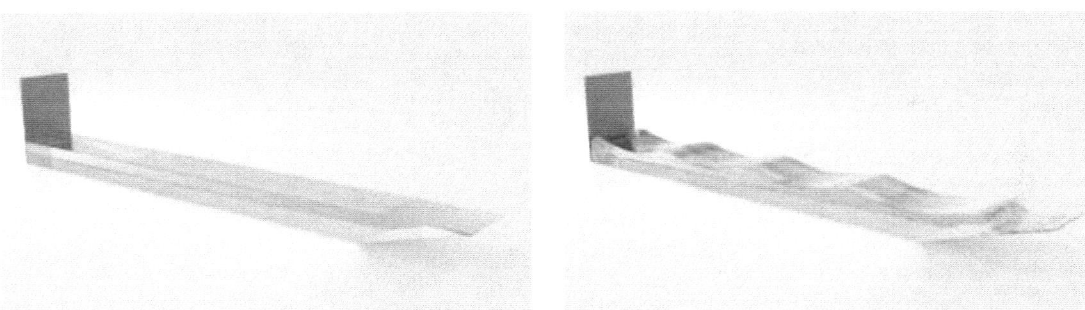

(b)

Figure 8. Post-processing results of wave propagation simulation of medium NWT with water depth *d = 0.4 m* in Paraview (a) and Blender (b)

4. Conclusion

Numerical Numerical simulation of three-dimensional wave propagation has been carried out using DualSPHysics ver. 5.0. The simulation results show that the large NWT with water depth d = 0.4 m does not have good accuracy and requires further parameter improvements, as seen from the significant phase difference. The wave elevation is not in accordance. In medium NWT with water depth d = 0.4 m, it has excellent accuracy. It has been validated with experimental data as seen from the absence of phase difference and accordance wave elevation.

The particle spacing (dp) affects the wave profile both in terms of height and the wave phase, where decreasing particle spacing will cause the number of particles to increase. Still, it also will increase computation times and the total number of particles.

The numerical simulations can be visualized through post-processing using Paraview software, which supports simple scientific visualization, and Blender software, which supports complex and realistic visualization.

This research has promising results, however, further studies are needed, especially for other parameter adjustments to produce a numerical simulation of three-dimensional regular waves in large numerical wave tanks (NWT).

Acknowledgments

Authors wish to thank laboratory of ship hydrodynamics Department of Naval Architecture Universitas Diponegoro for computer facilities and registration conference fee.

References

[1] M. A. Salvikran, A. Trimulyono, and E. S. Hadi, "Validation of multiphase water wave propagation using smoothed particle hydrodynamics," *IOP Conf. Ser. Earth Environ. Sci.*, vol. 1166, no. 1, p. 012032, 2023.

[2] A. Trimulyono and H. Hashimoto, "Experimental Validation of Smoothed Particle Hydrodynamics on Generation and Propagation of Water Waves," *J. Mar. Sci. Eng.*, vol. 7, no. 1, pp. 17, 2019

[3] R. A. Dalrymple and B. D. Rogers, "Numerical modeling of water waves with the SPH method," *Coast. Eng.*, vol. 53, no. 2–3, pp. 141–147, Feb. 2006, doi: 10.1016/j.coastaleng.2005.10.004.

[4] A. Trimulyono and H. Hashimoto, "Experimental Validation of Smoothed Particle Hydrodynamics on Generation and Propagation of Water Waves," *J. Mar. Sci. Eng.*, vol. 7, no. 1, pp. 17, 2019.

[5] A. Trimulyono, S. T. M. Owhaamsorrc Gultom, E. S. Hadi, and D. B. Purwanto, "Investigation of Sloshing with Vertical and Horizontal Baffle in the Prismatic Tank using Meshfree CFD," *CFD Lett.*, vol. 15, no. 6, pp. 115–129, 2023.

[6] A. Trimulyono, D. Chrismianto, H. Atthariq, and Samuel, "Numerical Simulation Low Filling Ratio of Sway Sloshing in the Prismatic Tank Using Smoothed Particle Hydrodynamics," *CFD Lett.*, vol. 14, no. 7, pp. 113–123, 2022.

[7] A. Trimulyono, A. Haikal, C. Deddy, and S. Samuel, "Investigation of Sloshing in The Prismatic Tank With Vertical And T-Shape Baffles," *Brodogradnja*, vol. 73, no. 2, pp. 43–58, 2022.

[8] A. Trimulyono, Samuel, and M. Iqbal, "Sloshing Simulation of Single-Phase and Two-Phase SPH using DualSPHysics," *Kapal: J. Ilmu Pengetah. dan Teknol. Kelaut.*, vol. 17, no. 2, pp. 50–57, 2020.

[9] K. Pan, R. H. A. Ijzermans, B. D. Jones, A. Thyagarajan, B. W. H. van Beest, and J. R. Williams, "Application of the SPH method to solitary wave impact on an offshore platform," *Comput. Part. Mech.*, vol. 3, no. 2, pp. 155–166, Apr. 2016, doi: 10.1007/s40571-015-0069-0.

[10] J.M. Domínguez, G. Fourtakas, C. Altomare, R.B. Canelas, A. Tafuni, O. García-Feal, I. Martínez-Estévez, A. Mokos, R. Vacondio, A.J.C. Crespo, B.D. Rogers, P.K. Stansby, M. Gómez-Gesteira, DualSPHysics: from fluid dynamics to multiphysics problems, *Comput. Part. Mech.* 9 (2022) 867–895. doi:10.1007/s40571-021-00404-2.

[11] A.J.C. Crespo, J.M. Domínguez, B.D. Rogers, M. Gómez-Gesteira, S. Longshaw, R. Canelas, R. Vacondio, A. Barreiro, O. García-Feal, DualSPHysics: Open-source parallel CFD solver based on Smoothed Particle Hydrodynamics (SPH), *Comput. Phys. Commun.* 187 (2015) 204–216. doi:10.1016/j.cpc.2014.10.004.

[12] C. Altomare, A. J. C. Crespo, J. M. Domínguez, M. Gómez-Gesteira, T. Suzuki, and T. Verwaest, "Applicability of Smoothed Particle Hydrodynamics for estimation of sea wave impact on coastal structures," *Coast. Eng.*, vol. 96, pp. 1–12, Feb. 2015, doi: 10.1016/j.coastaleng.2014.11.001.

[13] J. J. Monaghan, "Smoothed particle hydrodynamics and its diverse applications," *Annu. Rev. Fluid Mech.*, vol. 44, pp. 323–346, 2011, doi: 10.1146/annurev-fluid-120710-101220.

[14] R. A. Gingold and J. J. Monaghan, "Smoothed particle hydrodynamics theory and application to non-spherical stars," *Mon. Not. R. astr. Soc.*, vol. 181, pp. 375–389, 1977.
[15] J. J. Monaghan, "Simulating Free Surface Flows with SPH," *J. Comput. Phys.*, vol. 110, pp. 399–406, 1994.
[16] X. Y. Cao, F. R. Ming, and A. M. Zhang, "Sloshing in a rectangular tank based on SPH simulation," *Appl. Ocean Res.*, vol. 47, pp. 241–254, 2014, doi: 10.1016/j.apor.2014.06.006.

Saltwater Intrusion at Kapuas Besar Estuary Using 1-Dimensional Analytical Method

Danke Aziz * **and Ayi Tarya**

Department of Oceanography, Faculty of Earth Sciences and Technology, Bandung Institute of Technology, Bandung 40132, Indonesia

Email: dankeaziz@gmail.com

Abstract. The Kapuas Besar Estuary in West Kalimantan is vital for both the local community's ecological and economic well-being. One major issue faced in this estuary is saltwater intrusion. This study aims to investigate saltwater intrusion variation using a one-dimensional (1D) analytical model. The model input data includes estuary geometry, tides, salinity, and river discharge. Simulation results demonstrate the effectiveness of the 1D analytical model in predicting saltwater distance based on the Root Mean Square Error (RMSE), Percent Bias (PBIAS), and Nash-Sutcliffe Efficiency (NSE) values between the analytical model results and observed data ranging from 0.82-1.70 psu, 1.64-2.85%, and 0.96-0.99 respectively. The model proves applicable during the May-October period, subject to the condition that the river discharge remains below 3,200 m3/s. However, during November-April the model tends to underestimate compared to actual distance. Model results show saltwater intrusion increasing from December to June by 49 km, then decreasing from July to November by 38 km. The minimum distance of saltwater intrusion occurred in May (1 km from the estuary mouth) when river discharge of 8,853.82 m^3/s. Conversely, the maximum saltwater intrusion distance was observed in June (71 km from the estuary mouth) with a river discharge of 1,034.55 m^3/s.

1. Introduction

Estuaries, which are complex coastal ecosystems, are known as highly productive coastal ecosystems and sensitive to both human-induced and natural environmental pressures [3]. Estuaries are widely used to support human life activities such as water supply systems, agricultural irrigation channels to industrial matters. One common occurrence in estuaries is the mixing of water masses with different salinities, resulting in brackish conditions. This process is known as saltwater intrusion (SWI) [12]. saltwater intrusion has several factors that affect the amount of distance generated, including tides, river discharge, and the geometry of the estuary itself. These conditions ultimately affect the distance of saltwater intrusion and can reach deep into the upstream river resulting in salinity contamination of freshwater irrigation channels.

The saltwater intrusion phenomenon in the Kapuas Besar Estuary and watershed has several possibilities to occur with different distance variations at certain periods. The saltwater intrusion distance can be predicted by knowing several physical factors that occur when the phenomenon takes place with the support of adequate salinity field observation data both temporally and spatially [10]. To address the issue of seawater intrusion, Savenije (2012) developed a one-dimensional analytical modeling system that considers estuary geometry, river discharge, and curve fitting of field observation data. This modeling approach has been applied in numerous studies on seawater intrusion in 33 estuaries worldwide with various scenarios of river discharge flow in the study estuary area and can help predict seawater intrusion, especially during the dry season [9].

Consequently, this research aims to address the variations in saltwater intrusion distance during monthly periods in the Kapuas Besar Estuary using a one-dimensional analytical model. The predictions of saltwater intrusion distance obtained from this study can guide decisions related to raw water intake points, irrigation sources, ecosystem conservation, and land management in the Kapuas Besar Estuary and its surrounding watershed in West Kalimantan.

2. Materials and Methods

2.1. *Study Area*

This research study area is located on the Kapuas Besar estuary in West Kalimantan, Indonesia. The domain model is focused on the estuary geometry, including width, depth, and cross-sectional area. The continuous discharge of the Kapuas Besar River begins from the river source in Kapuas Hulu Regency and flows downstream into different regions, including Sungai Raya District, Kubu Raya District, and Rasau Jaya District, before reaching the Kapuas Besar estuary (Figure 1). The Kapuas River has a catchment area of 99,00 km2. The river discharge is influenced by the west and east monsoon cycle, ranging from 1,000 m3/s in the dry season to 10,000 m3/s in the wet season [6] and an average discharge of 5,500 m3/s [2]. The estuary mouth (x0) point will be placed at the Padu Empat trifurcation point, 11 km from the sea.

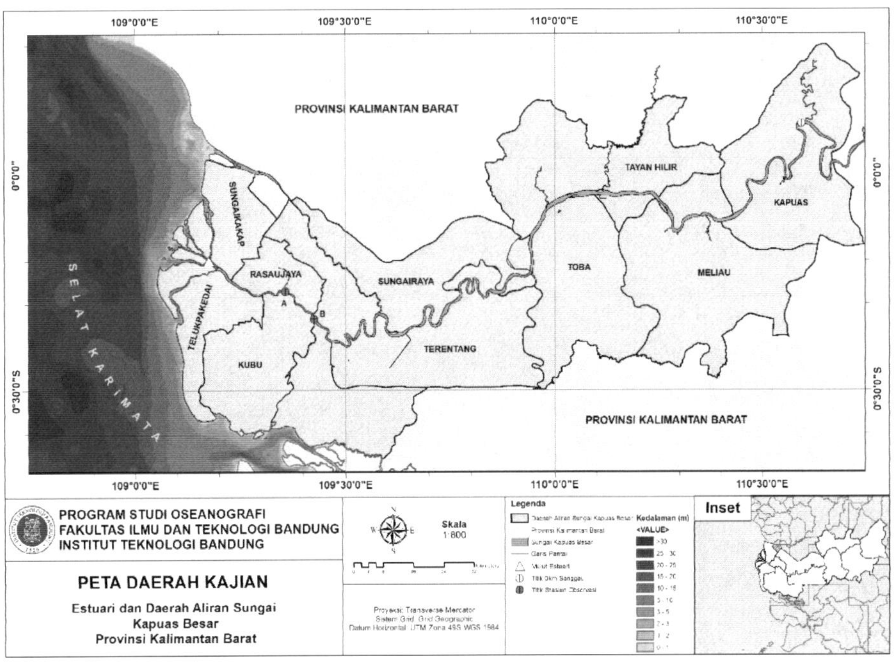

Figure 1. The study area and observation point of the model with the observation points (estuary mouth, point A, and point B) represent the distance of 0 km, 24 km, and 36 km.

2.2. *Data*

The data used in this research are estuary geometry, tides, salinity, and river discharge data from December 2011 to April 2015 (Tables 1 & 2). The estuary geometry data were obtained using estimated measurements from Google Earth with the average width and depth at the estuary mouth and every segment (1 km interval). The tides data were obtained from Tide Model Driver with the point of interest at the estuary mouth at the Padu Empat trifurcation point (0 km). The river discharge data were taken from the Regional Office of Kalimantan I River Basin, West Kalimantan Province, every day during the period 1998- 2013. The salinity data were obtained from the field observation data logger [6] with a depth of 4,6 m at 24 km (Rasau Jaya Station) and 36 km (Kubu Raya Station) from the estuary mouth every 30 minutes from August 2014 until April 2015. The field data has to be synchronized based on the tides (spring tide) to gain a slack water condition (high-water slack and low-water slack) when the current is near 0 m/s. Therefore, the salinity data doesn't need a tide correction in this model (Figure 2). The salinity data at the estuary mouth were obtained from several seasonal salinity research data: 20 psu (DJF) [1], 15 psu (MAM) [14], 24 psu (JJA) [7], and 15.8 psu (SON) [8].

Table 1. Salinity input at the estuary mouth (0 km).

Season	Salinity at the Estuary Mouth (psu)	Data retrieval time	Data Source
East monsoon (DJF)	20	December-February	[1]
First transitional (MAM)	15	May 2016	[14]
West monsoon (JJA)	24	June 2012	[7]
Second transitional (SON)	15.8	September 2016	[8]

Table 2. Salinity input at point A (24 km) and point B (36 km).

Season	Field Salinity Data		
	Distance (km)	Salinity HWS (psu)	Salinity LWS (psu)
East monsoon (DJF)	25	0.29	0.28
	36	0.29	0.19
First transitional (MAM)	25	0.04	0.03
	36	0.02	0.02
West monsoon (JJA)	25	15.50	12.78
	36	11.00	10.00
Second transitional (SON)	25	5.42	2.53
	36	1.83	1.83

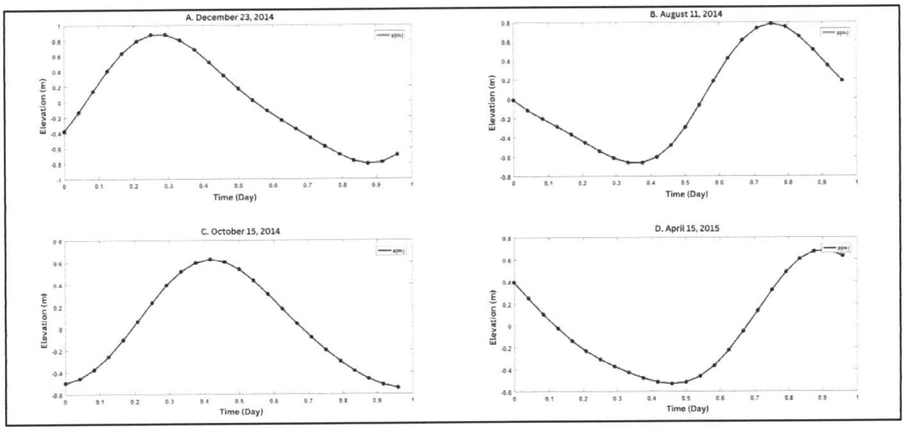

Figure 2. Tides for salinity synchronization data on (**A**) December, (**B**) August, (**C**) October, and (**D**) April.

2.3. Geometry analysis method

The geometry analysis method for the Kapuas Besar Estuary was based on the Savenije approach theory [11] for alluvial estuaries. The width, depth, and cross-sectional area are exponentially changing in this prismatic scheme. The value for A_0, B_0, and h_0 (Equations (1) to (3)) was obtained by fitting the simulated geometry characteristics to the observed data and will be used as boundary input data for the one-dimensional analytical model.

$$A = A_0 \exp\left(-\frac{x}{a}\right) \tag{1}$$

$$B = B_0 \exp\left(-\frac{x}{b}\right) \tag{2}$$

$$h = h_0 \exp\left(-\frac{x}{d}\right) \tag{3}$$

Where A, B, and h represent the cross-sectional area, width, and depth at the distance of x. A_0, B_0, and h_0 represent the cross-sectional area, width, and depth at the estuary mouth and a, b, and d are the cross-sectional area, width convergence length, and d can be found with $d=ab/(b-a)$. The Padu Empat trifurcation point is chosen to be the area of interest for this study referring to the geometry terms and condition approach on the alluvial estuary and the ideal estuary as can be seen in Figure 3 below.

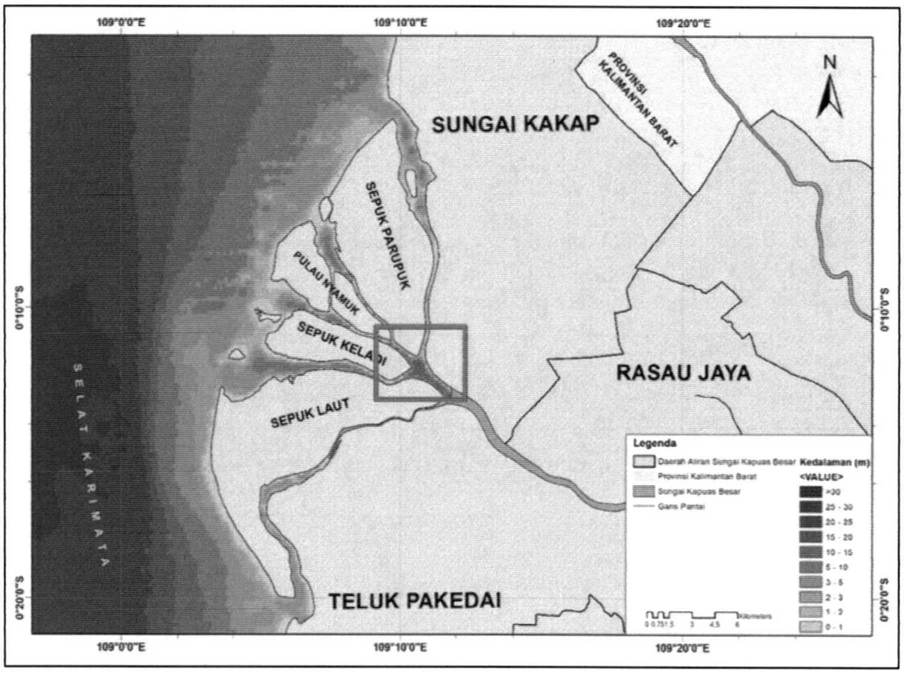

Figure 3. Point of interest for geometry analysis at Kapuas Besar Estuary.

2.4. *One-dimensional analytical model*

The model solves the saltwater intrusion distance prediction using Savenije's one-dimensional analytical approach for alluvial estuary [11]. The model is used to predict the saltwater intrusion distance with multiple scenarios during high, low, and average river discharge. In this model, the salinity distribution will be simulated under Tidal Average (TA) conditions and then converted into high-water slack (HWS) and low-water slack (LWS) conditions to determine the maximum and minimum saltwater intrusion conditions and determine the limits of saltwater intrusion in the study area. This analytical model assumes that the saltwater intrusion that occurs has no salinity stratification and uses the upstream prediction method towards the upstream of the river and the estuary is considered prismatic and shrinks upstream with a minimum width of the river width upstream. The calculation of salinity values in the Kapuas Besar Estuary will be done by segment with an average cross-sectional area of each segment. The distance between segments has a value of 1,000 m and the visualization of the geometry of measuring salinity values until it reaches a value of 0 psu (x_0, x_1, x_2, x_3, x_4, x_5, x_6, etc.) as it can be seen with Figure 4 below.

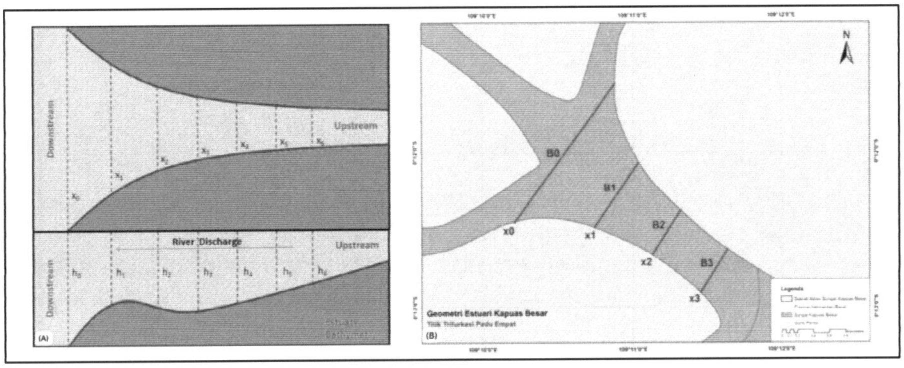

Figure 4. Geometry analysis application for the 1D model.

The equations applied in the 1-dimensional model by Savenije [x] are derived from the salt equilibrium equations for HWS, LWS, and TA as follows in Equation (4):

$$S_i - S_f = -\frac{A_i}{Q_f} D_i \frac{dS_i}{dx} \quad (4)$$

Where S_i is the salinity value sought at the x distance and S_f is the salinity limit for water flowing upstream which under the conditions in this study is 0 psu, D_i is the dispersion coefficient at the x distance, and A_i is the cross-sectional area at the x-point. Furthermore, the Van der Burgh coefficient (K) can help the calculation of the salt equilibrium equation as a shape factor coefficient for longitudinal salinity distribution in the estuary with an empirical method approach based on the average tidal dispersion under effective tidal dispersion conditions [13]. The coefficient is obtained from the results of consistent reduction of upstream tidal flow and the plot of effective dispersion against river discharge velocity with a narrowing cross-sectional area. The results of the Van der Burgh Coefficient can be written as in Equation (5) assuming 1 dimension on the x-axis.

$$\frac{dD_i}{dx} = -K\frac{Q_f}{A} \quad (5)$$

$$\frac{dS_i}{S_i - S_f} = \frac{1}{K}\frac{dD}{D} \quad (6)$$

D_i is obtained as a dispersion coefficient that will change exponentially upstream. Combining Equations (4, 5, & 6), the salinity and dispersion equations are expressed as Equation (7) below.

$$\frac{S - S_f}{S_0 - S_f} = \left(\frac{D}{D_0}\right)^{\frac{1}{K}} \quad (7)$$

S_0 and D_0 are the boundary conditions for salinity and longitudinal dispersion in the 1-dimensional analytical model, the longitudinal dispersion equation can be solved by substituting changes in the value of the dispersion coefficient D into Equation (8) as follows:

$$\frac{D}{D_{0i}} = 1 - \beta\left(\exp\left(\frac{x}{a}\right) - 1\right) \quad (8)$$

With β, the dispersion reduction for each x-distance is as follows,

$$\beta = \frac{KaQ_f}{D_0 A_0} \quad (9)$$

Then by substituting Equations (7)-(9), the saltwater intrusion distance equation can be simplified and written as in Equation (10). The value of the saltwater intrusion distance can be found with the condition where the salinity value sought has the same value as the freshwater salinity (0 psu) ($S_i = S_f$). Therefore when the seawater has a maximum distance upstream, it can be stated with $x = L$.

$$L = a\ln\left(\frac{1}{\beta} + 1\right) \quad (10)$$

The resulting saltwater intrusion distance (L) is in TA conditions, so it must be converted into HWS and LWS distance conditions with Equations (11) and (12), where E_0 is the horizontal tidal excursion distance obtained based on the results of curve-fitting calibration between the results of the model salinity curve and field salinity observation data on HWS and LWS conditions.

$$L_{HWS} = x + \frac{E_0}{2} \quad (11)$$

$$L_{LWS} = x - \frac{E_0}{2} \quad (12)$$

The calibrated parameters in this saltwater intrusion model are the Van der Burgh coefficient (K), dispersion coefficient at the mouth of the estuary (D_0), tidal excursion at the mouth of the estuary (E_0), and seawater salinity (S_0) [5]. The input parameters that can be changed in the saltwater intrusion model are river discharge (Q), tidal excursion (H), and saltwater salinity at the mouth of the estuary (S_0) which are adjusted based on the scenario conditions of the saltwater intrusion model in the seasonal period. Freshwater salinity (S_f) will be equated to a value of 0 in order to get the saltwater intrusion distance limit with a salinity of 0 psu.

2.5. *Error analysis*

Root Mean Squared Error (RMSE) error analysis method is used for the validation and performance of the 1D saltwater intrusion analytical model. Percentage Bias (PBIAS) and Nash-Sutcliffe Efficiency (NSE) coefficients are also used to determine the predictive power of the salt intrusion model. The equations of RMSE, PBIAS, and NSE are as follows:

$$RMSE = \sqrt{\frac{\sum_{i=1}^{n}(P_i - O_i)^2}{n}} \quad (13)$$

$$NSE = 1 - \frac{\sum_{i=1}^{n}\left(S_{pi} - S_{oi}\right)^2}{\sum_{i=1}^{n}(S_{oi} - S_{\bar{o}})^2} \quad (14)$$

$$PBIAS = \left[\frac{\sum_{i=1}^{n}\left(S_{pi} - S_{oi}\right)^2}{\sum_{i=1}^{n}(S_{oi} - S_{\bar{o}})^2}\right] \quad (15)$$

Where:
n = Data amount
S_p = Predicted salinity (psu)
S_o = Observed salinity
P_i = Predicted value
O_i = Observed value
$\bar{O}_i$ = Average observed value

3. Results and Discussion

3.1. Geometry result of the estuary

The start point (0 km) for the utilization of Equation (1)-(3) is at the Padu Empat trifurcation point before entering the Kapuas Delta (Figure 4.1). The geometry of the estuary in this area is similar to the ideal estuary proposed by Savenije [11]. Based on the observation data [6] the Kapuas Besar River has varying depths with a maximum depth of up to 17 m (from 0-76 km) and an average width of 600 m with the maximum width in the river reaching ±1,705 m at the bifurcation point branch [6].

The bathymetry of the Kapuas Besar watershed has silted up towards the downstream (estuary mouth) to an average depth of 8.8 m at the Padu Empat trifurcation point. Kastner's research [6] explained that there is a slope upstream so that the siltation that occurs in the Kapuas Besar Estuary has minimal effect on the river flow that flows towards the estuary mouth. Therefore, the resulting graph has been adjusted based on the ground elevation from the Padu Empat trifurcation point (1 m above sea level) to a distance of 76 km (9 m above sea level). The geometry characteristics of the Kapuas Besar Estuary that will be used as input data for the saltwater intrusion analytical model are shown in Table 3 as it follows.

Table 3. Geometry input value for the 1D model.

Kapuas Besar estuary geometry	Symbol	Value
Cross-sectional area at the estuary mouth	A_0	15,000 m^2
Cross-sectional area at the inflection point	A_1	8,000 m^2
Width at the estuary mouth	B_0	1,705 m
Width at the inflection point	B_1	800 m
Inflection point	x_1	1,000 m
Depth at the estuary mouth	h_0	8.8 m

3.2. Model calibration

The calibration carried out in this model is a curve-fitting process based on salinity data at 2 observation stations at a distance of 24 km and 36 km. The seasonal calibration results of the model in the Kapuas Besar Estuary have a bell-shaped saline intrusion curve.

Based on the calibration results, it is found that the difference in the resulting curve is largely influenced by the condition of the input parameters from each season, namely the dispersion coefficient at the mouth of the estuary (D_0) as the magnitude of salinity dispersion at each x-distance, then seawater salinity (S_0) as the initial value of salinity which affects the initial value of salinity at the mouth of the estuary and how much salinity will be dispersed at the x-distance, and the Van der Burgh coefficient (K) affected the slope of the curve in the salinity dispersion process until the salinity reaches a value of 0 psu at the x-distance point.

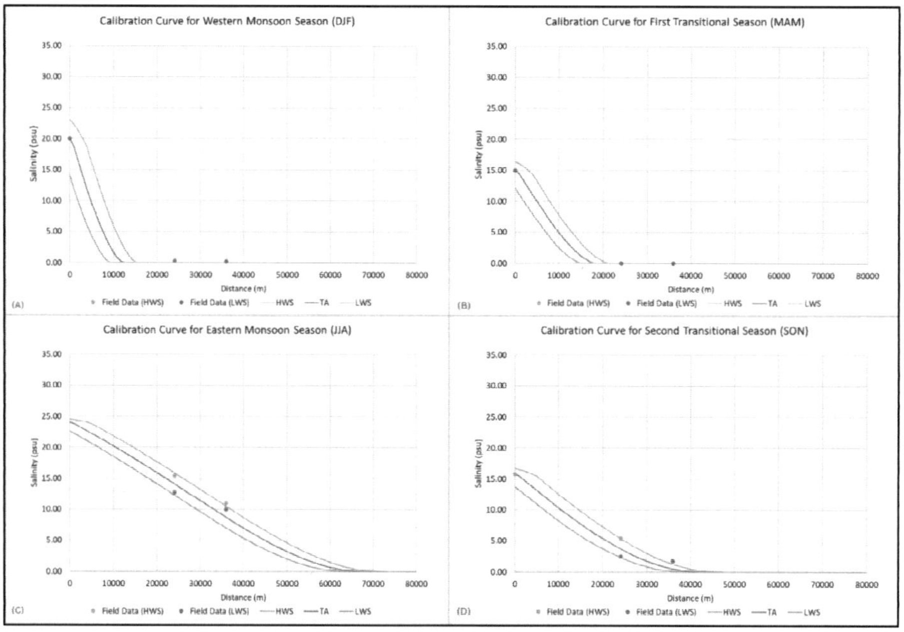

Figure 5. Seasonal salinity calibration curve for December-February (A), March-May (B), June-August (C), and September-November (D).

Each seasonal calibration result of the saltwater intrusion model (Figure 5) will be the main reference model for various river discharge scenarios (minimum, maximum, and average discharge) in the monthly scenario. The west monsoon season (DJF) calibration curve is used as the main reference for the model in December, January, and February, then the first transitional season (MAM) is used as the main reference for the model in March, April and May, then the east monsoon season (JJA) is used as the reference for the model in June, July, and August, and the second transitional season (SON) is used as the reference for the model in September, October and November.

3.3. Saltwater intrusion results

Based on the estuary number, the Kapuas Besar Estuary has a highly-stratified salinity stratification type from the Padu Empat trifurcation point (0 km) to a distance of 3.1 km upstream. While from a distance of 3.1 km to 73 km upstream has a well-mixed type of salinity stratification [4]. This is caused by the mixing process at a distance of 0-3.1 km then turned into a well-mixed after a distance of 3.1 km. With the result of well-mixed salinity stratification, the 1D saltwater intrusion model can be used in this study to predict the salinity value until the distance of 80 km.

3.3.1. Monthly saltwater intrusion distance prediction.

In general, the saltwater intrusion distance in December-February lengthens and shortens again in March, then lengthens again until June and shortens further until November. The range of saltwater intrusion distance differences in HWS and LWS conditions is 6,000 m (December-May) and 8,000 m (June-November), which is influenced by the calibration process for the four seasons. The largest margin value of the saltwater intrusion distance was obtained in July with a value of 45,000 m, while the smallest margin was obtained in April with a value of 7,000 m (see: Table 4).

Table 4. Saltwater intrusion prediction results.

No.	Bulan	Model Result					
		Minimum discharge scenario		Maximum discharge scenario		Average discharge scenario	
		Distance in LWS (m)	Distance in HWS (m)	Distance in LWS (m)	Distance in HWS (m)	Distance in LWS (m)	Distance in HWS (m)
1.	December	7,000	13,000	2,000	8,000	4,000	10,000
2.	January	9,000	15,000	2,000	8,000	6,000	12,000
3.	February	10,000	16,000	4,000	10,000	7,000	13,000
4.	March	7,000	13,000	1,000	7,000	3,000	9,000
5.	April	6,000	12,000	5,000	11,000	6,000	12,000
6.	May	16,000	22,000	7,000	13,000	11,000	17,000
7.	June	63,000	71,000	40,000	48,000	51,000	59,000
8.	July	62,000	70,000	25,000	33,000	37,000	45,000
9.	August	52,000	60,000	21,000	29,000	39,000	47,000
10.	September	38,000	46,000	16,000	24,000	26,000	34,000
11.	October	38,000	46,000	16,000	24,000	23,000	31,000
12.	November	17,000	25,000	10,000	18,000	13,000	21,000

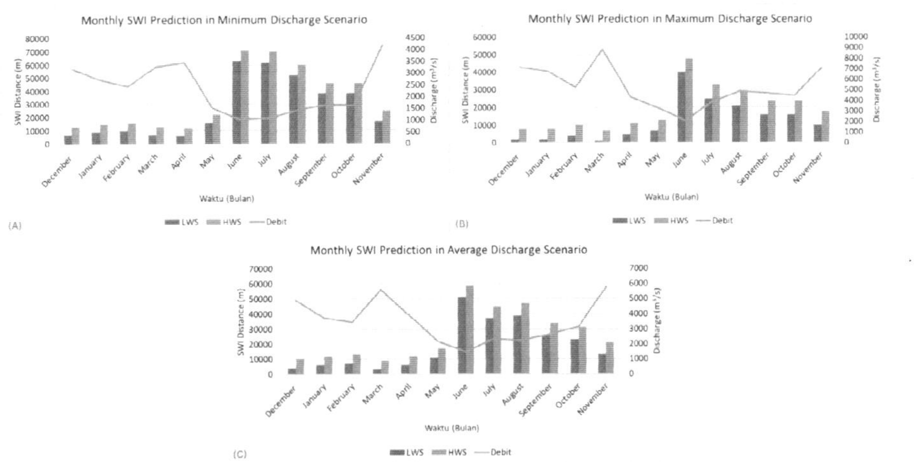

Figure 6. Monthly saltwater intrusion (SWI) distance prediction with several river discharge scenarios: minimum (A), maximum (B), and average (C).

In the monthly minimum river discharge scenario (Figure 6A & 7), the saltwater intrusion distance for the December-May period does not exceed 22,000 m with a stable increase or decrease, while for the June-October period has the maximum monthly saltwater intrusion distance is quite high up to 71,000 m and shortens until November. The reason was caused by the flow of river discharge during the June-October period (below 2,000 m^3/s). There is a drastic increase in saltwater intrusion distance at the May-June turnover and a drastic decrease at the October-November turnover, the reason it happened due to the influence of a very low river discharge in June of 1,034.55 m^3/s and a drastic increase in river discharge from 1,623.26 m^3/s in October to 4,207.14 m^3/s in November.

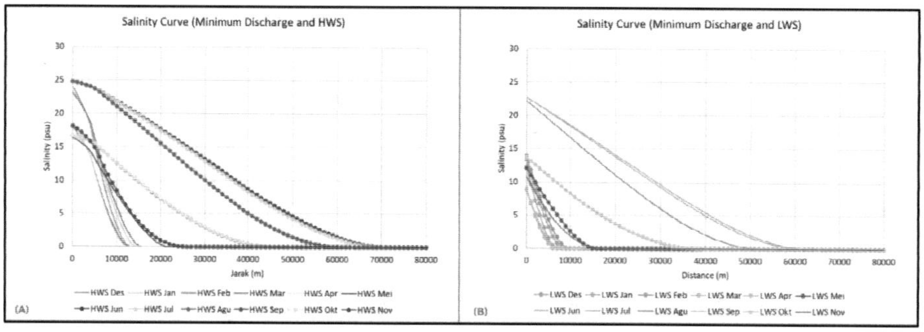

Figure 7. Saltwater intrusion model prediction results on minimum discharge scenario.

In the maximum monthly discharge scenario (Figure 6B & 8), most of the saltwater intrusion distances are quite low with values not exceeding 33,000 m with conditions of increasing and decreasing distance similar to the conditions of the minimum discharge scenario. However, it can be seen that there is an outlier in the June salinity curve with a distance of 48,000 m (HWS) and 40,000 m (LWS). It was influenced by maximum river discharge flow in June of only 2,109.94 m^3/s which is very different from other months with maximum discharges above 3,282.92 m^3/s.

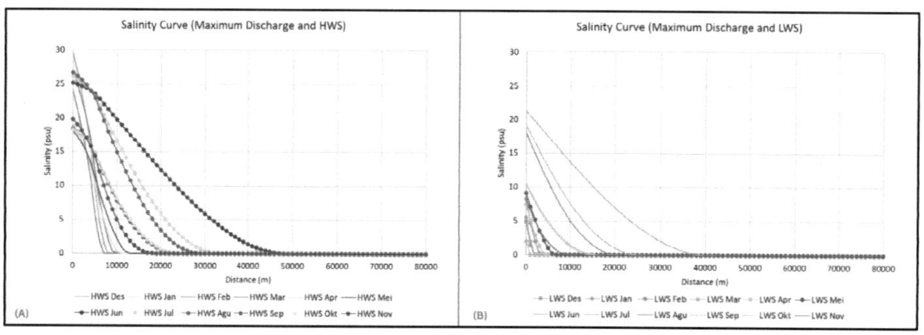

Figure 8. Saltwater intrusion model prediction results on maximum discharge scenario.

In the monthly average river discharge scenario (Figure 6C & 9) has a more detailed visualisation of the increased and decreased amount of river discharge between months. The graph generated in the average river discharge scenario looks smoother than other scenarios, this is due to the lack of inequality in the average river discharge between months. The distribution of the saltwater intrusion distance has similarities with the distribution at the minimum and maximum discharge with the saltwater intrusion distance being the median between the minimum and minimum distance each month. There is a fairly drastic increase in river discharge in March and November which results in outliers in the minimum and maximum saltwater intrusion distances in the monthly period.

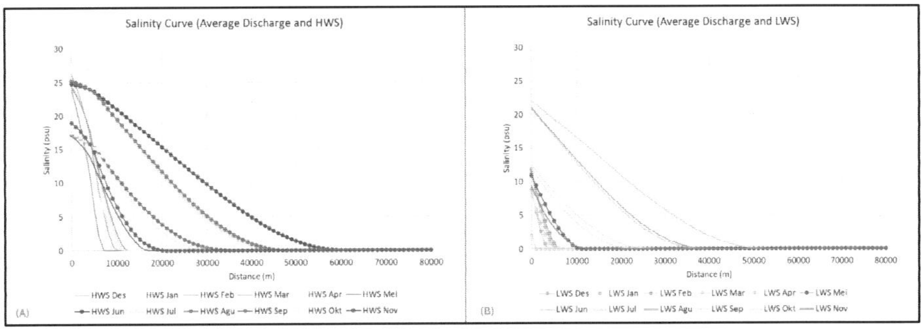

Figure 9. Saltwater intrusion model prediction results on average discharge scenario.

One of the factors that affect the saltwater intrusion distance in each model scenario is the rainfall that affects the amount of river discharge from the month before and after. In addition, the influence at the time of calibration of the saltwater intrusion graph also contributed to the resulting output, as for example in May and June which have a minimum river discharge that is not much different (1,520.08 m^3/s and 1,034.55 m^3/s) have much different results. One of the main reasons is that in May, the calibration curve will follow the first transitional season (MAM), whereas in June, it will follow the calibration curve for the east monsoon season (JJA) with distinct calibration input parameters.

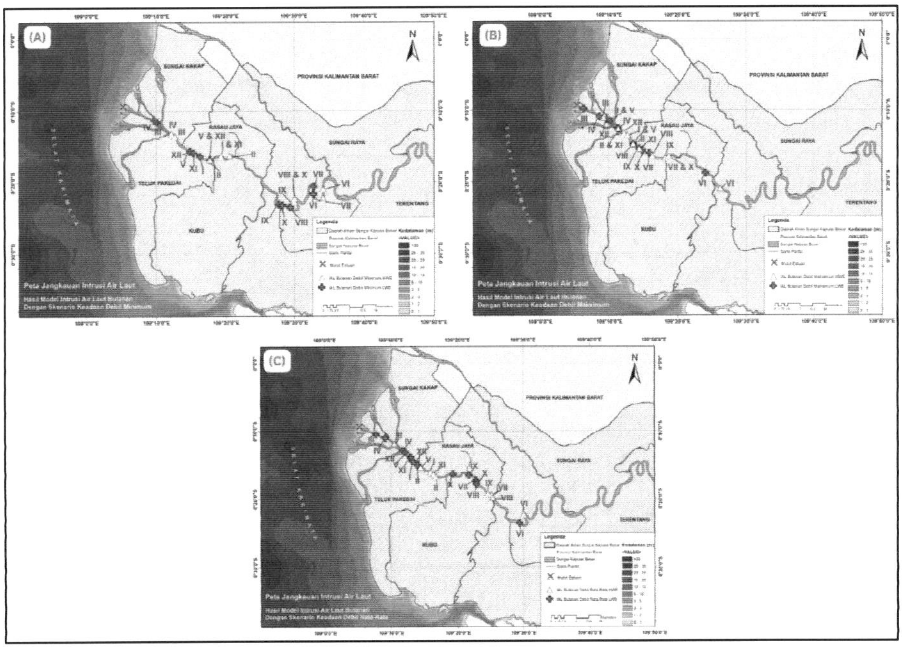

Figure 10. Spatial map for monthly saltwater intrusion distance prediction with several discharge scenarios (minimum (A), maximum (B), and average (C) from January (I) to December (XII) with HWS (red label) and LWS (blue label) conditions.

In the minimum river discharge scenario (Figure 10A), most of the saltwater intrusion has passed into Rasau Jaya Sub-district and is spread into 2 parts, namely in Rasau Jaya Sub-district and the next part has passed the observation station point of Kubu Raya Station to Sungai Raya Sub-district with the furthest point in June almost approaching the bifurcation point of the Kapuas Kecil River. The maximum river discharge scenario (Figure 10B) produced saltwater intrusion distances that were mostly spread from the Padu Empat bifurcation point to the Rasau Jaya sub-district boundary, but in this scenario, two saltwater intrusion points were obtained in June that had passed Rasau Jaya sub-district as outliers. In the scenario of average river discharge (shown in Figure 10C), the points of saltwater intrusion distance are evenly spread every month, with a reasonable distance between each point. The fairly even distribution of saltwater intrusion points starts from a distance of 9,000 m from the mouth of the estuary (0 km) to 59,000 m which has entered Sungai Raya Sub-district.

Reverse flow can be a factor in the long saltwater intrusion distances that occur in the Kapuas Besar River if the river discharge is <5000 m^3/s [6], where there will be changes in the direction of river discharge propagation from going downstream to the opposite direction. This condition happens because of the inability of the river discharge to withstand the mass of seawater entering upstream, resulting in water with high salt concentration entering far upstream of the Kapuas Besar watershed.

The results obtained in this study are in line with several previous studies, as stated by Kastner [6] that salinity is only detected at the Kubu Raya bifurcation point (data collection station) which is 36 km away with a very low river discharge and salinity never touches the Kapuas Kecil bifurcation point (85 km from the outer limit of the sea in the Kapuas Besar delta).

The results of this study can be used to determine the optimal location for installing water intake stations as a source of livelihood for communities around the Kapuas Besar watershed. The recommended location of the intake station installation point is to pass through the Kapuas Kecil bifurcation point as the furthest saltwater intrusion point, which is more than 71 km from the estuary mouth to avoid salinity contamination thus the irrigation water sources and raw water sources will be protected from seawater intrusion.

3.3.2. Maximum and minimum saltwater intrusion distance comparison.

The minimum saltwater intrusion distance was obtained in March with the maximum discharge scenario (8,853.82 m^3/s) and LWS conditions, resulting in a distance of 1,000 m located not far from the Padu Empat trifurcation point. Meanwhile, the maximum saltwater intrusion distance was obtained in June with the maximum discharge scenario (1,034.55 m^3/s) and HWS conditions, resulting in a distance of 71,000 m with a point in Sungai Raya District, as shown in Figure 11 below.

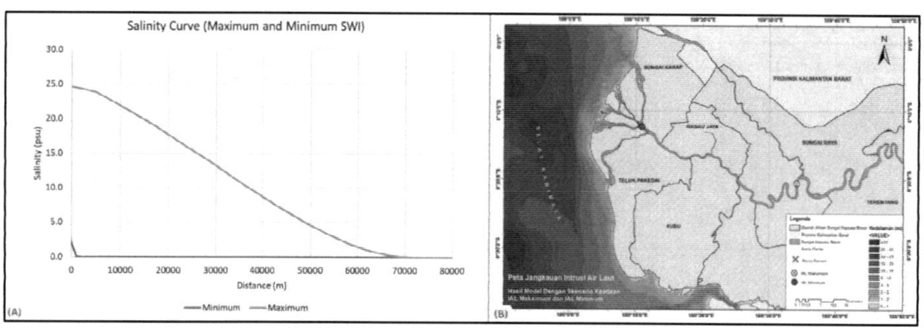

Figure 11. Maximum and minimum saltwater intrusion (SWI) distance comparison.

The range of saltwater intrusion distance between the maximum and minimum values is 70,000 m, which is mostly caused by differences in the initial value and river discharge in each month. The minimum saltwater intrusion is given a salinity input of 15 psu, a tidal excursion of 6,000 m, a Van der Burgh coefficient of 0.5, and a dispersion coefficient at the mouth of the estuary of 5,000 m^2/s, whilst the maximum saltwater intrusion is given a salinity input of 24 psu, a tidal excursion of 2,500 m, a Van der Burgh coefficient of 0.4, and a dispersion coefficient at the mouth of the estuary of 8,250 m^2/s.

3.4. Error analysis

The model has good performance with the minimum river discharge scenario based on the RMSE values of 0.82 psu (HWS) and 1.70 psu (LWS), which particularly still quite good due to the complexity of the calibration process for the 1D analytical saltwater intrusion model compared to the RMSE value of the average discharge scenario (3.58 psu for HWS and 5.11 psu for LWS) and the maximum discharge scenario (6.71 psu for HWS and 7.99 psu for LWS). The PBIAS and NSE values for HWS and LWS tidal conditions in 3 river discharge scenarios has the same effect as the RMSE values (Table 6). The minimum discharge scenario will result in smaller PBIAS value (1.64% for HWS and 2.85% for LWS) and NSE (0.99 for HWS and 0.96 for LWS) values than the average discharge scenario with PBIAS (4.67% for HWS and 9.27% for LWS) and NSE (0.82 for HWS and 0.63 for LWS) as well as the maximum discharge scenario with the highest PBIAS values (8.61% for HWS and 14.73% for LWS) and NSE (0.37 for HWS and 0.09 for LWS).

Table 5. RMSE value.

Time (seasonal)	Min. discharge scenario		Avg. discharge scenario		Max. discharge scenario	
	HWS (psu)	LWS (psu)	HWS (psu)	LWS (psu)	HWS (psu)	LWS (psu)
East monsoon (DJF)	1.06	1.70	2.95	5.49	5.43	8.72
First transitional (MAM)	0.97	1.55	2.14	3.98	5.08	7.51
West monsoon (JJA)	0.51	1.93	5.17	6.93	1.39	9.99
Second transitional (SON)	0.60	1.42	3.36	3.21	4.03	4.79
Average	**0.82**	**1.70**	**3.58**	**5.11**	**6.71**	**7.99**

Table 6. PBIAS and NSE value.

Tides condition	Min. discharge scenario		Avg. discharge scenario		Max. discharge scenario	
	PBIAS (%)	NSE (0-1)	PBIAS (%)	NSE (0-1)	PBIAS (%)	NSE (0-1)
HWS	1.64	0.99	-4.67	0.82	-8.61	0.37
LWS	2.85	0.96	-9.27	0.63	-14.73	0.09

The RMSE value (Table 5) shows that in the minimum river discharge scenario each season has a fairly good accuracy with a low error value compared to the average and maximum river discharge scenarios. The average RMSE value obtained in HWS (0.82 psu) and LWS (1.70 psu) conditions shows that the 1D analytical saltwater intrusion model is more accurate and suitable for use in discharge conditions <3,200 m^3/s in each month. The model worked quite well with an accuracy of 50% of the 1-year period, namely in May-October (6 months) with the average river discharge successively being 2,194.43 m^3/s, 1,467.58 m^3/s, 2,361.36 m^3/s, 2,246.63 m^3/s, 2,654.62 m^3/s, and 3,139.35 m^3/s.

Therefore, the greater the river discharge will produced the greater the RMSE value and likely underestimated the saltwater intrusion calculation with shorter distances due to the high error values of salinity in the Estuary and Kapuas Besar watershed.

4. Conclusions

In summary, the investigation of saltwater intrusion in the Kapuas Besar Estuary has been a successful endeavor and confirmed as appropriate for this case study. Moreover, successful salinity intrusion investigations have been carried out for the Kapuas Besar Estuary during both high-water slack (HWS) and low-water slack (LWS) conditions based on the resulting RMSE error values ranging from 0.82 psu to 1.70 psu, PBIAS values ranging from 1.64% to 2.85%, and NSE values ranging from 0.99 to 0.96.

The 1D saltwater intrusion analytical model is applicable during the May-October period with an average river discharge of less than 3,200 m^3/s, while in November-April it produces an underestimated model where model results are shorter than the actual distance. The saltwater intrusion distance increased in December-June by 49 km (due to a decrease in river discharge of 3,434.77 m^3/s) and decreased in the July-November period by 38 km (due to an increase in river discharge of 4,292.04 m^3/s). The minimum distance occurred in March (1 km from the estuary mouth) when the river discharge was 8,853.82 m^3/s, while the maximum distance occurred in June (71 km from the estuary mouth) when the river discharge was 1.034,55 m^3/s with an estimated point in the Teluk Bayur area, Kubu Raya District (10 km from the Kapuas Kecil bifurcation point).

In summary, the salt intrusion model proves to be highly dependable for investigating salt intrusion in the Kapuas Besar Estuary. Consequently, it is believed that the 1-D saltwater intrusion analytical model could serve as a valuable tool for determining the optimal location for installing water intake stations based on the salinity distribution of the Kapuas Besar Estuary.

References

[1] Blaber S J 2008 *Tropical estuarine fishes: ecology, exploitation and conservation* (Cleveland: John Wiley & Sons)

[2] Cohen S, Kettner A J, Syvitski J P, and Fekete B M 2013 WBMsed, a distributed global-scale riverine sediment flux model: Model description and validation *Computers & Geosciences* **53** 80-93.

[3] Dahuri R 1992 *Strategi penelitian estuari di Indonesia* Pros. Loka. Nas. Peny. Prog. Pen. Bio. Kelautan dan Proses Dinam. Pesisir. UNDIP, Semarang.

[4] Fitriyawita M, Jumarang M I, Apriansyah A, Sulistya W, and Saepudin M 2020 Hubungan Pola Garis Arus Angin (Streamline) dengan Distribusi Hujan di Kalimantan Barat. *Prisma Fisika.* **8(2)** 135-146

[5] Gisen J I, Adnan S S, Tajudin A A A, and Chan T X 2018 Investigation of salt intrusion condition in the Belat Estuary. *MATEC Web of Conferences*. Vol. 150:03005, EDP Sciences.

[6] Kästner K 2019 *Multi-scale monitoring and modelling of the Kapuas River Delta*. Doctoral Dissertation. Wageningen University and Research.

[7] Kusumaningtyas D I and Sukamto S 2015 Pengukuran Salinitas, Total Dissolved Solid (TDS) Dan Turbiditas Di Wilayah Pesisir Kabupaten Kubu Raya, Kalimantan Barat. *Buletin Teknik Litkayasa Sumber Daya dan Penangkapan* **13(1)** 43-49,

[8] Lestari A 2019 Rambatan Salinitas Di Sungai Kapuas Besar. *Jurnal Teknik Sipil*

[9] Nguyen A D 2008 *Salt Intrusion, Tides and Mixing in Multi-Channel Estuaries*. Doctoral Dissertation. UNESCO-IHE Institute Delft CRC Press, **153**.

[10] Savenije H H 1989 Salt intrusion model for high-water slack, low-water slack and mean tide on spread sheet. *Journal of Hydrology*. **107(1-4)** 9-18

[11] Savenije H H G 2005 *Salinity and tides in alluvial estuaries* (New York: Elsevier).

[12] Suprayogi I and Fatnanta F 2015 *Pengembangan Model Pengembangan Peramalan Intrusi Air Laut di Estuari Menggunakan Pendekatan Softcomputing*. Universitas Riau.

[13] Vd Burgh P 1972 Ontwikkeling van een methode voor het voorspellen van zoutverdelingen in estuaria, kanalen en zeeen. *Rijkwaterstaat Rapport*. **10-72**

[14] Widiastuti F and Rifanjani S 2017 Habitat Bekantan (Nasalis Larvatus Wurmb) di dalam dan Sekitar Areal Iuphhk-Ht Pt. Bina Silva Nusa Kecamatan Batu Ampar Kabupaten Kubu Raya Provinsi Kalimantan Barat. *Jurnal Hutan Lestari*. **5(3)**

Acknowledgments

Gratitudes are given to Mr. Karl Kästner to provide data for this research.

IOP Conf. Series: Earth and Environmental Science 1298 (2024) 012004 doi:10.1088/1755-1315/1298/1/012004

Analysis of deck house height and center of gravity in anti-capsize patrol boat

H F M Taqi [1], A Trimulyono[1*], O Mursid[1], M L Hakim[1]

[1]Department of Naval Architecture, Universitas Diponegoro, Indonesia

[*]Correspondence: anditrimulyono@live.undip.ac.id

Abstract. The design of patrol boats, especially in Indonesian waters which have extreme sea conditions, requires fairly good stability capabilities and special self-righting capabilities. Designing a self-righting ship will be closely related to the ship stability because the center of gravity (G) is affected by the load and the height of the deckhouse of a ship, which has implications for the ship's self-righting. Present study was carried out with experimental study for patrol boat design that has capability self-righting moment. The patroal boat has 4 deckhouse height variations with 2.01 m, 2.11 m, 2.21 m, and 2.31 m, respectively. There are 4 variations of the load conditions, i.e., the condition with 50% of the maximum amount of cargo, the condition 50% the passengers and cargo, full load condition without passengers, and the last one is the condition of the passenger baggage is only 50%. The results showed that a deckhouse with a height of 2.01 m has the worst GZ curve analysis, where in some conditions the results touch a negative number before heeling to 180°. The results showed by trial and error that the minimum deckhouse height is 2.07 m to have a self-righting moment in all conditions.

1. Introduction

The self-righting moment is a concept where the ship can return to its upright position after being upside down to a certain degree or even completely upside down (180°) without the help of an external force. The previous study of a patrol boat with an effective self-righting moment was performed using experimental and numerical methods [1,2]. The results showed that the numerical method has the same phenomenon as the experiment: the ship has self-righting in all condition load cases. Numerical methods such as CFD have become famous for marine use, such as fast ship design [3,4] and ship performance [5,6]. Another CFD scheme such as mesh-free CFD, i.e., smoothed particle hydrodynamics (SPH) that is suitable for marine engineering cases such as water wave propagation [7,8] and sloshing in LNG cases [9-11]. Designing a self-righting ship will closely affect the boat's stability because the load affects the centre of gravity (G). The height of the superstructure of a vessel has implications for the ship's self-righting conducted by Bai et al. [12] shows that the height of the centre of gravity affects the ship's self-righting moment, where there is a graph of the righting lever which shows that the higher GZ (height of the centre of gravity), the ship's self-righting moment will decrease its ability.

According to previous research, the addition of the height of the superstructure on the ship influences the stability of the ship because there is also an increase in the volume of the superstructure, which will have an impact on increasing the volume of the part of the ship that is submerged in water when the vessel is tilted at a certain degree when water superstructure. This influence can be seen from the area of the stability arm curve (GZ curve), where the elevation of the superstructure affects the maximum GZ magnitude and the angle at which the maximum GZ [13]. The present study is based on previous works [1], i.e., a patrol boat with four variations using different deckhouse heights where there are heights of 2.01 m to a height of 2.31 m with a difference of 0.1 m each. The analysis results showed that the condition of the ship with the lowest deckhouse height does not have an excellent self-righting moment compared to the other variations, whereas in a fully loaded condition, a patrol boat with a deckhouse of 2.01 m has a negative moment at an angle at 17°. Meanwhile, ships with the highest deckhouse height at 2.04 m find the best self-righting moments. This research was continued by validating and simulating using Star-CCM+. Patrol boats with a height of 2.04 m are simulated with maximum load conditions, able to return to an upright position in 0.7 seconds and produce stability in 3 seconds.

This research focuses on the modification of the height of the deckhouse and finding out the minimum deckhouse height that has a self-righting moment. Finally, the results showed that the minimum deckhouse height was found to be more than 2.0 m.

2. Methods

2.1. Ship Model

The reference ship model based on MV Barracuda which was designed in previous studies [2]. For the main dimension of the ship can be seen in Table 1 and for the three dimension model shows in Figure 1.

Table 1. Main Dimension Data (units in m)

No	Main Dimension	Dimension
1.	Length Over All	13.7
2	Length of Water Line (LWL)	13.0
3	Breadth Moulded (B)	4.20
4	Draught (T)	1.15
5	Height (H)	2.19

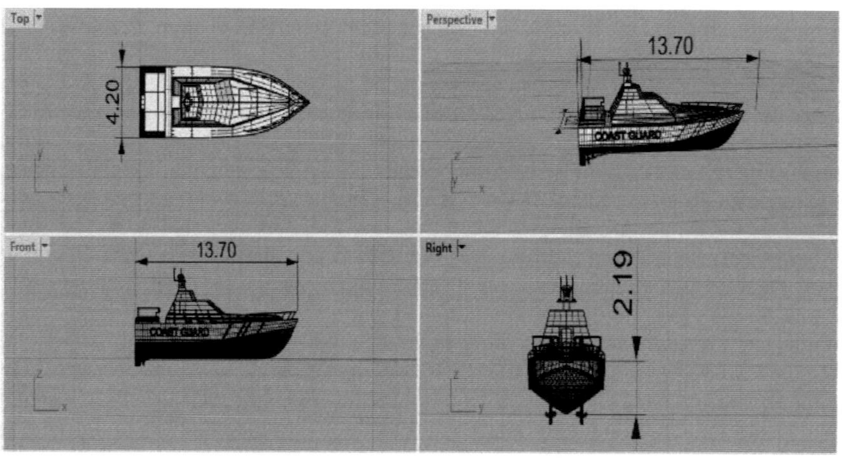

Figure 1. Three dimension of ship model

2.2. Deckhouse Variable

The reference variations for deckhouse heights were based on previous studies where four types of deckhouse patrol boats were used. The variations in the height of the deckhouse are illustrated in Figure 2, in which the change is 5%, 10% and 15% from the base model deckhouse height, so each model has an additional height of 0.1 m.

Therefore, the authors made variations in the weight of the component into four conditions by changing the percentage of the weight of the DWT component.

The four conditions are:

a. The ship's condition with 50% cargo weight, 0% passenger baggage, and 100% passengers totalling six people.
b. 50% total passengers with 0% passenger baggage and 50% cargo.
c. Only 50% of all passenger baggage.
d. 100% load, no passengers.

The analysis was carried out using the Maxsurf software to determine the GZ curve or stability calculation of moment koppel.

3. Results and Discussion

3.1. The Calculation of GZ curve

The load loading test was carried out out in 4 conditions with LWT data for four ship models as follows.

Table 3. LWT and CoG Data of Patrol Ship

Model	LWT (ton)	LCG (m)	TCG (m)	VCG (m)
Deckhouse 1	11.17	4.87	0	1.23
Deckhouse 2	11.18	4.85	0	1.23
Deckhouse 3	11.21	4.86	0	1.24
Deckhouse 4	11.27	4.87	0	1.25

Load case 1 is where each cargo weighs 50% of all and the condition of no passenger baggage but 100% of passengers. The results show that each variation of the ship's deckhouse has good stability, and no ship models touch the negative GZ to a 180° angle. This condition indicates that each ship model has a self-righting moment, therefore, the ship has a positive GZ curve (see Figure 3 (a)).

Load case 2 is a condition where the weight reduction is 50% of passengers, and the remaining cargo is 100%. The analysis shows that all deckhouses except deckhouse one experience good self-righting moments. Deckhouse 1 failed to return to the upright position at a heel angle of 170°, furthermore, deckhouse 1 has a negative GZ curve in this load condition (see Figure 3 (b)).

Load case 3 is a condition where the ship has three passengers who are placed on one side of the ship's passenger section, but the other side is left empty and also, for cargo conditions, is 50% of the total weight of the overall cargo. The analysis showed that each variation of the ship's deckhouse has good stability, and no ship models touch the negative GZ to heel angle of 180°. This condition indicates that all deckhouse heights have a positive GZ curve. However, this condition is unbalanced because the passenger load is only on one side of the ship, so there is a shift in the centre of gravity on the transverse axis (see Figure 3 (c)).

Load case 4 is a condition where all cargo is 100% fully loaded, but there are no passengers. The analysis showed that Deckhouse 1 failed to return to its original position at an angle of 170°. In contrast, the other deckhouses had good self-righting moments (see Figure 3 (d)). Therefore, deckhouse 1 did not have a self-righting moment.

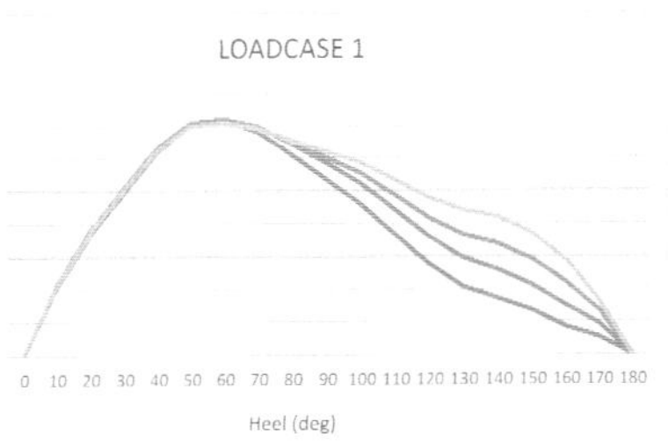

(a)

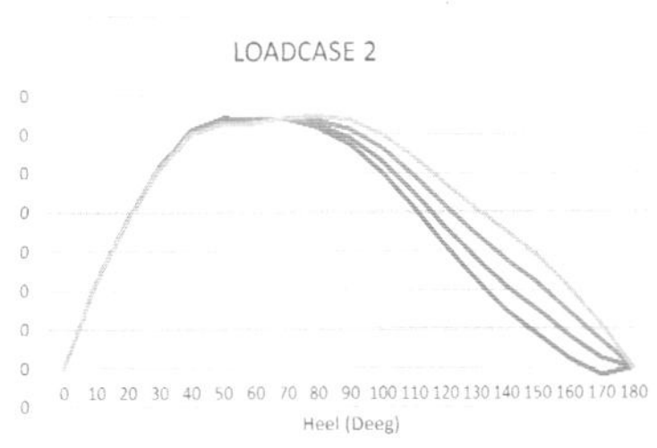

(b)

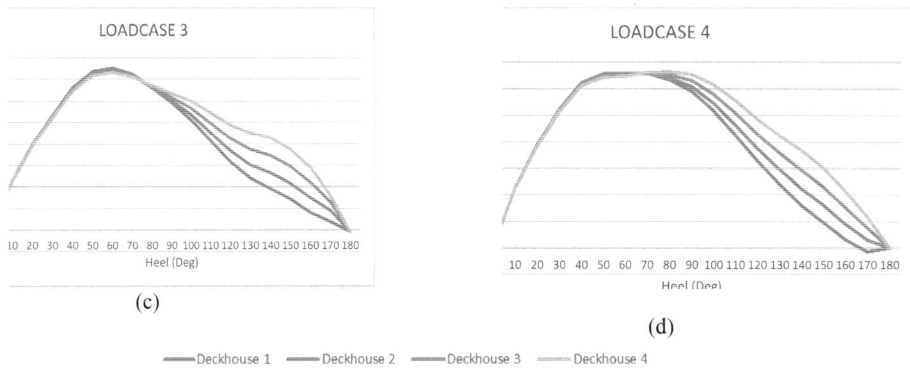

Figure 3. GZ curve for all loadcase condition

4. Conclusion

It was found the self-righting moment has linierly effect on deckhouse of patrol boat, nevertheless based on variations of deck-house in this study conclusions could be drawn as follows:

a. The distribution of cargo loads on patrol boats have significant influences to the stability of patrol boats. DWT distribution, especially for small ships, needs to be symmetrical as results ship have self-righting cabilities.
b. The minimum deckhouse height is 2.07 m with a ratio between the height and width of the ship of 4.2 m, the result is 1 m: 1,415 m.

Acknowledgments

The authors wish to thank laboratory of ship hydrodynamics Naval Architecture Department Universitas Diponegoro for computer facilities and funding conference registration fee.

References

[1] A. Trimulyono, A. F. Zakki, and M. A. Fuadi, "Experimental study of a ship with the self-righting moment in extreme condition," *IOP Conf. Series: Earth and Environmental Science*, 2022.

[2] A. Trimulyono, M. A. Fuadi, A. F. Zakki, O. Mursid, and M. Iqbal, "Design of Anti-Capsize Ship for Patrol Vessel with the Self-Righting Moment," *J. Mar. Sci. Eng.*, vol. 11, no. 1, 2023.

[3] Samuel, A. Trimulyono, P. Manik, and D. Chrismianto, "A numerical study of spray strips analysis on fridsma hull form," *Fluids*, vol. 6, no. 11, 2021.

[4] B. Utomo, S. Samuel, P. Manik, A. Chiquita, P. A. Syaiful Tambah, and Y. Zhang, "Hydrodynamic Analysis of High-Speed Vessels in Shallow and Deep Water Using CFD," *Kapal: Jurnal Ilmu Pengetahuan dan Teknolologi Kelautan*, vol. 20, no. 2, pp. 146–153, 2023.

[5] M. A. Asis, E. A. A. Dian, and R. R. Fadhil, "The Prediction of Hydrodynamic Performance of Pinisi Ship Muhammad," *Kapal: Jurnal Ilmu Pengetahuan dan Teknolologi Kelautan*, vol. 20, no. 2, pp. 134–145, 2023.

[6] D. A. Putri, M. H. N. Aliffrananda, S. Riyadi and I.K.P.A. Utama, "Numerical Analysis on Added Resistance of a Crew Boat with Variation of Wave Period," *Kapal: Jurnal Ilmu Pengetahuan dan Teknolologi Kelautan*, vol. 20, no. 12, pp. 16–26, 2023.

[7] M. A. Salvikran, A. Trimulyono, and E. S. Hadi, "Validation of multiphase water wave propagation using smoothed particle hydrodynamics," *IOP Conf. Ser. Earth Environ. Sci.*, vol. 1166, no. 1, pp. 012032, 2023.

[8] A. Trimulyono and H. Hashimoto, "Experimental Validation of Smoothed Particle Hydrodynamics on Generation and Propagation of Water Waves," *J. Mar. Sci. Eng.*, vol. 7, no. 1, pp. 17, 2019.

[9] A. Trimulyono, D. Chrismianto, H. Atthariq and S. Samuel, "Numerical Simulation Low Filling Ratio of Sway Sloshing in the Prismatic Tank Using Smoothed Particle Hydrodynamics," *CFD Lett.*, vol. 14, no. 7, 2022.

[10] A. Trimulyono. S.T.M.O. Gultom, E.S. Hadi and D.B. Purwanto, "Investigation of Sloshing with Vertical and Horizontal Baffle in the Prismatic Tank using Meshfree CFD," *CFD Lett.*, vol. 15, no. 6, 2023.

[11] A. Trimulyono., S. Utami, D. Chirsmianto., P. Manik, "Investigation of Sloshing in Different Tank Shapes using Smoothed Particle Hydrodynamics," *CFD Lett.*, vol. 15, no. 12, 2023.

[12] T. Bai, Z. Ding, X. Wang, Z. Zhang, and F. Zhang, "Theoretical Analysis of the Performance of a Self-Righting Boat," in *International Conference on Ocean, Offshore, and Arctic Engineering*, 2017.

[13] A. B. C. Putra and H. Hasanudin, "Desain Self-Righting Rescue Boat untuk BASARNAS," *J. Tek. ITS*, vol. 8, no. 2, 2019.

IOP Conf. Series: Earth and Environmental Science 1298 (2024) 012005 doi:10.1088/1755-1315/1298/1/012005

Tanker and SPM Motion Response Analysis due to Environmental Load

Y. S. Hadiwidodo[1], **F. Gymnastiar**[1*], **N. Syahroni**[1], **A. Rachman**[2]

[1] Ocean Engineering Department, Institut Teknologi Sepuluh Nopember, Surabaya, Indonesia

[2] Banner SPM Terminals Pty. Ltd., Australia

E-mail: farhangymm@gmail.com

Abstract. The use of tankers for oil and gas distribution requires auxiliary structures such as Single Point Mooring (SPM). In the process of loading cargo, tankers must maintain alevel of stability so that the load transfer process runs smoothly. Stability needs to be checked using inputs in the form of structural responses in free-floating conditions to get structural responses when moored. Analysis of the motion response of tankers and SPM in free-floating conditions showed that ballast tanker loads had a greater motion response than full loads with the largest movement on roll at 3.59 deg/m. The moored motion response of the tanker has the greatest amplitude in the surge movement towards the AP with a value of -0.2251 m.

1. Introduction

Indonesia is known as a country rich in valuable natural resources, especially oil and natural gas. These resources continue to be exploited to meet basic oil and gas needs. In offshore oil and gas extraction, there are two main types of facilities, namely fixed and floating structures. Apart from that, the construction of offshore platforms can be categorized into several types [1] [2], namely:

1) Floating structures: semi-submersibles, drilling vessels, tension leg platforms (TLP), and jack-ups.
2) Fixed structures: jacket platform, concrete gravity, and tripod.
3) Flexible or suitable structures, namely articulated and shielded towers.

Floating structures have extraordinary adaptability and are more efficient and effective, so they are suitable for operation in deeper waters compared to fixed structures [3]. However, floating structures are vulnerable to waves, which cause movement that can disrupt the comfort of the working crew's operations. Waves will cause dynamic responses in floating structures [4-15], so these movements need to be predicted early in the design.

Floating structures use Floating Storage and Offloading (FSO) or storage tankers as the primary means of transportation for oil and gas distribution. Among other things, shuttle tankers are essential in transporting oil to refineries or sending refined oil products for further distribution. This method of transportation offers greater flexibility and allows efficient oil distribution to different regions. To facilitate cargo transfer from shuttle tankers, a mooring system known as Single Point Mooring (SPM) can be used, simplifying and improving the transfer process.

SPM is an offshore floating structure designed as a mooring and interconnection point for loading and unloading gas and liquid products from tankers. Its effectiveness in the oil mining process makes it a component that is widely used in conjunction with FSOs or tanker units [16]. Overall, these structures are essential in facilitating efficient and smooth operations in the oil mining industry. Stability analysis of tankers during loading and unloading via SPM must be carried out to ensure the operation process runs smoothly. The analysis was carried out to obtain the response characteristics of the tanker ship and SPM motion, which then became the input for analysis of the tanker ship's motion response when moored at the SPM.

2. Research Methodology

2.1 Study of Literature and Data Collection

Study and collection of literature as a reference or study material and theoretical sources needed in research. The data used in the studies are as follows:
1) Shuttle tanker and SPM principal dimensions
2) Material properties of mooring lines and mooring hawsers used.
3) Environmental data is in the form of wind, waves, and currents.

2.2 Shuttle Tanker and SPM Modelling

The modeling of shuttle tankers and SPM is supported by response motion basis software, utilizing principal dimension data provided in Table 1 and dimension data for SPM from Banner SPM Terminals Pty. Ltd. Available in Table 2 [17].

Table 1. Shuttle tanker principal dimensions.

Parameter	Data	Unit
Ship Name	Elizabeth I.A	-
Class	ABS	-
IMO	9257149	-
LOA	332	m
LBP	320	m
Breadth	58	m
Height	31	m
Draft	23.5	m

Table 2. SPM principal dimensions.

Parameter	Data	Unit
Outer Skirt Diameter	17.885	m
Hexagonal Diameter	13.9	m
Buoy Hull Height	5.5	m
Design Draft	3.1	m
Displacement	408	ton

2.3 Model Validation

The modeled structure's validation will follow the ABS Rules for Building and Classing Mobile Offshore Units (MOU).

2.4 Tanker and SPM Motion Response Analysis

Upon successfully meeting the validation criteria, motion response analysis is performed as a Response Amplitude Operator (RAO) to facilitate graphic processing and interpretation. Mooring system modeling is done with mooring system analysis-based software. The tanker load condition scenario is applied to the analysis, which will be described in the next chapter. After the modeling is complete, proceed with analyzing the motion response of the tanker when tethered to the SPM.

3. Result and Discussion

3.1 Shuttle Tanker and SPM Modelling

Figure 1 displays the outcomes of the Elizabeth I. A. shuttle tanker in a lines plan model from body plan, breadth plan, and sheer plan. Figure 2 displays the SPM model from the top, side, and isometric view.

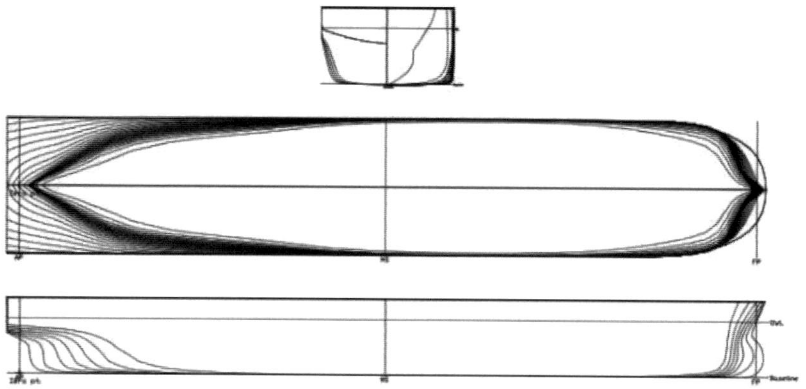

Figure 1. Tanker modeling

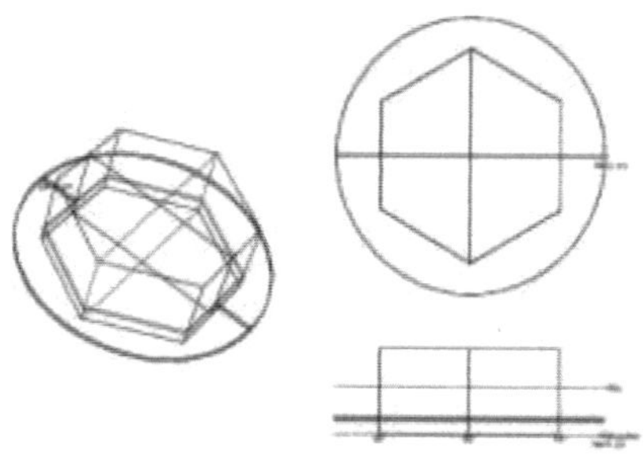

Figure 2. SPM modeling

3.2 Model validation

Validation involves comparing the model's results obtained from the software with existing data. By the ABS Rules for Building and Classing MOU [18], validation is conducted with a maximum allowable error of 2% for displacement and 1% for other parameters. The validation results for the tanker and SPM can be observed in Table 3 and Table 4, respectively. Because the validation results have been met, it can be continued in the following analysis.

Table 3. SPM model validation.

Parameter	Data	Software	Error
Outer Skirt Diameter	17.885	17.886	0.01%
Hexagonal Diameter	13.9	13.899	0.01%
Buoy Hull Height	5.5	5.5	0%
Design Draft	3.1	3.1	0%
Displacement	408	405.2	0.69%

Table 4. Tanker model validation.

Parameter	Calculation	Software	Error
Displacement	386416	389765	0.87%
Vol. Displacement	376991	380259	0.87%
LWL	329.6	329.996	0.12%
LCB	169.035	169.521	0.29%
VCB	12.33	12.39	0.48%
Draft	23.5	23.5	0%
CB	0.839	0.837	0.43%
CP	0.841	0.845	0.26%
CM	0.997	0.99	0.73%

3.3 Structure Motion Response Free Floating Condition

The structural analysis of the shuttle tanker and SPM in regular waves is performed using response motion-based software. The analysis is conducted under two tanker conditions: full load and ballast load. For the tanker, simulations are conducted with headings at 0°, 45°, 90°, 135° and 180°. However, due to the symmetrical shape, the SPM simulations for the SPM are carried out only at headings of 0°, 45°, and 90°. Table 5 presents the maximum RAO values for each tanker's six degrees of freedom under full load conditions. On the other hand, Table 6 shows the maximum RAO values for each of the tanker's six degrees of freedom movements under ballast load conditions.

Table 5. Maximum RAO full load condition.

Mode	Loading Direction					Maximum
	0°	45°	90°	135°	180°	
Surge (m)	0.81	0.62	0.05	0.61	0.81	0.81
Sway (m)	0	0.61	0.92	0.61	0	0.92
Heave (m)	0.86	0.93	1.44	0.94	0.86	1.44
Roll (deg)	0	1.35	1.85	1.35	0	1.85
Pitch (deg)	0.53	0.6	0.33	0.53	0.51	0.6
Yaw (deg)	0	0.26	0.02	0.26	0	0.26

Table 6. Maximum RAO ballast load condition.

Mode	Loading Direction					Maximum
	0°	45°	90°	135°	180°	
Surge (m)	0.84	0.63	0.01	0.63	0.84	0.84
Sway (m)	0	0.63	0.95	0.63	0	0.95
Heave (m)	0.87	0.94	1.17	0.94	0.87	1.17
Roll (deg)	0	1.72	3.59	1.7	0	3.59
Pitch (deg)	0.54	0.57	0.1	0.55	0.54	0.57
Yaw (deg)	0	0.27	0.02	0.27	0	0.27

Based on the analysis results under full load conditions, the highest value obtained in translational movement is in the heave motion at a heading of 90°, with a value of 1.44 m/m. Regarding rotational movement, the roll motion at a heading of 90° has the highest value, measuring 1.85 deg/m. Under ballast load conditions, the highest value in translational movement is observed in the heave motion at a heading of 90°, measuring 1.17 m/m. This value is smaller than the corresponding value in the full-load condition. However, in rotational movement, the roll motion at a heading of 90° exhibits the highest value, measuring 3.59 deg/m. This value is greater than the roll motion value in the full-load condition by 52%.

Despite the roll motion value increase in ballast load condition, it is relatively small compared to the ship's dimensions. These results indicate that the tanker possesses typical seakeeping characteristics. The RAO value for the primary vertical motion mode is not excessively high, which suggests that the ship has an excellent damping ability and stiffness factor, contributing to its stable behavior during motion at sea. Table 7 displays the maximum RAO values for each SPM's six degrees of freedom movement.

Table 7. Maximum RAO on SPM.

Mode	Loading Direction			Maximum
	0°	45°	90°	
Surge (m)	1.658	1.172	0.054	1.658
Sway (m)	0	1.172	1.658	1.658
Heave (m)	0.987	0.987	0.987	0.987
Roll (deg)	0	2.363	3.19	3.19
Pitch (deg)	3.323	2.329	0.127	3.323
Yaw (deg)	0.001	0.185	0.065	0.185

The SPM analysis revealed that the highest value obtained for translational movement is in surge at 0° heading and sway at 90° heading, with a value of 1.658 m/m. on the other hand, in rotational movement, the highest value is observed in pitch motion at 0° heading, with a value of 3.323 deg/m. The higher response values for the SPM are attributed to its small geometry and free-floating condition. However, in real-world applications, this issue has been effectively addressed by implementing mooring lines capable of withstanding the SPM's movements. The SPM's response can be significantly reduced by employing these mooring lines, leading to improved stability and functionality.

3.4 Mooring System Modelling Scenario

After obtaining a structural response in free-floating conditions, motion analysis will continue when the structure is moored. The analysis utilizes various input data, including the response of motion structures, such as displacement RAO, load RAO, wave drift, hydrodynamic drag, wind drag, stiffness of the structures, added mass, and damping. Additionally, the material properties of the moorings used in this case, the hawser used in this analysis is Double Braided mooring hawser with 18" Circ: 70 m length and 759-ton minimum breaking load. The chain studless mooring R3 type with a 4-inch diameter has the Minimum Breaking Load of 1.960 kips selected for the mooring line.

Moreover, the analysis requires essential environmental data, which includes sea depth, wave spectrum, wind speed, current speed, and their respective headings. All this environmental data is presented in Table 8 for reference during analysis. In this analysis, the wind speed considered is the 3-second gusts, which allows for capturing the maximum wind load value. Additionally, the environmental load applied in the analysis is collinear, meaning that it acts in the same direction. The environmental load includes the combined effects of wave, current, and wind load. This load scenario is applied at a sea depth of 35 meters. Due to the collinearity and alignment of the environmental load, it results in the highest impact on the structures. The combined effect of these forces amplifies their impact, making them the dominant factor affecting the structural response in this analysis. The analysis is performed using a time domain analysis, enabling the modeling of a non-linear matrix system and accurately representing all external loads. The analysis encompasses both full load and ballast load conditions for the tanker. Full-load is when the tanker has 100% of the cargo, while ballast load is when the tanker has 20%. The configuration used in the analysis is an In-Line configuration where the tanker is aligned with one of the mooring lines on the SPM. Based on the ABS Guide for Position Mooring System, the minimum duration for time domain analysis is 3 hours (10800 s) [19]. A cut-off method is applied, involving a selection of a 200-second time frame during which peak waves occur. It allows for the extraction of the maximum motion response of the structures.

Table 8. Environmental Data.

Parameter	Unit	Omni
Wind Speed		
1-minute winds	m/s	22.6
10-minute winds	m/s	20.2
1-hour winds	m/s	18.2
3-sec gusts	m/s	26.8
Wave		
Hmax	m	6.4
Associated Period	s	8
Hs	m	3.3
Crest Height	m	3.4
Current Speed		
Surface Current	cm/s	44
25% from surface	cm/s	42
50% from surface	cm/s	40
75% from surface	cm/s	36
1m from the bottom	cm/s	26
Storm		
Storm Surge	m	0.1

3.5 Tanker Motion Response Moored to The SPM

From the analysis, the distribution of motion responses in the structure during the 2000s is produced in Figures 3 and 4, and Table 9 shows the maximum value of movements that occur in full-load and ballast-load tankers, respectively, as follows. It can be seen in the table above that in the surge movement of ballast conditions and the resulting value is more significant than in full conditions. It is due to the cargo from the ship being lighter, with the damping and stiffness of the ship getting smaller, so the environmental load has a more significant impact. In movements other than surge, the resulting movement value is minimal due to the heading of the environmental load.

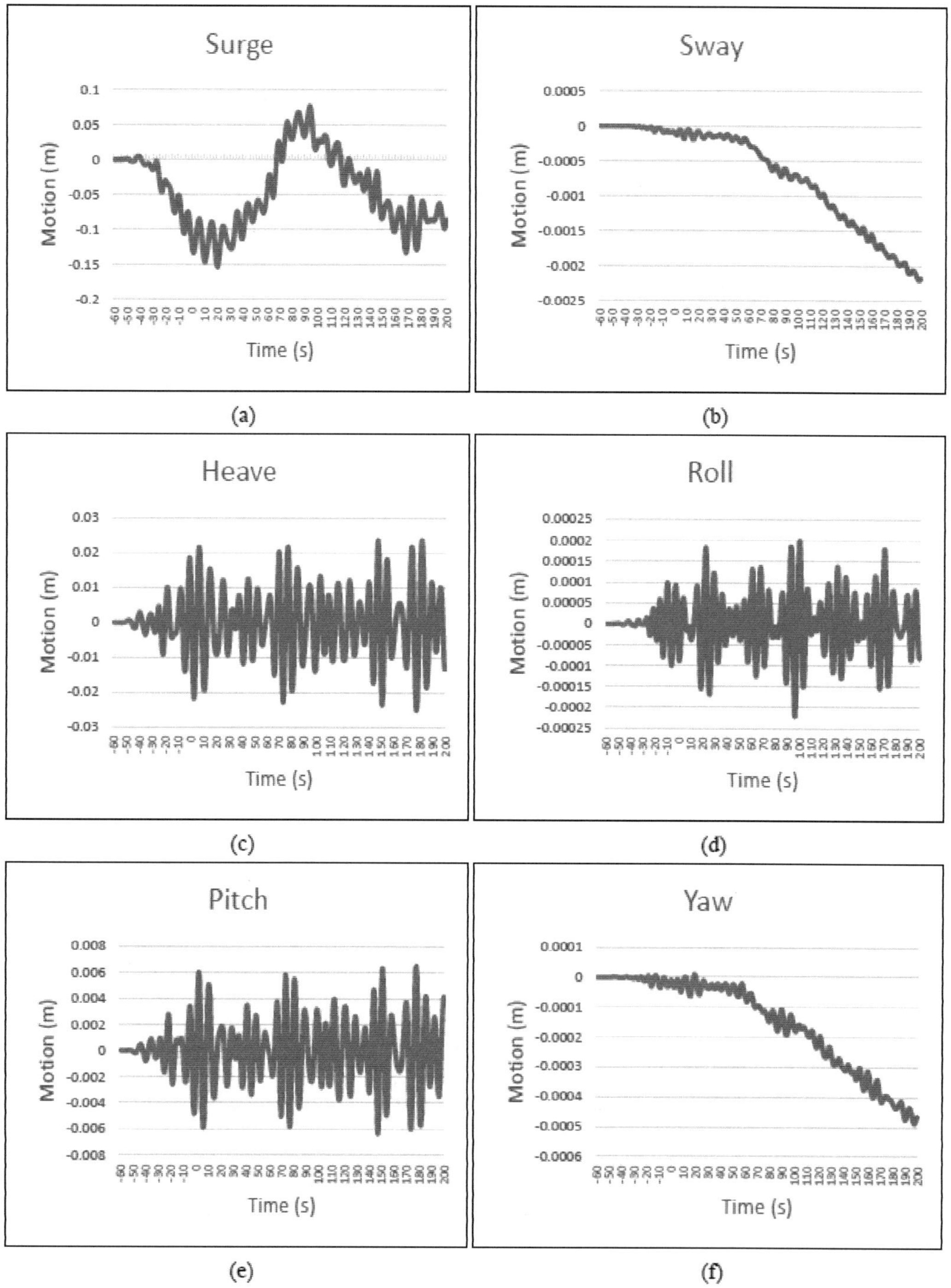

Figure 3. RAO Full load condition

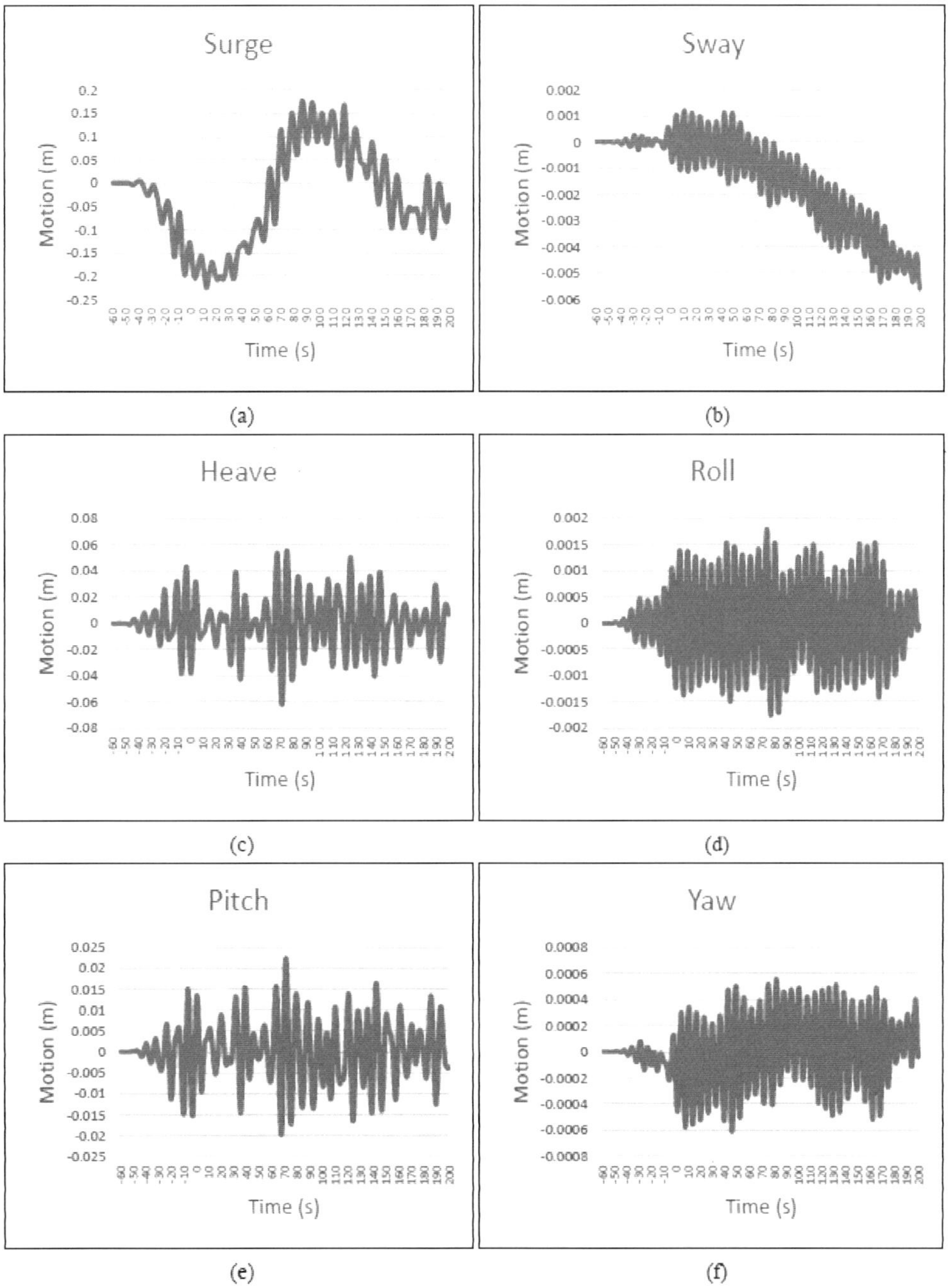

Figure 4. RAO ballast load condition

Table 9. Full-load and ballast-load maximum movement.

RAO	Subject	Full-Load	Ballast-Load
Surge (m)	FP	0.0768	0.1773
	AP	-0.1538	-0.2251
Sway (m)	PS	0.0000	0.0012
	SB	-0.0022	-0.0056
Heave (m)	Up	0.0238	0.0556
	Down	-0.0252	-0.0624
Roll (deg)	SB-Up	0.0002	0.0018
	PS-Up	-0.0002	-0.0018
Pitch (deg)	Bow-Up	0.0065	0.0224
	Stern-Up	-0.0005	-0.0019
Yaw (deg)	Bow-PS	0.0000	0.0006
	Bow-SB	-0.0005	-0.0006

4. Conclusions

1.M otion response free-floating condition

In a tanker, the maximum response occurs under ballast load conditions with the value for translational motion being 1.165 m/m in heave motion. In comparison, the maximum rotational motion occurs in roll movement with a value of 3.592 deg / m. In SPM, the maximum translational motion response is 1.658 m/m in surge and sway motion. At the same time, the maximum rotational motion is 3.323 deg/m in pitch motion.

2.M otion response moored condition

In full-load tankers, the farthest amplitude in translational motion is surge movement with a value of -0.1538 m towards the after perpendicular (AP) tanker. In rotational motion, the farthest amplitude of the tanker occurs in pitch movement with a value of 0.0065 degrees when the tanker's bow is going up. For ballast load conditions, the farthest amplitude of translational motion occurs in the surge movement with a value of -0.2251 m towards the AP direction of the tanker. For maximum rotational motion, the farthest amplitude occurs in pitch movement with a value of 0.0224 degrees when the tanker's bow goes up.

Acknowledgment

Researchers would like to thank the Department of Ocean Engineering, DRPM-ITS, and Institut Teknologi Sepuluh Nopember, Surabaya, Indonesia, for the financial support and facilities provided. And also, a special thanks to Mr. Sujantoko.

References

[1] Reddy, D. V. and Arockiansamy, M., 1991. *Offshore Structures*, Volume 1, Krieger Publishing Company, Malabar, Florida.

[2] Reddy, D. V. and Arockiansamy, M., 1991. *Offshore Structures*, Volume 2, Krieger Publishing Company, Malabar, Florida.

[3] Barltrop, N. D. P., and Adams, A. J., 1991. *Dynamics of Fixed Marine Structures*, Third Edition, Butterworth-Heinemann Ltd, Linacare House, Jordan Hill, Oxford OX2 8DP.

[4] Hales, L. Z, 1981. *Floating Breakwater: State of the art Literature Review*. Technical Report No. 81-1. U.S. Army Coastal Engineering Research Center, Fort Belvoir.

[5] McCartney, B. L., 1985. Floating breakwater design. *Journal of Waterway, Port, Coastal and Ocean Engineering,* 111 (2), pp. 304-318.

[6] Sujantoko, Djatmiko, E. B., Wardhana, W., Armono, H. D., Sholihin, and Ali, M. F., 2019. Experimental study on the effect of mooring angle on mooring tension of porous saw type floating breakwater. *Proceeding* of *The* 7th *International Seminars on Ocean and Coastal Engineering, Environmental and Natural Disaster Management,* Surabaya, Indonesia, Nov 6-7, pp. 94-99.

[7] Sujantoko, Djatmiko, E. B., Wardhana, W., Armono, H. D., and Wahyudi, 2019. Dynamic behavior analysis of porous saw floating breakwater under regular waves. *Proceeding of The* 7th *International Seminars on Ocean and Coastal Engineering, Environmental and Natural Disaster Management*, Surabaya, Indonesia, Nov 6-7, pp. 236-241.

[8] Sujantoko, Wardhana, W., Djatmiko, E. B., Armono, H. D., Putro, W. S., and Almuzaki, R. H., 2021. Study of wave characteristics of the floating breakwater for piling and tethered type. *Jurnal Teknik Hidraulik (Indonesian)*, 12 (1), pp. 39-51.

[9] Sujantoko, Djatmiko, E. B., Wardhana, W., Hidayatullah, A., 2021. Analysis of mooring tension on the saw-type floating breakwater with physical modeling. *Jurnal Teknik Sipil (Indonesian)*, 28 (3), pp. 289-300.

[10] Sujantoko, Djatmiko, E. B., Wardhana, W., 2022. Experimental investigation of the performance of porous-slope floating breakwater. *International Journal on Engineering and Applications*, 10(2), pp. 149-157.

[11] Sujantoko, Armono, H. D., Djatmiko, E. B., and Putra, R. D., 2022. Stability analysis of concrete block anchor on steep-slope floating breakwater. *Fluids*, 7 (8), 259.

[12] Sujantoko, Djatmiko, E. B., and Wardhana, W., 2022. Experimental study on mooring tension of porous-slope floating breakwater. *International Journal on Engineering and Applications*, 10 (5), pp. 345-352.

[13] Sujantoko, Djatmiko, E.B., Wardhana, W., and Firmansyah, I. G., 2023. Numerical modeling of hydrodynamic performance on porous slope type floating breakwater. *Nasemore*, 70 (2), pp. 149-157.

[14] Sujantoko, Djatmiko, E. B., Wardhana, W., and Mustain, M., 2023. Analysis of wave spectrum change of porous-saw floating breakwater. *International Journal on Engineering and Applications*, 11 (5).

[15] Sujantoko, Djatmiko, E. B., Wardhana, W., Mustain, M., and Ahidah, D. R., 2023. Analysis of wave spectrum change of porous-saw floating breakwater. *International Journal on Engineering and Applications*.

[16] Barltrop, N. D. P., 1998. *Floating Structures: A Guide for Design and Analysis*, CMPT.

[17] Rachman, A., 2022. *Deep Induction for Single Point Mooring Banner HEX 06,* Surabaya, practical work.

[18] ABS, 2023. *Rules for Building and Classing Mobile Offshore Units*, American Bureau of Shipping, Spring, Texas.

[19] ABS, 2022. *Guide for Position Mooring Systems,* American Bureau of Shipping, Spring, Texas.

Increasing Globalization Business and Ecotourism Branding on the Beach of Delegan, Gresik, East Java

Mahmud Mustain[1*] and Anggita Deva Ariyanti[1]

[1] Department of Ocean Engineering, Institut Teknlogi Sepuluh Nopember, Surabaya, Indonesia

[*]Email: mmustain@oe.its.ac.id

Abstract. Developing and managing ecotourism is an effort to utilize local resources optimally. Delegan Beach in Gresik, East Java, is an area for the management and development of ecotourism. This study aims to determine the potential resources in the Delegan beach tourism object using the Analysis Hierarchy Process (AHP) method. Based on the analysis, Delegan Beach has potential ecotourism because of the sloping beach topography, calm waves, and white sand. However, the development of Dalegan Beach as ecotourism has several obstacles, including the lack of supporting infrastructure for tourist attractions, cooperation between village communities and local governments, and poor management so that ecotourism development has not been maximized. The main priority of the policy regarding management studies and the potential of the Gresik coastal area, according to stakeholder perceptions, is the empowerment of human resources 51%, then followed by the development of tourism branding by 26% as the second priority, followed by the development of tourist destinations by 16%, and the last is ecotourism-based management. by 8%. The study's results also show that the determinants of the development of the Delegan Beach tourism area influence the forms of participation carried out by the local community. The resulting development directions include utilizing various resources, developing facilities as community business opportunities, empowering community skills, increasing area promotion, and efforts to preserve the environment.

Keywords: Delegan Gresik Beach, Ecotourism, Analytical Hierarchy Process, Stakeholders

1. Introduction

The influence of the role of globalization in the business world in any field feels very massive, including in the economic field. Increasing the speed of information also increases the speed of transportation in the world economy, namely in increasing the speed of transportation of goods and services. When talking about tourism services, it cannot be separated from beach and sea tourism. Even beach and sea tourism development is environmentally sound [1-4].

The beach is dynamic because it is a meeting place and interaction between land, sea and air [5-8]. These dynamic conditions make coastal areas vulnerable to coastal damage, namely erosion, sedimentation, abrasion and shoreline retreat [9-16]—many efforts to protect the coast naturally and with artificial structures. Coastal protection can naturally be done with mangrove and bamboo plants [2, 4, 17]. Another way is to build coastal structures, namely fixed breakwaters [18-24] and floating breakwaters [25-32]. The vast coastal area has a role as a potential resource which is an essential source of income for both the region and the country. The beach undergoes different adjustments every year due to natural factors. Almost all parts of the coastline in Indonesia have the potential to be developed, including the potential for fisheries, tourism, plantations and other potentials. However, this will increase the need for land and other infrastructure so new problems will arise in coastal areas. In this new normal era, the tourism sector is still trying to recover due to the COVID-19 pandemic, which has required the tourism sector to stop entirely for a while. People prefer to be in places away from crowds to avoid spreading the virus.

Coastal ecotourism is an approach to tourism development that aims to maintain biodiversity, preserve local culture, and improve the welfare of local communities [33]. In the context of developing the Delegan Gresik beach ecotourism, an in-depth understanding of the principles, strategies, and practices relevant to sustainable tourism management is required. The research conducted [3] identified several strategies for developing Delegans Gresik beach ecotourism. The management that has been carried out includes; empowering human resources, developing tourist destinations, tourism marketing, and developing the investment industry. Development efforts that have not been carried out in detail include; increasing accessibility, conserving biodiversity, and improving the skills of local communities. This effort aims to take concrete steps to balance the utilization of coastal natural resources. This condition is carried out while still paying attention to environmental sustainability and obtaining sustainable economic benefits, especially for the people around Delegan Beach. The ecotourism potential of Delegan Beach as a tourist attraction in Gresik includes the natural beauty, biodiversity, and cultural richness of Delegan Beach, which are essential factors in developing ecotourism. This potential creates opportunities to attract tourists and boost the local economy through sustainable tourism activities. Therefore, the importance of management in developing Delegan Gresik Beach ecotourism.

Effective management involves good planning, careful management of natural resources, close supervision, and community participation. In this context, good management will help ensure the sustainability of Delegan Beach ecotourism by maintaining a balance between economic, social, and environmental interests. Effective management also involves developing policies, working with stakeholders, and using appropriate technology to enhance the visitor experience. As a result, the ecotourism potential of Delegan Gresik Beach can be utilized as a source of developing local community economic income and a means of preserving and promoting existing natural and cultural wealth.

Delegan Beach is one of the tourism potentials in Gresik Regency, which is the area of this research. Figure 1 gives the research location. Delegan Beach has white sand with calm waves, so this beach has charm for tourists. To be precise, this beach is located in the waters of the Java Sea in the north of Gresik Regency. The distance from the city center of Gresik is approximately 53 km, with a distance of 1.5 hours by car. Access to this beach is quite good if we depart from downtown Gresik, can take directions to Sidayu through the Panceng teak forest, can follow signs to the beach area.

First, this study aims to identify important factors that influence the development of Delegan Beach ecotourism, including environmental challenges, financial sustainability, community participation, and effective management. Second, analyze the efforts that have been made in managing coastal ecotourism. Third, provide recommendations for further sustainability improvements in the future. Therefore, through this research, it is hoped that valuable solutions and recommendations can be found to develop Delegan Beach ecotourism and its surroundings. Thus, this research is expected to make a real contribution to environmental preservation, empowering local communities and improving the quality of the tourist experience on Delegans Beach.

Natural Resources Delegan Beach is one of the main assets in developing ecotourism. These resources include beautiful coral reefs, ecologically important seagrass beds, and abundant biodiversity around the coast [3] [33]. Tourism is a continuous process activity that has an impact, so there must be efforts to prevent or reduce risks in tourism activity. Then efforts are made to increase and utilize the potential based on the results of field observations and surveys. This research method uses the Analytical Hierarchy Process (AHP), where data collection is obtained through field surveys, interviews with local stakeholders, and direct observation of the condition of Delegan Beach. Secondary data, such as government reports and related documents, are also used to support analysis and draw conclusions.

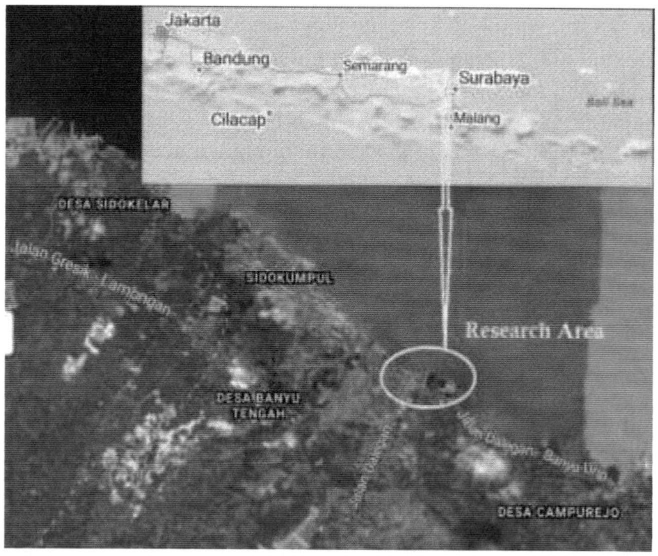

Figure 1. Research location at Delagan Beach

2. Methods

As mentioned in the introduction, this study uses the AHP method with more detailed specifications compared to previous research by Ikhwani et al [3, 33]. Questionnaire data obtained from respondents is processed through software expert choice to determine logical priorities and consistency. The most important thing is the calculation of weights on each of the necessary criteria. The steps performed in the AHP method are as follows:

a) Define the problem and determine the desired alternative solution.
b) Create a diagram of the hierarchical structure that begins with the primary goal.
c) Determine priorities on each criterion and alternative solutions that are carried out in pairs. The equation of the solving matrix calculates the weight and priority. This equation is used in the calculation of the geometric mean as follows.

$$a_w = \sqrt[n]{a_1 x\, a_2 \ldots\ldots x\, a_n} \quad (1)$$

With a_w = combined assessment, a_1 = 1st respondent assessment, and n = number of respondents. Pairwise comparisons start from the highest level of the hierarchy, where subjects or actors are used as a basis for making comparisons. The form of the pairwise comparison matrix is given in Table 1.

d) Measure Consistency
Consistency has 2 (two) meanings, first objects that are similar and grouped according to uniformity and relevance. Secondly, the degree of relationship between objects is based on specific criteria. Consistency is measured by calculating the consistency index and ratio using the following formula.

$$CI = (\lambda \max - n)/n \quad (2)$$

$$CR = CI \div IR. \tag{3}$$

With CR = consistency ratio, CI = consistency index, and IR = index random consistency. The random index value expresses the magnitude of the consistency of the reciprocal matrix that appears randomly on a scale of 1-7 and vice versa. Saaty [34] states that when random indices are expressed in matrices of order 1-10. Table 2 provides a random Index. The data judgment assessment must be corrected if the consistency index value exceeds 10%. However, if the Consistency Ratio is ≤ 0.1, the calculation results can be declared correct [35].

Furthermore, the interview method has been conducted to explore all potential information, problems, aspirations, and needs of the community and other stakeholders. Interviews were also conducted to communicate alternative strategies from this study by considering the conditions of the research site.

Table 1 Pairwise Comparison Matrix [34]

	A1	A2	...	An
A1	A1.A1	A1.A2	...	A1.An
A2	A2.A1	A2.A2	...	A2.An
...	...	...	...	...
An	An.A1	An.A2	...	An.An

Table 2 Random Index for Each Order of Matrix [34]

Order of the Matrix	1	2	3	4	5	6	7	8	9	10
Random Index	0.00	0.00	0.58	0.90	1.12	1.24	1.32	1.41	1.45	1.49

3. Results and Discussion

3.1 Data Compilation

3.1.1 Respondents Based on Perpetrators

Based on the results of a survey of actors participating in the management of the Delegans Gresik coastal area and stakeholders, it was determined that there were 30 respondents. The distribution of respondents was in the form of 20 local people (66.67%), the village government represented by the person in charge of the beach officers 2 people (6.67%), the beach manager Delegans 3 people (10%), and the tourists 5 people (16.67%). Figure 2 gives an idea of the distribution of perpetrators. As the primary user of marine resources, the surrounding community gets the largest percentage considering the success of the development of an area is influenced by community participation.

3.1.2 Respondents by Education and Employment Level

After the identification process of respondents based on the questionnaires distributed, it is known that the composition of respondents is based on education level, namely the distribution of education-based respondents. It is known that respondents are at the level of education the level; Bachelor (S1) 3 people (10%), Senior High School (SMA) 15 people (50%), Junior High School (SMP) education level 5 people (16.67%), and for elementary school level 7 people (23.33%). The job-based respondents only include traders 66.67% and others 33.33%.

3.2 The Process of Formulating the Ecotourism Development Strategy of Delegan Beach

The Analytical Hierarchy Process (AHP) structuring process begins with defining the problem or problem. Then decompose the problem by changing it into smaller components so that they cannot be separated into smaller ones. The next step in AHP is to determine its hierarchical structure. Figure 2 provides a hierarchical structure of the AHP diagram. The existence of a hierarchy allows solving complex or unstructured problems into sub-sub-problems. Then arrange it into a form of hierarchy. Namely, there are four levels or levels (hierarchy): goals, potentials, criteria, and alternative solutions.

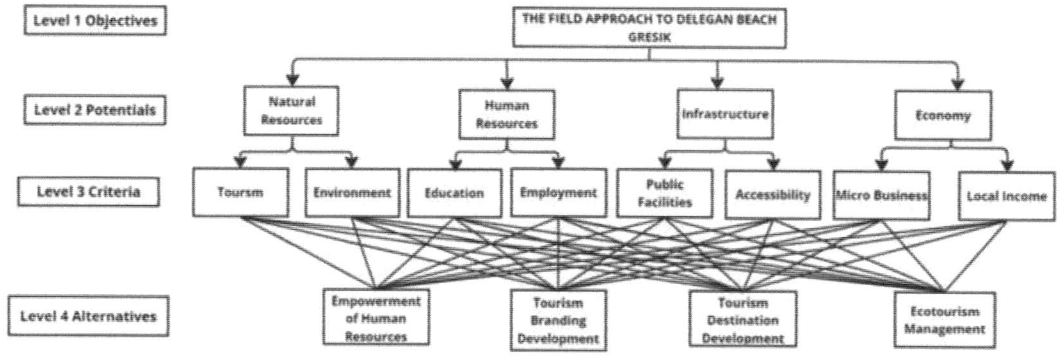

Figure 2. AHP Diagram

The first level of ecotourism development of Delegan Beach is goal setting and is related to the aspirations desired by all stakeholders. The goal is to obtain a Delegan Beach ecotourism development strategy following the actors, aspects, and alternative strategies. Based on the questionnaire data processing results, a paired comparison value is obtained, and a geometric average is generated. These results are then processed to produce priority factors.

Based on the calculation results, it was found that human resources occupy the priority in ecotourism development by 58.9%. Natural resources followed it at 22.3%, economy at 12.5%, and infrastructure at 6.3%. Table 3 provides Priority Values Based on Ecotourism Development Aspects. The consistency ratio value is 0.09, meaning that the opinions given by respondents have been consistent because the CR value still meets the requirements, namely CR = 0.1. The magnitude of the priority value of human resources shows that almost all respondents want human resources should be paid more attention to or prioritized to be improved in each of these coastal ecotourism developments. The high quality of human resources is an essential aspect in supporting the development of ecotourism. The public expects increased knowledge of various things recommended and prohibited in coastal areas to develop Delegan Beach ecotourism.

Table 3. Priority Values Based on Ecotourism Development Aspects

No	Aspect	Vactor Priority	Percentage (%)
1	Human Resources	0.589	58.9
2	Natural Resources	0.223	22.3
3	Economics	0.125	12.5
4	Infrastructure	0.063	6.3
Consistency Ratio (CR)		0.09	9 %

From the results of the recapitulation of respondents' perceptions, the priority of management policies and the potential of this coastal area is the strategy of empowering human resources (human resources) by 51%. It was followed by developing tourist destination branding at 26%, tourist destinations at 16%, and ecotourism-based management at 8%. Table 4 provides priority vector results from the recapitulation of respondents' perceptions. Stakeholders agreed that in developing Delegan Beach ecotourism, there should be assistance and provision of human resources first. In addition, there is also a need for cooperation between the government as a policyholder and the community as the subject of managing the potential and natural resources. Starting from planning to evaluating management so that beach tourism can continue growing during tourism competition in East Java, especially in Indonesia and the World.

Table 4. Priority Vactor Results from Respondent Perception Recapitulation Results

No	Aspects	Vactor Priority	%
1	Human Resource Empowerment	0.51	51
2	Development of Tourism Destination Branding	0.26	26
3	Development of Tourist Destinations	0.16	16
4	Ecotourism-Based Management	0.08	8
	Consistency Ratio (CR)	0.09	0.09

3.3 Confirm AHP Software Expert Choice Results on Manual Calculation.

A comparison of AHP results and manual calculations shows a minimal average error of less than 0.1%. This minimal error means that the results of the AHP analysis are very reliable with the standard reference manual calculation using the formula provided. Figure 3 gives the results of such a comparison. HR showed a figure of 58.9% in manual calculations, while in Software Expert Choice calculations, it was 61%, a difference of 2.1%. SDA shows a figure of 22.3% of manual mass while Software Expert Choice is 21.5%, a difference of 0.8%. On infrastructure, it showed a 6.3% and 5.3% difference, a difference of 1%. In economics, 12.5% and 11.9%, the difference is 0.6%.

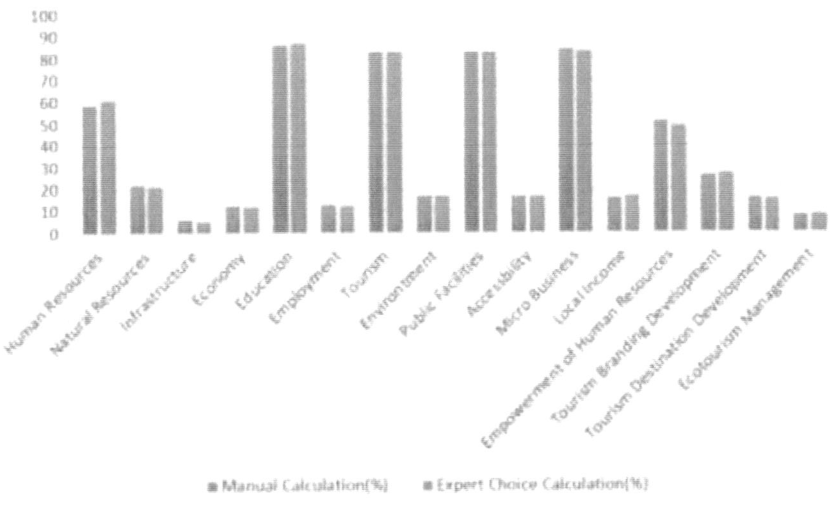

Figure 3. Comparison of manual calculation results with expert choice software

3.4 Ecotourism Development Strategy of Delegan Beach

Delegan Gresik Beach refers to all survey data in this study to prepare and determine the ecotourism development strategy, data of secondary, primary interview and questionnaire data, and AHP results to obtain the goals and results to be achieved. Some aspects that need to be considered in preparing this coastal ecotourism development plan are as follows:

a. Results of Secondary and Primary Data Studies:

From the results of this study, it is known that Delegans Village has the potential that be developed, namely:

a. Considerable human resource potential
b. Tourism potential from the beauty of the surrounding environmental conditions
c. Support from the Gresik Regency Government

d. The potential of facilities and infrastructure is relatively complete
e. Road access to paved tourist areas

b. AHP Study Results
Some aspects that become criteria in the development of Delegan Beach ecotourism are all towards improving the quality of human resources through formal and informal education and improving people's skills to create crafts that characterize this area. Refers to the results of AHP that the main priority of developing coastal ecotourism, according to respondents, is the empowerment of human resources first.

3.5 Discussion

In this coastal ecotourism development strategy, a collaboration between several elements is highly recommended from secondary data studies, primary or survey data, interviews, and AHP results. Indonesia has policies and regulations that define collaborative management. This condition is contained in the Regulation of the Minister of Forestry No. P19 / Menhut-II / 2004 concerning collaboration in developing Nature Reserve Areas (KSA) and Nature Conservation Areas (KPA). *Collaboration* is defined as the implementation of an activity or handling a problem to help improve the effectiveness of developing and managing a conservation area jointly and synergistically. This collaboration is carried out by the parties based on mutual understanding and agreement following applicable laws and regulations.

Figure 4. Potential and crowds captured in the condition and situation of Delegan Gresik-East Java beach tourism objects, Indonesia

The critical word stated in the paper's title in this study is the desire of researchers to concentrate on efforts to improve branding and globalization. It is highly recommended for branding interests to arrange internal aspects, namely the optimization of the five potentials owned. Globalization aims to collaborate with relevant stakeholders to expose to the outside world, for example, by creating a contemporary website. This condition can be done by creating a site containing content that attracts potential visitors domestically and internationally.

Concerning the priorities generated by AHP, the idea of branding and globalization is recommended to prioritize improving the quality of human resources by including other potentials. For example, with the increase in the quality of human resources, the physical appearance and service of this beach tourist attraction look upgraded at a higher level and can even skyrocket by having specific new icons. Thus, the purpose of branding and globalization can be realized so that, in the end, it can improve the welfare of the surrounding community and, generally, Gresik.

4. Conclusion

In addition to the tremendous human resource potential in the coastal area, Delegan Gresik is a natural beauty with the uniqueness of white sand. The surrounding community has taken the initiative to provide facilities for providing food, beverages, and unique souvenirs from Gresik. This beach also has paved road access, other tourist facilities, and good government support. The results of the Analytical Hierarchy Process (AHP) method provide that the priority order of this coastal ecotourism development policy is; HR empowerment at 51%, branding development at 26%, tourism destination development at 16%, and ecotourism-based management at 8%.

Associated with the brand and globalization of Delegan Beach, the recommended ecotourism development strategy directions include; Upgrading the tourism promotion system through the Internet network, developing facilities or facilities for tourist destinations, and ecotourism-based management. It is hoped that the direction of the strategy can help increase the attractiveness of domestic and international tourists globally, as well as the development of Delegan Beach's ecotourism potential in a sustainable manner.

Acknowledgment

Researchers would like to thank the Department of Ocean Engineering, DRPM-ITS, and Institut Teknologi Sepuluh Nopember, Surabaya, Indonesia, for the financial support and facilities provided. And also, a special thanks to Mr. Sujantoko.

References

[1] Mustain, M., Sujantoko, and Prasetyo, M. A., 2022. Empowerment and optimalization of Kenjeran beach tourism potential, Surabaya-East Java. *International Review of Civil Engineering*, 13 (4), pp. 246-254.
[2] Mustain, M., Garang, I. J., Ikhwani, H., and Sujantoko, 2022. Analysis and maximize the ecotourism potential of the Wonorejo mangrove in East Surabaya. *International Journal on Engineering Applications*, 10 (2), pp. 158-166.
[3] Ikhwani, H., Mustain, M., Satrio, D., Sujantoko, Hadiwidodo, S.H., Nasution, M.P.P., 2023, The Concept of Integrated Coastal Management for Eco-Tourism Development at Dalegan Beach, Gresik Regency, East Java, Indonesia, *10th International Seminar on Ocean and Coastal Engineering, Environmental and Natural Disaster Management (ISOCEEN 2022)*.
[4] Rosyid, D.M., Sujantoko, Armono, H.D., Djatmiko, E.B., Prastianto, R.W., Mulyadi, Y., Kurniati, N., Wardhani, M.K., and Wardhana, W., 2021. Kajian Pengelolaan Kawasan Mangrove Pesisir Selatan Kabupaten Bangkalan Berdasarkan UU No 1 Tahun 2014, *Sewagati*, 5 (3), pp.1-11.
[5] Dahuri, R. et al, 1996. Pengelolaan Sumberdaya Pesisir dan Lautan Secara Terpadu. Jakarta: PT. Pramadya Paramita.
[6] Mustain, M., 2017, The Model of Shoreline Change in Estuary of Porong River after Mud Volcano, *International Journal of Engineering and Applied Sciences (IJEAS)*, Vol. 4 (11).

[7] Mustain M., R. Leonard, 2019, Coastal Zone Management: The Field Approach to Wonorejo-East Surabaya, *International Journal on Engineering Applications (IREA)*, Vol 7 (4), pp. 145-151.

[8] Mustain, M., and Sujantoko, 2021. Simple Analysis of Determination of Tidal Types of Sea Water: A Case Study in Lamong Bay Waters, East Java. *Proceeding - IEEE Ocean Engineering Technology and Innovation Conference: Ocean Observation, Technology and Innovation in Support of Ocean Decade of Science, OETIC 2021*, 2021, pp. 13–17.

[9] Mustain, M., 2018. The model of the current pattern at the mouth of Porong River Sidoarjo, *International Journal of Civil Engineering and Technology*, 9 (1), pp. 688–701.

[10] Sujantoko and Natakusuma, D., 2003. Model simulasi interaksi gelombang dan arus di perairan dangkal, *Jurnal Teknik Sipil*, vol. 10 (3), 99-108.

[11] Sujantoko, 2009. The Analysis of wave refraction using SWAN model, *IPTEK The Journal of Technology and Science*, 20 (2), pp. 45-56.

[12] Sujantoko, Pratikto, W.A., Prastianto, R. W. Maulana, M. I. and Vebriyanti, A., 2021. Study of changes in coastal morphology due to utilization of the Surabaya city coastal area. *International Journal of Marine Engineering Innovation and Research*, 7 (1), pp. 26-32.

[13] Sujantoko, Pangestu, P. A. Saputro, D., Ekianto, B. P. P., Hasanudin, Kurniati, N., Kusumawardhani, R., and Hartanto, D., 2023. Hydrodynamic model simulation at the port of Tanjung Rhu Belitung. *International Journal of Marine Engineering Innovation and Research*, 8 (2), pp.141-149.

[14] Sujantoko, Mustain, M., Wahyudi, and Ikhwani, H., 2023. Hydrodynamic model due to reclamation in Lamong Bay. IOP Conference Series: Earth and Environmental Science, Volume 1198, 012004.

[15] Mustain, M., and Sujantoko, 2023. Spiral analysis of vertical currents in Lamong Bay. IOP Conference Series: Earth and Environmental Science, Volume 1166, 012024.

[16] Rudyani, F.P., Armono, H.D., and Sujantoko, 2013. Pemodelan Gelombang di Kolam Pelabuhan Perikanan Nusantara Brondong, *Jurnal Teknik ITS*, 2 (2), pp. G150-G154.

[17] Armono, H.D., Bromo, B.H., Sholihin, and Sujantoko, 2022, Numerical study of bamboo breakwater for wave reduction, *Fluids*, 7 (1), pp. 1-13.

[18] Sujantoko, Saputro, D., Pratikto, W.A., Mustain, M., and Ikhwani, H., 2023. Experimental Study of Wave Runup and Overtopping for Concrete Armor A-Jack on Seawall Structure. *International Journal on Engineering Applications (IREA).*

[19] Sujantoko, Pangestu, P.A., Pratikto, W.A., Mustain, M., Wahyudi, and Ikhwani, H., 2023. Experimental Study of Wave Reflection on A-Jack Armor Unit on Seawall Structure, *International Journal of Marine Engineering Innovation and Research (IJMEIR)*, 8 (2), pp. 232-238.

[20] Putra, A.O.P., Armono, H.D., and Sujantoko, 2013. Pengaruh Elevasi Muka Air Laut pada Koefisien Transmisi dan Refleksi Composite Breakwater, *Jurnal Teknik ITS*, 1 (2), pp. G47-G51.

[21] Armono, H.D., Winarto, A., Sujantoko, and Suastika, I.K., 2021. A laboratory study on wave transmission over hexagonal artificial reef, *Sustainability and Resilience of Coastal Management (SRCM)*, Surabaya.

[22] Daffa, M.Z., Armono, H.D., and Sujantoko, 2020. Model Numerik Pengaruh Artificial Reef dalam Reduksi Gelombang dan Perubahan Morfologi Pantai menggunakan DELFT3D, *Seminar Nasional Penelitian & Pengabdian Kepada Masyarakat (SNP2M) 2020 (Teknologi dan Sosial Sains)*, Makasar.

[23] Sujantoko, and Manurung, A., 2020. Analisis Hidrodinamika pada Perencanaan Jetty Siantan Selatan, Kepulauan Anambas, *Seminar Nasional Penelitian & Pengabdian Kepada Masyarakat (SNP2M) 2020 (Teknologi dan Sosial Sains)*, Makasar.

[24] Armono, H.A, Manurung, A., Sujantoko, and Suastika, I.K., 2022. Wave transmission model over hexagonal artificial reefs using DualSPHysics. IOP Conference Series: Earth and Environmental Science, Volume 1095, 012025.

[25] Sujantoko, Wardhana, W., Djatmiko, E. B., Armono, H. D., Putro, W. S., and Almuzaki, R. H., 2021. Study of wave characteristics of the floating breakwater for piling and tethered type. *Jurnal Teknik Hidraulik (Indonesian)*, 12(1), pp. 39-51.
[26] Sujantoko, Djatmiko, E. B., Wardhana, W., Hidayatullah, A., 2021. Analysis of mooring tension on the saw-type floating breakwater with physical modeling. *Jurnal Teknik Sipil (Indonesian)*, 28(3), pp. 289-300.
[27] Sujantoko, Djatmiko, E. B., Wardhana, W., 2022. Experimental investigation of the performance of porous-slope floating breakwater. *International Journal on Engineering and Applications*, 10(2), pp. 149-157.
[28] Sujantoko, Armono, H. D., Djatmiko, E. B., and Putra, R. D., 2022. Stability analysis of concrete block anchor on steep-slope floating breakwater. *Fluids*, 7(8), 259.
[29] Sujantoko, Djatmiko, E. B., and Wardhana, W., 2022. Experimental study on mooring tension of porous-slope floating breakwater. *International Journal on Engineering and Applications*, 10 (5), pp. 345-352.
[30] Sujantoko, Djatmiko, E.B., Wardhana, W., and Firmansyah, I. G., 2023. Numerical modeling of hydrodynamic performance on porous slope type floating breakwater. *Nasemore*, 70(2), pp. 149-157.
[31] Sujantoko, Djatmiko, E. B., Wardhana, W., and Mustain, M., 2023. Analysis of wave spectrum change of porous-saw floating breakwater. *International Journal on Engineering and Applications*, 11 (5).
[32] Sujantoko, Djatmiko, E. B., Wardhana, W., Mustain, M., and Ahidah, D. R., 2023. Numerical Modeling of Wave Hydrodynamic Performance with Various Layouts of Single-Slope Floating Breakwater Arrays, *International Journal on Engineering and Applications*, vol.11.
[33] Ikhwani, H., Satrio, D., Umari, W., Sujantoko, Hadiwidodo, Y.S., and Syarifudin, M.R., 2022. The Analysis of Coastal Society Vulnerabilities Against the Spread of Covid-19 in Surabaya Using the Analytical Hierarchy Process (AHP), *International Journal of Marine Engineering Innovation and Research (IJMEIR)*, 7 (4), pp.218-224.
[34] Saaty T.L., 1991. *Decision Making for Leaders; Analytic Hierarchy Process for Decision Making in Complex Situations.*
[35] Forman, H., & Gass, S.,2001. *The Analytical Hierarchy Process - An Exposition. Operation Research*, pp. 469-486

Simulation of a Wave Hydrodynamic Numerical Model on a Single Vertical Porous Breakwater

Sujantoko[1*], G. A. Pramestika[1], M. Mustain[1], H. Ikhwani[1], Y. S. Hadiwidodo[1]

[1] Department of Ocean Engineering, Institut Teknlogi Sepuluh Nopember, Surabaya, Indonesia

[*]Email: sujantoko@oe.its.ac.id

Abstract. One of the problems in the coastal area is a change in the coastline in the form of a coastline retreat which results in the loss or damage of infrastructure and settlement facilities caused by extreme sea wave activity. One way to anticipate this problem is to build a breakwater. This research is focused on the effect of porosity and freeboard on the transmission and reflection of waves in a single vertical porous breakwater. Numerical modeling was done using a 3D Numerical Wave Tank (NWT) based on Smoothed Particle Hydrodynamic (SPH). The numerical model results show that the average transmission and reflection coefficients are 0.58 and 0.42. In addition, it can be seen that the freeboard and porosity influence the value of the transmission and reflection coefficients. Generally, the smaller the freeboard, the greater the transmission coefficient and the smaller the reflection coefficient. Meanwhile, the greater the porosity, the greater the transmission coefficient, and vice versa, the smaller the reflection.

1. Introduction

As an archipelagic country, Indonesia has coastal areas with great potential to support human survival, so their existence is used intensively. The potential of human and coastal natural resources in Indonesia is an outstanding potential and a challenge that can be utilized as well as possible. One of these uses is the construction of a port. However, if the port construction is not carried out properly, it will impact the surrounding beaches, namely sedimentation and erosion [1-8]. Another problem about the coastal area is a change in the coastline in the form of a retreating coastline resulting in loss or damage to infrastructure facilities and settlements caused by high sea wave activity.

The solution to this problem can be done naturally, namely dampening the waves with mangrove plants [9-13] and bamboo [14] or adjusting the coastal border by implementing integrated coastal zone management (ICZM). Another way is to construct a coastal protection structure, such as a floating breakwater [15-26] or a fixed breakwater [27-32]. This type of fixed breakwater, horizontal and vertical, is a very effective structure in protecting the coast by dissipating wave energy before it reaches the beach. In waves that hit the wave damper building, some of the energy will be reflected, transmitted, and partly destroyed (dissipated) through the breaking of the wave, which depends on the characteristics of the incident wave. Transmission (H_t) and reflection (H_r) waves are measured by the transmission coefficient (K_t) and reflection (K_r) parameters. Several previous studies have proven that the value of the transmission and reflection coefficients is strongly influenced by the height and period of the waves.

Increasing the length and height of the waves decreases the transmission coefficient, while the reflection coefficient is directly proportional to the length and height of the waves.

This study will focus on the vertical structure of porous breakwaters. The pore-shaped structure has the advantage that it can dampen wave energy by holding its energy through the pore so that the wave energy received by the structure is reduced. Apart from that, this pore-shaped structure is designed to utilize small stones to be included in the porous box frame because, at the time of construction, not every project site had natural stone available with the weight according to the design. Many researchers have carried out numerical modeling using Smoothed Particle Hydrodynamics (SPH) DualSPHysics 5.0 software before [33-38]. This paper will apply an SPH-based analysis to obtain the transmission and reflection coefficients on a vertical porous breakwater.

2. Methods

2.1 Data and 3D models

The determination of the research variable is used to see its effect on what is being observed. In this study, numerical analysis will be carried out using the DualSPHysics software. After a literature study on previous studies, the impact of wave parameters on the structure will be reviewed further. Test data in this study used wave height and period, 0.1-0.18 m and 1.35 and 1.8 seconds, respectively.

In this study, the 3D models that will be made include flume tanks, vertical porous breakwater structures, and beds. The 3D model creation was designed using the open-source software FreeCAD 0.20 (Fig. 1) because DualSPHysics does not have its interface yet. FreeCAD 0.20 software can add user interfaces from other software via "macros" from DualSPhysics 5.0. The specification data for the 3D Numerical Wave Tank (NWT) to be used has dimensions of 30.0m in length, 0.5m in width, and 1.0m in height (Fig. 2). By using the DualSPHysics 5.0 user interface, FreeCAD 0.20 can be used to adjust the piston motion on the flume tank model to generate waves.

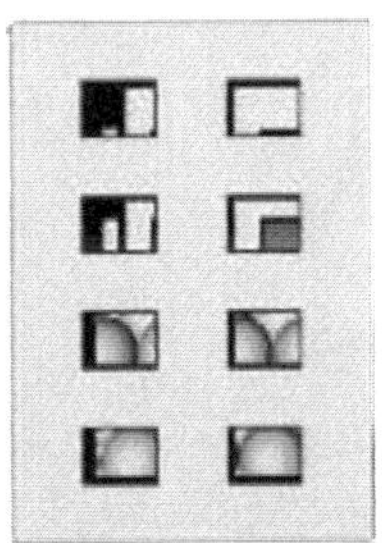

Figure 1. Model of a single vertical porous breakwater with FreeCAD 0.20

Figure 2. The flume tank model in FreeCAD 0.20

2.2 Model Validation

The validation process in this study is used to determine the dimensions of the numerical wave tank (NWT) before finally proceeding with the model simulation process. This validation ensures that the model dimensions of the numerical flume tank used follow the physical and theoretical conditions and are feasible for use in the research being conducted. In this study, validation was carried out by comparing the numerical results using DualSPHysics with the experimental results of the physical model [39]. The wave probe is placed according to the location determined in the physical model test. Wave probe 1 is placed 5 meters from the piston, which is used as wave input. Then wave probe 3 is placed 2 m away from the outermost structure, wave probe 2 is placed 3.5 m before wave probe 3, and wave probe 4 is 2 m behind the structure.

This numerical modeling was validated using sensitivity analysis using variations in the distance between different particles. This fact is done to get the optimum distance between particles in this research. The distance between the particles used is 0.03m, 0.02m and 0.01m. The results of the comparison of numerical model simulations and experiments are shown in Fig. 3 - 5. Based on numerical modeling, it can be seen that the distance between particles affects the accuracy of the model results. The smaller the distance between the particles, the better the result. A distance between particles of 0.01 m has the lowest error rate between the distance between particles of 0.02 m and 0.03 m (Table). However, the program simulation time on a computer with the same specifications will be longer. Good validation results show that the wave elevation between the numerical and physical models is not much different (Table 1). The error approach uses the mean absolute percentage error (MAPE) method [41].

Table 1. MAPE calculation results for numerical and physical models

Distance between particles (m)	Estimation of model running time (hours)	Result of MAPE calculation (%)				
		wp1	wp2	wp3	wp4	Average
0.01	72-94	0.03591	0.03679	0.02361	0.03017	0.03162
0.02	38-54	0.03687	0.04877	0.02566	0.03150	0.03570
0.03	5-10	0.03707	0.04931	0.02572	0.03170	0.03595

3. Results and Discussion

Analysis of the transmission coefficient K_t and K_r reflection was carried out based on the parameters of wave steepness and freeboard dimensions. The steepness of the waves shows a nondimensional variable due to the height and period of the waves expressed in H/gT^2. The freeboard is a nondimensional variable due to the effect of freeboard height and waves (Rc/H), where these two variables are variables that affect the transmission coefficient (K_t) and reflection (K_r). In this numerical model simulation, the freeboard is varied with a height of 0.0m and 0.10m relative to sea level elevation (SWL). Porosity in the model is set by filling a porous concrete box with gravel of a specific size, and then the volume filled with gravel is compared to the volume of the entire porous box. This study set the porosity at 5% and 10%.

Based on the numerical model simulation results, it was found that the porosity vertical breakwaters were 0%, 5%, and 10% on the same freeboard (Fig. 6), indicating that K_t in a 5% porous vertical breakwater has a greater K_t value than without pores (0%). At 10% porosity, the value of K_t will be even greater because, in this condition, much wave energy is not entirely damped compared to without porosity. As for the influence of the steepness of the wave, it can be observed that the greater the steepness of the wave, the value of the transmission coefficient K_t also increases (Fig. 7). Conversely, the higher the freeboard (Rc = 0.1m), the lower the K_t value because the wave energy will be damped by the freeboard which is 0.1m above the SWL.

The same thing also happens to the phenomenon of wave reflection. On a high freeboard (Rc=0.1m), the waves will be partially reflected to a greater extent than a structure with a freeboard in SWL (Rc=0.0m) (Fig. 8 and 9).

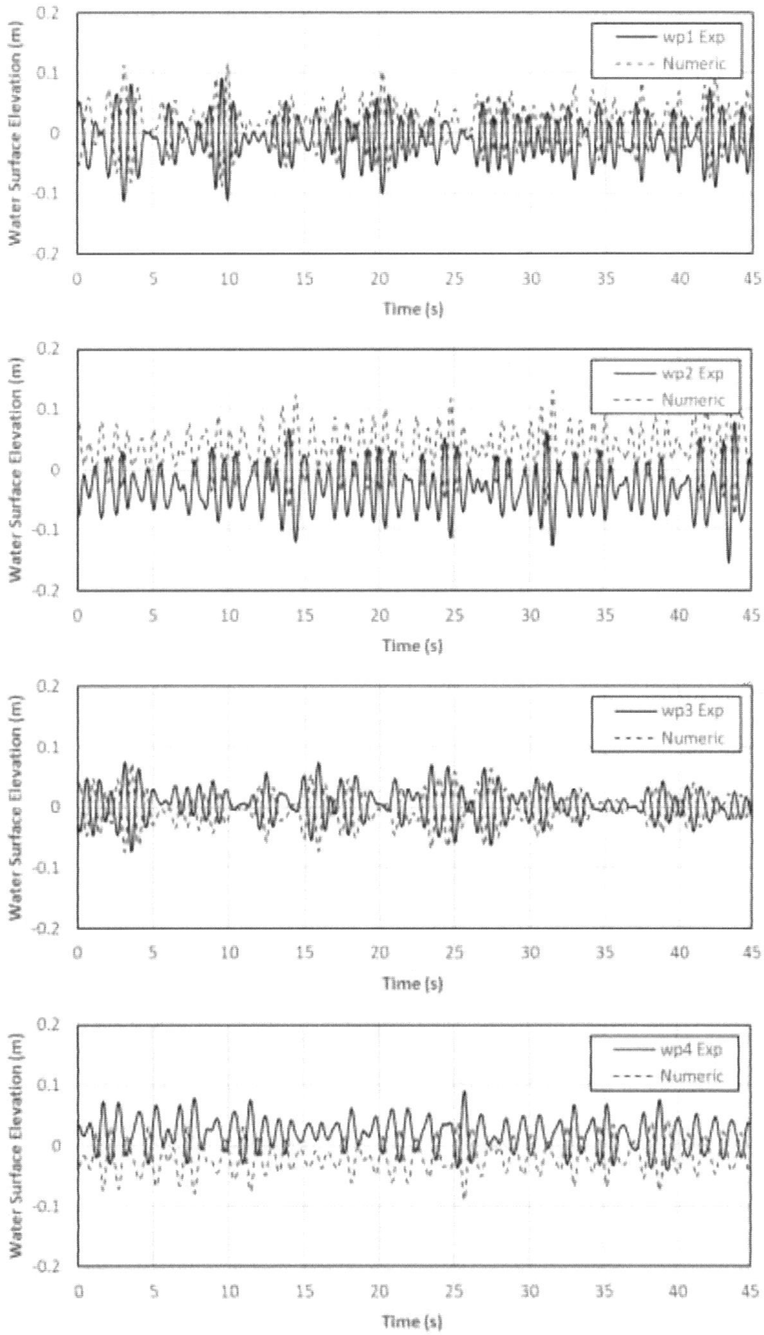

Figure 3. Results of the validation of numerical and experimental models with a distance between particles of 0.03m

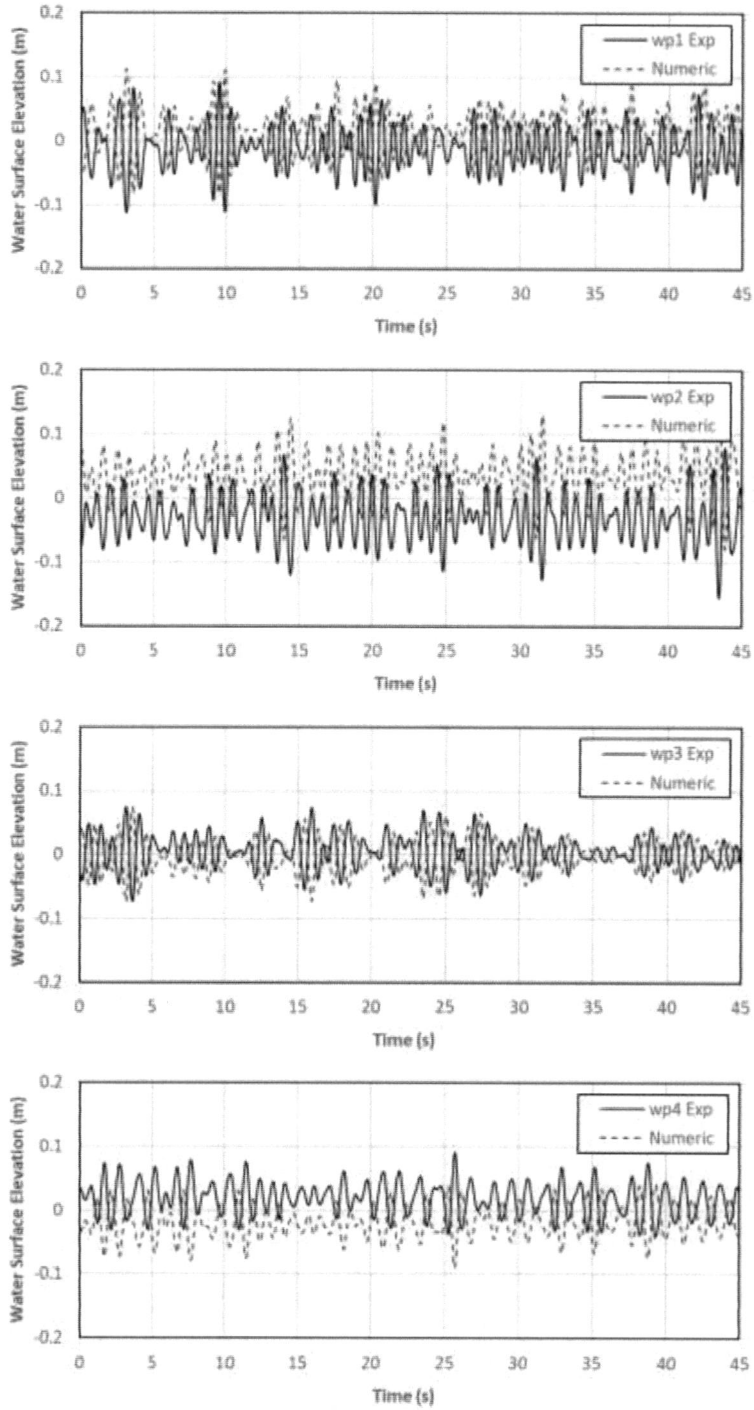

Figure 4. Results of the validation of numerical and experimental models with a distance between particles of 0.02m

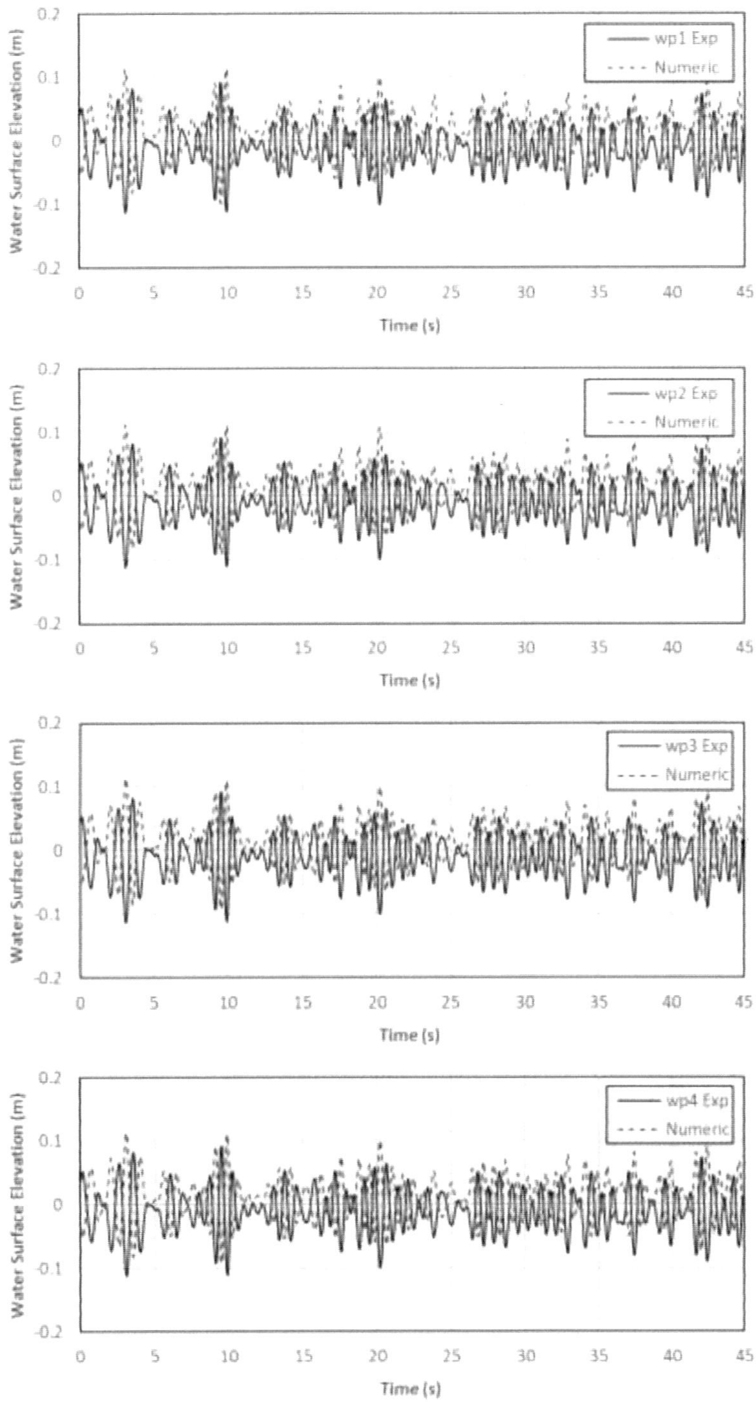

Figure 5. Results of the validation of numerical and experimental models with a distance between particles of 0.01m

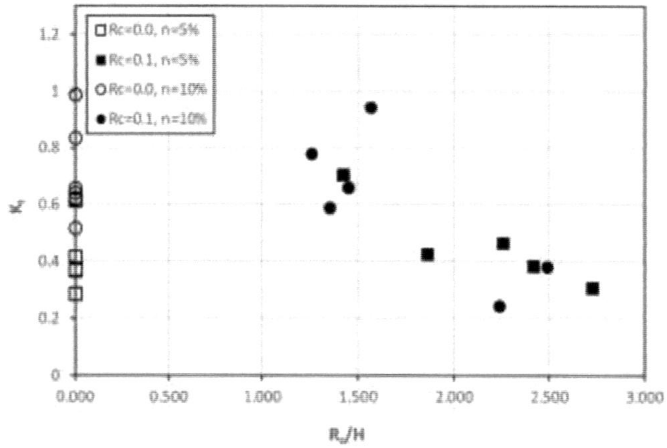

Figure 6. Effect dimensionless freeboard to K_t

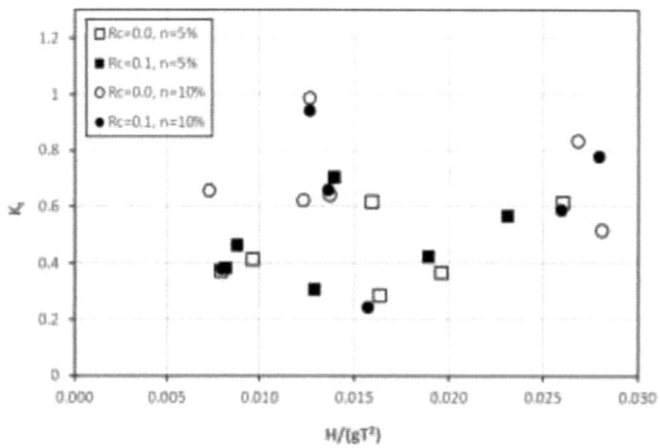

Figure 7. Effect of wave steepness on K_t

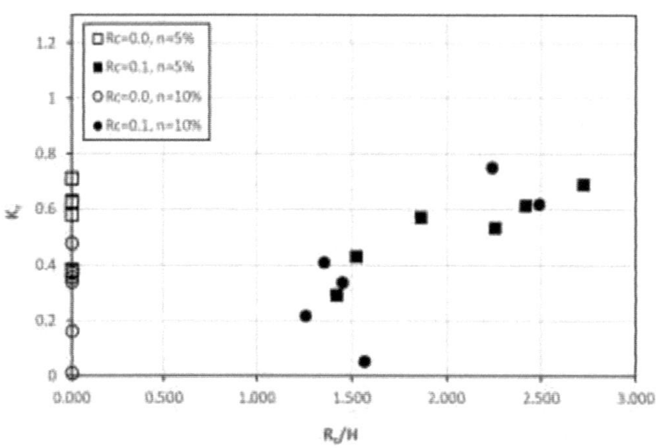

Figure 8. Effect dimensionless freeboard to K_r

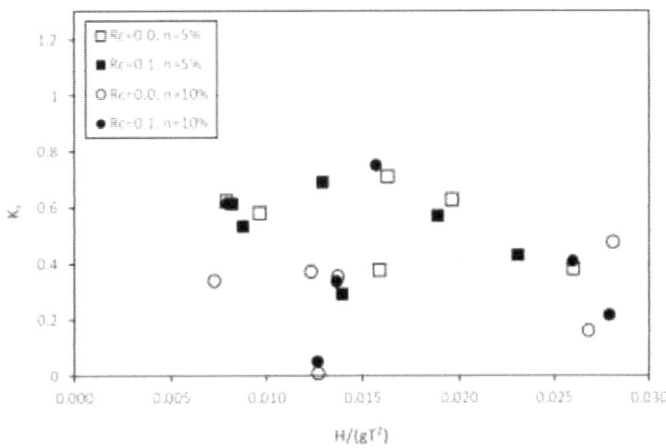

Figure 9. Effect of wave steepness on K_r

4. Conclusion

Based on the results of the numerical model simulation that has been done, it can be concluded that the greatest transmission coefficient is located at Rc = 0 and n = 10%, and the largest reflection coefficient is located at Rc = 0.1m and n = 5%, with an average of K_t and K_r of 0.58 and 0.42. Freeboard is very influential on K_t and K_r. The smaller the freeboard and the greater the porosity, the greater K_t. Conversely, K_r is greater if the freeboard is also larger. This fact proves that freeboard and K_t are inversely related. While porosity with Kt is directly proportional and K_r is inversely proportional.

Acknowledgment

Researchers would like to thank the Department of Ocean Engineering, DRPM-ITS, and Institut Teknologi Sepuluh Nopember, Surabaya, Indonesia, for the financial support and facilities provided.

References

Pustaka SPH Submerged Breakwater
[1] CERC, 1984. Shore Protection Manual. Coastal Engineering Research Centre, Dept. of the Army Waterways Experiment Station, US Army Corps of Engineers, Washington.
[2] Sujantoko and Natakusuma, D., 2003. Model simulasi interaksi gelombang dan arus di perairan dangkal, *Jurnal Teknik Sipil*, vol. 10 (3), 99-108.
[3] Sujantoko, 2009. The Analysis of wave refraction using SWAN model, *IPTEK The Journal of Technology and Science*, 20 (2), pp. 45-56.
[4] Sujantoko, Pratikto, W.A., Prastianto, R. W. Maulana, M. I. and Vebriyanti, A., 2021. Study of changes in coastal morphology due to utilization of the Surabaya city coastal area. *International Journal of Marine Engineering Innovation and Research*, 7 (1), pp. 26-32.
[5] Sujantoko, Pangestu, P. A. Saputro, D., Ekianto, B. P. P., Hasanudin, Kurniati, N., Kusumawardhani, R., and Hartanto, D., 2023. Hydrodynamic model simulation at the port of Tanjung Rhu Belitung. *International Journal of Marine Engineering Innovation and Research*, 8 (2), pp.141-149.
[6] Sujantoko, Mustain, M., Wahyudi, and Ikhwani, H., 2023. Hydrodynamic model due to reclamation in Lamong Bay. *IOP Conference Series: Earth and Environmental Science*, Volume 1198, 012004.
[7] Mustain, M., and Sujantoko, 2023. Spiral analysis of vertical currents in Lamong Bay. IOP Conference Series: Earth and Environmental Science, Volume 1166, 012024.

[8] Mustain, M., and Sujantoko, 2021. Simple Analysis of Determination of Tidal Types of Sea Water: A Case Study in Lamong Bay Waters, East Java. *Proceeding - IEEE Ocean Engineering Technology and Innovation Conference: Ocean Observation, Technology and Innovation in Support of Ocean Decade of Science (OETIC)*, pp. 13–17.

[9] Mustain, M., Sujantoko, and Prasetyo, M. A., 2022. Empowerment and optimalization of Kenjeran beach tourism potential, Surabaya-East Java. *International Review of Civil Engineering*, 13 (4), pp. 246-254.

[10] Mustain, M., Garang, I. J., Ikhwani, H., and Sujantoko, 2022. Analysis and maximize the ecotourism potential of the Wonorejo mangrove in East Surabaya. *International Journal on Engineering Applications*, 10 (2), pp. 158-166.

[11] Salem, M. E., and Mercer, D. E., 2012. The economic value of mangroves: a meta-analysis. *Sustainability*, 4, pp. 359–383.

[12] Winterwerp, J. C., Erftemeijer, P. L. A., Suryadiputra, N., Eijk, P.V., and Zhang, L., 2013. Defining eco-morphodynamic requirements for rehabilitating eroding mangrove-mud coasts. *Wetlands*, 33, pp. 515–526.

[13] Rosyid, D.M., Sujantoko, Armono, H.D., Djatmiko, E.B., Prastianto, R.W., Mulyadi, Y., Kurniati, N., Wardhani, M.K., and Wardhana, W., 2021. Kajian Pengelolaan Kawasan Mangrove Pesisir Selatan Kabupaten Bangkalan Berdasarkan UU No 1 Tahun 2014, Sewagati, 5 (3), pp.1-11.

[14] Armono, H.D., Bromo, B.H., Sholihin, and Sujantoko, 2022, Numerical study of bamboo breakwater for wave reduction, *Fluids*, 7 (1), pp. 1-13.

[15] Hales, L. Z, 1981. *Floating Breakwater: State of the art Literature Review*. Technical Report No. 81-1. U.S. Army Coastal Engineering Research Center, Fort Belvoir.

[16] McCartney, B. L., 1985. Floating breakwater design. *Journal of Waterway, Port, Coastal and Ocean Engineering,* 111 (2), pp. 304-318.

[17] Sujantoko, Djatmiko, E. B., Wardhana, W., Armono, H. D., Sholihin, and Ali, M. F., 2019. Experimental study on the effect of mooring angle on mooring tension of porous saw type floating breakwater. *Proceeding* of *The* 7th *International Seminars on Ocean and Coastal Engineering, Environmental and Natural Disaster Management,* Surabaya, Indonesia, Nov 6-7, pp. 94-99.

[18] Sujantoko, Djatmiko, E. B., Wardhana, W., Armono, H. D., and Wahyudi, 2019. Dynamic behavior analysis of porous saw floating breakwater under regular waves. *Proceeding of The* 7th *International Seminars on Ocean and Coastal Engineering, Environmental and Natural Disaster Management*, Surabaya, Indonesia, Nov 6-7, pp. 236-241.

[19] Sujantoko, Wardhana, W., Djatmiko, E. B., Armono, H. D., Putro, W. S., and Almuzaki, R. H., 2021. Study of wave characteristics of the floating breakwater for piling and tethered type. *Jurnal Teknik Hidraulik (Indonesian)*, 12(1), pp. 39-51.

[20] Sujantoko, Djatmiko, E. B., Wardhana, W., Hidayatullah, A., 2021. Analysis of mooring tension on the saw-type floating breakwater with physical modeling. *Jurnal Teknik Sipil (Indonesian)*, 28(3), pp. 289-300.

[21] Sujantoko, Djatmiko, E. B., Wardhana, W., 2022. Experimental investigation of the performance of porous-slope floating breakwater. *International Journal on Engineering and Applications*, 10(2), pp. 149-157.

[22] Sujantoko, Armono, H. D., Djatmiko, E. B., and Putra, R. D., 2022. Stability analysis of concrete block anchor on steep-slope floating breakwater. *Fluids*, 7(8), 259.

[23] Sujantoko, Djatmiko, E. B., and Wardhana, W., 2022. Experimental study on mooring tension of porous-slope floating breakwater. *International Journal on Engineering and Applications*, 10 (5), pp. 345-352.

[24] Sujantoko, Djatmiko, E.B., Wardhana, W., and Firmansyah, I. G., 2023. Numerical modeling of hydrodynamic performance on porous slope type floating breakwater. *Nasemore*, 70(2), pp. 149-157.

[25] Sujantoko, Djatmiko, E. B., Wardhana, W., and Mustain, M., 2023. Analysis of wave spectrum change of porous-saw floating breakwater. *International Journal on Engineering and Applications*, 11 (5).

[26] Sujantoko, Djatmiko, E. B., Wardhana, W., Mustain, M., and Ahidah, D. R., 2023. Numerical Modeling of Wave Hydrodynamic Performance with Various Layouts of Single-Slope Floating Breakwater Arrays, *International Journal on Engineering and Applications*, vol.11

[27] Sujantoko, Saputro, D., Pratikto, W.A., Mustain, M., and Ikhwani, H., 2023. Experimental Study of Wave Runup and Overtopping for Concrete Armor A-Jack on Seawall Structure. *International Journal on Engineering Applications (IREA)*.

[28] Sujantoko, Pangestu, P.A., Pratikto, W.A., Mustain, M., Wahyudi, and Ikhwani, H., 2023. Experimental Study of Wave Reflection on A-Jack Armor Unit on Seawall Structure, *International Journal of Marine Engineering Innovation and Research (IJMEIR)*, 8 (2), pp. 232-238.

[29] Putra, A.O.P., Armono, H.D., and Sujantoko, 2013. Pengaruh Elevasi Muka Air Laut pada Koefisien Transmisi dan Refleksi Composite Breakwater, *Jurnal Teknik ITS*, 1 (2), pp. G47-G51.

[30] Armono, H.D., Winarto, A., Sujantoko, and Suastika, I.K., 2021. A laboratory study on wave transmission over hexagonal artificial reef, *Sustainability and Resilience of Coastal Management (SRCM)*, Surabaya.

[31] Daffa, M.Z., Armono, H.D., and Sujantoko, 2020. Model Numerik Pengaruh Artificial Reef dalam Reduksi Gelombang dan Perubahan Morfologi Pantai menggunakan DELFT3D, *Seminar Nasional Penelitian & Pengabdian Kepada Masyarakat (SNP2M)*, Makasar.

[32] Sujantoko, and Manurung, A., 2020. Analisis Hidrodinamika pada Perencanaan Jetty Siantan Selatan, Kepulauan Anambas, *Seminar Nasional Penelitian & Pengabdian Kepada Masyarakat (SNP2M)*, Makasar.

[33] Islam, M.R., Hayano, K., and Nadeesha, A.M.K.N., 2013. Simulating the effect of flow volume on scouring process around breakwater under tsunami condition by SPH method, *Int. Journal of Geomate*, Vol. 5, No. 2 (Sl. No. 10), pp. 723-728.

[34] Armono, H.A, Manurung, A., Sujantoko, and Suastika, I.K., 2022. Wave transmission model over hexagonal artificial reefs using DualSPHysics. IOP Conference Series: Earth and Environmental Science, Volume 1095, 012025.

[35] Meringolo, D. D., Aristodemo, F, and Veltri, p., 2015. SPH numerical modeling of wave–perforated breakwater interaction, *Coastal Engineering*, Volume 101, pp 48-68.

[36] Rogers, B. D., 2009. Simulation of caisson breakwater movement using 2-D SPH, *Journal of Hydraulic Research*, 48(1), pp.135-141.

[37] Crespo, A.J.C., Domínguez, J.M., Rogers, B.D., Gómez-Gesteira, M., Longshaw, S., Canelas, R., Vacondio, R., Barreiro, A., and García-Feal, O., 2014. DualSPHysics: Open-source parallel CFD solver based on Smoothed Particle Hydrodynamics (SPH). *Computer Physics Communications*, 187, pp. 204-216.

[38] Liu, B. jin, Cheng, D., Sun, Zhao-Chen, Zhao, Xi-Zeng, Chen, Y., Lin, and Wei-dong, 2019. Experimental and numerical study of regular waves past a submerged breakwater. *Journal of Hydrodynamics*. 31, pp. 641–653.

[39] Aghaei, B., Arasteh, A.M., Lari, K. and Torabi Azad, M., 2021. Numerical modeling of wave reflection from sloped impermeable seawalls using the SPH method: Case Study of Chabahar Port, *Research Article Advances in Civil Engineering*, p.17.

[40] Tuan, T. Q, Luan M. T., and Cuong, L. N., 2022. Laboratory study of wave damping by porous breakwaters on mangrove mudflats in the Mekong River Delta. *Ocean Engineering*, 258. 111846.

[41] Lewis, C.D., 1982. *International and Business Forecasting Methods*. Butterworths, London.

Small Scale Plant of Desalination Using Multiple Tray Aerator and Two Different Membranes of Reverse Osmosis Using Low-Pressure Pump

M M Danial[1*], R Purnaini[2], F Hadary[3], A D Lestari[1], H Prasetya[4], D D Yhavez[4], E M Putri[4], Apriyantine[4]

[1]Lecturer, Ocean Engineering Department, Universitas Tanjungpura, Pontianak, Indonesia
[2]Lecturer, Environmental Engineering Department, Universitas Tanjungpura, Pontianak, Indonesia
[3]Lecturer, Electrical Engineering Department, Universitas Tanjungpura, Pontianak, Indonesia
[4] Undergraduate student, Ocean Engineering Department, Universitas Tanjungpura, Pontianak, Indonesia

E-mail: meddydanial@civil.untan.ac.id

Abstract. The seawater is an abundant source that still not be utilized as potential raw water source to overcome the deficit of drinking water. This paper aims to design the small-scale desalination plant with the capacity of 2000 gpd for treating the brackish and seawater. There are two split processes, i.e., multi-tray aerator and filtration for reducing the Fe and Mn, while the two different type reverse osmosis membranes were used for desalinating process. Parameters used to verify the performance of small-scale plant are salinity and total dissolved solid for membranes, while the parameters for examining the pre-treatment results are TSS, BOD, pH, Fe, and Mn. The test results will be evaluated under the Regulation of The Minister of Health, Number 32/2017. The influence of pump pressure on flow will also be analysed. The results show that multi-tray aerator needs some improvement to reduce the Mn, and Fe. The two types of membranes could significantly reduce the salinity level up to 91.7%, from 22,900 ppm to 1200 ppm. The pump pressure needs to be replaced into the high-pressure pump to reduce the salinity level significantly.

Keywords: Desalination, multiple tray aerator, two types of reverse osmosis membranes, low-pressure pump

1. Introduction

Water is essential and fundamental element for human life. However, only 0.8% of total water in the earth that could be consumed as freshwater [1]. In developing country, people living in coastal area and island mostly experience the lack of freshwater. Moreover, the highest freshwater demand is located in the coastal zone and delta estuary [2]. The difficulty of having freshwater in coastal community which is surrounding by the brackish and seawater is the irony amid modern era which is supported by advanced technology. Therefore, the abundant of seawater and brackish water near the coastal area should be utilized as raw water source. The characteristic of seawater and brackish water, containing of high level of salinity, can be desalinated by using the reverse osmosis membrane to get the freshwater [3,4].

Application of the reverse osmosis method as the cutting-edge technology for the water treatment has been investigated by researchers for many years [3,5]. One of the researchers studied the application of reverse osmosis membrane using low pressure pump (1.5–3.5 bar) for the brackish water well, located at the Sidoarjo resident area, in East Java [6]. Their pre-treatment used two media filters consisting of sand and activated carbon that is contained by FRP. Their result showed that the salinity only reduced by 58%. Moreover, another researcher carried out the combination of biofiltration, membrane ultrafiltration, and reverse osmosis for the treatment of river water, without being influenced by the salinity intrusion [7]. Their results show that the system can reduce parameters of TDS, Mn and Fe by 97.8%, 57–85%, 58–84%, respectively.

In coastal area, soil water mostly contains the iron (Fe). To reduce the iron, one of the most popular method that is used in pre-treatment process is multi-tray aerator [8,9]. The mechanism of aerator device is to interact between the air and the iron (Fe) as much as possible to. The number of trays is intended to increase the aeration process to reduce the Fe level. In the previous experiments, four trays have the reducing level of Fe at 98.2% [9]. On the other hand, the single tray aerator can only reduce the Fe level at 25% [10]. Those two experiments demonstrated that the multi-tray aerator is much more effective than compared to the single-tray aerator.

Based on the previous works, the investigation of applying reverse osmosis method using two low pressure pumps, to amplify the pump pressure, in order to reduce the salinity of seawater significantly, is important task.

This paper aims to design the desalination plant for converting sea water or brackish water to freshwater using sequential processes consisting of pre-treatment and core treatment. The pre-treatment stage applies multi tray aerator, whereas the core treatment process, consisting of ultrafiltration membrane (UF), and reverse osmosis (RO) membrane with two types of RO, i.e., ultra-low-pressure membrane (RO ULP) and brackish membrane (RO BW).

2. Materials and methods

2.1. Working principle of reverse osmosis

The working principle of reverse osmosis as a technology of desalination is by pressing the seawater using the high-pressure pump, where the pressure of pump must be greater than the pressure of seawater. This condition pushes freshwater through the RO membrane, whereas the residual water, containing the salinity, will be rejected by the RO membrane [11].

Aeration is widely used for treating water containing high iron and manganese [8]. Principle process of multiple tray aerator is based on the aeration process, which generates the oxidation process between the water and air. The longer the contact, the increasing the oxygen content. In addition, this process can also reduce the CO_2, H_2S, CH_4, and VOCs which could affect bad taste and flavour of the water.

2.2. Materials and installation design

There are two main installations of the desalination equipment, i.e., installation of pre-treatment and core treatment. The scheme diagram of desalination plant design can be shown in Fig. 1.

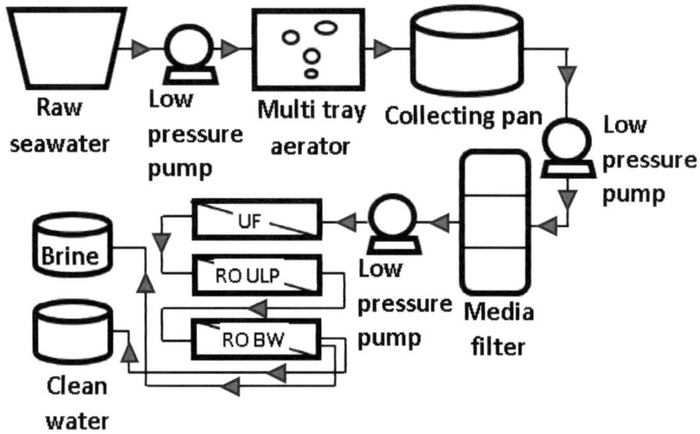

Figure 1. Scheme diagram of small desalination plant

Moreover, there are two components of pre-treatment that should be assembled, i.e., multi tray aerator and filter media, whereas the core treatment only has component of membranes. The material of multi tray aerator unit made by using galvanized iron frame, consisting of five trays with the total height is 4 m, and the dimension of tray is 0.94 m x 0.74 m x 0.17 m.

The last unit is the core treatment, which consist of three membranes, i.e., UF, RO ULP, and RO BW membranes which is assembled in galvanized irone frame. Figure 2 shows the design of pre-treatment and core treatment of desalination equipment.

Figure 2. Pre-treatment using multi tray aerator and alum (left), core-treatment using UF, ULPRO and BWRO membrane (right)

In this small plant, there are two types of RO membrane, which are implemented in two different situations, i.e., if the water condition is categorized as light brackish, then only the ULPRO membrane is operated. However, if the water condition is categorized as brackish or seawater, then both membranes are operated.

The other vital component of desalination equipment is high pressure pump, which must achieve a certain pressure in order to fulfil the requirement of reverse osmosis process, ranging from 75—350

Psi[12,13]. In addition, these pressures depend on the salinity content where the higher the salinity, the greater the pressure of pump is required [12].

2.3. Methods

The performance of desalination equipment design must be evaluated in real situation. There are two types of raw water sample, i.e., first, the existing of raw water in the river near the coastal area where the salinity level at 5000 ppm. Second, the seawater with the salinity level at 21 ppt.

In this research, tests for raw water samples are divided based on two purposes, the first one is to evaluate the performance of multiple tray aerator using river water sample, and the second one is to examine the performance of RO membrane using seawater sample. Thus, these examinations of both samples are not sequential, where the water sample from pre-treatment does not used as water input for the core treatment.

Figure 3 shows both raw water sample of river water and seawater. Due to always rain falls in Pontianak throughout the year 2022, in order to obtain the seawater salinity, we must take the samples from the ocean, assisting by the fisherman, transporting by small boat and bring it by the water gallon.

Figure 3. River water/slightly brackish (left), gallons of seawater sample (right)

In this research, for pre-treatment process, there are only seven parameters of water quality tested in the laboratory, namely biological oxygen demand (BOD), total dissolved solid (TDS), total suspended solid (TSS), iron (Fe), and Manganese (Mn). After aeration process, coagulation process is conducted by adding aluminum sulphate (alum) [14]. The dosages of alum are determined by jar tests, where the optimal dose is 125 gr for 1 m^3 of water sample, as presented in Fig. 4.

Figure 4. Jar test for determining coagulant dose

The test method for BOD and COD is using the azide winkler and closed reflux method, respectively. The parameter of Fe and Mn are tested using atomic absorption spectrometry (AAS), while the TSS is examined by using gravimetric method. In addition, for the core treatment stage, only two water quality parameters are evaluated, viz. TDS and salinity, which are measured by using conductivity meter. The result of water sample test will be evaluated using the quality standard, according to the regulation of Minister of health, Number 32/2017.

The pressure specification of pump used in this research is 4 bar or 58 Psi, while the pressure specification of membranes ULPRO and BW are 150 Psi and 225 Psi, respectively. Two low pressure of pumps is connected to amplify the pump pressure in order to gain the desalination process.

3. Result and analysis

Figure 5 shows the 5 water sample bottles, i.e., two inlet bottles before pre-treatment and two outlet bottles after the treatment; and the last bottle (the rightmost bottle) is after adding coagulant to the collecting pan. The outlet bottles show more brighter, indicating pre-treatment process works. However, reducing mineral of Fe and Mn are still not significant. However, after adding the coagulant, the water quality fit the requirements of the Minister of Health regulation. This means that the multi tray aerator need further improvement to reduce parameters TDS, TSS, BOD, and Fe, Mn. This problem also happen to the previous work conducted by researcher where the pre-treatment does not affect several parameters, particularly for the TDS and TSS [7].

Figure 5. Water sample bottles before and after the treatment of multi tray aerator. The rightmost bottle is the sample after being added the coagulant.

To further analysis whether the aeration process works or not, it can be measured using dissolved oxygen (DO) parameter and calculated using DO meter. The result of DO parameter can be shown in Figure 6, where the DO increases in increasing time. This reaction proves that the oxidation process occurred, where there is contact between the air and water during the aeration process. As a result, the solubility of oxygen contained in the raw water increases. The process of mixing the oxygen with the water is characterized by an increase in the value of DO in the colleting pan, after aeration process in multi tray aerator.

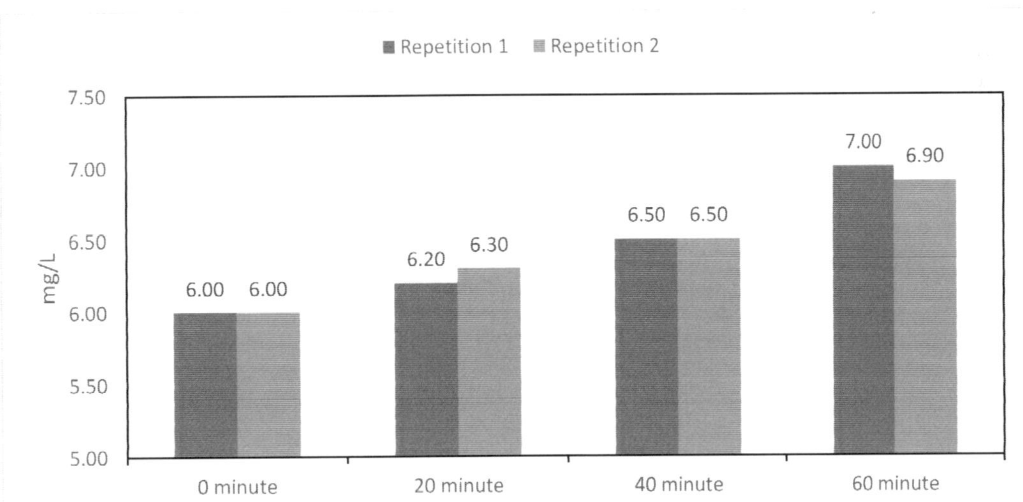

Figure 6. DO parameter test after multi tray aerator process.

The performance of desalination equipment can be seen from Table 1, demonstrating the result of water quality parameters for river water and seawater sample from the multi tray aerator and RO test, respectively.

From the Table 1, the pH parameter of water samples behaves differently between the adding coagulant and without the adding coagulant. The pH parameter of water sample, where the water samples were added with alum coagulant, decreases from 7.4 to 4.7. This outcome is consistent to the result from the previous study, conducted by another researcher [15].

Table 1. Results of water quality samples.

	Pre-treatment		Core treatment	
	Raw water from river water	Outlet multi tray aerator without coagulant/with coagulant	Raw water from seawater	Outlet RO Membrane (1 RO/2RO)
TDS (ppm)	844	852 / 838	18,000	1,236 / 1,137
TSS (mg/l)	18	25 / 2	-	-
Salinity (ppm)	0.08	0.08 / 0.03	22,900	2,000 / 1,200
pH	7.0	7.6 / 4.7	-	-
BOD (mg/l)	7.42	3.66 / 2.01	-	-
Fe (mg/l)	3.22	3.61 / 0.19	-	-
Mn (mg/l)	0.1	0.09 / 0.21	-	-

Figure 7 shows the comparison of clean water flow and pressure pump between the seawater and brackish conditions. It can be shown that for brackish condition, the pump pressure is lower than that for seawater condition. In contrast, the clean water flow for brackish is higher than that for the seawater condition. It means, the higher the salinity concentration, the higher the pump pressure. As a result, the clean water flow decreased.

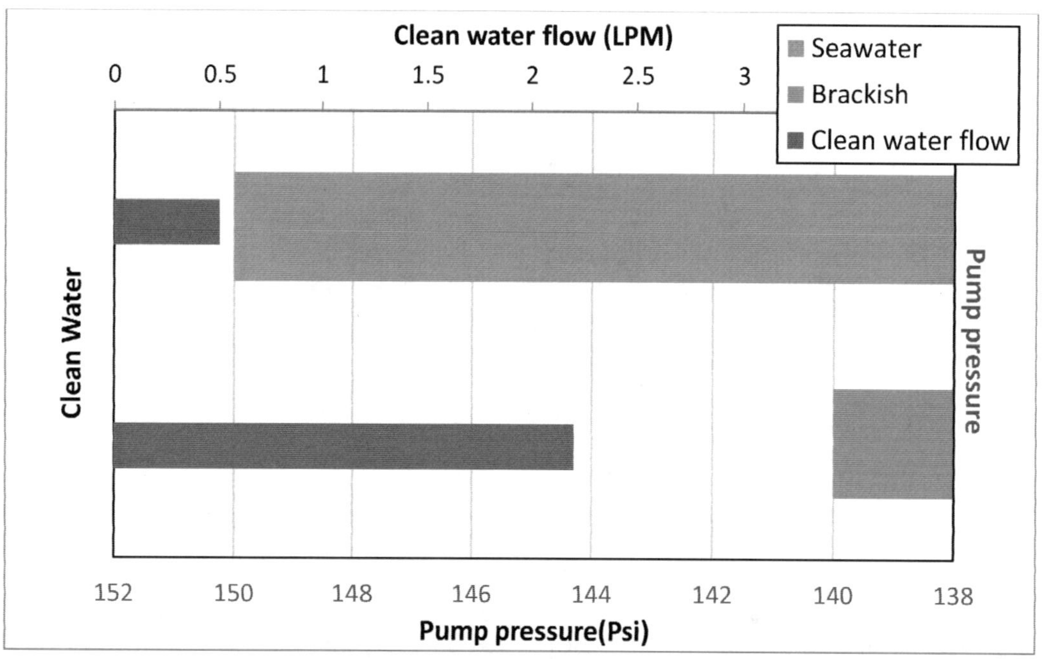

Figure 7. Comparison of clean water flow and pressure pump between the seawater and brackish conditions.

4. Discussion

Theoretically, the multi tray aerator can reduce the Fe and Mn significantly. The multi tray aerator affects the oxidation process, as evidence in decreasing the COD and BOD. Moreover, it can also be proved from the DO parameter analysis result, that the DO increases with increasing time the, implying the aeration process of multi tray aerator is running well. However, the oxidation process is only able to create colloid, not floc as expected. Therefore, to accelerate the formation of large floc, adding alum coagulant is needed. However, alum can result in lower pH due to the chemical process related to salt ions [15]. The multi tray aerator probably can be more effective if the settlement time is more than 4 hr, as revealed in the previous paper by another researcher [7]. Moreover, the filter component inside the tray of aerator can also be replaced by another filter to be able to improve the decreasing of iron content, such as aquadine filter mat combined with sand and gravel, or combined with silica sand, zeolite, and activated carbon [16].

In the desalination process, the goal of applying multiple membranes is to assure the process of converting seawater to freshwater can work well. As a result, the UF, RO ULP, and BW membrane can effectively reduce the salinity, where the pressure gauge indicates that the total pump pressure can reach 150 Psi. This pressure can be achieved because the experiment test uses the series of two pumps, which is able to amplify the total pressure of pump.

As proven in the experiment test, applying both RO membranes can be more effective in reducing the salinity, compared to applying only 1 RO membrane (See Table 1). This also means that the reverse osmosis process remain works although applying the two low pressure of pump on the reverse osmosis membranes. This happens, because the type of low-pressure membrane of reverse osmosis enabling the low-pressure pump to remain work to obtain desalination process, though must work with heavy loads.

Nevertheless, the flow water rate of clean water for seawater condition is too minor compared to the brackish water condition, where from the flowmeter measurement, it can only reach ~0.5 l/min. It means that the higher the concentration of salinity, the lower the flow can produce. As the comparison to the previous works by other researcher, reducing their salinity concentration by using only one low pressure pump is much more limited, which can only reach at 58% [6,17]. Therefore, employing of two series pumps is necessary if a greater reduction in salinity concentration is desired.

5. Conclusion

Design and construction of small-scale desalination device using multi-tray aerator and reverse osmosis membranes using low-pressure pump has been implemented. The performance of multi-tray aerator can increase oxidation process, implying the aeration process is running well. However, the additional alum coagulant is still needed to accelerate the process of flocculation to meet the quality of water that match with the regulation of Minister of health, Number 32/2017.

The three membranes consisting of UF, RO ULP, and RO BW can reduce the salinity by 91.7%, from 22,900 ppm to 1,200 ppm, though need further improvement. In addition, the discharge of clean water for seawater needs to be increased. To optimally reduce the salinity level and increase the flow rate of clean water, the pump must be replaced that having the pressure around 24 bar or 340 Psi.

Acknowledgment
This research paper is part of Penelitian Inovatif (PINOV) 2023, supported by the DIPA of Universitas Tanjungpura.

6. References

[1] Greenlee L F, Lawler D F, Freeman B D, Marrot B and Moulin P 2009 Reverse osmosis desalination: Water sources, technology, and today's challenges *Water Res.* **43** 2317–48
[2] Jeuken A, Termansen M, Antonellini M, Olsthoorn T and van Beek E 2017 Climate Proof Fresh Water Supply in Coastal Areas and Deltas in Europe *Water Resour. Manag.* **31** 583–6
[3] Misdan N, Lau W J and Ismail A F 2012 Seawater Reverse Osmosis (SWRO) desalination by thin-film composite membrane-Current development, challenges and future prospects

Desalination **287** 228–37
[4] Kurihara M and Takeuchi H 2018 SWRO-PRO system in "Mega-ton Water System" for energy reduction and low environmental impact *Water (Switzerland)* **10** 1–15
[5] Dhakal N, Salinas-Rodriguez S G, Hamdani J, Abushaban A, Sawalha H, Schippers J C and Kennedy M D 2022 Is Desalination a Solution to Freshwater Scarcity in Developing Countries? *Membranes (Basel).* **12** 1–15
[6] Etikasari Y, Tuhu Agung R and Rudy L 2010 Pengaruh Tekanan Reverse Osmosis Pada Pengolahan Air Payau Menjadi Air Bersih *Envirotek J. Ilm. Tek. Lingkung.* **2** 78–87
[7] Said N I 2009 Uji Kinerja Pengolahan Air Siap Minum Dengan Proses Biofiltrasi, Ultrafiltrasi Dan Reverse Osmosis (Ro) Dengan Air Baku Air Sungai *J. Air Indones.* **5** 144–61
[8] Ronny and Hasim A H 2018 Effectiveness of multiple tray-aerators in reducing iron (Fe) water wells in Gowa Regency, Indonesia *Ecol. Environ. Conserv.* **24** 22–5
[9] Al Kholif M, Sugito S, Pungut P and Sutrisno J 2020 Kombinasi Tray Aerator Dan Filtrasi Untuk Menurunkan Kadar Besi (Fe) Dan Mangan (Mn) Pada Air Sumur *ECOTROPHIC J. Ilmu Lingkung. (Journal Environ. Sci.* **14** 28
[10] Utomo B P B, Yuliansyah A T and Prasetya A 2020 Perbaikan Kualitas Air Yang Memiliki Konsentrasi Besi (Fe) Tinggi Menggunakan Single Tray Aerator *Media Ilm. Tek. Lingkung.* **5** 76–82
[11] Said N I 2003 Aplikasi Teknologi Osmosis Balik Untuk Memenuhi Kebutuhan Air Minum *J.Tek.Ling. P3TL-BPPT* **4** 16–35
[12] Said N I 2008 Pengolahan Payau Menjadi Air Minum dengan Teknologi Reverse Osmosis *Teknol. Pengolah. Air Minum* 443–501
[13] Sagle A and Freeman B 2004 Fundamentals of membranes for water treatment *Futur. Desalin. Texas* 1–17
[14] Malik Q H 2018 Performance of alum and assorted coagulants in turbidity removal of muddy water *Appl. Water Sci.* **8** 1–4
[15] Thammasane S and Kaosol T 2018 Impact of Chemical Coagulants for Oscillatoria sp. Removal from Raw Water on Chemical Coagulation Process *Am. J. Environ. Sci.* **14** 257–65
[16] Nainggolan, A. H., Tarigan, A. P. M. Khair A 2017 Saringan Pasir , Karbon Aktif , Dan Zeolit Dalam Menyisihkan Parameter Fe Dan Mn Dari Air Tanah Di Pesantren Ar-Raudhatul Hasanah *J. Tek. Lingkung.* **3** 1–12
[17] Putro E P, Widodo E, Fahruddin A and Iswanto I 2020 Analisis Head Pompa Sentrifugal Pada Rangkaian Seri Dan Paralel *Media Mesin Maj. Tek. Mesin* **21** 46–56

Processed Seafood Determination for Fishermen Prosperity Using Macro Ergonomic and Analytical Hierarchy Process

Eko Nurmianto[1], S H Kusuma[2], S Soehardjoepri[3], S R Nadhiroh[4], S Rahmaningsih[5], F N Fauziah[6], G S Trikurniawan[2]

[1]Department of Industrial and System Engineering, Institut Teknologi Sepuluh Nopember, Surabaya 60111, Indonesia
[2]Department of Urban and Regional Planning, Institut Teknologi Sepuluh Nopember Surabaya 60111, Indonesia
[3]Department of Actuarial, Institut Teknologi Sepuluh Nopember, Surabaya 60111, Indonesia
[4]Department of Nutrition, Universitas Airlangga, Surabaya 60115, Indonesia
[5]Department of Fisheries, Universitas Ronggolawe, Tuban 62381, Indonesia
[6]Department of Development Studies, Institut Teknologi Sepuluh Nopember, Surabaya 60111, Indonesia

E-mail: nurmi@ie.its.ac.id; ekonurmianto1@gmail.com

Abstract. The high rate of stunting in children in East Java Province is a problem that has become a major focus in the national scope. Based on these problems, the researcher intends to increase the intake of animal protein from fish for fishermen's families, by prioritizing fish processing in coastal areas of East Java and providing education and knowledge for fishermen in fish processing methods in coastal areas of East Java Province. An analysis was carried out using the macro ergonomic and Analytical Hierarchy Process (AHP) method to determine the priority of processed food from fish ingredients to reduce stunting rates. The criteria in this analysis are Community Preferences, Community Knowledge, Ergonomic process, Commodity Production, and Fish Processing Tools. The results of the analysis found that priority choices in percentage include Fish Nuggets, Fish Sticks, Fish Meatballs, and Fish Crackers. The final result of this study was the priority of processed fish food ingredients to reduce stunting rates in children in coastal areas of East Java Province. With these results, it can help SMEs and coastal communities to process fish food ingredients with the aim of developing communities and coastal areas and increasing fishermen's family interest in consuming processed fish ingredients to reduce stunting.

1. Introduction

Community empowerment in coastal areas that has been carried out previously in Lamongan dan Probolinggo [1] found stunting cases in coastal areas of East Java. Stunting is the impaired growth and development that children experience from poor nutrition, repeated infection, and inadequate psychosocial stimulation. Children are defined as stunted if their height-for-age is more than two standard deviations below the WHO Child Growth Standards median [2]. This is interesting because areas that have high stunting rates are areas that have abundant resources in terms of foodstuffs, especially the coast of East Java, an area with abundant food resources but in which there is still a high stunting problem. This is because many small children are bored with eating fish. The daily life of his parents who are fishermen and the lifestyle that sees fish every day, makes these little children bored and do not want to eat fish. This is why in the coastal areas of East Java although natural resources are abundant, stunting cases are also high in this region then potential to be coastal eco-tourism [3]. The fish processing process can be carried out with the support of renewable energy sources from the ocean [4–6].

Previous research [7] on Various Fish and Clams Processed Business in Kalirejo Village, Kraton District, Pasuruan Regency this time improved fish and clams processing management which includes training and assistance in the process of processing fish and clams into various processed fish and clams, financial management, capital, packaging methods and brands, sales techniques and marketing management and entrepreneurship for product processors sea area of Kraton District to improve their welfare.

Overcoming stunting in coastal areas of East Java is one of the challenges to solve it [8]. It is necessary to consider the right way to overcome the problem of stunting with the availability of existing resources. Processed fish products were chosen as one way to overcome stunting in coastal areas of East Java. This processed product is a way out to overcome the boredom of young children about fish. This processed product is processed in such a way that the shape of the fish and the taste of it can be different and become a new food choice for them. These processed products include fish and clam's crackers, fish meatballs, fish and clams' nuggets, fish mangut, and smoked fish and smoked clams. Fish mangut is one type of Javanese cuisine made mainly from smoked fish with a taste like goulash but thinner and spicier. These processed products are the main solution in overcoming stunting in coastal areas of East Java.

In its development, these processed products are products that can be processed from raw fish ingredients. Of course, for better and more targeted production, it is necessary to prioritize which processed products are the main interests of coastal communities. By using macro ergonomic it produces a comprehensive analysis of the causes and effects of the problems faced which in this case is stunting. This analysis provides a diagram on how each stakeholder plays an important role in the problem of stunting on the coastal area of East Java. Furthermore, it is combined with Analytical Hierarchy Process (AHP) analysis can be found the order of priority ranking which preparations are of interest to coastal communities, starting from the first, second, and so on. The Analytical Hierarchy Process (AHP) offers a thorough framework for dealing with such issues. It is a method for representing the elements of any problem in a methodical manner. It organizes basic reasoning by breaking down a problem into smaller constituent parts and then developing priorities in each hierarchy using simple pairwise comparison judgments [9]. Several criteria in determining alternatives are also involved according to the existing conditions of the community such as community preferences, community knowledge, commodity production, fish processing tools [10], and ergonomic process. This AHP analysis involved 33 respondents from several coastal areas in East Java such as Surabaya, Tuban, Pasuruan, and Probolinggo to obtain valid results.

Therefore, the research objectives are:

1. Determine the best processed seafood foods among the processed foods chosen by respondents.
2. Determine cause-and-effect relationships between stakeholders in the problems that occur.
3. Choose the criteria of good, processed seafood food.
4. Reducing stunting rates in children in coastal areas of East Java.

2. Literature Review

Community empowerment in coastal areas that has been carried out previously in Lamongan and Probolinggo [1]. In some of these empowerment activities, there is a phenomenon that coastal communities in East Java are tired of eating fish because almost every day they meet the same commodity. Boredom of these fish results in a lack of animal protein intake, especially fish, giving rise to one of the indicators of stunting. Over time, the stunting phenomenon is like an iceberg so that it becomes a pandemic that spreads to several cities in the coastal area of East Java. The reduction in stunting can be overcome by one of them returning to eating fish with various preparations that are managed into various preparations [7]. In addition to solutions to reduce stunting in the form of processing fish foodstuffs, previous studies conducted an analysis in the form of a combination of AHP and Macro Ergonomic which produced outputs in the form of the best scenario decisions in which there was a stakeholder analysis to present cause-and-effect relationships. The process of the

3. Methodology

The selection of a multi-criterion method for dealing with a particular problem is a multi-criterion decision problem without a clear solution. Choosing one method over another is often done arbitrarily, due to the lack of standard rules [12]. Macro Ergonomic is used as an analysis that determines the causal relationship between each stakeholder involved in this study. To prioritize and choose health infrastructure projects, the Analytical Hierarchy Process (AHP) was employed. It is suggested for this type of ranking problem with scores [13,14].

The method known as AHP has caught the interest of many academics, due to its good mathematical nature and the ease with which the essential input data may be obtained [15]. Its simplicity is defined by comparing alternative pairs based on specific criteria [16]. Furthermore, the emergence of various software tools has popularized and simplified this procedure. Expert Choice is the most well-known [17] and Super Decisions [18]. In this study, we used Expert Choice program version 11, which is a decision-making software based on multi-criteria decision making, and we implemented AHP by calculating relative weights rather than counting by hand. Modern fish smoking devices are used as one method to produce some processed fish products. Organoleptic tests [19] are also carried out to determine the level of relevance between product quality and consumer taste. This method is applied to be able to adjust the target consumers to be targeted by this processed fish product so that it can be well accepted in the community.

4. Result and Discussion

Figure 1 shows the causal relationship between each stakeholder involved. Macro Ergonomic is a concept to make a friendly relationship to an environmental system.

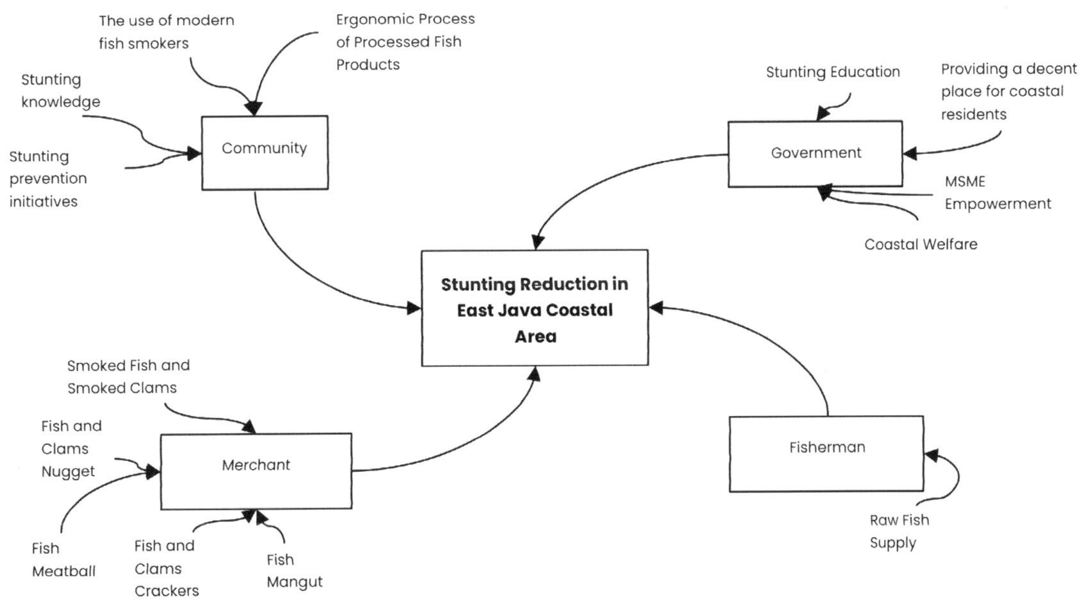

Figure 1. Causal relationship analysis diagram between stakeholders

In figure 1, the decrease in stunting cases in the coastal areas of East Java occur if each stakeholder gives their own role as the diagram shows. Environmental systems that support each other would be one of the factors in reducing stunting rates in coastal areas of East Java.

Figure 2 shows the hierarchical structure of prioritization to compile an analytical hierarchy process (AHP) analysis framework.

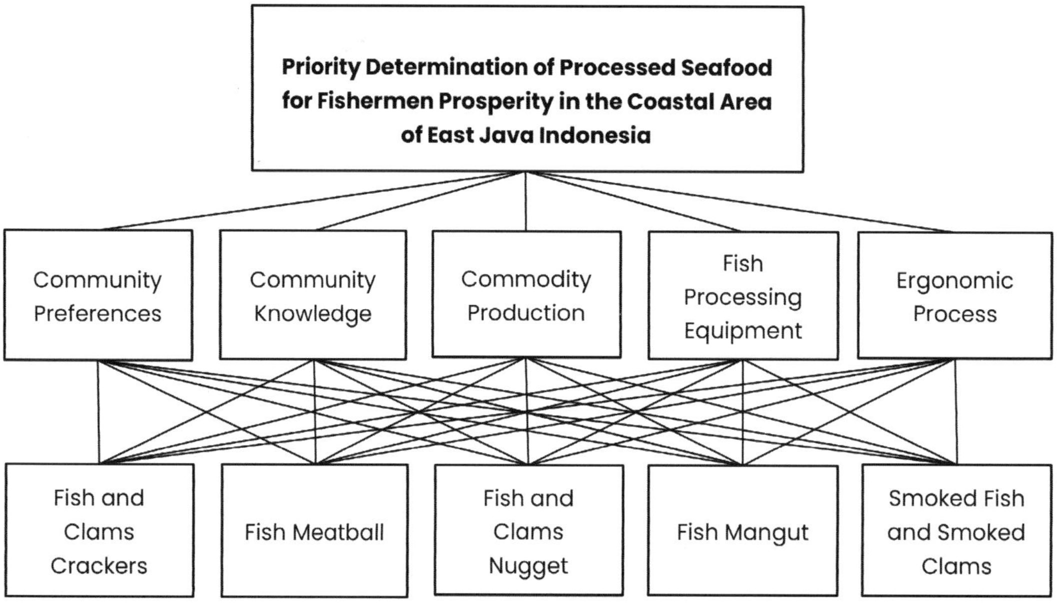

Figure 2. Hierarchy Structure of Prioritization

In figure 3 the results of the AHP analysis are displayed which is a visualization of respondent data that has been combined in such a way by the Expert Choice 11 application. Figure 3 visualizes priority data on criteria that influence the selection of alternative processed food products, with the results of community preference being the highest priority criterion with a value of 0.538 with an inconsistency of 0.09. Then figure 4 shows the result of the priority of alternative processed fish products where the highest priority of processed fish products is fish mangut with a value of 0.403 with an inconsistency value of 0.08.

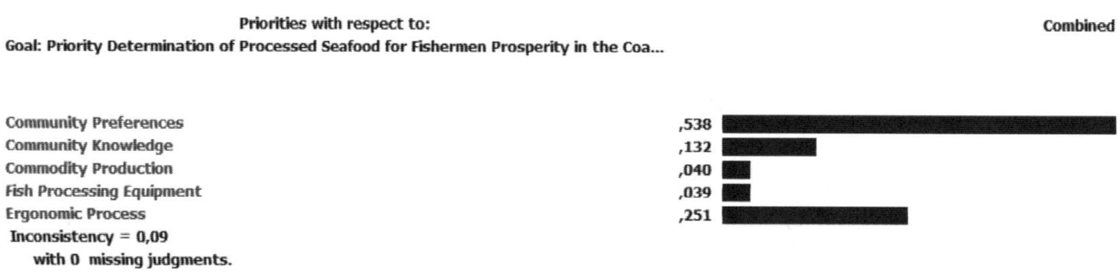

Figure 3. Priority of Criteria for Determining Processed Food Ingredients

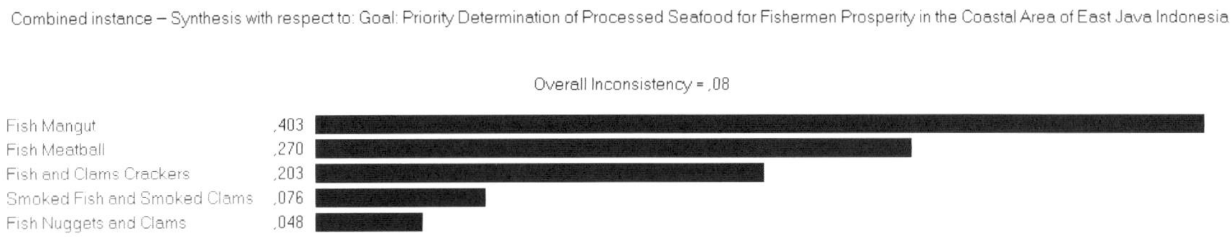

Figure 4. Alternative Priorities for Determining Processed Food Ingredients

In figure 3, Community Preferences is ranked the highest priority with a value of 0.538 which means that community preferences are the most dominant factor in choosing alternative priorities for processed fish products. The Ergonomic Process criteria were ranked second with a value of 0.251 followed by community knowledge with a value of 0.132, commodity production with a value of 0.040, and finally the fish processing equipment criteria with a value of 0.039.

Followed by figure 4, where this image is a graphic visualization of the results of prioritizing processed fish products. The first rank is occupied by fish mangut with a value of 0.403 followed by fish meatball with a value of 0.270, fish and clams crackers with a weight of 0.203, smoked fish and smoked clams with a weight of 0.076, and finally fish nuggets and clams with a weight of 0.048.

Table 1. Criteria Weights (Combined)

Criteria	Weight
Community Preferences	0.538
Community Knowledge	0.132
Commodity Production	0.040
Fish Processing Equipment	0.039
Ergonomic Process	0.251
inconsistency	0.090

Table 2. Criterion Weights – Alternatives (Combined)

Weights	Community Preferences	*Community Knowledge*	Commodity Production	*Fish Processing Equipment*	Ergonomic Process
Fish Mangut	0.551	0.307	0.0307	0.0309	0.307
Fish Meatball	0.242	0.300	0.292	0.289	0.281
Fish and Clams Crackers	0.113	0.258	0.262	0.262	0.263
Smoked Fish and Smoked Clams	0.064	0.075	0.076	0.080	0.091
Fish and Clams Nugget	0.031	0.060	0.063	0.060	0.057
inconsistency	0.070	0.060	0.060	0.06	0.080

Table 1 presents qualitative data on the weights of each criterion compared to each other. Then in table 2 qualitative data is presented about the weight of each alternative on each criterion. In this data, fish mangut is the most popular choice for the public in each criterion. All are done with an inconsistency level below 0.1. Table 3 shows the final result of AHP. In the table, fish mangut has the highest weight of 0.403 followed by other alternatives, namely fish meatballs with a weight of 0.270, fish crackers and clam crackers with a weight of 0.203, smoked fish and smoked clams with a weight of 0.076, and finally fish and clam's nugget with a weight of 0,048.

Table 3. Priority End Results of Processed Food Ingredients

Rank	Criteria	Weight
1	Fish Mangut	0.403
2	Fish Meatball	0.270
3	Fish and Clams Crackers	0.203
4	Smoked Fish and Smoked Clams	0.076
5	Fish and Clams Nugget	0.048
6	Inconsistency	0.080

The following is some documentation about processed fish products that are the choice of coastal communities and documentation about the discussion process with local residents.

Figure 5. Processed Fish Products (Nuggets, Meatballs, Crackers, and Sticks)

Figure 6. Discussion of Socialization Atmosphere

Figure 7. Modern Fish Smoking Devices

In figure 5, figure 6, and figure 7, documentation can be seen about real steps in conducting socialization and training to introduce processed fish products as well as taking samples of interest in selecting criteria and processed fish products in coastal communities of East Java.

4.1. Sensitivity Test

Figure 8 shows that this is the initial scenario of the sensitivity test with the ideal mode visualized by the performance model.

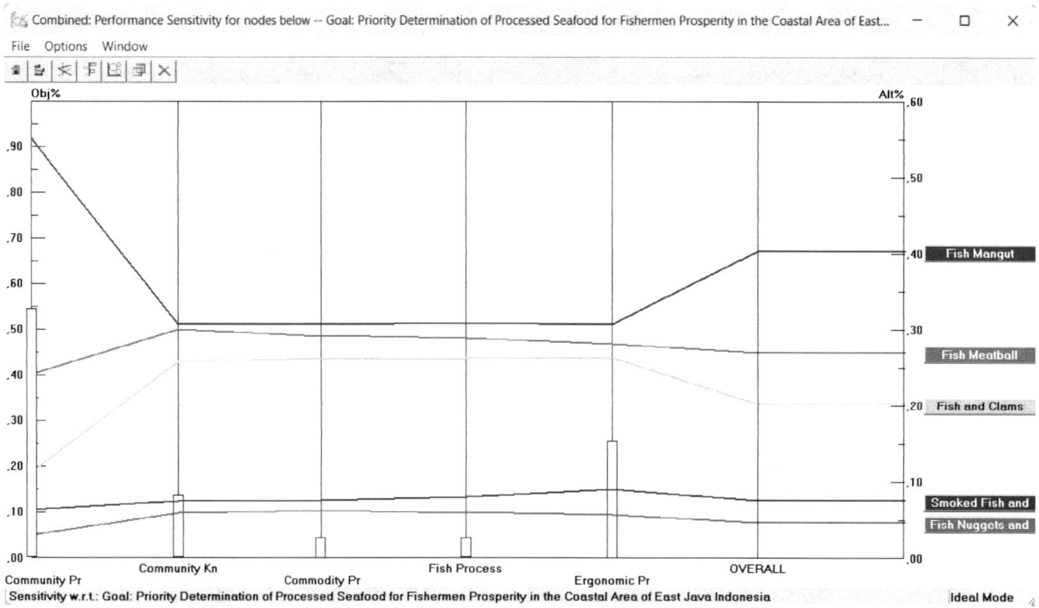

Figure 8. Sensitivity Analysis (Performance) – Initial Scenario

A sensitivity analysis was performed to help understand the effect of changes in the weight of key criteria on alternative ratings. It is useful to perform a "what will happen if" analysis to see how the final result will change if the weight of the criteria is different [18]. This sensitivity analysis will show how much change will occur if each criterion does not fit the initial scenario.

There are two scenarios studied and become a benchmark for sensitivity to influence criteria with alternative results:

- Scenario 1: Sensitivity analysis of ergonomic process criteria
- Scenario 2: Sensitivity analysis of community knowledge criteria

Table 4 is the weight of each criterion if changes are made to find out how much influence the criteria are changed.

Table 4. Sensitivity Analysis Scenario Weighting

Criterion	Initial Scenario Weighting	Scenario 1	Scenario 2
Community Preferences	0.538	0.009	0.017
Community Knowledge	0.132	0.283	0.538
Commodity Production	0.040	0.086	0.163
Fish Processing Equipment	0.039	0.084	0.159
Ergonomic Process	0.251	0.538	0.122

As the data shown in table 4, by conducting two test scenarios on the criteria of ergonomic process and community knowledge, results with different weights were obtained. Both scenarios result in different priorities of determining criteria. In the initial scenario, the priority of the highest determining criterion is community preferences. However, when testing scenario 1, the results changed so that ergonomic process was the highest determining criterion as well as what happened in scenario 2 where community knowledge was the highest priority criterion. This is to test how strong and how consistent the alternative results of processed fish products have been selected before.

Scenario 1: Sensitivity analysis of Ergonomic Process criteria
In scenario 1 visualized in figure 9, it can be seen that there is no change at all in the changes in the weight of the criteria made earlier.

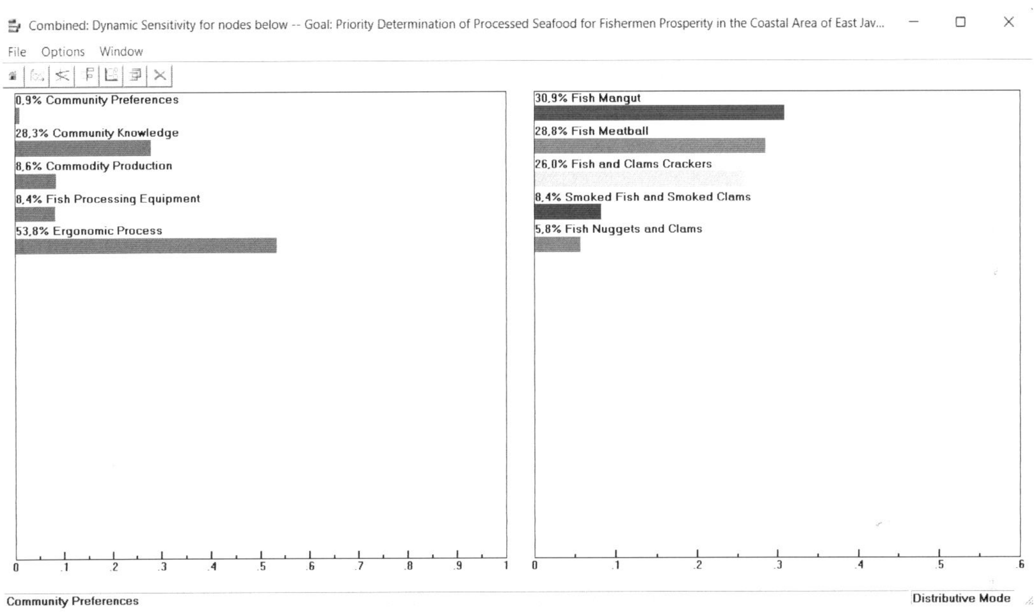

Figure 9. Sensitivity Analysis Scenario 1

By varying the weight of the ergonomic process criteria to reach a value of 53.8%, there was no change in the priority ranking of processed fish products in comparison to the first scenario. Mangut ikan still ranked first with 30.9% then followed by fish meatballs 28.8%, fish and clams crackers 26%, smoked fish and smoked clams 8.4%, and fish nuggets and clams 5.8%.

Scenario 2: Sensitivity analysis of community knowledge criteria
Just like before, in scenario 2 visualized in figure 10, there is no change at all in the changes in the weight of the criteria made earlier.

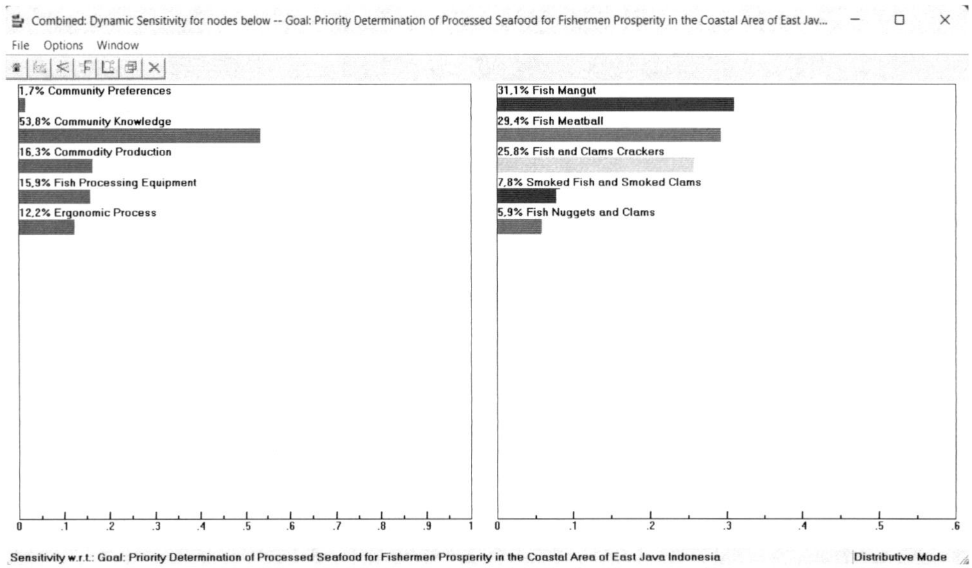

Figure 10. Sensitivity Analysis Scenario 2

By varying the weight of the ergonomic process criteria to reach a value of 53.8%, there was no change in the priority ranking of processed fish products in comparison to the first scenario. Now, fish mangut 31.1%, fish meatballs 29.4%, fish and clams crackers 25.8%, smoked fish and smoked clams 7.8%, and finally fish nuggets and clams 5.9%.
By making changes using 2 scenarios, no results were found about alternative ranking changes. This shows that the ranking of selected alternatives is a strong and consistent alternative.

4.2. Originality and Research Value

This study is a study that applies macro ergonomic and AHP in the process of determining and selecting processed products from fish as an effort to overcome stunting cases in coastal areas of East Java. This study provides decision makers with an approach that takes criteria into consideration as an implementation from the point of view of fish smoking device innovation and community habits to produce a targeted ranking according to the purpose for which the study was made.

5. Conclusion

Research Conclusions

1. The best processed fish products selected based on the analysis were fish nuts with a value (0.403), followed by fish balls (0.270), fish crackers and clam crackers (0.203), smoked fish and smoked mussels (0.076), and fish nuggets and clam nuggets (0.048) being the last choice.
2. The role of each stakeholder is clearly visible and illustrated to be able to create a supportive environment for the reduction of stunting cases in coastal areas of East Java.
3. The criteria that dominate processed seafood in this analysis are community preference with a value of (0.538), followed by public knowledge (0.132), commodity production (0.040), fish processing equipment (0.039), and ergonomic process (0.251) being the last.
4. With this ranking, it will make the stunting rate in children decrease because they can eat a varied diet with healthy, tasty, and nutritious fish-based ingredients.

6. Acknowledgments

The authors gratefully acknowledge financial support from the Institut Teknologi Sepuluh Nopember for this work, under project scheme of the Publication Writing and IPR Incentive Program (PPHKI) 2023. And also, a special thanks to Dr. Dendy Satrio.

References

[1] Nurmianto E, Wessiani N A, Negoro N P and Agustin T I 2012 Ergo-Desain Alat Pengasapan Ikan yang Mobile dan Portable Untuk Meningkatkan Nilai Ekonomi pada UMKM Perikanan di Wilayah Pesisir Lamongan *Proceeding SENTA 2012 Semin. Nas. Teor. dan Apl. Teknol. Kelaut.*
[2] World Health Organization 2015 Stunting in a nutshell
[3] Ikhwani H, Putro M H M, Mustain M, Satrio D, Sujantoko, Hadiwidodo Y S and Nasution P H O 2023 The Concept of Integrated Coastal Management for Eco-Tourism Development at Dalegan Beach, Gresik Regency, East Java, Indonesia *IOP Conference Series: Earth and Environmental Science* vol 1198 (Surabaya: IOP Publishing Ltd) pp 1–9
[4] Satrio D, Adityaputra K A, Suntoyo S, Silvianita S, Dhanistha W L, Gunawan T, Muharja M and Felayati F M 2023 The Influence of Deflector on the Performance of Cross-flow Savonius Turbine *Int. Rev. Model. Simulations* **16** 27–34
[5] Satrio D, Suntoyo S, Erwandi E, Albatinusa F, Tuswan T and Hakim M L 2023 The Benefit Using a Circular Flow Disturbance on the Darrieus Turbine Performance *Int. J. Eng. Appl.* **11** 55–63
[6] Satrio D, Suntoyo S, Albatinusa F, Junianto S, Musabikha S, Madi M and Setyawan F O 2023 Effect of Positioning a Circular Flow Disturbance in Front of the Darrieus Turbine *IOP Conference Series: Earth and Environmental Science* (Surabaya: IOP Publishing Ltd) pp 1–8
[7] Soehardjoepri, Nurmianto E and Anzip A 2021 PKM Usaha Aneka Olahan Ikan dan Kerajinan Kerang *J. Pengabdi. Kpd. Masy. Bina Darma* **1** 173–9
[8] Has E M M, Prakhasita R C, Krisnana I, Wahyuni S D, Arief Y S and Nursalam 2020 Risk Factors for Stunting and Severe Stunting among Children (12-59 Months Old) in the Coastal Area of Surabaya, East Java, Indonesia *Int. J. Psychosoc. Rehabil.* **24**
[9] Saaty T L 1986 Axiomatic Foundation of the Analytic Hierarchy Process *Manage. Sci.* **32**
[10] Nurmianto E 2018 Ergonomics smoke machine for indigenous people in Indonesia ed S Ma'mun, H Tamura and M R A Purnomo *MATEC Web Conf.* **154** 01103
[11] Nurmianto E, Iswandy D, Armono H D, Damanhuri, Agustin T I and Negoro N P 2012 Program Hi-Link : Budidaya dan Pengolahan Rumput Laut Untuk Peningkatan Kesejahteraan Ekonomi Masyarakat Pesisir Kab. Probolinggo *Proceeding SENTA 2012 Semin. Nas. Teor. dan Apl. Teknol. Kelaut.*
[12] Al-Shemmeri T, Al-Kloub B and Pearman A 1997 Model choice in multicriteria decision aid *Eur. J. Oper. Res.* **97**
[13] Ishizaka A and Nemery P 2013 *Multi-Criteria Decision Analysis* (John Wiley & Sons, Ltd)
[14] Ikhwani H, Satrio D, Umari W, Sujantoko S, Hadiwidodo Y S and Syarifudin M R 2022 The Analysis of Coastal Society Vulnerabilities Against the Spread of Covid-19 in Surabaya Using the Analytical Hierarchy Process (AHP) *Int. J. Mar. Eng. Innov. Res.* **7** 218–24
[15] Triantaphyllou E and Mann S H 1995 Using the Analytic Hierarchy Process for Decision Making in Engineering Applications: Some Challenges *Int. J. Ind. Eng. Theory, Appl. Pract.* **2**
[16] Viana Vargas R 2010 Using the Anaytic Hierarchy Process (AHP) To Select and Prioritize Projects in a Portfolio *PMI Glob. Congr. 2010 - North Am.* 1–24
[17] Saaty T L and Forman E 2022 Expert Choice
[18] Mu E and Pereyra-Rojas M 2018 *Practical Decision Making Using Super Decisions v3: an Introduction to the Analityc Hierarchy Process*
[19] Nurmianto E, Anzip A and Negoro N P 2022 Evaluasi Desain Ergonomi Alat Pengasapan Ikan Untuk Pemberdayaan Masyarakat *J. Pengabdi. Kpd. Masy. Bina Darma* **2** 25–37

IOP Conf. Series: Earth and Environmental Science 1298 (2024) 012010 doi:10.1088/1755-1315/1298/1/012010

Use of Ergonomics and Analytical Hierarchy Process Model for Selection of Cities of Small Medium Enterprises

Eko Nurmianto[1], S H Kusuma[2], M Nurif[3], M D Kurniawan[4], F N Nurmianto[3], K A Nugraha[2]

[1]Department of Industrial and System Engineering, Institut Teknologi Sepuluh Nopember, 60111, Indonesia
[2]Department of Urban and Regional Planning, Institut Teknologi Sepuluh Nopember, 60111, Indonesia
[3]Department of Development Studies, Institut Teknologi Sepuluh Nopember, 60111, Indonesia
[4]Department of Industrial Engineering, Universitas Muhammadiyah, 61121, Indonesia

E-mail: nurmi@ie.its.ac.id; ekonurmianto1@gmail.com

Abstract. Fish Smoking is a traditional method that has long been used in fisheries. The use of fish smoking devices can have a negative impact on human health which results in the entry of smoke into the eyes, lungs, and skin. In order not to be harmful to human health, ergonomic and healthy fish smoking devices are used. This study analyzes the prioritization of Small Medium Enterprises (SMEs) groups in the use of fish smoking devices to reduce the risk of health problems. The approach method used in this analysis uses the Ergonomics and Analytical Hierarchy Process (AHP) method to determine the priorities of SMEs groups in the use of fish smoking devices. The criteria in this analysis are Ergonomics (Safe, Comfortable, Healthy), Health, Education Level, Community Participation, Motivation, Habits, and Resistance. The results of the analysis found that priority choices in percentage include E, A, B, D, and C regencies. It is hoped that with the holding of Diffusion Technology Results of Innovation Products (DTHPI), the use of modern fish smoking devices that are more ergonomic, healthy, and practical can help improve the welfare of SMEs.

1. Introduction

Indonesia is recognized as one of the countries rich in the diversity of fish and marine resources. Its strategic geographic location between the Pacific and Indian Oceans positions Indonesia with abundant potential for marine resources, surrounded by islands that offer beautiful landscapes. The estimated total potential fish resources across the 11 Fisheries Management Areas of the Republic of Indonesia are projected to reach 12.01 million tons per year, with a permitted total allowable catch (TAC) of about 8.6 million tons per year (Kepmen KP, 19/2022). With abundant fish resources, Indonesia holds the potential for a significant fishing community. The Ministry of Home Affairs records that the number of fishermen in Indonesia is approximately 1.27 million people by the end of 2022. This demonstrates that from Sabang to Merauke, many individuals have chosen fishing as their main occupation to provide for their families. Beyond fishing, coastal communities also engage in trade to sell their catches.

The captured results are sold in both fresh and processed conditions. Coastal fishermen have tools for cooking these fish. They use traditional smoking tools that facilitates easy selling of their products. However, these traditional smoking tools have both positive and negative aspects.

Traditional smoking tools is a method that has been long used in the fishing industry. The fish smoking process takes a considerable amount of time, up to 12 hours per day. This can be considered inefficient as it doesn't add significant value for the fishermen. Moreover, traditional smoking tools has adverse effects on human health due to the production of substances that can potentially cause illnesses over time. Therefore, there is a need to further address traditional smoking methods by adopting safer and more modern approaches for the sake of health. The development of modern smoking tools is aimed at assisting Micro and Small Enterprises (MSEs) in coastal areas and increasing efficiency. The innovation behind this modern smoking tools involves reducing smoke by integrating several smoke-reducing functions to ensure safety, comfort, and health for users, while also enhancing the efficiency of the fish smoking process.

Research Objectives:

a. To determine which MSE among the MSEs perform well in several cities.
b. To select the criteria for identifying successful MSEs.
c. To enchance the performance of MSEs for better outcomes.

2. Literatur Review

Local Economic Development Based on Fisheries for the Self-Sufficiency of Poteran Island, Sumenep Regency, which was previously conducted by Adjie Pamungkas, Aries Sulisetyono, and Eko Nurmianto in 2016 [1]. The earlier study explained that many fishermen still use traditional smoking tools [2]. They have grown accustomed to this traditional method as it has been utilized for years. However, this tools has negative health implications, as it can lead to various health issues for humans. The smoke generated by this tools can hinder breathing and visibility on the streets. Consequently, the previous researchers developed an innovation in the form of modern smoking tools to assist MSEs in several cities in using this tool.

3. Methodology

The selection of a Multi-Criteria Decision Method to address specific issues is a problem of multi-criteria decision-making without a clear-cut solution. Choosing one method over another is often done arbitrarily due to the lack of standardized rules[3]. The Analytical Hierarchy Process (AHP) has been used to determine the priority of existing criteria and alternatives. It is recommended for ranking problems with scores[4].

The AHP method has piqued the interest of numerous researchers, mainly due to its sound mathematical nature and the fact that the required input data is relatively easily obtainable[5]. Its simplicity is marked by comparing pairs of alternatives according to specific criteria[6]. Furthermore, this method has gained popularity and is user-friendly due to the development of various software programs. The mostrenowned ones are Expert Choice [7] and Super Decisions [8]. In this study, we utilized Expert Choiceversion 11 software, which is a decision-making tool based on multi-criteria decision-making and implements AHP by calculating relative weights instead of manual calculations.

Additionally, there exists a method that can be employed to assess ergonomic risk factors, namely the Rapid Entire Body Assessment (REBA) method [9].

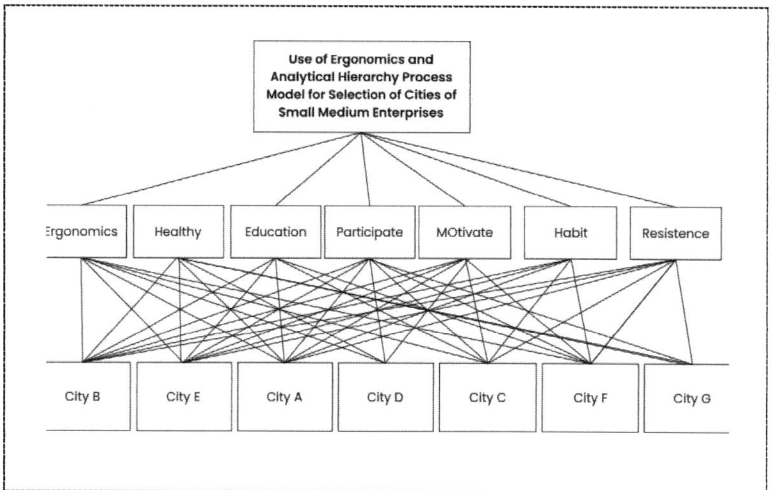

Figure 1. Presents the AHP Hierarchy Structure for Determaining the Best MSEs in Various Cities

As depicted in Figure 1, the AHP Hierarchy is explained as follows:

1. The application of Ergonomics and Analytical Hierarchy Process in Prioritizing MSEs with the use of Fish Smoking Tools serves as the Goal or Objective in the AHP method.
2. Ergonomics, Healthy, Education, Participation, Motivate, Habit, and Resistance serve as criteria in the AHP method.
3. The cities of A, B, C, D, E, F, and G are identified as Alternatives in the AHP method.

4. Result

In the selection of criteria in this AHP analysis, it was found that the criteria with the highest results are Ergonomics, Healthy, Education, Motivate, Habits, and Resistance. Figure 2 and Table 1 illustrate the combined results of the AHP analysis based on responses from the participants. From Figure 2, it can be concluded that the criterion with the highest priority is Ergonomics, with a result of (0.419), which is the highest value among the other criteria. Figure 3 indicates that the alternative with the highest priority is E, with a result of (0.392), which is the highest value among the other alternatives.

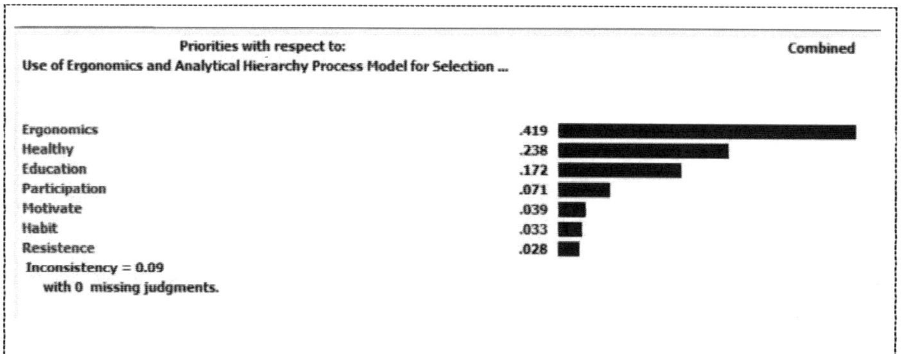

Figure 2. Prioritization Results of Fish Smoking Tools Criteria

Table 1. Results of Each Critera

Criteria	Score
Ergonomics	0,419
Healthy	0,238
Education	0,172
Participation	0,071
Motivate	0,039
Habits	0,033
Resistence	0,028

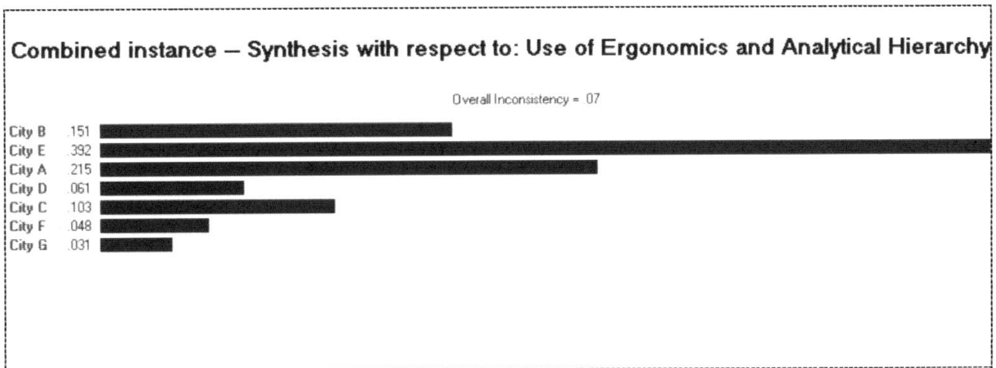

Figure 3. Alternative Prioritization Results of Fish Smoking Tools

Table 2. Criteria - Alternative (Combined)

	Ergonomis	Healthy	Education	Participation	Motivate	Habits	Resistensi
City B	0,159	0,138	0,156	0,124	0,128	0,180	0,153
City E	0,395	0,423	0,330	0,425	0,452	0,390	0,407
City A	0,218	0,180	0,245	0,232	0,233	0,163	0,225
City D	0,055	0,065	0,069	0,059	0,050	0,068	0,069
City C	0,093	0,116	0,116	0,099	0,082	0,124	0,069
City F	0,050	0,049	0,050	0,032	0,028	0,039	0,049
City G	0,030	0,030	0,034	0,028	0,027	0,036	0,027

Table 2. shows the final AHP results, where the Alternative has the highest score.

Table 3. Final AHP Results for Fish Smoking Tools Alternative Prioritazion Final Result Priority

Alternative	Score
City E	0,392
City A	0,215
City B	0.151
City C	0.103
City D	0.061
City F	0.048
City G	0.031

4.1. Sensitive Analysis

The final determination of alternative priorities is influenced by the weights assigned to criteria. Sensitivity analysis is conducted to analyze "what if" scenarios, aiming to assess how the final results would change with varying criterion weights [8]. Sensitivity analysis is crucial in understanding the effects of changing the weights of main criteria on alternative rankings. The sensitivity analysis results are depicted in Figure 4.

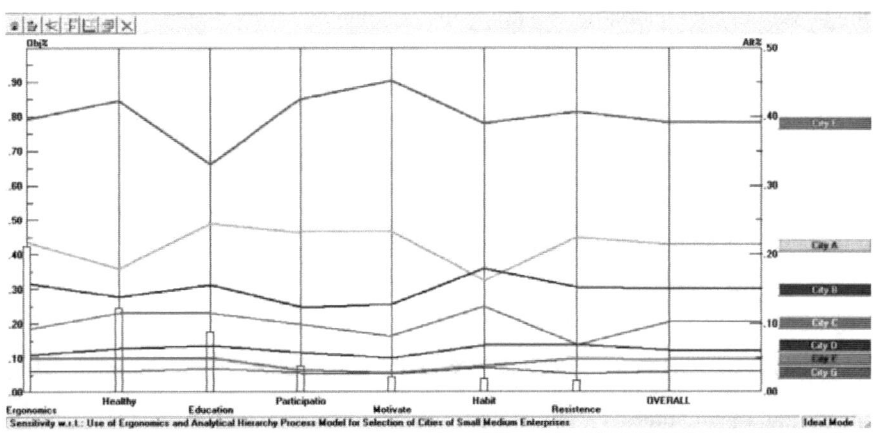

Figure 4. Sensitivity analysis - initial scenario (data provided by Expert Choice software)

It's evident that there are three distinct scenarios, follows:

a. Scenario 1 : Sensitivity analysis concerning the Education
b. Scenario 2 : Sensitivity analysis concerning the Participate
c. Scenario 3 : Sensitivity analysis concerning the Habit

Table 4 illustrates how the scores of each criteria change to assess the extent of impact when those criteria undergo modifications.

Table 4. Scenario Sensitivity-Analysis

	Score First Scenario	Scenario 1	Scenario 2	Scenario 3
Ergonomic	41,9	32,6	22,3	14,6
Healthy	23,8	18,5	12,6	8,3
Education	17,2	35,5	24,2	15,9
Participation	7,1	5,5	35,5	23,3
Motivate	3,9	3	2,1	1,3
Habit	3,3	2,6	1,8	35,5
Resistance	2,8	2,2	1,5	1

IOP Conf. Series: Earth and Environmental Science 1298 (2024) 012010 doi:10.1088/1755-1315/1298/1/012010

In Scenario 1, the most influential criterion is Education, indicating its significant role compared to the other criteria with relatively lower values. Similarly, in Scenario 2, the most influential criterion is Participation. In Scenario 3, the most influential criterion is Habits

4.1.1. Scenario 1: Sensitivity analysis concerning the Education

In Scenario 1, it can be observed that in terms of relative importance to the goal, the alternatives remain unchanged. City E still ranks first in determining the priority of MSEs in various cities, with a value of 38.1%. City A retains the second position with a value of 22%, followed by City B at 15.2% for the third position, City C maintaining the fourth position with a value of 10,6%, City D remaining at the fifth position with 6.2%, followed by City F at 4.8%, and finally, City G at the last position with a value of 3.2%.

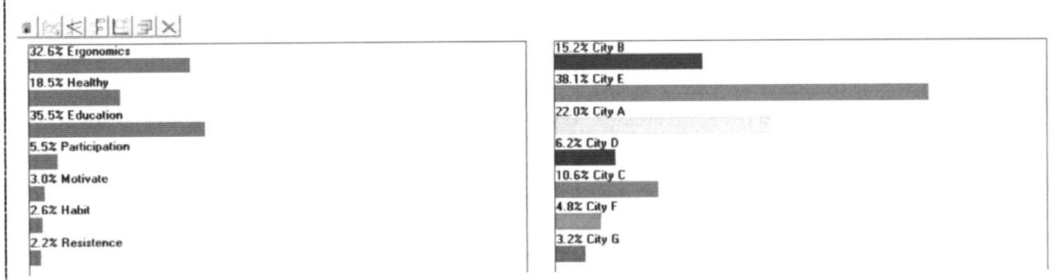

Figure 5. Sensitive Analysis Scenario 1

4.1.2. Scenario 2 : Sensitivity analysis concerning the Participate

In Scenario 2, it can be seen that in terms of relative importance to the goal, the alternatives remain unchanged. City E still ranks first in determining the priority of MSEs in various cities, with a value of 39.5%. City A retains the second position with a value of 22,4%, followed by City B at 14,3% for the third position, City C maintaining the fourth position with a value of 10.4%, City D remaining at the fifth position with 4.3%, followed by City F at 4,3%, and finally, City G at the last position with a value of 3.1%.

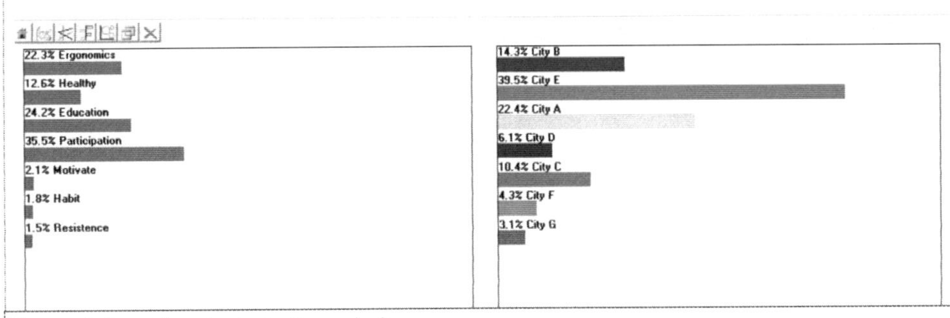

Figure 6. Sensitive Analysis Scenario 2

4.1.3. Scenario 3 : Sensitivity analysis concerning the Habit

In Scenario 3, it can be observed that in terms of relative importance to the goal, the alternatives remain unchanged. City E still ranks first in determining the priority of MSEs in various cities, with a value of 38.2%. City A retains the second position with a value of 20.8%, followed by City B at 15.7% for the third position, City C maintaining the fourth position with a value of 11.3%, City D remaining at the fifth position with 6.5%, followed by City F at 4.2%, and finally, City G at the last position with a value of 3.3%.

This indicates that after conducting 3 scenarios, there have been no changes in the alternative rankings.

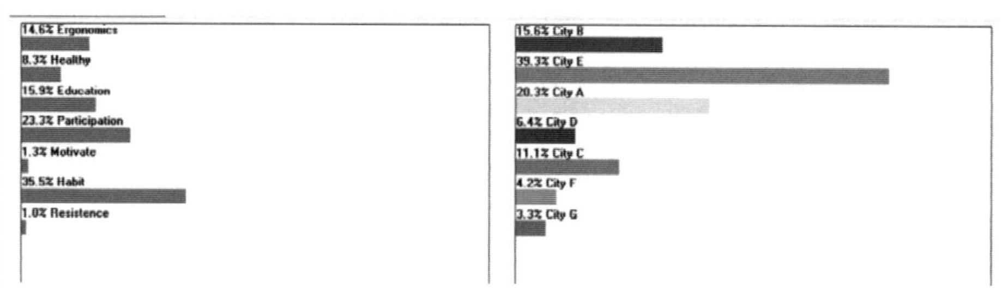

Figure 7. Sensitive Analysis Scenario 3

4.1.4. Original and Research Values

This study represents the first endeavor to apply Ergonomics and AHP in the determination and selection of MSEs using fish smoking equipment across various cities in Indonesia. As a pioneering implementation, this study furnishes decision-makers with a methodology that takes into account criteria from the perspective of fish smoking equipment innovation and societal habits, resulting in accurate and targeted rankings in accordance with the objectives of this study.

5. Conclusion

1. Good MSEs are located in various cities: City E (0.392), City A (0.215), City B (0.151), City C (0.103), City D (0.061), City F (0.48), and G (0.031)
2. The criteria for good MSEs include Ergonomics (0.419), Healthy (0.238), Education (0.172), Participation (0.071), Motivate (0.030), Habits (0.033), and Resistence (0.028).
3. Enhancing the performance of MSEs for the better: Through the ranking provided by the AHP analysis results, the performance of MSEs can be further improved.

6. Acknowledge

The authors gratefully acknowledge financial support from the Institut Teknologi Sepuluh Nopember for this work, under project scheme of the Publication Writing and IPR Incentive Program (PPHKI) 2023. And also, a special thanks to Dr. Dendy Satrio.

7. References

[1] A. Pamungkas, E. Nurmianto, V. K. Siswanto, and A. Sulistyono, "Appropriate technologies for local economic development based on fisheries products in Poteran Island," *IOP Conf. Ser.Earth Environ. Sci.*, vol. 202, no. 1, 2018, doi: 10.1088/1755-1315/202/1/012016.

[2] E. Nurmianto, A. Anzip, and N. P. Negoro, "'C-MOX' Alat Pengasapan Ikan yang Mobile dan Ergonomis," *Hibah Inov. Dana Lokal ITS Tahun 2021*, 2021.

[3] T. Al-Shemmeri, B. Al-Kloub, and A. Pearman, "Model choice in multicriteria decision aid," *Eur. J. Oper. Res.*, vol. 97, no. 3, 1997, doi: 10.1016/S0377-2217(96)00277-9.

[4] A. Ishizaka and P. Nemery, *Multi-Criteria Decision Analysis*. John Wiley & Sons, Ltd, 2013. doi: 10.1002/9781118644898.

[5] E. Triantaphyllou and S. H. Mann, "Using the Analytic Hierarchy Process for Decision Making in Engineering Applications: Some Challenges," *Int. J. Ind. Eng. Theory, Appl. Pract.*, vol. 2, no. No. 1, 1995.

[6] R. Viana Vargas, "Using the Anaytic Hierarchy Process (AHP) To Select and Prioritize Projects in a Portfolio," *PMI Glob. Congr. 2010 - North Am.*, pp. 1–24, 2010, [Online]. Available: http://www.ricardo-vargas.com/pt/articles/analytic-hierarchy-process/#english

[7] T. L. Saaty and E. Forman, "Expert Choice," 2022. http://www.expertchoice.com (accessed Aug. 16, 2023).

[8] E. Mu and M. Pereyra-Rojas, *Practical Decision Making Using Super Decisions v3: an Introduction to the Analityc Hierarchy Process*, no. 9. 2018.

[9] E. Nurmianto, A. Anzip, and D. E. Kusrini, "REBA: an Ergonomics Assessment for Machine Operator of Porang (Amorphophallus Muelleri)," *IPTEK J. Proc. Ser.*, vol. 0, no. 2, p. 19, 2021, doi: 10.12962/j23546026.y2020i2.9021.

8. Appendices

a. City E

b. **City A**

c. City B

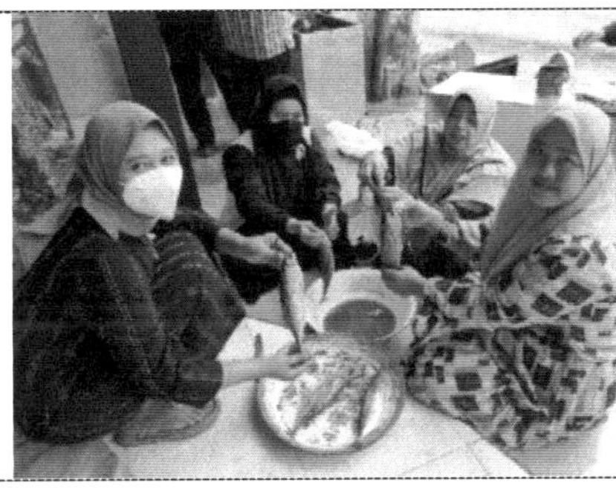

Analysis of container loading-unloading performance at Teluk Lamong Terminal Surabaya

W L Dhanistha[1*], M F Farabi[1], I Rohmawati[2]
[1]Department of Ocean Engineering, Institut Teknologi Sepuluh Nopember, Surabaya, Indonesia
[2]Department of Mechanical Engineering, Meiji University, Japan

*Corresponding author: wimala.danistha@gmail.com

Abstract. PT. Teluk Lamong Terminal (TTL) is a container terminal whose primary business is to serve the loading and unloading of containers. As a result, shipping by water is necessary and will become more so with each passing year. This was seen between 2017 and 2021, when the arrival of ships and the flow of containers at PT. Teluk Lamong Terminal increased. The goal of this study is to determine how strong the association between the factors investigated is the loading and unloading performance (container flow) at Teluk Tamong Terminal Port. This research employs both qualitative and quantitative approaches, as well as primary and secondary data sources. Using the regression technique, the container flow of Teluk Lamong Terminal in 2031 is 1,357,208 TEUs, with a growth rate of 4.49%, falling short of the UNCTAD criterion. The results revealed a substantial positive association between the variables Idle Time (X1), Not Operation Time (X2), ASC Utility (X3), and Container Flow (Y), with Fcount 276.730 > Ftable 19.2 and a significance of 0.044 0.05 accepting Ho and Ha. The Adjusted R Square score for the association between Idle Time (X1), Not Operation Time (X2), ASC Utility (X3), and Container Flow (Y) is 0.995, indicating a strong relationship.

1. Introduction

Indonesia is a maritime country with a large enough ocean so that sea lanes are a means of transportation and distribution between provinces in Indonesia. Under these conditions, the maritime sector is of great significance to the goods and passenger transportation system for both domestic and international shipping. As a maritime country, in supporting inter-island activities, it must have adequate infrastructure. One of the supports for inter-island activities is the availability of ports.

A port is a facility that functions as a place to switch modes of transportation and move goods, cargo and passengers in it [1]. In loading and unloading services, obstacles often occur in operations. In addition, loading and unloading productivity is not optimal due to lack of maintenance tools so that it has a direct impact on the productivity of loading and unloading equipment. These obstacles can result in losses for the company, ineffective loading and unloading equipment.

Furthermore, in research conducted by [2] where in the conditions of loading and unloading equipment and adequate human resources, the BCH standard is 25 boxes but the realization only reaches an average of 18 boxes. Each type of tool has a capacity with advantages and disadvantages of each that can affect loading and unloading productivity. The results showed that there was a partially significant positive relationship between the variables Crane (X_1) and Berthing Time (Y), where tcount 4.839> ttable 2.03224 and significantly 0.000 <0.05, then H_o was rejected and Ha was accepted. Judging from the correlation coefficient between Productivity Crane (X_1) and Berthing Time (Y), 0.639 indicates a strong relationship because it is within the interval (0.600-0.799).

IOP Conf. Series: Earth and Environmental Science 1298 (2024) 012011 doi:10.1088/1755-1315/1298/1/012011

2. Materials and methods

2.1 Loading unloading

Loading and unloading activities are activities of unloading goods from on the ship by using the crane and ship sling to the nearest land in edge of the ship which is commonly called the wharf, then from the wharf with using lorries, forklifts loaded and arranged into the nearest warehouse appointed by the harbor master. While loading activity is the opposite activity. Loading and unloading operations from/to ships.

2.2 Five whys analysis

Five why analysis is a simple root cause analysis tool and can be used to analyze system failures and can work well in identifying the causes and effects of an event [3]. 5 why analysis is used to dig deeper to the real root of the problem, the root cause can be known by asking "why" repeatedly until it reaches a point where the answer to the question has shown a root problem [4].

2.3 Multiple linear regression

The multiple linear regression method is a probabilistic model which states a linear relationship of more than one variable where one variable is considered to influence the other variables [5]. The following is the formula for the multiple linear regression method:

$$Y = \beta 0 + \beta 1X1 + \beta 2X2 + \dots + \beta nXn \qquad (1)$$

Which

Y = Dependent Variable

$\beta_0, \ldots, \beta n$ = Regression Coefficient

$X_1, \ldots, X_n$ = Independent Variable

2.4 Classical assumption test

The classical assumption test is a statistical data test that is used as a prerequisite to provide certainty that the regression equation obtained has accuracy in estimation, is not biased and is consistent.

1) Normality test
   Normality test aims to determine whether the data distribution deviates or not from the normal distribution. Data that is good and feasible to prove the research models is data that has a normal distribution.
2) Heteroscedasticity test
   A heteroscedasticity test was conducted to analyze whether a linear model has a different variance from the residual of one observation to another or not. Heteroscedasticity indications can be seen from the difference in the residual variance of each independent variable. To indicate heteroscedasticity, the Spearman test and scatter plot graphs were carried out [6].
3) Multicollinearity test
   The multicollinearity test can be determined by carrying out the Variance Inflating Factor (VIF) test. VIF is an easy method for analyzing the multicollinearity of data. In the VIF test it can be observed when the t (tolerance) value is above 0.1 and the VIF value is below 10, it can be said that there is no multicollinearity between the variables being analyzed or vice versa [6].
4) Autocorrelation test
   The autocorrelation test is analyzed to see whether there is a correlation between the error in the period (t) and the error in the previous period (t-) in a regression model or not. The autocorrelation test is carried out using the Durbin – Watson test or the DW test.

3. Results and discussion

3.1 Loading-unloading performance
In this study the loading and unloading performance is used as the object of research to analyze the factors that affect loading and unloading performance. The data used in the analysis process are Idle Time, Not Operation Time, and the Automated Stacking Crane Utility.

3.2 Five whys analysis
Then a search for the root causes of the delay in the loading and unloading process and not achieving performance standards at Teluk Lamong Terminal was found by brainstorming and interviews, then processed using 5 Whys Analysis.

3.3 Classical assumption test
The classical assumption test is a statistical data test that is used as a prerequisite to provide certainty that the regression equation obtained has accuracy in estimation, is not biased and is consistent.

1) Normality test
In the normality test the software is used to obtain significant values as follows.

Table 1. Result of normality test.

Variable	Sig
Residual	0.2

The Kolmogorov-Smirnov test above shows the significance of all variables around 0.200 where the value is above the limit_(standard error), 0.05 or 5%. This shows that the variable data tested is normally distributed and can be continued for other classical assumption tests.

2) Heteroscedasticity test
Spearman's rank test can be done using SPSS with regression of all independent variables on residual values. If there is a significant independent variable influence (Sig <0.05) then it can be said that there is a heteroscedasticity problem. To analyze whether there are signs of heteroscedasticity, the significance of each variable can be used by regressing the residual variance of each independent variable. If the table significance value of each independent variable is above 0.05, it can be said that there is no heteroscedasticity. If the table significance value of each independent variable is below 0.05, it can be said that there is an indication of heteroscedasticity.

Table 2. Result of heteroscedasticity test.

Variable	Sig
IT	0.351
NOT	0.309
Utility of ASC	0.425

Based on the table above, the significance value of each independent variable shows a value above 0.05 (5%). This shows that there is no sign of heteroscedasticity among the independent variables to be analyzed, so that the data is homoscedastic.

3) Multicollinearity test
The multicollinearity test can be determined by using the Variance Inflating Factor (VIF) test. The VIF test is an easy test method for analyzing data multicollinearity. To analyze indications of multicollinearity, t and VIF values can be observed if the t (tolerance) value is above 0.1 and the VIF value is below 10. It can be said that there is no multicollinearity between the variables being analyzed and vice versa.

Table 3. Result of multicollinearity test.

Variable	Tolerance	VIF
IT	0.584	1.713
NOT	0.172	5.804
Utility of ASC	0.192	5.207

The multicollinearity test above shows that the VIF of each variable is above 10 and the tolerance value is above 0.1, so that each variable is independent of multicollinearity.

4) Autocorrelation test
The autocorrelation test was carried out to analyze whether there is a correlation between the t-period error and the previous period error (t-) in the regression model or not. The autocorrelation test was carried out using a run test. The results of the run test are shown in the table below.

Table 4. Result of run test.

Sig. (2-tailed)	0.230

Based on the table above, it can be concluded that there is no autocorrelation problem in the regression model.

3.4 Multiple linear regression
Multiple linear regression tests were carried out on the variables Idle Time (X1), Not Operation Time (X2), and ASC Utility (X3) which can be seen in the table below.

Table 5. Result of multiple linear regression.

Variable	Unstandardized B	Coefficients Std. Error
Constant	-60635.460	73183.817
IT (X1)	5156.832	8787.786
NOT (X2)	2785.210	5321.348
Utility of ASC (X3)	25838.957	1940.496

Based on the results of the table above, the equation to form a linear regression model can be seen in the Unstandardized Coefficients table in column B. Based on the analysis results, a linear regression model can be formed as follows.

$$Y = -60635.460 + 5156.832X1 + 2785.210X2 + 25838.957X3 \quad (2)$$

Based on the results of the multiple linear regression above, it can be seen that:
1. The regression coefficient value of Idle Time (X1) 5156.832 has the understanding that every 1 hour increase of Idle Time (X1) will affect an increase in Flow (Y) 5156.832 TEUs assuming other variables are constant.
2. The regression coefficient value of Not Operation Time (X2) 2785.210 has the understanding that every 1 hour increase in Not Operation Time (X2) will affect an increase in Flow (Y) 2785.210 TEUs assuming other variables are constant.
3. The regression coefficient value of ASC Utility (X3) 25838.957 has the understanding that every 1% increase in ASC Utility (X3) will affect an increase in Flow (Y) 25838.957 TEUs assuming other variables are constant.

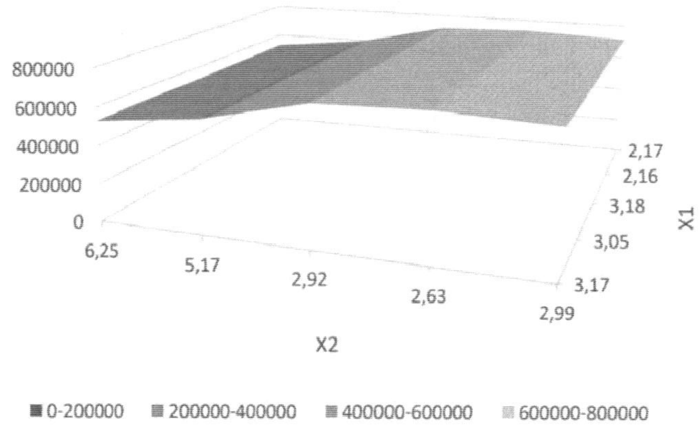

Figure 1. Multiple linear regression of X1 and X2.

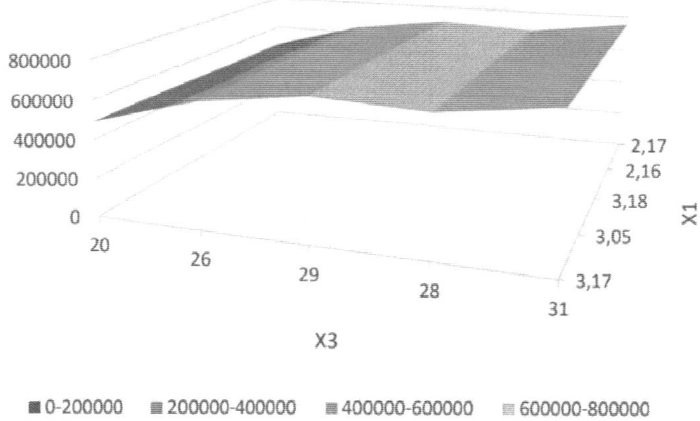

Figure 2. Multiple linear regression of X1 and X3.

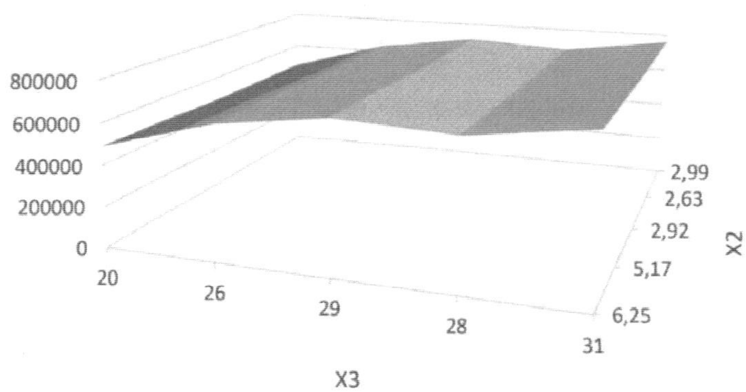

Figure 3. Multiple linear regression of X2 and X3.

The linear equation obtained in equation 2 can be visualized through figures 1, figure 2, and figure 3, where the y-axis represents the sum of TEUs. And the conclusion wherein figures 2 and figure 3 exhibit a similar pattern, culminating in an increase at the end.

4. Conclusions

From the results of the analysis that has been carried out regarding loading and unloading performance, the following conclusions are obtained:

- Factors that influence the failure to achieve loading and unloading performance standards are due to human, machine and environmental factors.
- From the results of the Classical Assumption Test with 3 independent variables including IT, NOT, and ASC Utility, the results show that the data used is normally distributed, there are no indications of heteroscedasticity, multicollinearity, and autocorrelation.
- In the Multiple Linear Regression results, the most influential variable on loading and unloading performance is the ASC Utility with a constant value of 25838.957.

Acknowledgements

The authors are deeply indebted to the Department of Ocean Engineering and DRPM ITS and PT. Lamong Surabaya which has sincerely made the data available. Gratitude and great appreciations without the support from those two institutions this study might not be successfully carried out.

References

1. Antiko W K D 2021 Analisis Kinerja Pada Fasilitas Dan Peralatan Bongkar Muat Petikemas di PT. Terminal Teluk Lamong. Institut Teknologi Sepuluh Nopember Surabaya
2. Darunanto D, Wahyuni, Rr E, and Saidah D 2020 Produktivitas alat bongkar muat (crane) terhadap berthing time *Jurnal Manajemen Bisnis Transportasi Laut Dan Logistik* **6** 2
3. Dogget A M 2005 Root Cause Analysis: A Framework for Tool Selection *The Quality Management Journal* **34**
4. Ohno T 1988 Toyota Production System
5. Cheng C L, Shalabh, and Garg G 2014 Coefficient of determination for multiple measurement error models *Journal of Multivariate Analysis* **126** 137–152
6. Prihananto A H 2014 Optimizing waiting time at container terminal in port of Tanjung Perak Indonesia *World Time University*

Surface Current Measurement Using Large Scale Particle Image Velocimetry (LSPIV) in Ender Estuary, Cirebon

A Tarya[1] and M R Aqdam Prasetyo[1]

[1] Department of Oceanography, Faculty of Earth Sciences and Technology, Bandung Institute of Technology, Ganesha Street No. 10, Bandung, 40132, Indonesia

Email: ayi.tarya@itb.ac.id

Abstract. The Large-Scale Particle Image Velocimetry (LSPIV) method is an image- based technique that uses a non-intrusive approach or avoids direct physical contact with the water being studied to measure the velocity of water flow. The present study aims to identify surface velocity using LSPIV captured by Unmanned Aerial Vehicle (UAV) in Ender estuary, Cirebon. Field observation for surface velocity was conducted at three locations during flood and ebb tides. The surface current results of LSPIV compared with in situ velocity measurement by a current meter. The results of the LSPIV method are strongly influenced by the data acquisition process such as lighting, interrogation area, and video recording angle. Based on comparing the surface flow between results of LSPIV and observation data, the coefficient correlation (R) is 0.84 for velocity magnitude and R = 0.92 for flow direction. The Root Mean Square Error (RMSE) of 0.024 m/s for velocity magnitude and 4.81 degree for flow direction, and the Mean Absolute Error (MAE) range from 0.002 to 0.059 m/s (1 to 12 degree) for the velocity magnitude and the flow direction, respectively. Furthermore, the Mean Absolute Percentage Error (MAPE) value of 10.41% (8.84%) for magnitude and direction, respectively. Therefore, LSPIV was applied successfully to obtain surface velocity measurements.

1. Introduction

Ocean currents play a vital role in estuaries, facilitating processes such as sediment and nutrient transport, water body development, and supporting various human activities such as transportation and tourism [1]. Generally, velocity observations are limited to using intrusive measurement method including drifter, current meter, and Acoustic Doppler Current Profiler (ADCP). However, many studies have used photogrammetric methods to measure the surface velocity of rivers in order to lower the risk of field observations. There are more image-based approaches that can be used for surface velocity measurement as a result of the advancement of technology. This related state-of-the-art technology has been continuously developed over many years [2,3].

River surface velocities can be calculated using photos using photogrammetry. Unmanned aerial vehicles (UAVs), which possess highly maneuverable qualities, can overcome terrain constraints, have orthographic capabilities, and provide high economic benefits, are preferred by an increasing number of researchers [4,5]. UAVs have been extensively used in agricultural and ecological surveys [6], environmental monitoring projects [7] and disaster monitoring efforts [8], due to their ability to navigate terrain obstacles and conduct aerial photography operations.

Among the optical techniques available, large-scale particle image velocimetry (LSPIV) uses digital images taken from a fixed location to measure remote surface flow [2,9]. The field's surface velocity can be obtained by using LSPIV. Additionally, LSPIV offers a flexible and affordable flow measurement technique that can efficiently get around the shortcomings of river flow monitoring networks. The results of the surface velocity data obtained by LSPIV and the results obtained by acoustic Doppler current profilers (ADCPs) are comparable, according to a number of studies based on field experiments [10]. As a result, it can be concluded that LSPIV techniques are accurate enough to be relied upon for determining river surface velocities.

The LSPIV method has been successfully used to measure surface current velocity in rivers such as the Houlong River, Taiwan [11], Alpine River, Austria [12], East Central Illinois River, USA [13], and Murg River, Switzerland [5]. These previous studies were carried out in the area of uniform current velocity. Meanwhile, in this study the LSPIV method will be used to identify current circulation in river mouths, where the conditions of river mouths are more complex due to the influence of tides. The present study aims to measure the surface velocity of the Ender Estuary in Cirebon, West Java, using UAV and LSPIV instruments and an intrusive current meter observation to verify the accuracy of the LSPIV measurements.

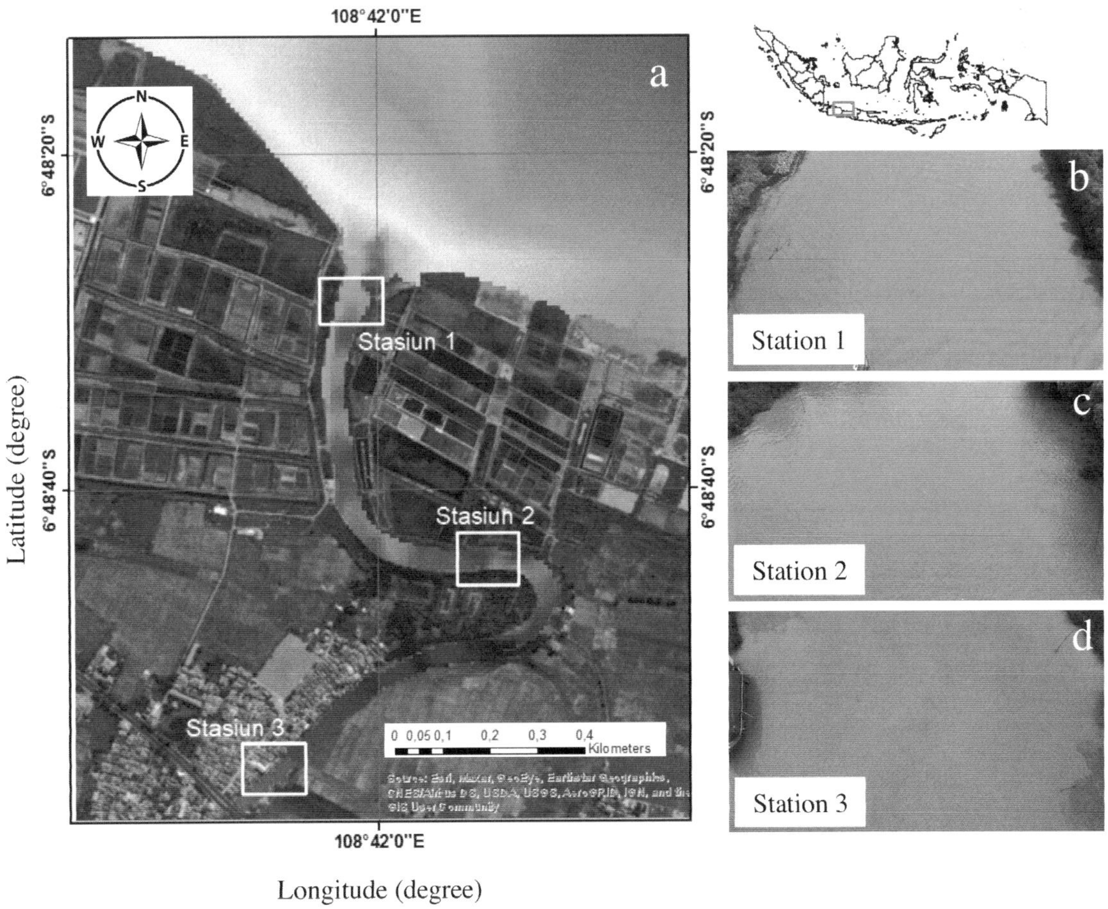

Figure 1. Study area of the Ender Estuary, Cirebon, West Java (a). Yellow box is the area of investigation for UAV measurement at three locations (Station 1 (a), Station 2 (b) and Station 3 (c)).

2. Methods

2.1 Field site and observation

This research was conducted in the estuary and river area in Ender, Pangenan District, Cirebon Regency, as shown in Figure 1a. The Ender River has a length of approximately 18 km, with its river upstream located in Kuningan, West Java. The mouth of the river has a width of 127 m, and the average width of the river is about 40 m. Ender is a shallow estuary with average depth of 2.5 m. The tide is predominantly semidiurnal with the tidal range varies from 0.2 m during neap tides to 0.8 m during spring tides. The Ender estuary area has land use dominated by salt and fish ponds, as well as residential areas.

In this study, three field campaigns were conducted on 26 January, 6 March, and 8 March in 2023 to measure the surface velocity. Measurements were conducted at three stations during flood and ebb tides condition: the first station was located at the river mouth, the second station was situated in the meandering section of the river representing the river body, and the third station was at the point where the ship was docked representing the furthest distance from the river mouth. At the first station, a transect measuring 24 x 41 m in a square shape was established at the river mouth Figure 1b. For the second station, situated in the river's meandering bend, a transect measuring 24 x 25 m was created Figure 1 (c). Lastly, at the third station, located at the ship docking area, a transect measuring 12 x 17 m was set up Figure 1d. At each of the three transects, three disks were placed along each bank. Real Time Kinematic Global Navigation Satellite System (RTK-GNSS) surveying equipment was used to determine 4 points to serve as georeference control points, measuring and marking the georeference control points x, y, z coordinates and marking the position for the drone.

Observation method used in this study involved capturing video footage of the surface flow using a drone (UAV) with the transect plot method. The model of the unmanned aerial vehicle used in this study is DJI Mavic 2 which consists of a drone body, a remote control, a gimbal camera, and the DJI Mavic 2 program. Footage of the event was collected at 1920 × 1080 pixel resolution at an acquisition rate of 30 frames per second (FPS). In addition to the UAV methods combined with the LSPIV method to measure surface velocity, this study also uses traditional surface velocity measurement methods to verify the measurement results of the LSPIV method. A Valeport Current Meter used to measure surface velocity at the three stations. Instantaneous measurements of current meter were taken simultaneously with the UAV recording. About 6-7 locations of surface velocity were observed at each station.

2.2 Large Scale Particle Image Velocimetry

Non-intrusive measurement procedures have rapidly advanced in the last few decades due to the development of scientific technology, including computer vision, Large Scale Particle Image Velocimetry (LSPIV) methods, and image-based measurements. [15] used the same PIV method to measure the flow velocity of rivers. Subsequent applications were carried out in the hydrological sector using the LSPIV method [3,14], which has developed angle correction calculations for image acquisition.

Large Scale Particle Image Velocimetry (LSPIV) is an optical remote sensing with spatiotemporal dimensions. Figure 2 show illustration of LSPIV measurement sequence. LSPIV operates on the fundamental principle of using traceable particles (seeding flow) introduced into the flowing water. These particles are then illuminated and recorded in photos or videos in a sequential manner. The seeding process is a concept in fluid dynamics to explain the influence of foreign objects on fluid flow. The output of LSPIV provides a 2D velocity vector field on the surface of the flow, covering a relatively large area. In simple terms, LSPIV captures fluid flow patterns over time using an optical sensor, such as recording videos with a camera or unnamed area vehicle (UAV) [11]. Then, the displacement of patterns in the x and y coordinates within the image matrix is analysed with cross-correlation matrices of paired patterns. Entire matrix in the image is processed to obtain velocity vectors on the measurement plane. By assuming the actual distance and pixel distance in the camera image, along with the time difference between two images, the displacement of patterns in the image sequence is converted into units of distance per time, such as m/s.

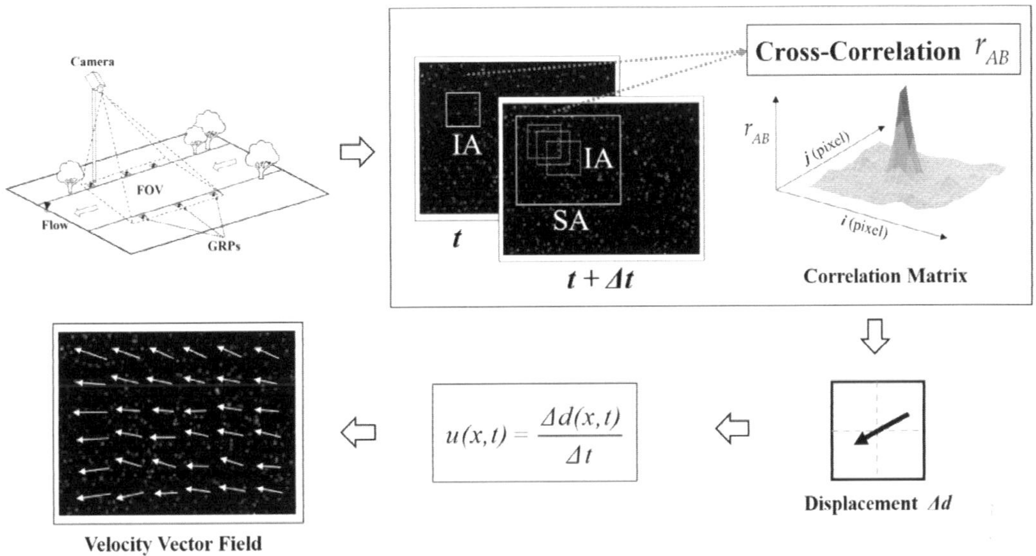

Figure 2. Flow diagram of the Large-Scale Particle Image Velocimetry (LSPIV) (FOV is the Field of View; GRPs is the Ground Reference Points; IA is the Interrogation Area; SA is the Searching Area) [11].

3. RESULTS AND DISCUSSION

3.1 Performance of LSPIV method

In order to quantity the accuracy of the LSPIV method, the results of LSPIV for velocity magnitude and direction compared with the observations. LSPIV performance is quantified by the Mean Absolute Error (MAE), the Root Mean Square Error (RMSE), the correlation coefficient (R), the Mean Absolute Percentage Error (MAPE) as shown in Table 1 and 2. The MAE between LSPIV and observation is ranging from 0.002 to 0.059 m/s for magnitude and 1 to 12 degrees for direction. RMSE value is 0.024 m/s for velocity magnitude and 4.81 degree for direction. Furthermore, the correlation between LSPIV and observation is 0.84 for velocity magnitude and 0.96 for velocity direction. Percentage error (MPAE) between LSPIV and observation is 10.41% for magnitude and 8.84% for direction. These imply that the LSPIV method provides a reasonably accurate estimation of current magnitude and direction.

3.2 Pattern of surface currents

The surface currents obtained from LSPIV at the station 1 as shown in Figure 3. It can be seen that the surface current towards the river mouth during flood tides (Figure 3a). Meanwhile, at low tide, the direction of surface flow is towards upstream of the river (Figure 3b). At high tide conditions, the magnitude of the current ranges from 0.2 – 1 m/s with an average magnitude of 0.73 m/s. At low tides, the surface velocity is ranges from 0.4 – 1 m/s with an average magnitude of 0.74 m/s. At station 2 (Figure 4), the direction of surface velocity flows to upstream during flood tides with an average velocity of 0.39 m/s. Whereas, the surface currents toward downstream when ebb tides with an average flow of 0.48 m/s.

Table 1. Quantification of validation between LSPIV and observation for velocity magnitude.

Date	Time	Velocity magnitude (m/s)		MAE (m/s)	MAPE (%)	RMSE (m/s)	R
		LSPIV	Current meter				
26 Jan 23	9.33	0.239	0.268	0.029	10.82	0.024	0.84
	9.33	0.292	0.278	0.014	5.04		
	9.38	0.319	0.298	0.021	7.05		
	9.38	0.308	0.287	0.021	7.32		
	10.59	0.212	0.198	0.014	7.07		
	10.59	0.206	0.178	0.028	15.73		
	11.02	0.210	0.176	0.034	19.32		
	11.02	0.242	0.213	0.029	13.62		
	12.23	0.124	0.147	0.023	15.65		
	12.23	0.112	0.125	0.013	10.40		
6 Mar 23	11.00	0.130	0.142	0.012	8.45		
	11.00	0.122	0.120	0.002	1.67		
	15.06	0.195	0.200	0.005	2.50		
	15.06	0.195	0.210	0.015	7.14		
	15.25	0.279	0.315	0.036	11.43		
	15.25	0.310	0.330	0.020	6.06		
	15.36	0.321	0.287	0.034	11.85		
	15.36	0.361	0.342	0.019	5.56		
	15.50	0.385	0.356	0.029	8.15		
	15.50	0.306	0.310	0.004	1.29		
	15.51	0.285	0.300	0.015	5.00		
	15.51	0.288	0.305	0.017	5.57		
8 Mar 23	7.15	0.471	7.15	0.048	11.35		
	7.15	0.509	7.15	0.031	6.49		
	7.32	0.349	7.32	0.005	1.41		
	7.32	0.385	7.32	0.043	12.57		
	7.46	0.175	7.46	0.04	29.63		
	7.46	0.121	7.46	0.011	10.00		
	11.27	0.418	11.27	0.027	6.07		
	11.27	0.459	11.27	0.036	8.51		
	11.58	0.326	11.58	0.059	22.10		
	11.58	0.25	11.58	0.04	13.79		
	12.33	0.089	12.33	0.011	11.00		
	12.33	0.114	12.33	0.014	14.00		

Table 2. Quantification of validation between LSPIV and observation for velocity direction.

Date	Time	Velocity direction (degree)		MAE (degree)	MAPE (%)	RMSE (degree)	R
		LSPIV	Current meter				
26 Jan 23	9.33	10	12	2	16.67	4.81	0.92
	9.38	9	10	1	10.00		
	9.48	11	5	6	22.22		
	9.52	13	14	1	7.14		
	10.54	15	13	2	15.38		
	10.59	15	17	2	11.76		
	11.02	60	57	3	5.26		
	11.07	63	66	3	4.55		
	12.23	67	64	3	4.69		
	12.28	72	68	4	5.88		
6 Mar 23	11.00	38	35	3	8.57		
	11.05	35	39	4	10.26		
	15.06	32	40	8	20.00		
	15.11	39	45	6	13.33		
	15.25	47	42	5	11.90		
	15.30	31	43	12	27.91		
	15.36	41	47	6	12.77		
	15.46	38	43	5	11.63		
	15.50	59	52	7	13.46		
	15.55	42	38	4	10.53		
	16.11	40	49	9	18.37		
	16.16	43	42	1	2.38		
8 Mar 23	7.15	150	153	3	1.96		
	7.20	154	159	5	3.14		
	7.32	155	160	5	3.13		
	7.38	154	142	12	8.45		
	7.46	160	152	8	5.26		
	7.53	162	169	7	4.14		
	11.27	340	335	5	1.49		
	11.32	356	353	3	0.85		
	11.50	350	346	4	1.16		
	11.55	345	340	5	1.47		
	12.33	340	348	8	2.30		
	12.38	338	342	4	1.17		

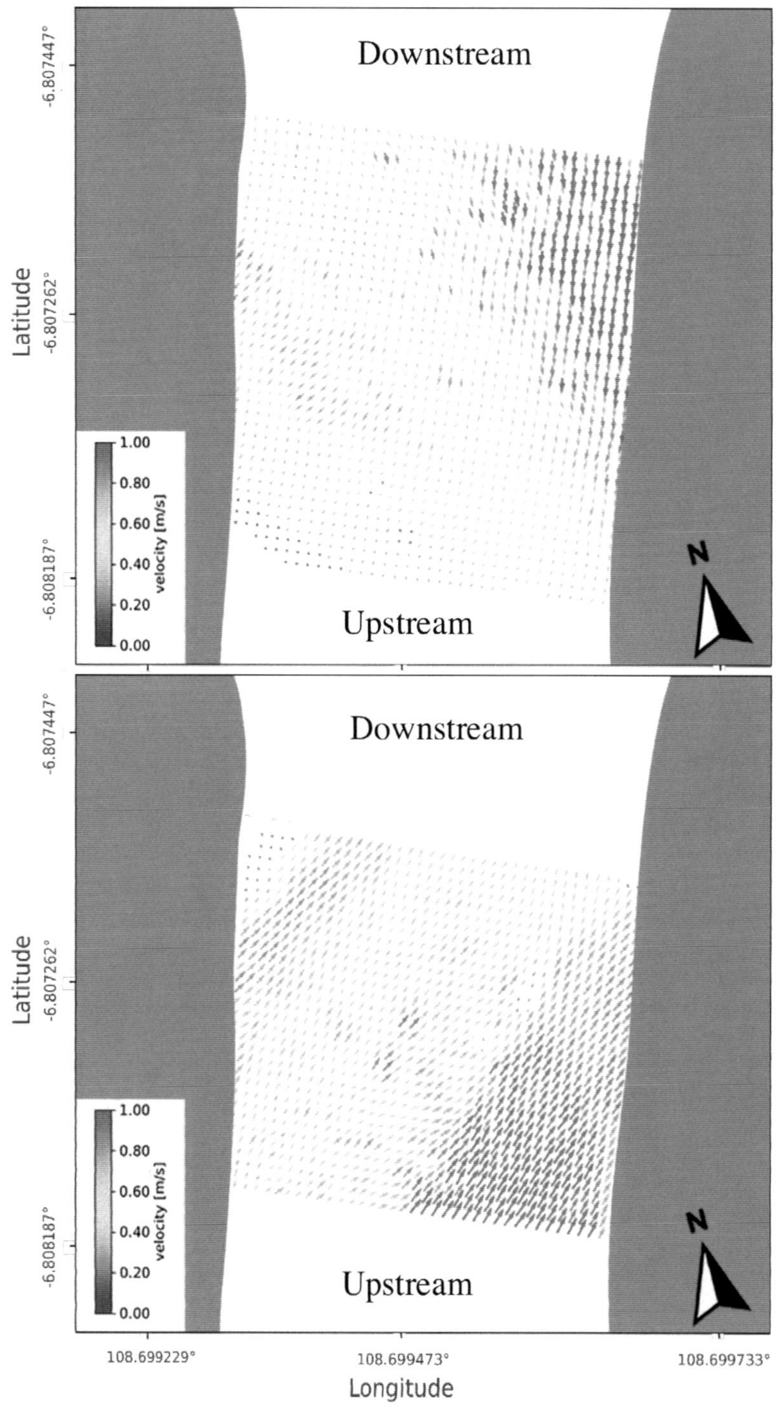

Figure 3. Surface current patterns obtained by LSPIV method at Station 1 during flood tides (a) and ebb tides (b).

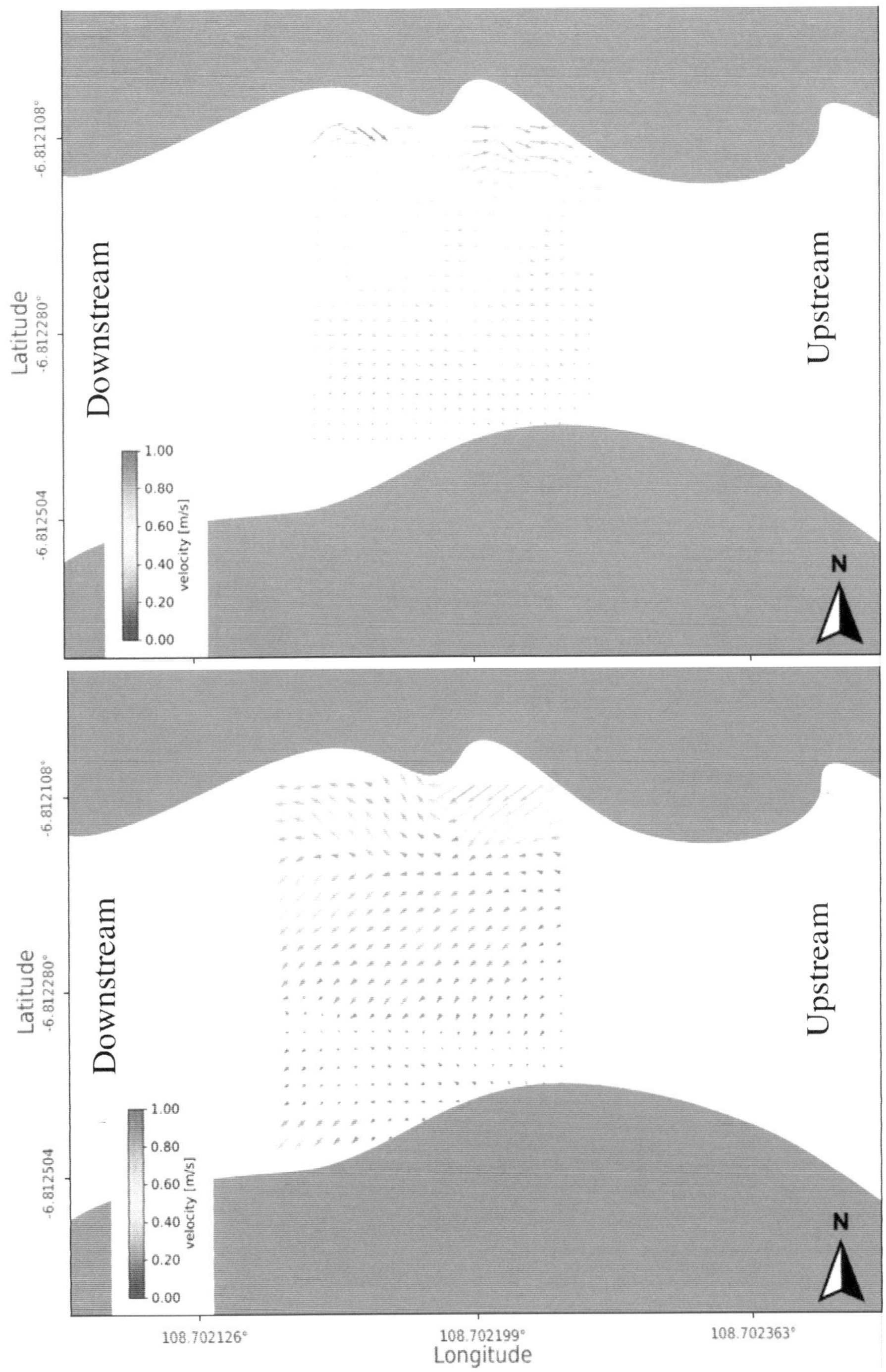

Figure 4. Surface current pattern obtained by LSPIV method at Station 2 during flood tides (a) and ebb tides (b).

Figure 5 is the results of surface current at station 3 during flood and ebb tides. The direction of surface currents is towards the estuary during flood and ebb condition. The current speed ranges from 0.3 – 1 m/s with an average magnitude of 0.52 m/s during low tides, while during high tides the surface velocity ranging from 0.2 – 0.8 m/s, with average velocity of 0.31 m/s.

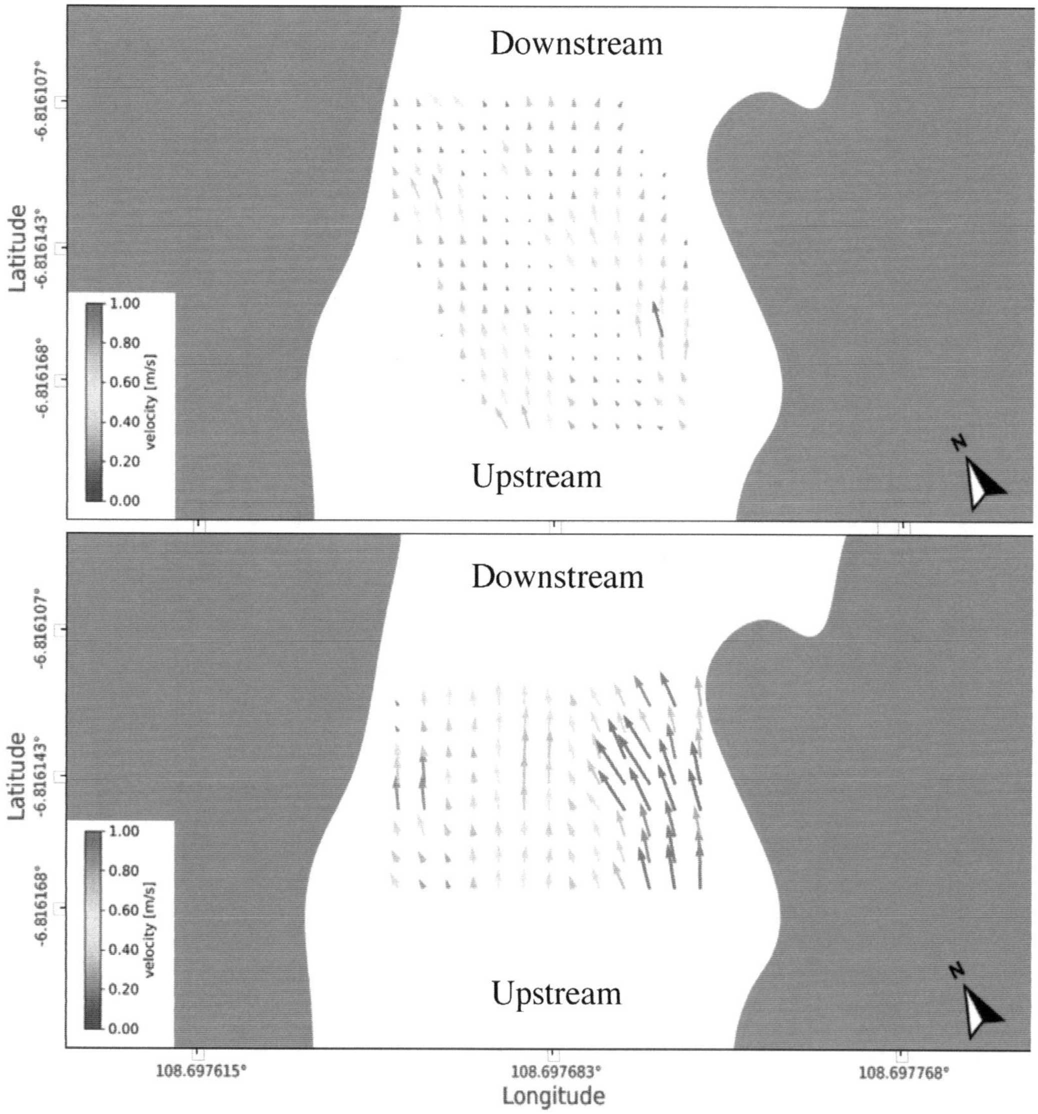

Figure 5. Surface current pattern obtained by LSPIV method at station 3 during flood tides (a) and ebb tides (b).

3.3 Effect of interrogation area

The LSPIV method describes the surface velocity field from the Eulerian perspective by determining flow velocities over a regular grid. Interrogation areas (IAs) are used to describe each cell in the regular grid. Figure 6 show surface current pattern at station 2 during ebb tides for two interval of interrogation area (IA). In order to identify the effect of IA, the results of surface velocity with pre-processing stage uses a resolution grid of 0.125 cm (Figure 6a) compared to the grid of 1.25 cm (Figure 6b). The grid is denser in the estimation results using an IA of 0.125 cm, making it challenging to determine the current direction. While an IA 1.25 cm produces a sparser grid, making the direction of velocity more obvious. The results of the present study is similar with [16]. They found that when the IA is too small, image matching from frame to frame becomes poor, resulting in less accurate surface velocity estimation. On the other hand, when the IA is too large, the calculated surface velocities for each IA become uniform.

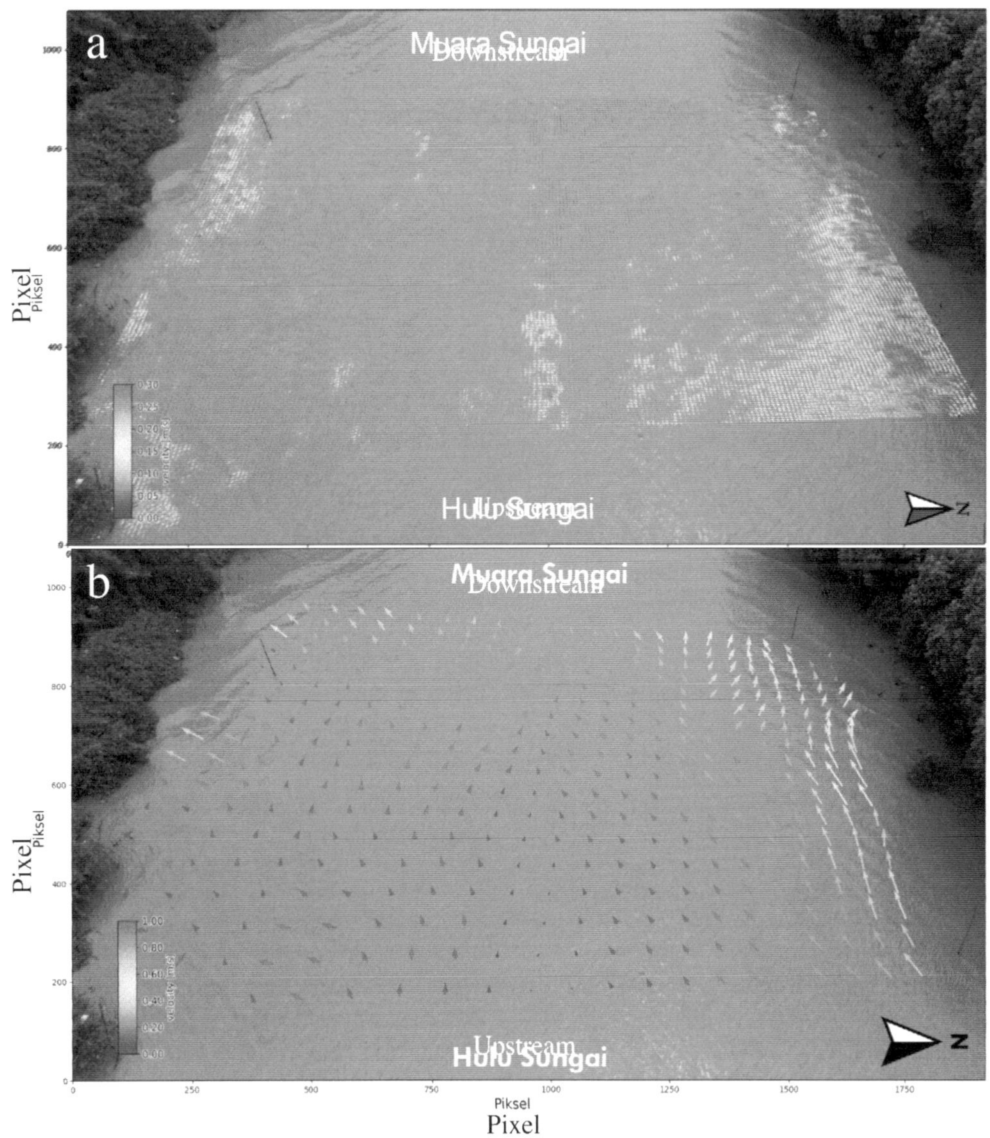

Figure 6. Surface current pattern at station 2 for Interrogation Area (IA) 0.125 cm (a) and Interrogation Area (IA) 1.25 cm (b).

4. CONCLUSIONS

The LSPIV method was implemented to determine the surface velocity at the Ender Estuary in Cirebon. Field measurements of surface velocity was conducted at three locations during flood and ebb tides. The surface velocity of the LSPIV method was compared with the velocity observed by a current meter. According to statistical analysis, the coefficient of correlation (R) for velocity magnitude is 0.84 and 0.92 for flow direction. The velocity magnitude and flow direction had the root mean square error (RMSE) of 0.024 m/s (4.81°) and the mean absolute error (MAE) ranges from 0.002 to 0.059 m/s (1 to 12 degree), respectively. Furthermore, the mean absolute percentage error (MAPE) for magnitude and direction is 10.41% (8.84%). As a result, LSPIV was effectively used to obtain the surface velocity, making it a promising technique for further research and development.

Acknowledgments

The writing manuscript was funded by Indonesian Ministry of Education, Culture, Research and Technology (Kemdikbudristek) under the research program of Penelitian Fundamental Research (PFR) 2023 Research Grant Contract No. 110/E5/PG.02.00.PL/2023. The fieldwork data was funded by Bandung Institute of Technology under the program Research, Community Service and Innovation (PPMI–ITB 2022).

References

[1] Zheng C., Xiao Z., Zhou W., Chen X., and Chen X. (2018). Ocean Current Characteristics. In: 21st Century Maritime Silk Road: A Peaceful Way Forward. *Springer Oceanography*. Springer, Singapore.

[2] Fujita, I., Muste, M., and Kruger, A. (1998). Large-scale particle image velocimetry for flow analysis in hydraulic engineering applications. *J. Hydraul. Res*., 36, 396–414.

[3] Muste, M., Fujita, I., and Hauet, A. (2008). Large-scale particle image velocimetry for measurements in riverine environments. *Water Resour. Res*., 44, W00D19.

[4] Tauro, F., Pagano, C., Phamduy, P., Grimaldi, S. and Porfiri, M. (2015). Large-scale particle image velocimetry from an unmanned aerial vehicle. *IEEE/ASME Trans. Mechatron*., 20, 3269–3275.

[5] Detert, M., Johnson, E.D., and Weitbrecht, V. (2017). Proof-of-concept for low-cost and non-contact synoptic airbone river flow measurement. *Int. J. Remote Sens*., 38, 2780–2807.

[6] Yang, M.D., Huang, K.S., Kuo, Y.H., Tsai, H.P., and Lin, L.M. (2017). Spatial and spectral hybrid image classification for rice lodging assessment through UAV imagery. *Remote Sens*., 9, 583.

[7] Hemmelder, S., Marra, W., Markies, H., and De Jong, S.M. (2018). Monitoring river morphology & bank erosion using UAV imagery—A case study of river Buech, Hautes-Alpes, France. *Int. J. Appl. Earth Obs. Geoinf*., 73, 428–437

[8] Kastridis, A., Kirkenidis, C., and Sapountzis, M. (2020). An integrated approach of flash flood analysis in ungauged Mediterranean watersheds using post-flood surveys and unmanned aerial vehicles. *Hydrol. Process*., 34, 4920–4939.

[9] Huang, W.C., Young, C.C., and Liu, W.C. (2018). Application of an automated discharge imaging system and LSPIV during typhoon events in Taiwan. *Water*, 10, 280.

[10] Kantoush, S.A., De Cesare, G., Boillat, J.L., and Schleiss, A.J. (2008). Flow field investigation in a rectangular shallow reservoir using UVP, LSPIV and numerical modelling. *Flow Meas. Instrum*., 19, 139–144.

[11] Liu, W.-C., Lu, C.-H., and Huang, W.-C. (2021). Large-Scale Particle Image Velocimetry to Measure Streamflow from Videos Recorded from Unmanned Aerial Vehicle and Fixed Imaging System. *Remote Sensing*. 13, 2661.

[12] Strelnikova D, Paulus G, Käfer S, Anders K-H, Mayr P, Mader H, Scherling U, and Schneeberger R. (2020). Drone-Based Optical Measurements of Heterogeneous Surface Velocity Fields around Fish Passages at Hydropower Dams. *Remote Sensing*. 12(3):384.

[13] Lewis, Q.W., Lindroth, E.M., and Rhoads, B.L. (2018). Integrating unmanned aerial systems and LSPIV for rapid, coast-effective stream gauging. *J. Hydrol*., 560, 230–246.

[14] Fox, J. F., and Patrick, A. (2008). Large-Scale Eddies Measured with Large Scale Particle Image Velocimetry, *Flow Measurement and Instrumentation*, 19(5), 283–291.
[15] Fujita, I., and Komura, S. (1994). Application of Video Image Analysis for Measurements of River Surface Flows, *Proceedings of Hydraulic Engineering*, 38, 733–738.
[16] Fujita, I. and Kunita, Y. (2011). Application of Aerial LSPIV To The 2002 Flood of The Yodo River Using A Helicopter Mounted High Density Video Camera. *Journal of Hydro-environment Research*, 5(4):323–331.

IOP Conf. Series: Earth and Environmental Science 1298 (2024) 012013 doi:10.1088/1755-1315/1298/1/012013

Mapping mangrove above ground carbon stock of benoa bay bali using sentinel-1 satellite imagery

Z Hidayah[1,2*], R Y S Utama[1], A R As-Syakur[3]
H A Rachman[1], D B Wiyanto[3]
[1]Department of Marine Science and Fisheries, Faculty of Agriculture Universitas Trunojoyo Madura
[2]Centre of Research and Community Development, Universitas Trunojoyo Madura Jl. Raya Telang 02 Kamal Bangkalan Madura 69162 East Java, Indonesia
[3]Department of Marine Science, Faculty of Marine and Fisheries, Universitas Udayana Jl. Raya Kampus Unud, Jimbaran, Badung 80361, Bali Indonesia

*Corresponding author E-mail : zainulhidayah@trunojoyo.ac.id

Abstract. Mangrove forest ecosystems distributed in tropical countries play an important role in carbon sequestration. The presence of mangrove forests is estimated to be capable of reducing CO_2 levels caused by air pollution. The process of photosynthesis absorbs CO_2 gas from the atmosphere and stores it as organic matter in the form of plant biomass. The amount of carbon stock stored in mangrove forests can be estimated using a regression model based on satellite imagery pixel values and above ground biomass (AGB) measurements. This study used the backscattering value of SAR Sentinel 1 images and field AGB measurements to map carbon stocks in the mangrove forest ecosystem of Benoa Bay, Bali. The processed backscattering value is the result of VV and VH polarization in the SAR Sentinel 1 image acquired in 2022. Meanwhile, AGB is calculated using the allometric equation based on the mangrove's species found at the study site. The biomass at the study location was estimated to be in the range of 64.18 Mg/Ha to 175.24 Mg/Ha, with an average of 115.34 ± 25.33 Mg/Ha. Therefore, the carbon stock of Benoa Bay can be estimated to have values in the range of 30.16 MgC/Ha to 82.36 MgC/Ha, with an average of 54.21 ± 11.09 MgC/Ha. In terms of estimating AGB biomass based on SAR pixel values, the VH polarization produces a better model ($R^2 = 0.635$) than the VV polarization.

1. Introduction

Mangrove ecosystems, like other coastal ecosystems, absorb and store carbon. These functions have been considered to play an important role in reducing CO_2 levels in the Earth's atmosphere. Carbondioxide (CO_2) absorbed during photosynthesis is used for the growth and development of mangrove plants [1,2]. Photosynthesis absorbs CO_2 gas from the atmosphere and stores it as organic matter in the form of plant biomass. The amount of organic matter stored in mangrove biomass per unit area and per period of time is the basis of mangrove forests' carbon-storage productivity [3]. Despite accounting for less than 2% of marine environments, mangroves are responsible for 10 to 15% of global carbon burial [4]. Mangroves are carbon-absorbing machines that absorb CO_2 while growing and sink it to the seafloor when they die.

IOP Conf. Series: Earth and Environmental Science 1298 (2024) 012013 doi:10.1088/1755-1315/1298/1/012013

Remote sensing sensors acquire electromagnetic waves reflected on the Earth's surface by vegetation and non-vegetation objects. Using a vegetation index, a subsequent examination of the images produced by the sensors is able to differentiate vegetation from other objects. Furthermore, a mathematical equation that combines several remote sensing bands to produce a single index value yields the vegetation index value. Other functions of the vegetation index include calculating percentages of vegetation cover and plant biomass, as well as estimating carbon sequestration and photosynthetic capacity.

From a remote sensing standpoint, above ground biomass (AGB) is recognized as an important ecological indicator for determining the environmental condition of mangrove forests. AGB is defined as the dry weight of organic matter produced and stored in above-ground living vegetation, including stems, branches, bark, seeds, and leaves [5]. Estimation of AGB has received attention in recent decades as a result of growing awareness of global warming caused by climate change and the critical role of mangrove forest biomass in the absorption and release of greenhouse gas carbon due to deforestation [6]. In addition, AGB can be estimated using destructive sampling methods, in which trees are felled and weighed, or non-destructive methods such as estimation based on allometric equations developed using dimensional trees such as diameter at breast height (dbh) and tree height. Recent research has focused on developing mathematical models that can be used to estimate AGB based on the satellite image pixel values.

The most widely used method is to extract pixel values of Visible-Infra Red (V-IR) bands (EM wave lenght 600 nm-950 nm) from medium resolution satellite images such as Landsat ETM 7, Landsat 8 OLI, ALOS, or the most recent Sentinel 2A/2B [7–9]. However, detection via the above mentioned passive satellites remote sensing is frequently limited by weather and thus cannot penetrate cloud cover. Other satellite sensors for example as SAR (Synthetic Aperture Radar) images offer an alternative for mangrove mapping. Synthetic Aperture Radar (SAR) is an active remote sensing sensor that generates its own energy to transmit microwave (radio) signals in order to image a specific scene. It is not dependent on sunlight and can thus operate at any time of day or night. In addition, SAR is regarded as an all-weather imaging sensor because it is not affected by cloud cover and can penetrate dense canopy cover. Radar sensors work in the microwave portion of the electromagnetic spectrum, with frequencies ranging from 0.3 GHz to 300 GHz, or 1 m to 1 mm in wavelength [10,11].

Sentinel-1 provides SAR images with a medium to high geometric resolution (5-20 m on the ground) in the C-band. It can send a signal in either horizontal (VH) or vertical (VV) polarisation and receive in both. Complex values can be found in Dual Polarisation Level-1 Single Look Complex (SLC) products. The inter-channel phase information, in addition to the backscatter intensity that can be measured from each single polarisation, allows for a more in-depth analysis of backscattering properties [12]. Several studies have demonstrated the use of Sentinel-1 SAR images to model the estimation of mangrove biomass, with varying results. A previous study in the Philippines found that VH polarisation has a better correlation with biomass than VV polarisation (Pearson correlation = 0.48-0.51) [13]. Meanwhile, another study in the West African dryland discovered that Sentinel-1 had insufficient accuracy in mapping AGB biomass, with RMSE = 78.6 and MAE = 25.6 [14]. Furthermore, several studies in China and Indonesia have attempted to estimate mangrove biomass using a combination of Sentinel 1 and Sentinel 2 data with several classification algorithms. The results obtained were representative with a coefficient of determination (R^2) = 0.787 and (R^2) = 0.821 respectively [15,16].

Questions have been raised regarding the accuracy and validity of using SAR images to estimate vegetation biomass. However, very few studies have investigated the association between SAR data of Sentinel-1 and specific vegetation biomass in particular mangroves. Previously published studies are limited to non-mangroves vegetation. Therefore this study aims to measure the accuracy of Sentinel-1 SAR data in mapping AGB biomass in a mangrove ecosystem. Furthermore, the results will be used to estimate the amount of carbon stock in the study area.

2. Methods

2.1 Study Location

This study was conducted in August to December 2021. The location of the study is the mangrove forest of Tahura Ngurah Rai Benoa Bay. Ngurah Rai Tahura is administratively located in Kuta District Badung Regency South Denpasar Bali Province, Indonesia. The geographic coordinate of the location is 115°9'31'' to 115°14'07'' East and 08°42'15'' to 08°47'27''South with an estimated area of 1,373.50 Ha. Thirty sampling plots measuring 10x10 m for field measurement of biomass and carbon stock were established. The plots were distributed using a purposive sampling method based on representation and access to the area (Figure 1).

Figure 1. Study Location the Mangrove Forest of Ngurah Rai Tahura Benoa Bay Bali

2.2 Image Processing

The Sentinel-1A SAR image used in this research was acquired on October 27, 2021 with Ground Area Detected (GRD) processing level. The image was downloaded from https://search.asf.alaska.edu. The specification of Sentinel-1 imagery data is explained in Table 1. The image was captured in Interferometric Wide Swath mode with VV and VH polarization and a pixel size of 10 m. The Sentinel application (SNAP) was used for image pre-processing, which included image sub-setting, orbit file application, radiometric calibration, speckle reduction, and Range-Doppler terrain correction. Radiometric image calibration converts the SAR image's pixel values to those that represent radar

IOP Conf. Series: Earth and Environmental Science 1298 (2024) 012013 doi:10.1088/1755-1315/1298/1/012013

backscatter from reflecting surfaces, as well as incident angle effects. This process converts the SAR image's pixel values into radar intensity backscattering coefficients (σ). Moreover, speckle reduction using the Refined Lee Filter is performed to reduce the effect of speckles on the image to enable better backscatter analysis and interpretation. Finally, Range-Doppler terrain correction is used to re-project the SAR image into a map projection.

Table 1. Specification of Sentinel-1 imagery data

No	Characteristics	Description/ Value
1	Swath width	250 km
2	Incidence angle range	29.1°– 46.0°
3	Azimuth Resolution	20 m
4	Ground Range Resolution	5 m
5	Azimuth and range looks	Single
6	Polarization options	Dual HH+HV, VV+VH Single HH, VV
7	Maximum Noise Equivalent Sigma Zero (NESZ)	-22 dB
8	Radiometric stability	0.5 dB (3σ)
9	Radiomatric accuracy	1 dB (3σ)
10	Phase error	5°

2.3 Field Survey

The non-destructive sampling method was applied to measure biomass of mangrove vegetations. A total of 30 sampling points were established for biomass measurement. At each sampling point, identification of mangrove species and measurement of the circumference of the stem at breast height (GBH) were carried out, which were then converted into stem diameters (DBH) (Figure 2). Biomass estimation was carried out using equations according to the type of mangrove found (Equation 1-5). Furthermore, biomass values were converted into carbon stock values (Equation 6).

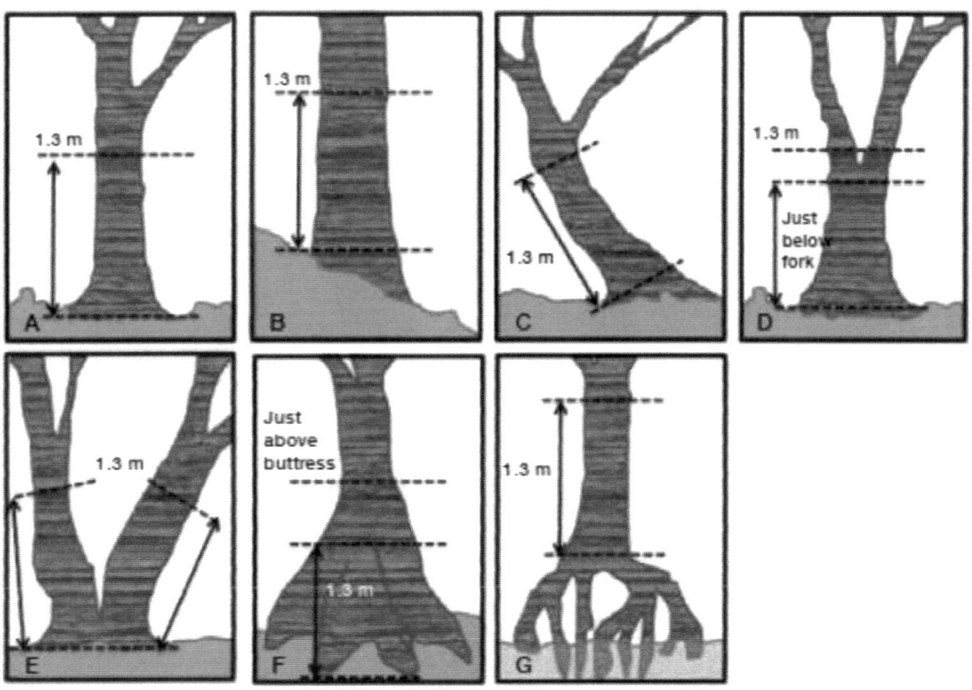

Figure 2. DBH Measurement of Mangrove Vegetations
(Suardana et al, 2022 *modified* from Pearson et al, 2005)

Sonneratia alba biomass (kg) = $0.3841*D^{2.101}*\rho$..........................(1)
Rhizophora apiculate biomass (kg) = $0.043*D^{2.63}$........................(2)
Rhizophora mucronata biomass (kg) = $0.128*D^{2.6}$........................(3)
Bruguiera gymnorrhiza (kg) = $0.0754*D^{2.505}*\rho$...................(4)
$C = B \times 0.47$..(5)

Note : D = Diameter at breast height (DBH) cm; ρ = wood specific density (gr cm^{-3}), C = carbon stock (kg); B = biomass (kg).

3. Results and Discussions

3.1 Above Ground Biomass and Carbon Stock

Benoa Bay is a tidal waterway on Bali's southern coast. Due to the reclamation of Serangan Island, the bay's mouth has narrowed by up to 75%, resulting in a semi-enclosed bay typology. Because it is a shallow estuary ecosystem with a number of rivers flowing into it (Tukad Punggawa, Tukad Balian, Tukad Badung, Tukad Mati, Tukad Soma, Tukad Mumbul, and Tukad Bualu), Benoa Bay is a unique ecosystem area. This esthetic condition distinguishes Benoa Bay from other shallow coastal waters, where a number of ecologically important communities, particularly mangrove communities, seagrass beds, macrozoobenthos, and infauna components with high abundance and diversity exist.

This location is strategically situated at the epicenter of Bali's golden triangle, which includes the Sanur - Kuta - Nusa Dua area as a tourism area that has developed and advanced as a centre for economic growth based on tourism, trade, and services. The beaches in Benoa Bay's coastal area have nearly identical physical characteristics, with the exception of the mangrove beach, which is distinguished by mangrove vegetation and its associations in the form of alluvial and muddy deposits. On the coasts of Benoa and Serangan, there are also sandy sloping beaches. In general, the coastal area of Benoa Bay is lowland with a slope of 0 - 2%. Mangrove forests grow on the sides of Benoa Bay from Tukad Loloan to Tanjung Benoa. The mangrove forest area in Benoa Bay accounts for 62.9% of Bali's total mangrove forest area. According to Minister of Forestry Decree Number: 544/Kpts-II/1993 dated September 25, 1993, the mangrove forest area totals was 1373.5 Ha [17].

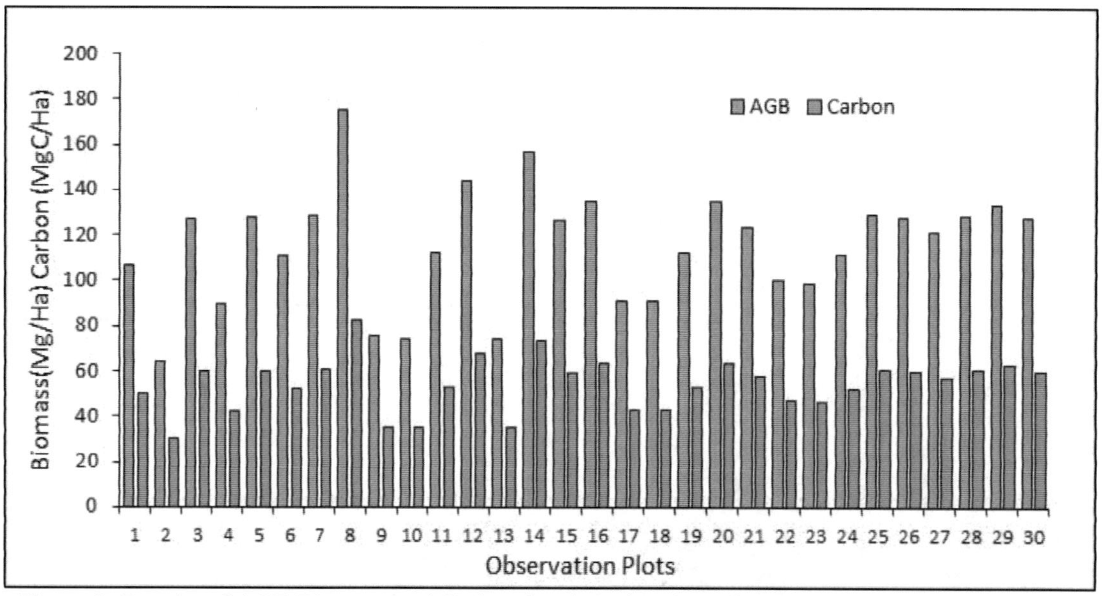

Figure 3. Results of AGB Biomass and Carbon Stock Estimation of Mangrove Forest of Benoa Bay Bali

The findings of this study's field survey identified several dominant mangrove species, including *Rhizophora mucronata*, *Rhizophora apiculata, Sonneratia alba* and *Bruguiera gymnorhiza*. The mangrove species discovered during the survey are dominant species that play an important role in the mangrove ecosystem. To adapt to the high salinity tidal environment, mangrove vegetations develop morphological specializations such as aerial roots and other special physiological mechanisms, particularly for salt excretion [18]. Taxonomically, true mangrove species are distinct from land plants because they are only found in mangrove forests and form pure stands, never mixing with land plants [19].

The above-ground biomass (AGB) of mangrove vegetation was measured in order to estimate carbon stocks and calculate CO_2 sequestration potential. The substrate in the mangrove growing area influences the amount of biomass produced by mangroves. The biomass content of tree stands increases with age. The biomass at the study location was estimated to be in the range of 64.18 Mg/Ha to 175.24 Mg/Ha, with an average of 115.34 ± 25.33 Mg/Ha based on the measurement results of 864 mangrove stands from 30 sampling plots (Figure 1). As a result, the carbon stock of Benoa Bay can be estimated to have values in the range of 30.16 MgC/Ha to 82.36 MgC/Ha, with an average of 54.21 ± 11.09 MgC/Ha.

The estimated AGB carbon stock of mangrove forest in Benoa Bay was relatively higher compare to several rehabilitated mangrove forest in Indonesia, for example in Tanakeke South Sulawesi (37.6 MgC/Ha), Deli Serdang North Sumatra (40-50 MgC/Ha) and Bregasmalang Central Java (30.27 MgC/ha) [20–22]. The carbon stock in the mangrove forests of Benoa Bay Bali, however, is significantly lower when compared to natural mangrove forests in relatively undisturbed environments. Figure 4 compares carbon stocks from other locations in Indonesia and other Asian countries [3,20,21,23–28]

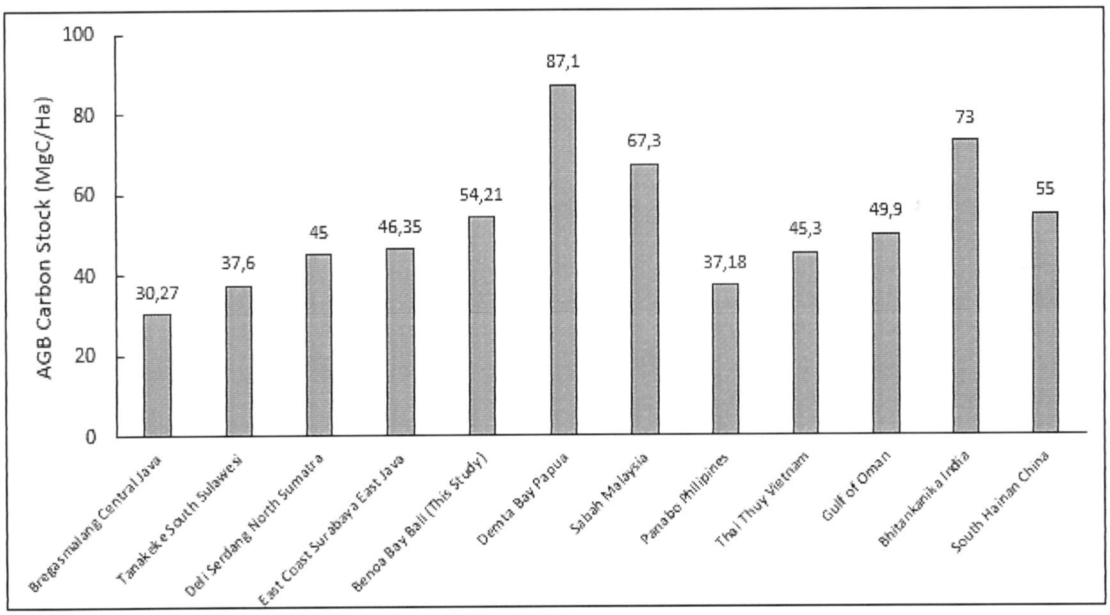

Figure 4. Comparison of Estimated Mangrove's Carbon Stock (MgC/Ha)

3.2 Image Processing and AGB Carbon Stock Model

The processing of Sentinel-1 satellite imagery was the next analysis performed in this study. Backscattering values on the vertical (VV) and horizontal (VH) polarizations can be obtained after the pre-processing stage. The VV and VH pixel values were then extracted from 30 points according to the coordinates of the carbon stock measurement plot. The average backscatter value at VV polarization was -7.86 dB which for the highest value was -5.10 dB and the lowest value is -9.99 dB. Whereas the average value of backscatter in VH polarization was -13.43 dB which for the highest value was -11.21 dB and the lowest value is -16.44 dB. Figure 5 represents the pixel values obtained from the sampling plots.

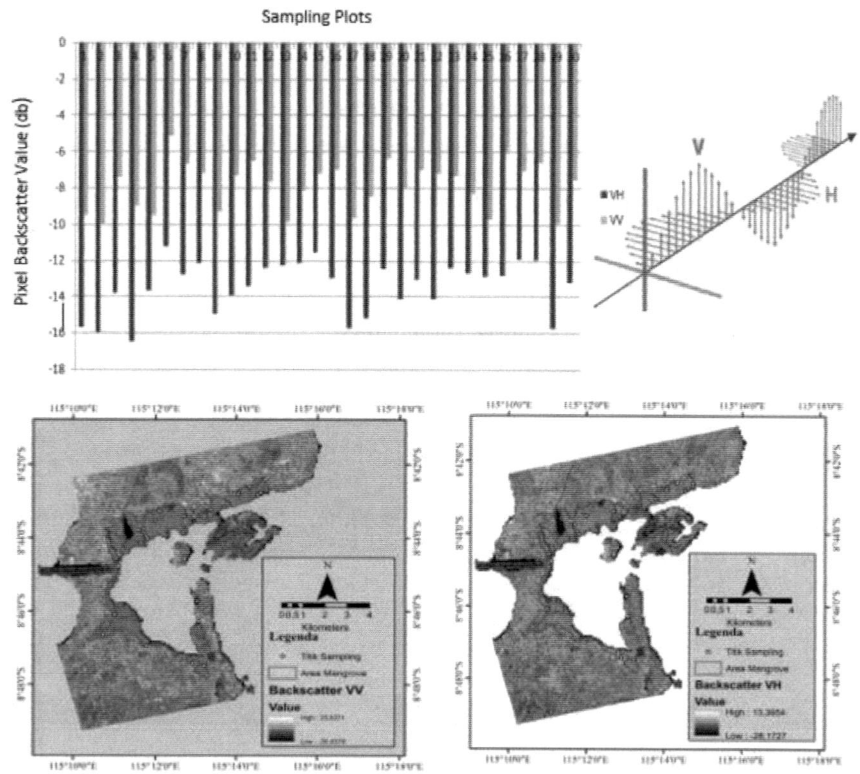

Figure 5. Pixel Value of VV and VH Polarization of Sentinel-1 SAR Data Benoa Bay Bali

The Sentinel-1 SAR instruments operate in single (HH or VV) and dual polarisation (HH+HV or VV+VH), with one transmit chain (switchable to H or V) and two parallel receive chains for H and V polarisation.VV is a mode that transmits and receives vertical waves to generate the SAR image, whereas VH is a mode that transmits and receives horizontal waves to generate the SAR image. Sentinel-1 VV and VH backscatter, in particular, are known to be sensitive to changes in vegetation conditions as a result of natural or anthropogenic disturbance such as storm surge and land conversion [29,30].

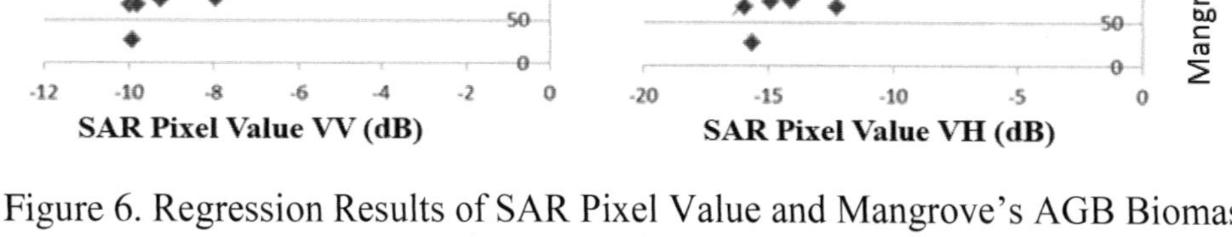

Figure 6. Regression Results of SAR Pixel Value and Mangrove's AGB Biomass of Benoa Bay Bali

Moreover, AGB modeling with Sentinel-1 image SAR pixel values was carried out by applying linear regression to find statistical correlations. The findings are presented in the form of a mathematical equation that can be used to estimate the amount of AGB and carbon stock in mangrove forests. The linear regression model produced by the VV polarization is AGB (kg)=38.206*(pixel value)+469.3 with R^2=0.521. Meanwhile, the VH polarization results in the regression equation AGB (kg) = 36.953* (pixel value) + 665.63, with R^2= 0.635 (Figure 6). These results confirm that SAR pixel value of Sentinel-1 satellite imagery is sufficient to estimate the amount of AGB biomass and carbon stock of mangrove forest. The VH polarization produce better model compare to its counterpart. A similar result was obtained from previous study in the Philippines where R^2 of VH-AGB model (0.42) was higher than VV-AGB model (0.33) [31]. To estimate the amount of AGB biomass and carbon stock of the study area, an estimation map was produced using the best model (VH-AGB). Based on the map, AGB biomass of Benoa Bay Bali was estimated at the average of 160.53 Mg/Ha, with maximum value of 267.16 Mg/ha and minimum value of 76.44 Mg/ha. Following the results, therefore the estimated carbon stock in Benoa Bay Bali is 75.44 MgC/Ha in average (Figure 7).

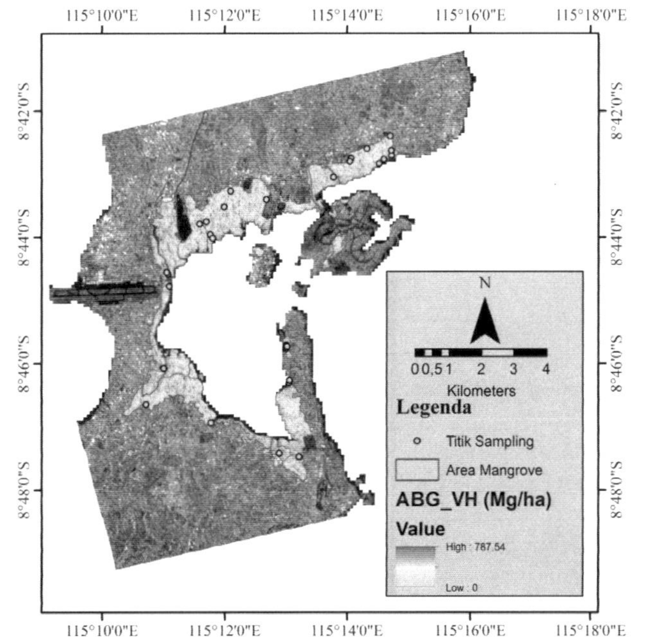

Figure 7. AGB Biomass Map of Benoa Bay Bali Produced from VH-AGB Model (R^2=0.635)

In 2021, a research at the same location using ALOS PALSAR-2 (2020) was conducted. The results for calculating the average AGB obtained were 133.74 Mg/Ha, which was lower than the results of this study, which were 160.53 Mg/Ha. This difference may be due to differences in spatial resolution between ALOS PALSAR imagery (7 meters) and Sentinel-1 SAR (10 meters). However, compare to its other predecessor satellites such as ERS-1, ERS-2, JERS, SIR-C/X-SAR, RADARSAT, SRTM, EnviSAT-ASAR, RADARSAR-II, LIGHTSAR [10], Sentinel-1 SAR offers better resolution thus become more reliable to be used for medium scale vegetation mapping.

The short revisit time (temporal resolution) of Sentinel-1 images is advantageous in forest change detection applications. Furthermore, the aforementioned images are free to use and easily accessible. Longer wavelength SAR (e.g., ALOS PALSAR products) is more suitable for forest applications because it penetrates deeper than shorter wavelength SAR (e.g., Sentinel-1 SAR products). Below-canopy information such as tree trunks/stems and underlying undergrowth vegetation, deadwood (including standing deadwood), and soil organic matter all contribute significantly to the amount of aboveground biomass and carbon pools in tropical forests [32,33].

Mangrove forests play a crucial role in the global carbon cycle by sequestering and storing large amounts of carbon in their biomass and soils. These coastal forests are highly efficient at capturing and storing carbon dioxide from the atmosphere, making them important carbon sinks. Studies have shown that mangrove carbon stocks vary across different regions and countries [3,34]. Global patterns indicate that mangrove carbon stocks are higher in the tropics compared to temperate regions. This is due to the favorable conditions for mangrove growth, such as high temperatures and nutrient availability. However, there are also variations within tropical regions, with differences in carbon stocks observed between hemispheres and latitudes.The loss of mangrove forests has significant implications for carbon stocks and the global carbon cycle. Deforestation and degradation of mangroves result in the release of stored carbon into the atmosphere, contributing to greenhouse gas emissions. A study estimated that between 2000 and 2012, global mangrove deforestation led to the loss of approximately 122 million metric tons of carbon and potential emissions of 316 million metric tons of CO_2 [35,36]. The consequences of mangrove carbon stock loss extend beyond the carbon cycle. Mangroves provide numerous ecosystem services, including coastal protection, habitat for biodiversity, and support for local communities. The destruction of mangroves not only disrupts these services but also leads to the loss of valuable carbon sinks.

The use of multi-temporal satellite images for example Landsat TM-8, Sentinel-1 and Sentinel-2 are important for mangrove monitoring. Image processing can detect various spatial environmental change parameters over multiple time periods. Other parameters, such as carbon content, require field data. This research demonstrates how a combination of field data and image processing results can be used to create a spatial database with high accuracy. To produce a more representative model of estimated biomass and carbon stock, sample selection must be varied to account for differences in density or canopy cover.

4. Conclusions

In conclusion, mangrove carbon stocks are an important component of the global carbon cycle. Their loss contributes to greenhouse gas emissions and has negative implications for coastal ecosystems and communities. This study has demonstrated the application of SAR data to identified the extent of mangrove forest, and with the combination of field survey to estimate AGB and carbon stock of Benoa Bay Bali. The biomass at the study location was estimated to be in the range of 64.18 Mg/Ha to 175.24 Mg/Ha, with an average of 115.34 ± 25.33 Mg/Ha. Therefore, the carbon stock of Benoa Bay can be estimated to have values in the range of 30.16 MgC/Ha to 82.36 MgC/Ha, with an average of 54.21 ± 11.09 MgC/Ha. In terms of estimating AGB biomass based on SAR pixel values, the VH polarization produces a better model (R^2 = 0.635) than the VV polarization. Sample selection must be varied to account for differences in density or canopy cover in order to produce a more representative model of estimated biomass and carbon stock. Additional research is required to better understand mangrove carbon stocks and their role in climate change mitigation. Mangrove forest protection and restoration are critical for preserving their carbon sequestration potential and the valuable ecosystem services they provide.

References

[1]. Guo Y, Liao J, Shen G. Mapping large-scale mangroves along the maritime silk road from 1990 to 2015 using a novel deep learning model and landsat data. Remote Sens. 2021;13(2):1–20.

[2]. Sharifi A, Felegari S, Tariq A. Mangrove forests mapping using Sentinel-1 and Sentinel-2 satellite images. Arab J Geosci [Internet]. 2022;15(20). Available from: https://doi.org/10.1007/s12517-022-10867-z

[3]. Hidayah Z, Rachman HA, As-Syakur AR. Mapping of mangrove forest and carbon stock estimation of east coast Surabaya, Indonesia. Biodiversitas. 2022;23(9):4826–37.

[4]. Bouillon S, Borges A V., Castañeda-Moya E, Diele K, Dittmar T, Duke NC, et al. Mangrove

production and carbon sinks: A revision of global budget estimates. Global Biogeochem Cycles. 2008;22(2):1–12.

[5]. Hu T, Zhang YY, Su Y, Zheng Y, Lin G, Guo Q. Mapping the global mangrove forest aboveground biomass using multisource remote sensing data. Remote Sens. 2020;12(10).

[6]. Gouvêa LP, Serrão EA, Cavanaugh K, Gurgel CFD, Horta PA, Assis J. Global impacts of projected climate changes on the extent and aboveground biomass of mangrove forests. Divers Distrib. 2022;28(11):2349–60.

[7]. Pham TD, Le NN, Ha NT, Nguyen LV, Xia J, Yokoya N, et al. Estimating mangrove above-ground biomass using extreme gradient boosting decision trees algorithm with fused sentinel-2 and ALOS-2 PALSAR-2 data in can Gio biosphere reserve, Vietnam. Remote Sens. 2020;12(5).

[8]. Sulistiyono N, Tarigan AA, Patana P. Application of landsat 8 sattelite imagery for estimated distribution of above ground carbon in Percut Sei Tuan forest landscape. IOP Conf Ser Earth Environ Sci. 2020;454(1).

[9]. Darmawan S, Sari DK, Takeuchi W, Wikantika K, Hernawati R. Development of aboveground mangrove forests' biomass dataset for Southeast Asia based on ALOS-PALSAR 25-m mosaic. J Appl Remote Sens. 2019;13(04):1.

[10]. Jay Labadisos Argamosa R, Conferido Blanco A, Balidoy Baloloy A, Gumbao Candido C, Bart Lovern Caboboy Dumalag J, Lee Carandang DImapilis, Lady, et al. Modelling Above Ground Biomass of Mangrove Forest Using Sentinel-1 Imagery. ISPRS Ann Photogramm Remote Sens Spat Inf Sci. 2018;4(3):13–20.

[11]. Ghorbanian A, Zaghian S, Asiyabi RM, Amani M, Mohammadzadeh A, Jamali S. Mangrove ecosystem mapping using sentinel-1 and sentinel-2 satellite images and random forest algorithm in google earth engine. Remote Sens. 2021;13(13):1–18.

[12]. Zhang B, Wdowinski S, Gann D, Hong SH, Sah J. Spatiotemporal variations of wetland backscatter: The role of water depth and vegetation characteristics in Sentinel-1 dual-polarization SAR observations. Vol. 270, Remote Sensing of Environment. 2022. 1–70 p.

[13]. Castillo JAA, Apan AA, Maraseni TN, Salmo SG. Estimation and mapping of above-ground biomass of mangrove forests and their replacement land uses in the Philippines using Sentinel imagery. ISPRS J Photogramm Remote Sens. 2017;134:70–85.

[14]. Forkuor G, Benewinde Zoungrana JB, Dimobe K, Ouattara B, Vadrevu KP, Tondoh JE. Above-ground biomass mapping in West African dryland forest using Sentinel-1 and 2 datasets - A case study. Remote Sens Environ [Internet]. 2020;236(November 2018):111496. Available from: https://doi.org/10.1016/j.rse.2019.111496

[15]. Huang Z, Tian Y, Zhang Q, Huang Y, Liu R, Huang H, et al. Estimating mangrove above-ground biomass at Maowei Sea, Beibu Gulf of China using machine learning algorithm with Sentinel-1 and Sentinel-2 data. Geocarto Int. 2022;37(27):15778–805.

[16]. Nuthammachot N, Askar A, Stratoulias D, Wicaksono P. Combined use of Sentinel-1 and Sentinel-2 data for improving above-ground biomass estimation. Geocarto Int. 2022;37(2):366–76.

[17]. Suardana AAMAP, Anggraini N, Aziz K, Nandika MR, Ulfa A, Wijaya AD, et al. Biomass estimation model and carbon dioxide sequestration for mangrove forest using SENTINEL-2 in Benoa bay, Bali. Int J Remote Sens Earth Sci [Internet]. 2022;19(1):91–100. Available from: https://jurnal.lapan.go.id/index.php/ijreses/article/view/3797

[18]. Su CJ, Hsieh SY, Chiang MWL, Pang KL. Salinity, pH and temperature growth ranges of Halophytophthora isolates suggest their physiological adaptations to mangrove environments.

Mycology [Internet]. 2020;11(3):256–62. Available from: https://doi.org/10.1080/21501203.2020.1714768

[19]. Quadros AF, Helfer V, Nordhaus I, Reuter H, Zimmer M. Functional traits of terrestrial plants in the intertidal: A review on mangrove trees. Biol Bull. 2021;241(2):123–39.

[20]. Cameron C, Hutley LB, Friess DA, Brown B. Community structure dynamics and carbon stock change of rehabilitated mangrove forests in Sulawesi, Indonesia. Ecol Appl. 2019;29(1).

[21]. Harefa MS, Nasution Z, Mulya MB, Maksum A. Mangrove species diversity and carbon stock in silvofishery ponds in Deli Serdang District, North Sumatra, Indonesia. Biodiversitas. 2022;23(2):655–62.

[22]. Sugiatmo, Poedjirahajoe E, Pudyatmoko S, Purwanto RH. Carbon stock at several types of mangrove ecosystems in Bregasmalang, Central Java, Indonesia. Biodiversitas. 2023;24(1):182–91.

[23]. Bai J, Meng Y, Gou R, Lyu J, Dai Z, Diao X, et al. Mangrove diversity enhances plant biomass production and carbon storage in Hainan island, China. Funct Ecol. 2021;35(3):774–86.

[24]. Indrayani E, Kalor JD, Warpur M, Hamuna B. Using Allometric Equations to Estimate Mangrove Biomass and Carbon Stock in Demta Bay, Papua Province, Indonesia. J Ecol Eng. 2021;22(5):263–71.

[25]. Bhomia RK, Mackenzie RA, Murdiyarso D, Sasmito SD, Purbopuspito J. Impacts of land use on Indian mangrove forest carbon stocks: Implications for conservation and management. Ecol Appl. 2016;26(5):1396–408.

[26]. Suhaili NS, Fei JLJ, Sha'ari FW, Idris MI, Hatta SM, Kodoh J, et al. Carbon stock estimation of mangrove forest in sulaman lake forest reserve, Sabah, Malaysia. Biodiversitas. 2020;21(12):5657–64.

[27]. Alimbon JA, Manseguiao MRS. Species composition, stand characteristics, aboveground biomass, and carbon stock of mangroves in panabo mangrove park, philippines. Biodiversitas. 2021;22(6):3130–7.

[28]. Savari A, Khaleghi M, Safahieh AR, Hamidian Pour M, Ghaemmaghami S. Estimation of biomass, carbon stocks and soil sequestration of Gowatr mangrove forests, Gulf of Oman. Iran J Fish Sci. 2020;19(4):1657–80.

[29]. Sosa L, Justel A, Molina Í. Detection of crop hail damage with a machine learning algorithm using time series of remote sensing data. Agronomy. 2021;11(10):1–18.

[30]. De Souza Moreno GM. Deep Semantic Segmentation of Mangrove Combining Spatial, Temporal and Polarization from Sentinel-1 Time Series in the Brazilian Territory. University of Brazilia; 2022.

[31]. Makinano-Santillan M, Bolastig CG, Santillan JR. Aboveground biomass estimation of mangroves in Siargao Island, Philippines using Sentinel-1 image. In: 40th Asian Conference on Remote Sensing, ACRS 2019: Progress of Remote Sensing Technology for Smart Future. 2020. p. 1–10.

[32]. Vashum KT, Jayakumar S. Methods to Estimate Above-Ground Biomass and Carbon Stock in Natural Forests - A Review. J Ecosyst Ecography. 2012;02(04).

[33]. Hidayah Z, De Oliveira Vieira L, Rahma S, Rachman HA, As-Syakur AR. Mapping Sea Grass Coverage of Tanjung Benoa Bali Using Medium Resolution Satellite Imagery Sentinel 2B. Sains Malaysiana. 2023;52(4):1315–32.

[34]. Giri C, Long J, Abbas S, Murali RM, Qamer FM, Pengra B, et al. Distribution and dynamics of

mangrove forests of South Asia. J Environ Manage. 2015;148(48):101–11.

[35]. Macreadie PI, Costa MDP, Atwood TB, Friess DA, Kelleway JJ, Kennedy H, et al. Blue carbon as a natural climate solution. Nat Rev Earth Environ. 2021;2(12):826–39.

[36]. Hidayah Z, De Oliveira L, Safitri R, Rachman HA, As-Syakur AR. Mapping sea grass coverage of Tanjung Benoa Bali using medium resolution satellite imagery Sentinel 2B. Sains Malaysiana. 2023;52(4):1315-32.

Comparison of peak ground acceleration using deterministic and probabilistic approach in the coastal area of Loh Buaya, Rinca Island, Indonesia

Heryawan Kurnia Rakhman[1], Sito Ismanti[1], Trias Aditya[2]

[1] Department of Civil and Environmental Engineering, Faculty of Engineering, Universitas Gadjah Mada, Indonesia
[2] Department of Geodetic Engineering, Faculty of Engineering, Universitas Gadjah Mada, Indonesia

Corresponding author email: sito.ismanti@ugm.ac.id

Abstract. An area located near an active volcano, subduction zone, and/or active fault has a risk of an earthquake. Earthquakes can cause damage to buildings with a large number of losses. One of the earthquake-related parameters required in planning an earthquake-resistant building is the Peak Ground Acceleration (PGA). The objective of this study is to compare the PGA value at the coastal area of Loh Buaya, Rinca Island, Indonesia, using Probabilistic Seismic Hazard Analysis (PSHA) and Deterministic Seismic Hazard Analysis (DSHA). The DSHA is calculated based on earthquake data series from the United States Geological Survey (USGS) in the period between 1922 to 2022. Also, it is calculated using 10 data of Ground Motion Prediction Equation (GMPE). Data from the USGS is constrained with magnitude $\geq$ 5.0 M_W and earthquake distance $\leq$ 200 km from the study site. The PSHA is based on the Indonesian earthquake hazard map and the de-aggregation map from the National Earthquake Study Centre of Indonesia with a probability exceeding 2% in 50 years. Based on the parameters of PGA maximum, distance, and magnitude, the DSHA by Kanno method has the closest result to the PSHA. It is concluded that this area has a high earthquake risk with a PGA maximum of more than 0.5 g.

1. Introduction

Indonesia is one of the countries frequently affected by earthquakes. Earthquakes are disasters that can occur in areas close to volcanoes, subduction zones, and active faults. Shock motion is detected on the Earth's surface due to seismic waves. It caused damage to structures; subsidence and liquefaction, and slope failure; and tsunami [1]. In the last two decades, there have been several major earthquakes in Indonesia, including the Banda Aceh earthquake with 9.2 M_W in 2004 [2]; Nias with 8.7 M_W in 2005 [3]; Pangandaran with 7.7 M_W in 2006 [4]; Yogyakarta with 6.3 M_W in 2009 [5]; Padang with 7.6 M_W in 2009 [6]; Lombok with 7.0 M_W in 2018 [7]; Palu with 7.5 M_W in 2018 [8]; dan Flores with 7.4 M_W in 2021 [9]. One of the main earthquake parameters used to measure ground motion strength is Peak Ground Acceleration [10].

Peak ground acceleration (PGA) is the maximum ground acceleration that occurred during earthquake shaking at a site. It is equal to the amplitude of the maximum acceleration measured on an accelerogram at a location during an earthquake [11]. Determining PGA is an important requirement in planning earthquake-resistant buildings [12]. Earthquakes with identic magnitude can offer different results of PGA. Even moderate-magnitude earthquakes can produce larger PGA than larger-magnitude earthquakes [13].

There is no Ground Motion Prediction Equation (GMPE) to calculate PGA in specific locations in Indonesia yet [14]. Indonesian seismic hazard analysis was adapted GMPE from other locations with similar geology, tectonics, and earthquake mechanisms [15]. Probabilistic Seismic Hazard Analysis (PSHA) method and de-aggregation map from the Indonesian Earthquake Study Center (Pusgen) have been able to provide information on the magnitude of the PGA value at a location, but the information on earthquake magnitude and distance provided is still in the form of a range [16]. Knowing the specific magnitude and distance of an earthquake to produce a PGA value can improve accuracy in earthquake risk planning.

This research has two objectives. First, to compare the PGA, magnitude, and distance of earthquakes by Deterministic Seismic Hazard Analysis (DSHA) through existing GMPEs from various countries with PSHA from Pusgen. Second, to find out the factors that make the research location prone to earthquakes. The research is located in Loh Buaya, Rinca Island, Indonesia. The research was conducted at Loh Buaya, Rinca Island, Indonesia. It is expected that this research can provide an overview of which GMPE is closest to the PSHA results and the cause of the earthquake.

2. Material and Method

2.1. *Peak Ground Acceleration (PGA)*

PGA is one of the variables in estimating ground motion intensity by considering independent variables such as magnitude, source-to-site distance, site class, and fault type [17]. While earthquake magnitude is used to estimate the losses caused by an earthquake, PGA is more widely used concerning the damage to buildings caused by the earthquake [14]. The value of PGA will be directly proportional to the damage to the building.

There are several methods to determine the PGA value, which include the probabilistic and deterministic approach. The probabilistic approach to determine PGA is also called Probabilistic Seismic Hazard Analysis (PSHA) method whereas deterministic approach is also called Deterministic Seismic Hazard Analysis (DSHA) method. PSHA method in Indonesia can refer to the Indonesian standard in SNI 1726:2019 [12] and has been updated by Pusgen 2022 [16] to determine the dominant magnitude and epicenter distance. Meanwhile, the DSHA method uses Ground Motion Prediction Equation (GMPE) equations based on earthquake records at specific locations to predict PGA. From 1964-2010, 289 GMPEs from all around the world were formulated based on their characteristics of location, geology, fault type, fault mechanism, and location respectively [18].

2.2. *Probabilistic Seismic Hazard Analysis (PSHA)*

Probabilistic approach or PSHA method is a method of analyzing earthquake hazards by considering the uncertainty of location, magnitude, and time of occurrence using the probability theorem. This research uses the calculation of PGA from SNI 1726-2019 which is the determination of PGA. In planning earthquake-resistant buildings using a return scale of 2,500 years or probability exceeding 2% of 50 years [12]. The PGA value is obtained from the response spectra (RSA) Cipta Karya website by selecting the site coordinates. The PGA value is the PGA at the bedrock, so it needs to be multiplied by the site class coefficient according to equation (1). Methods for determining the classification of site classes according to (Table 1).

According to [16], De-aggregation Map complements the probabilistic method of RSA Cipta Karya to provide information on representative magnitude (M) and epicenter distance (R) from the site to the reviewed earthquake source. In determining the magnitude and epicenter distance, a logic tree scheme

IOP Conf. Series: Earth and Environmental Science 1298 (2024) 012014 doi:10.1088/1755-1315/1298/1/012014

with various GMPEs was used to provide quantification of uncertainties in the model and its input parameters (Table 2).

$$PGA_M = F_{PGA}\, x\, PGA \tag{1}$$

Table 1. F_{PGA} coefficient [12]

Site Class	PGA (g)					
	≤ 0.1	0.2	0.3	0.4	0.5	≥ 0.6
SA (dense gravel)	0.8	0.8	0.8	0.8	0.8	0.8
SB (Gravel)	0.9	0.9	0.9	0.9	0.9	0.9
SC (Hard soil)	1.3	1.2	1.2	1.2	1.2	1.2
SD (medium soil)	1.6	1.4	1.3	1.2	1.1	1.1
SE (soft soil)	2.4	1.9	1.6	1.4	1.2	1.1
SF (specific soil)	Site Specific					

Table 2. The equation used in the logic tree in the De-aggregation map [16]

Source Model	Ground Motion Prediction Equation
1. Subduction (megathrust)	a. BC Hydro 2012 b. Youngs 1997 c. Atkinson-Boore 2003
2. Fault and shallow background	a. Boore-Atkinson NGA 2014 b. Campbell-Bozorgnia NGA 2014 c. Chiou-Youngs NGA 2014
3. Deep background	a. Abrahamson 2018 b. Zhao 2006 c. Atkinson-Boore 2003

2.3. Deterministic Seismic Hazard Analysis (DSHA)

The calculation of PGA in deterministic approach or DSHA is carried out to calculate the level of ground acceleration that occurs due to earthquakes based on fault mechanism, hypocenter distance, earthquake depth, magnitude, and local soil conditions. GMPE is DSHA method that is formulated based on a collection of seismic data at a location and then approached with the characteristic equation in the form of geological, tectonic conditions and fault mechanism. That are also consideration in the application of GMPE to different locations. Because of limited ground motion records, there is no GMPE in specific sites available in Indonesia [14]. The following GMPE is used to predict PGA in the Loh Buaya, Rinca Island:

2.3.1. GMPE by Donovan [19]. The GMPE equation from Donovan used earthquake data from 214 earthquake records located in San Fernando, California, USA. The proposed equation is according to equation (2):

$$y = b_{D_1} e^{b_{D_2} M} (R + 25)^{-b_{D_3}} \tag{2}$$

with y is PGA (gal); b_{D_1} is 1080; b_{D_2} is 0.5; R is distance; M is magnitude.

2.3.2. GMPE by Bornstein and Donovan [20]. The GMPE equation from Donovan and Bornstein was used to characterize earthquake records in California with the review of soil types of rock and hard soil. The proposed equation is according to equation (3)-(6):

$$y = b_1 e^{b_2 M} + (R + 25)^{b_3} \tag{3}$$

$$b_1 = c_1 R^{-c_2} \tag{4}$$

$$b_2 = d_1 + d_2 logR \quad (5)$$

$$b_3 = e_1 + e_2 logR \quad (6)$$

with y is PGA (gal); c_1 is 2,154,000; c_2 is 2.10; d_1 is 0.046; d_2 is 0.445; R is distance; M is magnitude.

2.3.3. GMPE by Mcguire [21]. This GMPE equation used characteristics of 11 earthquake records on rocky ground and 69 earthquake records on alluvial ground. The magnitude under review is M < 6.5. The proposed equation is according to equation (7):

$$\ln x = b_{G_1} + b_{G_2}M + b_{G_3}\ln R + b_{G_4}Y_s \quad (7)$$

with x is PGA (cm/s2); b_{G_1} is 3.40; b_{G_2} is 0.89; b_{G_3} is -1.17; b_{G_4} is -0.20; Y_s is 0 for gravel or 1 for soil; M is magnitude; R is distance.

2.3.4. GMPE by Campbell [22]. The GMPE equation from Campbell used earthquake data characteristic of the western United States located 50 km from the fault. The magnitudes of the earthquakes reviewed were between 5.0-7.7 M. The earthquakes were shallow earthquakes without considering the influence of subduction. The proposed equation is according to the equation (8):

$$PGA = 0.0159 \exp(0.868M)\,[R + 0.0606 \exp(0.700M)]^{-1.09} \quad (8)$$

with M is magnitude and R is distance.

2.3.5. GMPE by Youngs [23]. The GMPE equation from Youngs used the characteristics of 389 earthquake records from Alaska, Chile, Peru, Japan, Mexico, and Solomon. The proposed equation is according to (9):

$$\ln a_{max} = C_1 + C_2M_w + C_3 \ln[R + C_4\exp(C_5M_w)] + BZ_t \quad (9)$$

with a_{max} is PGA (g); C_1 is 19.16; C_2 is 1.045; C_3 is 4.738; C_4 is 205.5; C_5 is 0.0968; B is 0.54; Z_t is 0 for gravel or 1 for soil; M_W is magnitude; R is distance.

2.3.6. GMPE by Fukushima and Tanaka [24]. The GMPE equation from Fukushima and Tanaka uses data characteristic of earthquakes in Japan between 1960-1990 with hypocenter depth, H < 100 km, and epicenter distance, R < 300 km. The proposed equation is according to equation (10).

$$\log a = 0.41M - \log(R + 0.032.\,10^{0.41M}) - 0.0034R + 1.30 \quad (10)$$

with a is PGA (cm/s^2); R is distance; M is magnitude.

2.3.7. GMPE by Youngs [25]. This GMPE equation considered hypocenter depths between 10-229 km. A total of 98 records of interface earthquakes (earthquakes of hypocenter depth < 50 km) and 66 records of intra-slab earthquakes (earthquakes of hypocenter depth > 50 km). The proposed equation is according to equation (11).

$$\ln(y) = -0.6697 + 1.438M + C_{Y_1} + C_{Y_2}(10 - M)^3 + C_{Y_3}\ln(R + 1.097e^{0.617M}) + 0.00648H + 0.3643Z_t \quad (11)$$

with y is PGA (g); C_{Y_1} is 0 ; C_{Y_1} is 0 ; C_{Y_3} is -2.329; C_{Y_4} is 1.45; C_{Y_5} is -0.1; Z_t is 0 for gravel or 1 for soil; M is magnitude and R is distance.

2.3.8. GMPE by Garcia [26]. The GMPE equation from Garcia used earthquake data of hypocenter depth 35 < H < 138 km, magnitude M_W > 5.0, and epicenter distance R < 400 km. The proposed equation is according to equations (12)(14):

$$\log Y = C_{G_1} + C_{G_2}M_w + C_{G_3}R_{cld} - C_{G_4}logR_{cld} + C_{G_5}H \tag{12}$$

$$R_{cld} = \sqrt{R^2 + \Delta^2} \tag{13}$$

$$\Delta = 0.00750 \times 10^{0,507M} \tag{14}$$

with y is PGA (cm/s2); C_{G_1} is -0.2; C_{G_2} is 0.59; C_{G_3} is -0.0039; C_{G_4} is 1.00; C_{G_5} is 0.008; M is magnitude; R is distance.

2.3.9. GMPE by Kanno [27]. The GMPE equation is based on 91,371 earthquake records from 4,967 earthquakes in Japan, California, and Turkey. For earthquakes D ≤ 30 km, according to equation (15) while for earthquakes D > 30 km, according to equation (16).

$$\log y = a_1M_w + b_1X - \log(X + d_1 10^{0,5M_w}) + c_1 \tag{15}$$

$$\log y = a_2M_w + b_2X - \log(X) + c_2 \tag{16}$$

with y is PGA (cm/s^2); a_1= 0.56; a_2= 0.4; b_1= -0.0031; b_2= -0.0039; c_1= 0.26; c_2= 1.56; d_1= 0.0055. There is a PGA correction factor based on shear wave velocity (V_{S30}). According to equations (17) and (18).

$$G = plog(V_{S30}) + q \tag{17}$$

$$\log pre_{cor} = logpre + G \tag{18}$$

with G is correction to V_{S30} (g); p = -0.55; q = 1.35 and $\log pre_{cor}$ is the corrected PGA (g). The V_{S30} value obtained from converting the SPT blows according to equation (19) from Iyisan [28].

$$V_s = 51.5N^{0.516} \tag{19}$$

with N is SPT blows.

2.3.10. Lin & Lee [29]. This GMPE equation used earthquake data characteristic of hypocenter depth 3.94 < H < 161 km. Earthquake records are from Taiwan and the addition of subduction earthquakes from Mexico, the western United States, and New Zealand. The proposed equation is according to equation (20).

$$\ln(y) = C_{L_1} + C_{L_2}M + C_{L_3}\ln R + C_{L_4}e^{C_{L_5}M} + C_{L_6}H + C_{L_7}Z_t \tag{20}$$

With y is PGA (g). At site class D & E of soil types, the following C_{L_1}: -0.9; C_{L_2}: 1.00; C_{L_3};-1.90; C_{L_4}: 0.9918; C_{L_5}: 0.5263; C_{L_6}: 0.0004; C_{L_7}: 0.31.

3. Result and Discussion

3.1. Geology and seismic review

Rinca Island is located between several active faults: Flores back arc thrust in the north, Sape strike-slip in the west, and the Bondowatu fault in the south. Among the three active faults, the Flores back arc thrust fault is the longest fault. The Flores back arc thrust fault consists of 6 segments stretching from the north of Bali Island to Wetar Island in Maluku Province (Table 3). Flores back arc thrust is a reverse fault, while Bondowatu fault is a normal fault. Flores back arc thrust has the highest slip rate compared to Bondowatu fault and Sape strike-slip, in the range of 7.0-18.3 mm/yr. This results in the Flores back arc being more dominant as an earthquake trigger. Several large earthquakes have occurred due to the Flores back arc thrust activity, such as the Lombok earthquake in 2018 with a magnitude of 6.9 M_W and with hypocenter depth < 40 km [30], the Flores earthquake on December 12, 1992, with 7.7 M_W [31] and the Flores earthquake on December 14, 2021, with 7.4 M_W [9].

Besides being caused by active faults, Rinca Island is located near the subduction zone between the Australian plate and the Eurasian plate. The subduction zone extends from the western part of Sumatra Island to the southern part of the Nusa Tenggara islands. The NTB and Bali segments are the closest segments to Rinca Island (Table 4). The potential for earthquakes due to plate subduction in both segments is up to 9.0 M_W with a historical earthquake magnitude of 7.0 M_W recorded in 1977 [32]. Loh Buaya Rinca Island is included in an area that has the potential to experience strong earthquakes with an intensity of more than VIII Modified Mercalli Intensity (MMI) or PGA of 0.34-0.65g [33] This intensity can cause ground cracks, landslides, and ground shifts.

Table 3. Fault segments around Rinca Island [32]

No.	Structure Name		Slip-Rate (mm/yr)	Type	Top	Bottom	Length (km)	M_{max}
	Main	Segment						
1.	Flores *Back Arc Thrust*	Bali	7.0	R45N	3	18	84	7.4
2.	Flores *Back Arc Thrust*	Lombok Sumbawa	9.9	R45N	3	18	310	8.0
3.	Flores *Back Arc Thrust*	West Nusa Tenggara	11.6	R45S	3	18	217	7.5
4.	Flores *Back Arc Thrust*	Middle Nusa Tenggara	11.6	R45S	3	18	173	7.4
5.	Flores *Back Arc Thrust*	East Nusa Tenggara	5.6	R45S	3	18	236	7.5
6.	Flores *Back Arc Thrust*	Wetar	18.3	R45S	3	18	216	7.5
7.	Sape *Strike-slip*	-	0.5	SS	3	18	54	6.7
8.	Bondowatu *fault*	-	0.5	Norm60S	3	18	44	6.5

Table 4. Sumba subduction zone around Rinca Island [32]

No.	Segment	Length (km)	Width (km)	Slip rate (cm/yr)	M_{max}	Historical Earthquake
1.	Bali	500	200	4.0	9.0	-
2.	NTB	400	200	4.0	8.9	M_w = 7.0 in 1977

Based on earthquake data from the USGS, it is found that 330 earthquakes have occurred in the range of 1922-2022 with magnitude limits > 5.0 M_W and epicenter distance < 200 km (Figure 1). Earthquakes that have occurred are dominated by shallow earthquakes with a total of 191 events (Table 5). Generally, Shallow earthquakes generate stronger PGA than intermediate and deep earthquakes because the energy is released closer to the surface.

Table 5. Earthquakes occurred in 1922-2022 with the boundary of M ≥ 5.0 M_w and distance R ≤ 200 km from Loh Buaya on Rinca Island [34]

Earthquake Depth	Magnitude (M)	Event	Total Event
Shallow (0-70 km)	5.0 – 5.4	117	191
	5.5 – 5.9	40	
	6.0 – 6.4	25	
	6.5 – 6.9	9	
Intermediate (70-300 km)	5.0 – 5.4	81	92
	5.5 – 5.9	9	
	6.0 – 6.4	2	
Deep (>300 km)	5.0 – 5.4	31	47
	5.5 – 5.9	10	
	6.0 – 6.4	4	
	6.5 – 6.9	1	
	7.0 – 7.5	1	
		Total	**330**

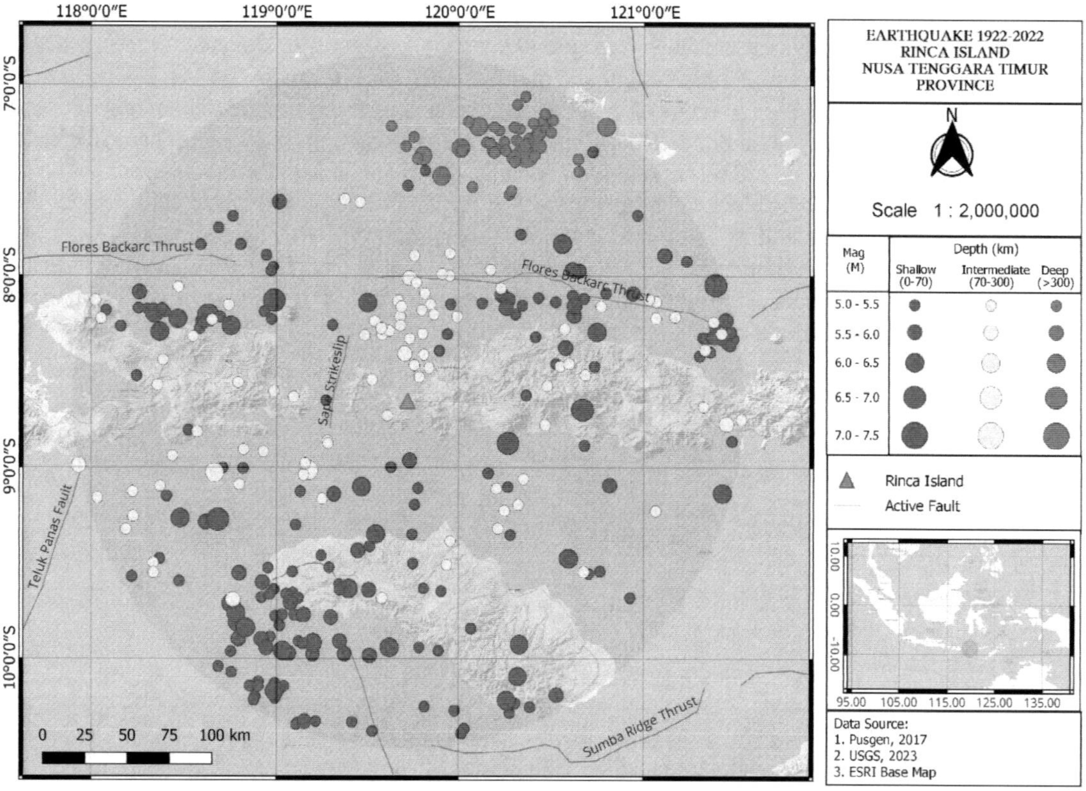

Figure 1. Earthquake epicenter points around Loh Buaya, Rinca Island for the period 1922-2022

3.2. Geotechnical investigation

Geotechnical investigation is required to calculate PGA by PSHA or DSHA. In the PSHA method, it is required to determine the site class classification. While in DSHA as in GMPE Kanno, Geotechnical investigation is required to calculate the correction factor. Standard Penetration Test (SPT) data was used as the basis for finding the V_{S30}. SPT was conducted on soil dominated by sandy soil. SPT results up to 28m obtained average SPT blows are 7.34 and V_{S30} conversion value using equation (18) are 144.06 m/s (Figure 2). Based on the following results, the site class classification is SE. This type of Ground conditions can result in PGA amplification so that the impact of shocks due to the earthquake will be more pronounced. In SPT data, the depth of data has a limitation of only 28m depth. To increase the accuracy of the correlation from SPT to V_{S30}, it would be better to collect data up to 30m in depth.

3.3. Comparison of PSHA and DSHA

A comparison of maximum PGA calculations using PSHA Pusgen and DSHA with 10 GMPE is provided in (Table 6). This comparison aims to determine which GMPE is the closest to the results of PSHA Pusgen. PSHA uses earthquake sources from a combination of shallow crustal, Benioff, and megathrust. The PSHA method found that the PGA_m was 0.59 g, magnitude 64-66 km, and earthquake distance 70-80 km. From 10 GMPEs, various results were obtained. The largest PGA was obtained using GMPE from Donovan and Bornstein of 0.85 g and the smallest from Youngs of 0.06 g. By comparing with PSHA, GMPE Kanno is the closest based on PGA, magnitude, and distance criteria. The difference in PGA is 0.01 g, the difference in distance is 5.19 km and the difference in magnitude is 0.3 M_w.

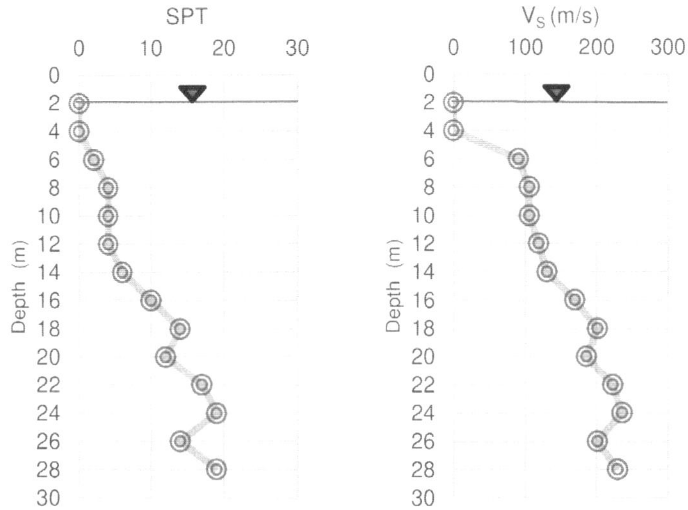

(a) (b)

Figure 2. SPT test results with sandy soil conditions with groundwater level (a) SPT blows (b) Shear wave velocity (V_s)

Table 6. Comparison of earthquake variable results between PSHA and DSHA

No.	GMPE Method	PGA *max* (g)	R (km)	M_w	PSHA Pusgen (*All Source*)		
					PGA_m	M_w	R (km)
1.	Donovan, 1973	0.08	64.91	6.8			
2.	Bornstein and Donovan, 1978	0.85	34.17	5.9			
3.	Mcguire, 1978	0.07	64.91	6.8			
4.	Campbell, 1981	0.05	64.91	6.8			
5.	Youngs, 1988	0.06	56.29	6.3	0.59	6.4 – 6.6	70-80
6.	Fukushima & Tanaka, 1990	0.10	64.91	6.8			
7.	Youngs, 1997	0.35	15.69	5.4			
8.	Garcia, 2005	0.60	118.43	5.1			
9.	Kanno, 2006	0.58	64.91	6.8			
10.	Lin & Lee, 2008	0.27	15.69	5.4			

The factor that caused GMPE Kanno close to Pusgen's PSHA results is the use of 91,371 earthquake records from 4,967 earthquakes in Japan, California, and Turkey. Employing many earthquake records will make the GMPE equation better. In addition, there is a correction factor based on the soil type at the location under review based on V_{S30} data like Kanno. This correction factor causes the PGA amplification based on soil type to be considered. In another study, the GMPE from Kanno had the best reliability using the Euclidean Distance Ranking (EDR) method by comparing four GMPEs with the PGA recorded from the accelerograph at the Mataram Geophysical Station [35]. Based on active faults and subduction zones, the Loh Buaya, Rinca Island, and Mataram City earthquakes were triggered by the Flores back arc fault and the Sumba subduction zone of the Bali and West Nusa Tenggara segments.

4. Conclusion

The earthquake that occurred around Loh Buaya, Rinca Island was influenced by faulting and subduction activities. The Flores back arc is the dominant fault that causes earthquakes because it has the largest slip rate among other faults. In addition, the subduction zone in the south of Rinca Island has the potential to produce large earthquakes. Soil conditions included in site class SE cause the PGA would be amplified so that the impact of the earthquake will be increased. Using probabilistic approach or PSHA method, the PGA value from Pusgen resulted in 0.59 g. By deterministic approach or DSHA method, GMPE Kanno is the closest to PSHA Pusgen results compared to the other 10

GMPE. The result based on the variables of earthquake distance, magnitude, and PGA. PGA values from both methods show PGA values of more than 0.5 or equivalent to VIII MMI. By considering the PGA value and soil conditions, the earthquake that occurred has the potential to cause another disaster, such as liquefaction. Further research needs to be conducted regarding the potential for liquefaction disasters.

Acknowledgment
We would like to express our sincere gratitude to the Ministry of Public Works and Housing which has supported the data and all parts of this article.

References

[1] Hunt R E 2007 *Geologic hazards: a field guide for geotechnical engineers* (Boca Raton, FL: CRC/Taylor & Francis)
[2] Jalil A, Fathani T F, Satyarno I and Wilopo W 2020 *Int. J. GEOMATE* **18** 147-55
[3] Hsu Y-J, Simons M, Avouac J-P, Galetzka J, Sieh K, Chlieh M, Natawidjaja D, Prawirodirdjo L and Bock Y 2006 *Science* **312** 1921–6
[4] Mori J, Mooney W D, Afnimar, Kurniawan S, Anaya A I and Widiyantoro S 2007 *Seismol. Res. Lett.* **78** 201–7
[5] Diambama A D, Anggraini A, Nukman M, Lühr B-G and Suryanto W 2019 *Geophys. J. Int.* **216** 439–52
[6] Wilkinson S M, Alarcon J E, Mulyani R, Whittle J and Chian S C 2012 *Nat. Hazards* **63** 521–47
[7] Wang C, Wang X, Xiu W, Zhang B, Zhang G and Liu P 2020 *Seismol. Res. Lett.* **91** 733–44
[8] Mason B 2019 *The 28 September 2018 M7.5 Palu-Donggala, Indonesia Earthquake* (GEER Association Report - 61)
[9] Hakim A R, Saputro A H, Rohadi S, Gunawan M T and Kardoso R 2022 *IOP Conf. Ser. Earth Environ. Sci.* **1047** 012019
[10] Wang A, Li S, Lu J, Zhang H, Wang B and Xie Z 2023 *Geophys. J. Int.* **234** 12–24
[11] Douglas J 2003 *Earth-Sci. Rev.* **61** 43–104
[12] BSN 2019 SNI 1726:2019 *Tata cara perencanaan ketahanan gempa untuk struktur bangunan gedung dan nongedung*
[13] Minson S E, Baltay A S, Cochran E S, McBride S K and Milner K R 2021 *Seismol. Res. Lett.* **92** 460–8
[14] Hardiyatmo H C 2022 *Rekayasa gempa untuk analisis struktur dan geoteknik* (Yogyakarta: UGM Press)
[15] Asrurifak M, Irsyam M, Budiono B, Triyoso W and Hendriyawan 2010 *Civ. Eng. Dimens.* **12** 52–62
[16] Pusgen 2022 *Peta Deagregasi Bahaya Gempa Indonesia Untuk Perencanaan Dan Evaluasi Infrastruktur Tahan Gempa* (Bandung: Ministry of Public Works and Housing)
[17] Sucuoğlu H, Sucuoğlu H and Akkar S 2014 *Basic earthquake engineering: from seismology to analysis and design* (Cham: Springer)
[18] Douglas J 2011 *Ground-motion prediction equations 1964 - 2010* (Pacific Earthquake Engineering Research Center University of California, Berkeley, CA)
[19] Donovan N C 1973 *Proc. Fifth World Conf. Earthq. Eng.* **1** 1252–61
[20] Bornstein A E and Donovan N C 1978 *Am. Soc. Civ. Eng.* **104** 869–87
[21] McGuire R K 1978 *J. Geotech. Eng. Div.* **104** 481–90
[22] Campbell K W 1981 *Bull. Seismol. Soc. Am.* **71** 2039–70
[23] Youngs R R, Day S M and Stevens J L 1988 *Earthquake Engineering and Soil Dynamics II—Recent Advances in Ground-Motion Evaluation* (ASCE) pp 445–62
[24] Fukushima Y and Tanaka T 1990 *Bull. Seismol. Soc. Am.* **80** 757–83
[25] Youngs R R, Chiou S-J, Silva W J and Humphrey J R 1997 *Seismol. Res. Lett.* **68** 58–73

[26] García D, Singh S K, Herráiz M, Ordaz M and Pacheco J F 2005 *Bull. Seismol. Soc. Am.* **95** 2272–82
[27] Kanno T 2006 *Bull. Seismol. Soc. Am.* **96** 879–97
[28] Iyisan R 1996 *Tech J Chamb. Civ. Eng Turk.* **6** 1187–99
[29] Lin P-S and Lee C-T 2008 *Bull. Seismol. Soc. Am.* **98** 220–40
[30] Yang X, Singh S C and Tripathi A 2020 *Geophys. J. Int.* **221** 758–68
[31] Okal E A 2019 *Pure Appl. Geophys.* **176** 2771–93
[32] Pusgen 2017 *Peta sumber dan bahaya gempa Indonesia tahun 2017* (Bandung: Ministry of Public Works and Housing)
[33] Suantika G and Hidayati S 2013 *Earthquake Hazard Zone Map of East Nusa Tenggara Province* (Bandung: Indonesian center of volcanology and geological hazard mitigation)
[34] United States of Geological Survey 2022 *Search Earthquake Catalog* [Accessed: February 16, 2022]
[35] Taruna R M and Setiadi T A P 2020 *JST J. Sains Dan Teknol.* **9** 20–9

IOP Conf. Series: Earth and Environmental Science 1298 (2024) 012015 doi:10.1088/1755-1315/1298/1/012015

The application of OREDA (offshore reliability data) in developing valve criticality analysis criteria

D Priyanta[1], M B Zaman[1], Semin[1]

[1] Department of Marine Engineering, Faculty of Marine Technology, Institut Teknologi Sepuluh Nopember, Surabaya 60111, INDONESIA

E-mail: priyanta@its.ac.id

Abstract. In oil and gas processing plants, extensive networks of hundreds to thousands of valves are vital in controlling the entire process. However, not all valves have the same importance in process control. Therefore, identifying and prioritizing critical valves is essential for efficient and safe operation. Assessing the importance of valves requires a risk-based approach. However, quantifying the associated risks becomes challenging, especially when limited field data makes it difficult to model the probability of valve failure. One possible solution is to use Offshore Reliability Data (OREDA) as a valuable failure probability data source. OREDA provides valuable insight into historical failure and performance data related to oil and gas equipment such as valves. To go one step further, the new proposal presented in this paper includes implementing a modified OREDA approach that considers the cumulative operating time of the valve. By considering cumulative operating hours, a more comprehensive and accurate assessment of failure probability can be achieved, enabling more informed decision-making regarding valve criticality and prioritization of maintenance.

1. Introduction

An oil and gas processing plant is a complex facility that contains a complex network of pipelines, tanks, compressors, pumps, valves, and other critical components that jointly extract, process, and refine valuable hydrocarbon resources. This equipment should be managed so that the integrity of all equipment can work properly. The integrity of all equipment or assets should be managed through asset integrity management (AIM).

The AIM ensures that plant operations, including process control, remain uninterrupted by maintaining equipment efficiency. AIM can be defined as the management system for ensuring the integrity of assets throughout the life cycle of the assets[1]. The AIM involves identifying potential risks, implementing measures to manage and mitigate them, and regularly monitoring and assessing the asset's condition to ensure it performs its intended function safely and effectively.

Of the equipment mentioned above, valves are vital in controlling the entire process. Failure of offshore oil and gas valves can have significant negative consequences and risks. Consequences include loss of property, loss of production due to plant shutdowns, health, safety, and environmental (HSE) issues such as oil and gas spills, air pollution, and human losses[2]. However, not all valves have the same importance in process control. Therefore, identifying and prioritizing critical valves is essential for efficient and safe operation. Once the critical valves are identified, the management can focus on developing maintenance strategies to keep the critical valves functioning and prevent the asset from failing.

Preventive maintenance aims to prevent equipment failures, such as valve failures. Maintenance is also referred to as the failure defense task (FDT). Therefore, the development of maintenance tasks should consider the possible failure modes of the asset. Risk-based maintenance (RBM) is a methodology to develop maintenance based on the assets' risk profile. The methods used to create maintenance tasks depend on the equipment failure mode[3]. In the context of AIM, RBM helps guide decision-making by identifying critical assets, assessing potential risks, and identifying appropriate maintenance strategies. RBM is a key tool in the AIM approach to more proactive and efficient asset management.

2. Criticality Analysis

The initial step to implement RBM is identifying the assets register and identifying the critical item. Criticality analysis refers to assigning a criticality rating to an asset based on a set of criteria. In the oil and gas industry, it is commonly used to determine the criticality of equipment before prioritizing maintenance strategies. Identifying critical equipment is known as the equipment criticality analysis (ECA). Once the critical items are identified, the maintenance task of the critical item can be developed based on its dominant failure mode[3]. There are several approaches to determining the criticality of equipment. Failure Mode Effects and Criticality Analysis (FMECA) is considered the earliest structured method in criticality analysis and then followed by other methods.

2.1. MIL-STD-1629A

The MIL-STD- 1629A is one of the earliest official guideline documents for criticality analysis[4]. This standard aims to identify potential modes of failure, their associated effects, and the criticality of allocating resources and effort to mitigation and prevention.

The FMECA Table documents the identification and definition of the system or equipment under analysis. The FMECA document includes identifying potential modes of failure for the equipment or systems, their associated causes, and the severity of each mode's impact on the system/equipment. Every failure mode can be assigned by criticality number (C_m), as shown in equation (1).

$$C_m = \beta \alpha \lambda_p t \tag{1}$$

C_m is the criticality number for a failure mode, β is the condition probability of mission loss, α is the failure mode ratio, λ_p is the part failure rate, and t is the duration of the applicable mission phase.

In the case of equipment or a system having more than one failure mode, the criticality of it can be represented by criticality numbers (C_r). The criticality number (C_r) is the sum of the failure mode criticality number (C_m), as shown in equation (2).

$$C_r = \sum_{n=1}^{j} \left(\beta \alpha \lambda_p t\right)_n \tag{2}$$

Cr is a criticality number for equipment or a system under investigation. The number of failure modes in the equipment or a system that fall under a particular criticality classification is represented by n, which can take values from 0, 1, 2, ..., j. The last failure mode in the equipment or a system under the criticality classification is represented by j. The criticality number shows the criticality of the equipment or a system relative to others.

2.2. Risk Priority Number (RPN)

An alternative metric to measure the criticality is a risk priority number (RPN), as found in [5]. This document describes the approaches contained in MIL-STD-1629A. RPN metric is practicably useful when quantitative failure rate data is unavailable. Examples of the use of RPN can be found in[6][7][8] [9]. Equation (3) shows the formula to calculate RPN.

$$RPN = S \times P \times D \tag{3}$$

S is the severity of the effect of failure, P is the probability of failure, and D is the ease of detection.

The severity of the equipment failure can also determine the asset's criticality. NORSOK Z-008 helps identify critical equipment for maintenance through its consequence classification process. The process considers the potential consequences of equipment failure, such as safety, environmental, and production impacts. Equipment with higher criticality levels requires more frequent or specialized maintenance tasks [10].

2.3. Consequence-Based Criticality Approach

A consequence-based criticality is a systematic approach to assessing the criticality of an asset based on its potential risks. It is a method of evaluating equipment's criticality based on its failure consequences. The approach involves identifying the potential consequences of equipment failure and then assigning a criticality rating based on the severity of those consequences. Equipment with higher criticality levels requires more frequent or specialized maintenance tasks. NORSOK Z-008 is one of the official reference codes for determining the consequence-based criticality of the equipment[10]. The use of this standard in determining the critical item and maintenance strategies in subsea production systems can be found in [11], while in a gas central processing plant can be found in [12].

2.4. Risk-Based Criticality Approach

Risk is also often used as the criticality metric of equipment. In this approach, the higher the risk, the more critical the equipment is. In terms of AIM, risk can be interpreted as the event of the failure of the equipment. The formal definition of failure is the loss of function of a system, structure, asset, or component to perform its required or intended function(s). In the case of risk-based inspection (RBI), failure represents a loss of containment[13].

Risk is expressed as the combination of the equipment's probability of failure (PoF) and the consequence of failure (CoF). Mathematically, the risk is the multiplication of PoF and CoF, as seen in equation (4).

$$Risk = PoF \times CoF \tag{4}$$

A risk matrix or iso-risk line is commonly a risk presentation. Examples of risk as criticality criteria in oil and gas industries can be found in [14][15].

For the screening purpose, a 3x3 risk matrix will be used to assess the criticality of the valve, as in Figure 1. The corporate policy sets the matrix dimension, probability ratings, consequence ratings, and risk level. The risk level can be determined by combining the probability and consequence rating of the assessed valve. The risk level corresponds to the criticality of the valve. The high, medium, and low risk corresponds to the equipment criticality's C1, C2, and C3, respectively.

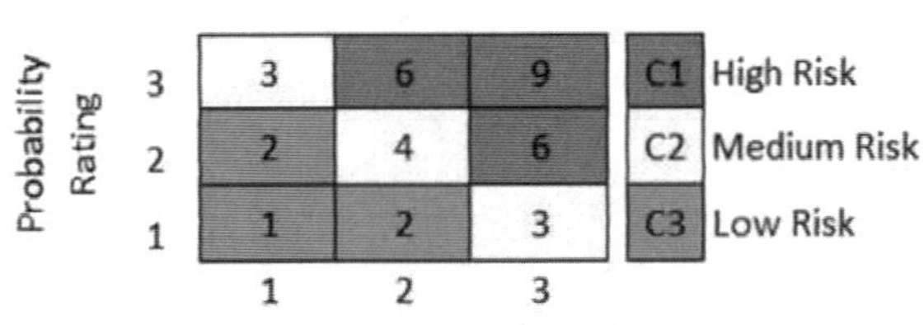

Figure 1. Risk matrix and the equipment criticality.

The probability ratings are determined based on the number of failures per year. Since the probability ratings are hard to determine based on real historical data, the secondary data will refer to the reputable source of failure database. The OREDA (offshore and onshore reliability data) is considered the most reputable reliability and failure database source. The consequence ratings are measured based on failure impact concerning the ultimate impact on plant operation, personnel injury, environmental impact, and financial loss. Table 1 and Table 2 show the probability and consequence criteria to determine the assessed valve's risk level and criticality.

Table1. Probability rating.

Rating	Descriptor	Criteria (Failures/year)
1	Improbable	$> 10^{-2}$
2	Occasional	10^{-4} to 10^{-2}
3	Probable	$< 10^{-4}$

Table2. Consequence rating.

Rating	Descriptor	Ultimate impact	Personal injury	Environmental impact	Financial lost
1	Insignificant	No disruption	First aid	Oil Spill < 10 barrel	< $ 1 M
2	Moderate	Unit/train shutdown	Hospitalized, permanent disability	Oil spill 10 – 100 barrel	$ 1M – $5 M
3	Catastrophic	Plant/total shutdown	One or more fatalities	Oil spill > 100 barrel	> 5 M

2.5. Safety Critical Equipment

Safety critical equipment (SCE) refers to equipment required for the safe operation of a facility. The equipment in the SCE list, if it is damaged, could result in a serious accident or death. In asset integrity management, SCE is identified and managed to ensure that it is fit for purpose and will continue to function safely throughout its life.

SCE regulations in Europe emphasize functionality, availability, and reliability while surviving the fire/explosion incidents they should prevent and mitigate. As a result of this consequence-based approach, SCE is often defined to include most items on the master equipment list. It is not surprising since the origins of the regulations stem from the Piper Alpha disaster[1].

On the other hand, asset integrity management may require a narrower definition of safety-critical equipment based on risk instead of consequence. Equipment criticality aims to identify the subset of equipment critical to managing major incident hazards. Before and during a major incident, this equipment ensures the prevention, control, and mitigation of major hazards. Much of this safety-critical equipment will require preventive maintenance and calibration to ensure its reliability to function on demand. Some equipment may become critical or increase in criticality during its life cycle due to deterioration and increased risk of failures, such as a corroded pressure vessel subject to a fitness-for-service assessment requiring more frequent inspections and a greater monitoring effort.

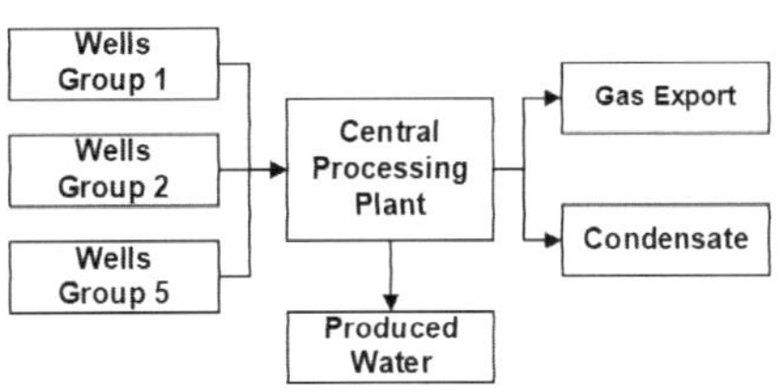

Figure 2. Block area layout of CPP.

3. The object of the study

The object of the study is the central processing plant (CPP) and connecting systems to CPP. The block area layout of the systems is shown in Figure 2. Sour gases from wells groups 1, 2, and 5 are sent to the CPP for further processing via trunkline. Wells group 3 is for future development, while wells

IOP Conf. Series: Earth and Environmental Science 1298 (2024) 012015 doi:10.1088/1755-1315/1298/1/012015

group 4 is abandoned. The CPP will deliver sweet gas to customers via a gas metering station through the pipeline. The stabilized condensate is also sent to the tanker ship in the condensate jetty station through the pipeline. The produced water is sent to the produced water tank through the pipeline. The produced water will be reinjected into the wells by the production water injection pump.

The CPP consists of an acid gas removal unit (AGRU), a gas dehydration unit, a condensate stabilization system, a produced water treatment system, and other necessary supporting systems. The gas from the well is normally wet and contains impurities. The gas needs to be purified and dried. The gas being purified is known as sour gas. An AGRU is a unit that removes hydrogen sulfide and carbon dioxide from sour gas. Amine solution is widely used to purify sour gas. The purifying process normally occurs in an amine scrubber. Lean amine is the condition of amine before it absorbs hydrogen sulfide and carbon dioxide in the gas-sweetening process. Rich amine refers to an amine's condition when it is rich in hydrogen sulfide and carbon dioxide. Rich amine is recycled by removing hydrogen sulfide and carbon dioxide from the regenerator unit.

Dehydrating the sour gas is essential after removing hydrogen sulfide and carbon dioxide. Gas dehydration units remove water vapor from natural gas. Triethylene glycol (TEG) dehydration is one of the oil and gas industry's most widely used dehydration systems because of its low operating costs and low CAPEX.

Stabilizing natural gas condensates reduces the vapor pressure to a value that ensures safety during transportation and storage, as well as separating lighter from heavier condensates. Stabilized natural gas condensate removes C_1-C_4 hydrocarbons or is fractionated in stages sequentially, removing ethane (C_2), propane (C_3), and butane (C_4). Stabilized LPG/gasoline is stripped of lighter ends, making it less flammable.

In oil and gas refineries, it also produces water. Produces water needs to be treated before it is utilized for other purposes. Hydrocyclone unit is often used in produced water treatment since it separates solids from liquids with different densities. Hydrocyclone is also able to separate liquids with different densities.

4. Valve Criticality Analysis

A risk-based criticality approach will be adopted to assign the criticality of the valves in the gas processing plant. However, quantifying the associated risks becomes challenging, especially when limited field data makes it difficult to model the probability of valve failure. The analysis is started by registering all valves. It is recommended to assess which valves are categorized as safety critical valves. The safety critical valve will be categorized as C1 or the most critical or high-risk. The other valves will be assessed further to determine whether they will be categorized as C1, C2, or C3. Figure 3 shows the proposed flowchart to conduct a risk-based valve criticality analysis.

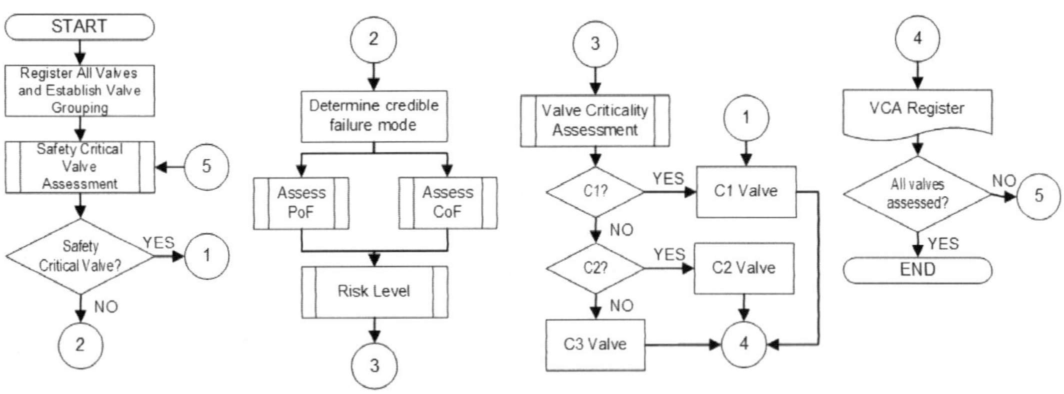

Figure 3. Flowchart to determine valve critical equipment.

Table 3. Distribution of type of valves in CPP.

Area	Ball Valve	BDV	Butterfly Valve	Check Valve	Control Valve	Gate Valve	Globe Valve	Needle Valve	Safety Valves	SDV	Total
Train 1:											
AGRU	294	3	4	19	11	18	6	0	8	13	376
Condensate Stabilization	1094	9	7	96	58	283	145	22	49	24	1787
Gas Dehydration Unit	172	0	0	11	5	10	16	0	10	3	227
Hydrocyclone Unit	37	0	0	2	4	3	7	11	1	0	65
Train 2:											
AGRU	188	1	0	14	6	54	64	21	11	1	360
Condensate Stabilization	423	5	0	21	20	27	44	0	10	13	563
Hydrocyclone Unit	34	0	0	2	4	3	7	11	1	0	62
Production and Test Manifold	109	4	0	6	0	0	15	0	4	7	145
Sales Gas Export	57	1	0	1	1	1	4	0	5	1	71
Sales Gas Separator	101	1	0	4	0	5	10	0	4	0	125
Metering at Jetty	47	0	0	0	6	4	1	0	1	0	59
Wells Group 1	69	0	0	1	7	1	0	0	2	2	82
Wells Group 2	70	0	0	10	7	1	0	0	2	2	92
Wells Group 5	50	0	1	2	7	1	1	0	2	2	66
Total	2745	24	12	189	136	411	320	65	110	68	4080

4.1. *Valves register*

The valve criticality analysis is started by documenting all the valves operating in the CPP. As many as 4080 valves are identified in the CPP. The types of identified valves are ball valves, blowdown valves, shutdown valves, butterfly valves, control valves, gate valves, globe valves, needle valves, and safety valves. Table 3 shows the detailed distribution of valves by area in the CPP.

4.2. *Safety critical valve*

A safety critical valve, as in SCE, is essential for the facility's safe operation. Figure 3 depicts the flowchart to identify the safety-critical valve (SCV). The flowchart is developed based on the one discussed in [1]. Once the valve is identified as the SCV, the valve is grouped as the most critical. Table 2 summarizes the safety critical valve in the CPP facility. The SCV is determined by using the flowchart shown in Figure 4.

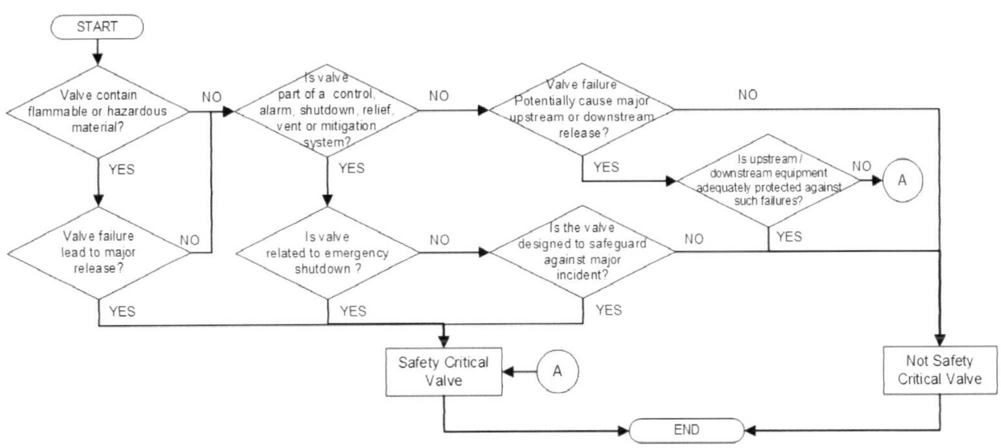

Figure 4. Flowchart to determine valve critical equipment.

Table 4. Distribution of SCV based on the type of valves.

	Ball Valve	BDV	Butterfly Valve	Check Valve	Control Valve	Gate Valve	Globe Valve	Needle Valve	Safety Valves	SDV	Total	Percentage
SCV	0	24	0	0	136	0	0	0	110	68	338	8.28%
Non-SCV	2745	0	12	189	0	411	320	65	0	0	3742	91.72%

Of the total valves, 338 (8.28%) are categorized as SCV, while the remaining 3742 (91.72%) are categorized as non-SCV. Table 4 summarizes the distribution of SCV and non-SCV based on the type of valves.

4.3. Credible failure modes

A failure can be defined as the termination or degradation of an item to perform its required function. An item's failure can occur due to one or more failure modes. On the other hand, a failure mode may happen due to one or more causes known as failure cause(s). In this case, the item denotes any hardware level in the asset hierarchy.

The OREDA (offshore and onshore reliability data) handbook is one of the most credible reliability databases of the oil and gas industry equipment. The data provided in the database covers both topsite and subsea installation.

4.3.1. The OREDA. There is a project in the oil and gas industry to collect data on equipment reliability, including failure rates and failure mode distribution. Phase I of the project was initiated in 1981 and called the OREDA (offshore reliability data) project. Phase I of data collection was initiated in 1981. Up to 1998, reliability data has been collected for approximately 24,000 units of offshore equipment, including approximately 33,000 failures. [16]. The program finished phase XIII in 2020.

In the beginning, the OREDA project collected only topside equipment of equipment in offshore platforms. The database is now extended to collect the equipment of the subsea as well as onshore installation in the oil and gas industry. The OREDA, now becoming offshore and onshore reliability data, provides valuable insight into historical failure and performance data related to oil and gas equipment such as valves.

The 6th edition OREDA handbook organizes the data based on the system and equipment class. The equipment class is the subcategory of the system. Table 5 summarizes the systems and equipment class of the OREDA handbook.

Table 5. Systems and equipment classes in the OREDA handbook[17].

Installation	Topsite				Subsea
System	Machinery	Electric Equipment	Mechanical Equipment	Control and Safety Equipment	Subsea
Equipment Class	1.1. Compressors 1.2. Gas Turbines 1.3. Pumps 1.4. Combustion Engines 1.5. Turbo Expanders 1.6. Steam Turbines	2.1. Electric Generators 2.2. Electric Motor 2.3. Battery and UPS	3.1. Heat Exchangers 3.2. Vessels 3.3. Heaters and Boilers	4.1. Fire & Gas Detector 4.2. Input Devices 4.3. Control Logic Units 4.4. Valves [a] 4.5. Valves [b]	5.1. Control System 5.2. Flowlines 5.3. Manifolds 5.4. Pipelines 5.5. Risers 5.6. Running Tools 5.7. Templates 5.8. Wellheads & X-mas Trees

a. Categorized by application
b. Categorized by taxonomy code

4.3.2. Valve failure modes. A valve failure can be total or partial, known as total failure and partial failure. The OREDA classified the severity of failure within one of the following categories: critical failure, incipient failure, degraded failure, and unknown. Critical failure causes an immediate and complete loss of an equipment unit's capability to provide output. Degraded failure is not critical but prevents an equipment unit from providing its output within specifications. Incipient failure is a failure that does not immediately cause loss of a unit's capability of providing its output but which, if not attended to, could result in a critical or degraded failure shortly. An unknown failure is a failure whose severity was not recorded or deduced.

For valve critical analysis, only critical failure will be referred. The following failure modes of valves lead to valve failure: Abnormal instrument reading, Delayed operation, External leakage - Process medium, External leakage - Utility medium, Fail to close on demand, Fail to open on demand, Fail to regulate, High output, Internal leakage, Valve leakage in a closed position, Low output, Other, Unknown.

4.4. Assessment of the valve probability of failure

The assessed valve's failure probability will be referred to the OREDA database. Databases provide valve failure data at various levels of detail. At the top level, the database summarizes all collected valve failure populations. The following levels provide more detailed information about the valve's type, application, and dimensions. Every level provides the following information: the number of valve failures, the operational time (both calendar and operation time), the failure rate, and active repair hours. For the assessment of valve PoF, only the detailed failure information matched to the assessed valve will be considered.

Table 6 shows samples of detailed information on valve reliability data for critical failure taken from the OREDA handbook. The mean, upper, and lower limit of the failure rate was estimated using equation (5), (6), and (7).

Table 6. Samples of detailed information on OREDA valve reliability data.

Section	Information	Aggregate time in service (10^6 hours)		Failure rate (per 10^{-6} hours)			Active Repair (hrs)	No. of Failure
		Calendar Time	Operational Time	Lower	Mean	Upper		
4.4.2.1.1	Application; Blowdown; Ball; 1.1-5 inch	4.258	4.2005	0.03	10.27	40	51	51
4.4.5.1.3	Application; ESD; Ball; 10.1-20 inch	0.5874	0.5867	0.001	2.28	9.68	32	1
4.5.1.13	Taxonomy code; Ball; Oil Export	0.375	0.3746	1.56	17	46.73	51	5
4.5.2.1	Taxonomy code; Butterfly; Cooling Systems	0.2425	0.2423	1		4.13	3	1

The failure rate in OREDA is estimated based on the assumption that the equipment is in a useful life. Therefore, the failure of the equipment is also assumed to be random, with a constant failure rate (λ). Statistically, the time to failure of the equipment will follow the exponential distribution. The maximum likelihood estimate (MLE) theory can estimate the failure rate of the failure data. In the case homogenous sample, that is, the failure data come from identical items operating under the same operation and environmental condition, then the maximum likelihood estimator of the parameter λ is given by equation (5)

$$\hat{\lambda} = \frac{n}{\tau} \tag{5}$$

In equation (5), n and τ represent the number of failures and aggregated time in service, respectively. The uncertainty of the estimate $\hat{\lambda}$ is presented with a 90% confidence interval. Mathematically, it is shown in equation (6).

$$P(\lambda_L \leq \lambda \leq \lambda_U) = 90\% \tag{6}$$

In equation (6), λ_L and λ_U denote the lower and upper bound of the failure rate, respectively. For exponential distribution, the upper and lower bound of equation (6) can be rewritten in terms of chi-square distribution as in equation (7).

$$P\left(\frac{1}{2\tau} z_{0.95,2n} \leq \lambda \leq \frac{1}{2\tau} z_{0.05,2(n+1)}\right) = 90\% \quad (7)$$

In equation (7), $z_{0.95,2n}$ and $z_{0.05,2(n+1)}$ symbolize the upper 95% and 5% percentile of the chi-square distribution with 2n and (2n+1) degrees of freedom, respectively.

It is suggested that the reliability data of the assessed equipment is also calculated in the same way as in the OREDA handbook. Unfortunately, most field data is not ideal. In this study, the number of failures of each valve cannot be obtained, and only the accumulating operating hours are available. It is found that the accumulative operating hours of the valve can be either more or less than that of the OREDA handbook for a similar valve. It is recommended not to disregard the accumulative operating hours of the existing valve when assessing the PoF of the valve. Therefore, the PoF represented by the valve's failure rate will be calculated by combining the failure mode listed in the OREDA handbook and the accumulative operating hour of the existing valve. The failure rate for the existing valve will be calculated based on equation (5), but it is modified to equation (8).

$$\widehat{\lambda^*} = \frac{n}{\tau^*} \quad (8)$$

In equation (8), $\widehat{\lambda^*}$, n and τ^* represent the calculated failure rate based on the actual accumulative operating hours of the analyzed valve, the number of failures of the selected valve in the OREDA handbook similar to the analyzed valve, and the actual aggregate operating hours of the analyzed valve. The upper and lower bound of the calculated failure rate can be obtained using equation (7).

Table 7. Samples of calculated failure rate based on cumulative operating hours.

Section	Information	No. of Failure	Population	Aggregated service time (10^6 Hours)	Failure rate (per 10^9 hours) Lower	Mean	Upper	Failure rate (per 10^{-3} years) Lower	Mean	Upper	Probability Rating
4.4.5.1.1	ESD, Ball, 1.1-5 inch	1	45	2.76	1.86	362	1090	16.3	3.17	9.51	2
4.4.12.1.1	Relief, Ball, 1.1-5 inch	1	372	22.8	2.25	43.8	131	0.0197	0.384	1.15	2
4.5.1.2	Ball, Condensate Processing	2	560	34.3	10.3	58.2	138	0.0907	0.51	1.21	2
4.5.1.10	Ball, Gas Processing	1	588	36.1	1.42	27.7	83.1	0.0125	0.243	0.73	2

Table 7 shows the samples of calculated failure rates based on cumulative operating hours of the homogenous sample of valves installed on the field. The plant started operating in November 2015 and has operated for 61320 hours. It means if there are 45 homogenous valves, the aggregate operating hours of the homogenous valve are 2.75 x 10^6 hours. The failure rate mean of the existing valves is estimated based on equation (8). The failure rate lower and upper bound of the actual valves is estimated based on equation (7) by substituting the appropriate parameters.

This study sets the probability ratings based on the number of failures per year. Therefore, the estimated valve failure rate unit shall be converted from failure per hour to failure per year. Only the mean failure rate will be referred to determine the probability rating of the valve. The probability rating of the assessed valve is determined based on the mean failure rate as in Table 1.

4.5. Assessment of the valve consequence of failure

Four categories of valve failure consequences should be determined. Those four categories include the ultimate impact on the plant operation, the impact on personnel health, the environmental effect, and the financial loss. Each category is governed by specific criteria outlined comprehensively in Table 2. This structured approach helps to measure the severity of the consequences of failure. The final consequence rating is determined by selecting the highest rating from the four categories analyzed. Of those four consequence ratings, the ultimate impact seems to have a more severe impact than others.

The ultimate impact of valve failure can be predicted by studying the plant operation, safety chart, process flow diagrams, piping and instrument diagrams, and other supporting documents. The impact of valve failure on the personnel is determined by examining the number of personnel working around the facilities. The environmental effect is estimated based on the released hydrocarbon when the leak has occurred. The financial loss is estimated based on production loss and repair costs. Table 8 shows the sample of consequence estimation of valve failure.

Table 8. Samples of estimated consequences of failure and ratings.

Section	Valve Tag	Ultimate Impact	Rating	Environ-mental	Rating	Injury	Rating	Financial	Rating	Final Rating
4.4.5.1.1	V2778	SCE	3	No oil spill	1	Not lead to accident	1	$4041	1	3
4.4.12.1.1	V2479	No disruption	1	No oil spill	1	Not lead to accident	1	$6730	1	1
4.5.1.2	V2767	No disruption	1	No oil spill	1	Not lead to accident	1	$7732	1	1
4.5.1.10	V2433	Unit Shutdown	2	No oil spill	1	Not lead to accident	1	$8233	1	2

4.6. Assessment of risk level and criticality of the valve

The risk level assessed valve can be determined by combining the probability and consequence ratings in a risk matrix. This study uses a 3x3 matrix, as in Figure 1. The risk level of the valve will determine the criticality of the valve. Table 9 summarizes the risk level evaluation and the criticality of the sample valves discussed in the previous sections.

Table 9. The risk level and criticality of sample valves.

Section	Valve tag	Probability rating	Consequence rating	Risk Level	Criticality
4.4.5.1.1	V2778	2	3	High risk	C3
4.4.12.1.1	V2479	2	1	Low risk	C2
4.5.1.2	V2767	2	1	Low risk	C1
4.5.1.10	V2433	2	2	Medium risk	C2

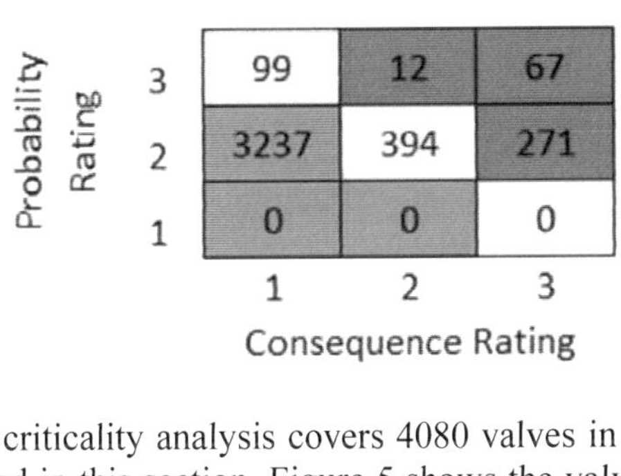

Figure 5. Valve criticality distribution based on a combination of probability and consequence ratings.

The criticality analysis covers 4080 valves in the plant. The valves are assessed similarly to those discussed in this section. Figure 5 shows the valve criticality distribution grouped by the combination of its probability and consequence ratings. As many as 3237 (79%) valves are categorized as low risk or having criticality C3. The other 493 (12%) valves are categorized as medium risk or criticality C2, and the other 350 (9%) valves are categorized as high risk or criticality C1.

5. Discussion and Conclusions

Valve criticality analysis can be determined using various methodologies. One of the proposed methodologies can be found in [18]. The methodology proposed was intended to identify critical valves in the company's water supply network based on their potential impact on customer service levels. The project uses an all-main hydraulic model of each water pressure zone to assess the criticality of thousands of valves.

Other methodologies may refer to [19]. The proposed method classified the valves based on a formally agreed priority ranking criteria described qualitatively. Three valve categories have been proposed: 1, 2, and 3. The most critical valve is designated Category 1, while the least critical valve is designated Category 3.

The first methodology mentioned above is more operational-based and designed to respond to the decreased service level. This methodology considers the impact of valve failure on the overall system and identifies valves critical to maintaining the system's service level. The second methodology is more for quick screening purposes. It is used to identify valves that require further assessment and analysis. This methodology is useful when there are many valves to assess, and a quick screening is required to prioritize the valves for further analysis.

The proposed methodology for valve criticality assessment is a risk-based valve criticality assessment that considers the operation history of the valves in running hours and the valves' operation risk. This methodology aims to identify valves at high risk of failure due to age, usage, and other factors. The results of the proposed methodology will be utilized to determine appropriate maintenance strategies for the valves.

Assessing 4080 valves using the proposed risk-based criticality methodology has resulted in valuable insights into their significance within the operational framework. The categorization of valve criticality into C1, C2, and C3 classifications serves as a comprehensive means of prioritizing maintenance efforts. Among these, C1 valves, representing the highest criticality, are of utmost concern due to their pivotal role in safety.

This study identified a notable distribution of criticality, with 9% categorized as C1, 12% as C2, and 79% as C3 valves. Identifying all safety-critical valves falling under the C1 category underscores the methodology's accuracy in recognizing potential hazards. The recommendation to prioritize high-priority maintenance for C1 valves aligns to ensure safety and operational continuity. Similarly, C2 valves, though less critical, merit significant attention.

The integration of reliability-centered maintenance (RCM) for C1 and C2 valves, considering the nature of their failure modes, is a prudent step toward optimizing maintenance strategies. Lastly, the flexibility in extending maintenance intervals for C3 valves within regulatory limits recognizes their lower impact on safety and operational integrity. This study establishes a robust framework for valuing valve criticality and paves the way for informed decision-making in resource allocation and maintenance planning within the broader operational landscape.

The discussion presented here reinforces the risk-based criticality methodology's significance in effectively managing valve assets. The clear categorization of valves based on their criticality aids in streamlining maintenance efforts and resource allocation. The prominence of C1 valves in the safety context necessitates their prioritized attention. Moreover, applying reliability-centered maintenance (RCM) to C1 and C2 valves offers a structured approach to addressing potential failure modes. While C2 valves are of lower criticality, they remain integral to operational efficiency and warrant diligent upkeep. The measured approach to C3 valves recognizes their lower impact and permits maintenance intervals to align with regulatory requirements. This study underscores the importance of adopting a holistic and systematic methodology for assessing and managing valve criticality, ensuring safe and efficient operations in the oil and gas industry.

6. References

[1] American Institute of Chemical Engineers. Center for Chemical Process Safety., *Guidelines for Asset Integrity Management.* American Institute of Chemical Engineers, 2016.

[2] K. Sotoodeh, 'A Review and Analysis of Industrial Valve Material Failures Due to Corrosion and Proposals for Prevention Measures Based on Industrial Experiences in the Offshore Sector of the Oil and Gas Industry', *Journal of Failure Analysis and Prevention*, 2020, doi: 10.1007/s11668-020-01064-9.

[3] D. Priyanta, M. B. Zaman, and Semin, 'The development of a risk-based maintenance flowchart to select the correct methodology to develop maintenance strategies of oil and gas equipment', *IOP Conf Ser Mater Sci Eng*, vol. 1052, no. 1, p. 012042, Jan. 2021, doi: 10.1088/1757-899x/1052/1/012042.

[4] United States Department of Defense, *MIL-STD-1629A MILITARY STANDARD PROCEDURES FOR PERFORMING A FAILURE MODE, EFFECTS AND CRITICALITY ANALYSIS*. 1977.

[5] Robert Borgovini, S. Pemberton, and M. Rossi, 'Failure Mode, Effects and Criticality Analysis (FMECA)', Rome, NY, 1993.

[6] B. S. DHILLON, 'Maintainability Tools', in *Engineering Maintainability*, Gulf Professional Publishing, 1999, pp. 50–81. doi: 10.1016/B978-088415257-6/50004-1.

[7] S. Medhekar, J. Moalli, and R. Caligiuri, 'Practical Risk Analysis - As a Tool for Minimizing Plastic Product Failures', in *Plastics Failure Analysis and Prevention*, William Andrew Publishing, 2001, pp. 289–295. doi: 10.1016/B978-188420792-1.50040-8.

[8] R. Kent, 'Quality Management in Plastics Processing Chapter 8 Design quality management', in *Quality Management in Plastics Processing*, Elsevier, 2016, pp. 227–262. doi: 10.1016/B978-0-08-102082-1.50008-3.

[9] D. R. Kiran, 'Failure Modes and Effects Analysis', in *Total Quality Management Key Concepts and Case Studies*, Butterworth-Heinemann, 2017, pp. 373–389. doi: 10.1016/B978-0-12-811035-5.00026-X.

[10] NORSOK, 'NORSOK Z-008:2017 Risk-based maintenance and consequence classification', 2017. [Online]. Available: www.standard.no/petroleum

[11] E. E. Uyiomendo and T. Markeset, 'Subsea maintenance service delivery: Mapping factors influencing scheduled service duration', *International Journal of Automation and Computing*, vol. 7, no. 2, pp. 167–172, May 2010, doi: 10.1007/s11633-010-0167-7.

[12] D. Priyanta, N. Siswantoro, M. B. Zaman, and F. Prasetyo, 'The Implementation of Norsok Z-008 for Equipment Criticality Analysis of Gas Central Processing Plant', *International Journal of Marine Engineering Innovation and Research*, vol. 4, no. 1, pp. 2548–1479, 2019.

[13] American Petroleum Institute, *Risk-based Inspection API Recommended Practice 580*. 2016.

[14] A. K. Attou and Q. Ahmed, 'Asset Management Practices at Qatargas', in *Proceedings of the 1st Annual Gas Processing Symposium*, Elsevier, 2009, pp. 286–296. doi: 10.1016/b978-0-444-53292-3.50036-5.

[15] D. Priyanta, 'The Development of Equipment Criticality Analysis (ECA) Protocols of Offshore Carbon Steel Static Mechanical Equipment', *Asian Journal of Applied Sciences*, vol. 4, no. 6, pp. 2321–0893, Dec. 2016, [Online]. Available: www.ajouronline.com

[16] H. Langseth, K. Haugen, and H. Sandtorv, 'Analysis of OREDA data for maintenance optimisation', *Reliab Eng Syst Saf*, vol. 60, no. 2, pp. 103–110, May 1998, doi: 10.1016/S0951-8320(98)83003-2.

[17] SINTEF and NTNU, *OREDA Offshore and Onshore Reliability Data Handbook Volume 1-Topside Equipment*, 6th ed., vol. 1. DNV GL, 2015.

[18] R. Casey, P. F. Boulos, C. H. Orr, and C. M. Bros, 'VALVE CRITICALITY MODELING', in 8th Annual Water Distribution Systems Analysis Symposium, Cincinnati, Ohio, USA, 2006. doi: http://dx.doi.org/10.1061/40941(247)32.

[19] Energy Institute (Great Britain). Technical Team, 'Report 454 : human factors engineering in projects.', London, Jun. 2020.

IOP Conf. Series: Earth and Environmental Science 1298 (2024) 012016 doi:10.1088/1755-1315/1298/1/012016

Reliability Analysis of a Four-legged Jacket Offshore Platform: A Case Study

Hastuti[1]*, RW Prastianto[1]*, D M rosyid, F Syalsabila[1], M R Syarifudin[1].
[1]Department of Ocean Engineering, Faculty of Marine Technology, Institut Teknologi Sepuluh Nopember, Surabaya 60111, Indonesia.

*E-mail address: rudiwp@oe.its.ac.id

Abstract. Offshore oil and gas exploration and exploitation requires reliable infrastructure. One crucial issue to consider is the reliability of the structure. Structural designs generally use deterministic or fixed values in their calculations. However, the factors involved in the design process may be random variables, such as environmental loads, materials, dimensions, and operational conditions, which may vary. Therefore, when calculating the reliability of a structure, these random variables must be treated with the science of statistics and probability. This research uses probabilistic and statistical methods via the Monte Carlo method to analyze the reliability of offshore jacket structures. The jacket platform is destined for the four-legged Java Sea around Gresik, Indonesia. The research results show that the Jacket platform has a high level of reliability for operating loads of K = 0.9883 and extreme loads of K = 0.9963 based on reliability analysis using the Monte Carlo Simulation method. Although the reliability figure does not exceed 1, this can be taken as a sign that the Jacket platform is declared safe in the analyzed situations. This reliability analysis ensures that the jacket platform is reliable and safe in dealing with the various environmental conditions and loads encountered during its operational life.

1. Introduction

Oil production offshore requires high technology. One type of offshore surfing platform is floating (semisubmersible) and fixed (jacket structure and gravity structure). In Indonesia's shallow waters, the commonly used platform is the fixed type, specifically the jacket structure. The platform uses a configured member and pile configuration to withstand loads of wind, current, and waves. Designing a jacket structure requires design criteria as a reference in designing the structure. The strength of the basic jacket structure on an offshore platform depends on the strength of the tube joints connecting the jacket legs and the retaining wire. Accuracy in calculating statically the strength of this tube connection has a very important meaning to ensure that the structure remains intact and functions properly during operation [1]. The American Petroleum Institute has issued regulations on the planning, designing, and constructing of a jacket platform called the API RP 2A – Working Stress Design (WSD) [2].

Design the jacket structure and find the initial structural configuration for the foundation jacket structure to maximize its structural rigidity. It can also provide sufficient resistance to extreme design loads and improve their dynamic performance [3]. In addition, it requires the ability to achieve the desired platform goals in operational and extreme conditions throughout its operating life, namely the reliability of the structure. The reliability of the jacket is measured by the reliability index and the probability of failure [4]. Deterministic analysis using a security factor approach is no longer sufficient consideration. Especially in offshore structures that face the irregular or random nature of the type of load. A probabilistic analysis must be presented to deal with this problem and provide the necessary information for optimal design [5].

Statistical uncertainty is indeed introduced in parameter estimates which are often involved in probabilistic models used for reliability analysis. When these parameters are not known with certainty, the probability of failure and the reliability index are also variables whose value cannot be ascertained [6].

The reliability assessment of the platform is conducted comprehensively, taking into account the deformation and energy of the jacket platform. Some of the random variables in this study include Young's modulus, the yield strength, the strain hardening rate; the diameter and thickness of members; the water depth; the wave height and period; the wave velocities on the surface, middle, and base of the platform; the drag coefficient; the inertial coefficient; and the horizontal concentrated scaled loads. It determines how performance functions are affected by these variables and identifies variables that are considered in platform reliability and performance analysis [7]. Wave height, current velocity, and wind speed are factors that influence the uncertainty of environmental loads in offshore platform design. Environmental load factors suitable for Indonesian waters are required to design structures that can withstand the specific conditions in those waters. That requires a structural reliability analysis. The reliability index indicates the probability that the criterion structure performance is met, known as structural reliability [8].

The basic concept of reliability analysis states that capacity and load factors, along with the assumptions used in the analysis, are statistical quantities with average values and standard deviations, and follow certain probability distributions such as normal, log-normal, or other probability distributions. Reliability is the opposite of the probability of failure. In reliability analysis, it is to evaluate the extent of this reliability to be able to understand how safe or reliable a structure or system is and, in many cases, to ensure that the risk of failure remains within acceptable limits [9].

Using a higher target reliability index might result in a greater degree of confidence in system security. However, this approach can also lead to heavier structures and result in unnecessarily higher costs. Therefore, adjustments were made to establish a reasonable target value of the reliability index sufficient to achieve the desired confidence level without sacrificing structural efficiency [10]. For a given configuration, member reliability, material, and loading statistics are designed standard deviation graphs that can be constructed for various structural index values of the model. For accurate and effective load estimation, stochastic procedures are based on reliability methods and Monte Carlo simulation (MCS) [11].

This research aims to evaluate the level of reliability in jacket structure design. The critical factor to consider is system reliability, essential in determining the optimal structural configuration. When the system being studied involves variables or parameters that have random values or experience random fluctuations, the simulation method is an appropriate choice. Through the application of Monte Carlo simulation methods, the reliability of structures can be assessed by considering uncertainties that may arise in variables or parameters during the design process. This approach helps find a more reliable and optimal solution to achieve the desired jacket structure.

2. Research Methods

The methodology used in this study is shown in the flowchart in **Figure 1**.

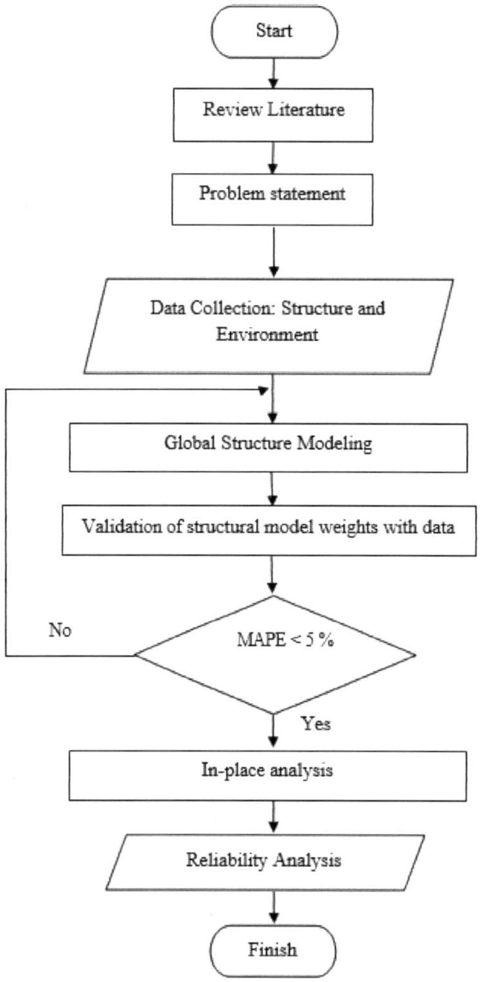

Figure 1. Research flowchart

In this research, modeling, and loading were carried out on the structure of the jacket platform operating in the Java Sea around Gresik, Indonesia which has four legs. The modeling was conducted using finite element method-based software and analyzed using the API RP 2A WSD code. In modeling consider environmental loads such as currents, waves, and wind. Structural inspection is carried out using inplace analysis which is validated according to the standards used.

In the next process, the examination of the results of the inplace analysis is a critical member. Reliability analysis will be carried out on this member to obtain reliability under the conditions of member stress performance in inplace analysis. Reliability analysis using the Monte Carlo simulation method is a powerful approach to determine the reliability of the structure. Through reliability analysis using the Monte Carlo simulation method, we can obtain more robust information about the reliability and performance of structures in various scenarios and load variations. This helps in identifying and mitigating potential problems or risks of failure, as well as designing safer and more reliable structures.

Structural data in this study are presented in **Table 1** below:

IOP Conf. Series: Earth and Environmental Science 1298 (2024) 012016 doi:10.1088/1755-1315/1298/1/012016

Table 1. Platform Structure Data

Description	Appurtenance
Structure Type	Wellhead Production
Location	East Java Sea
Number of Decks	4
Structure Elevation	Jacket Plan EL (-) 30 m
	Jacket Plan EL (-) 19 m
	Jacket Plan EL (-) 8 m
	Jacket Walkway EL (+) 4m
	Sub-Cellar Deck T.O.S EL (+) 9.5 m
	Cellar Deck T.O.S EL (+) 13.5 m
	Wellhead Access T.O.S EL (+) 16.9 m
	Main Deck T.O.S EL (+) 20.3 m
Leg Batter	No Batter
Platform Orientation	(+) 45°

3. Result and Discussion

3.1. Modeling of the jacket offshore platform

The jacket structure to be modeled is a platform operating in the Java Sea around Gresik, Indonesia with four legs and a water depth of about 30 m. The structure analyzed in this case will use the code, namely API RP 2A WSD. The location is at coordinates 6° 47' 50.470" S;112° 29' 18.783" E. This geometric modeling is the first step in analyzing the structure globally. In the jacket, the topside and jacket are modeled according to the existing data. The modeling results in this study can be seen in **Figure 2** below:

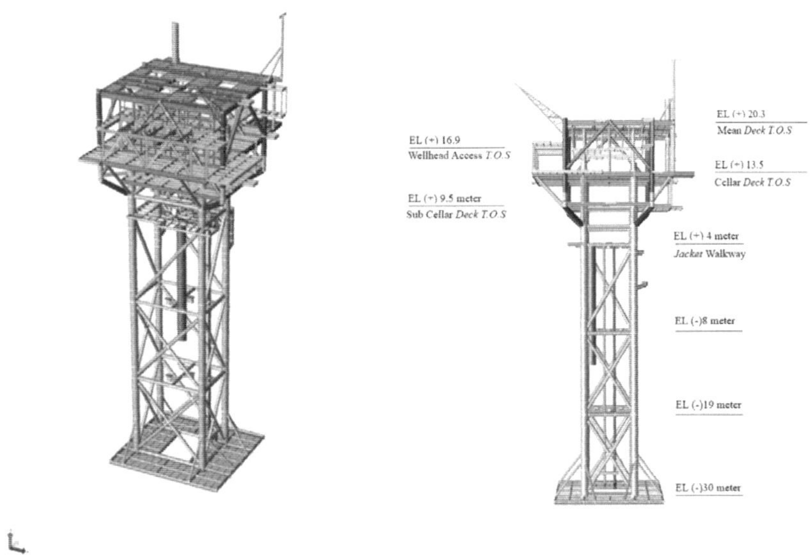

Figure 2. Modeling results for a four-legged platform jacket

3.2. Inplace Analysis

The results of in-place analysis of a structure are critical information that influences the design and operational safety of that structure. Using the unity check as a guide, the results of this analysis provide an overview of the extent to which stresses in the structure approach or exceed the allowable limits.

IOP Conf. Series: Earth and Environmental Science 1298 (2024) 012016 doi:10.1088/1755-1315/1298/1/012016

Practically, if the unity check (UC) exceeds 1, this is an important warning that the structural component cannot safely support the load.

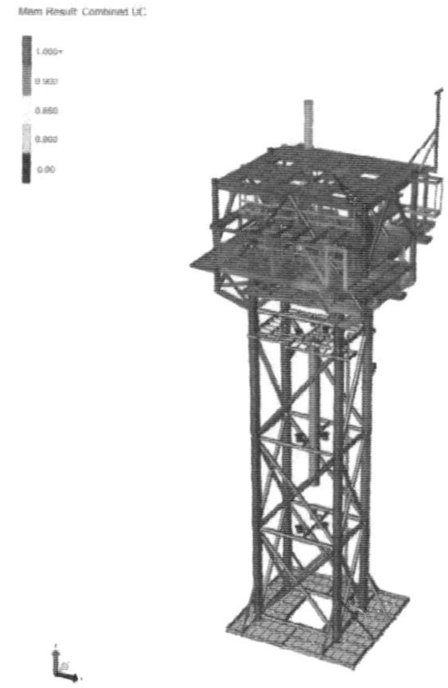

Figure 3. Inplace Analysis Results

By looking at Figure 3, which shows the results of in-place analysis, it can be concluded that no structural components have a unity check exceeding 1. Therefore, it can be concluded that the structure will not fail when operating under existing gravity loads and environmental loads. This analysis is an essential basis for ensuring the overall reliability and safety of the structure.

3.3. Reliability Analysis

Reliability analysis plays a central role in this research to evaluate the level of potential failure in the structure being studied. With an emphasis on failure patterns, this research focuses on specific parts of the structure that can potentially undergo full plasticity. Determining the failure or success of the part is based on the value of the reliability factor (MK). More specifically, some structures are categorized as "failed" if MK < 0 or MK > 1, while others are categorized as "successful" if the MK value is in the range 0 to 1. The failure mode used is equation 1, a combination of axial stress (tension or compression) and bending stress (API 2A-WSD).

$$\frac{f_a}{F_a} + \frac{\sqrt{f_{bx}^2 + f_{by}^2}}{F_b} \leq 1.0 \qquad (1)$$

In determining the failure mode above, it is essential to understand the random variables involved in the failure mode equation. The variables determined in the Monte Carlo simulation are fa, fbx, fby, and Fb, with each assumed Coefficient of Variance (COV) value (Moses, 1986), which can be seen in **Table 2** and **Table 3**.

Monte Carlo simulation transforms the Random Number Generator (RNG) for each variable, and making it a probability density function of failure is an important thing in carrying out this simulation. Transformation of random numbers into random variables in MS. Excel can be done using the following function:

- Lognormal distribution
  = LOGNORMAL.INV(random_number; mean; standard_deviation)
- Normal Distribution
  = NORMINV(random_number; mean; standard_deviation).

Table 2. Determination of environmental operating condition variables

Variable	Mean	Standard Deviation	Coefficient of Variant	Distribution
fa	4.237	0.294	0.30	lognormal
fbx	4.972	0.294	0.30	lognormal
fby	4.244	0.294	0.30	lognormal
Fa	450	67.500	0.15	normal
Fb	345	51.750	0.15	normal

Table 3. Determination of Extreme Environmental Condition Variables

Variable	Mean	Standard Deviation	Coefficient of Variant	Distribution
fa	4.334	0.294	0.30	lognormal
fbx	4.813	0.294	0.30	lognormal
fby	3.985	0.294	0.30	lognormal
Fa	450	67.500	0.15	normal
Fb	345	51.750	0.15	normal

In Monte Carlo Simulation, experiments are carried out by calling random numbers. Repetition is carried out until the target number of n simulations is reached, and a record of the number of failed structures is obtained. So, the PoF (probability of failure) equation is as follows:

$$PoF = \frac{number\ of\ failure}{number\ of\ simulation\ (n)} \quad (2)$$

The reliability of the structure can be determined by the equation below:

$$K = 1 - PoF \quad (3)$$

Monte Carlo simulations are carried out using tabulations to make it easier, as in **Table 4** and **Table 5**. To obtain accurate results, 10,000 simulations are carried out. To determine the accuracy of the number of simulations, the PoF value is recorded for each particular number so that a reliability value tends to be obtained constant.

Table 4. Reliability in Environmental Operating conditions

Σdata	ΣSucces	ΣFail	PoS	PoF	K
100	99	1	0.99	0	1.0
500	496	4	0.992	0	1.0
1000	983	17	0.9830	0	0.9830
5000	4929	71	0.9858	0.0142	0.9858
10000	9876	124	0.9876	0.0124	0.9876

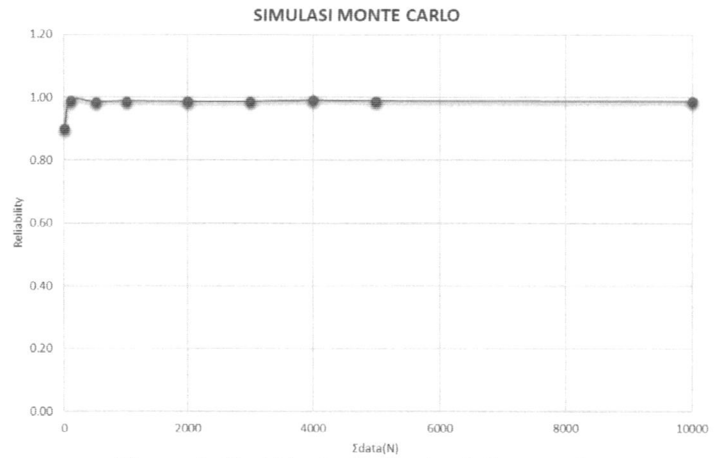

Figure 4. First Monte Carlo simulation results

Table 5. Reliability in Extreme Environmental Conditions

Σdata	ΣSucces	ΣFail	PoS	PoF	K
100	99	1	0.99	0	1.0
500	498	2	0.996	0	1.0
1000	994	6	0.9940	0	0.9940
5000	4979	21	0.9958	0.0042	0.9958
10000	9961	39	0.9961	0.0039	0.9961

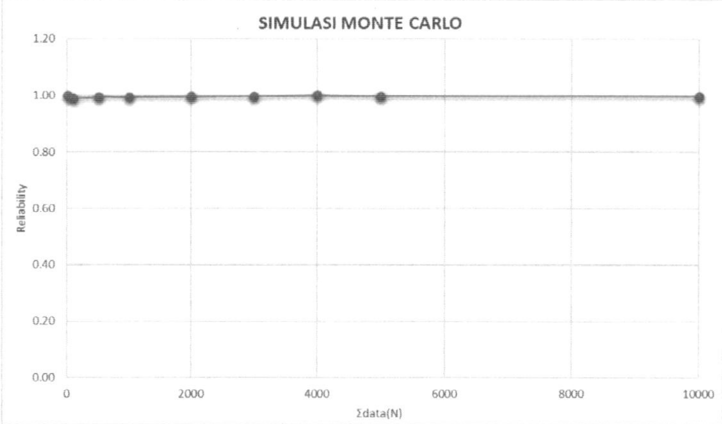

Figure 5. Second Monte Carlo simulation results

4. Conclusion

Based on the results of analysis carried out on four-legged platform jackets in the Java Sea around Gresik, Indonesia. In this research, the jacket platform has a very high level of reliability. Monte Carlo simulations were carried out 10,000 times, covering operational situations and extreme environmental conditions. The results show that the probability of failure under environmental operating conditions, the probability of failure (PoF) is 0.0039 with a reliability factor (K) of 0.9883, while in extreme environmental conditions, the probability of failure (PoF) is 0.0037 with a K of 0.9963. With these findings, the jacket platform has proven reliable in dealing with various conditions and loads that may occur. This conclusion is critical because it indicates that the structure can maintain optimal safety and performance, especially in a challenging maritime or offshore environment. Therefore, platform jackets can be relied on as a sturdy and reliable choice for various applications.

5. Recommendation

Recommendations can be given by developing a structural reliability analysis that takes into account the potential for failure due to fatigue, so it is necessary to analyze the combined ultimate and fatigue limit reliability.

6. References

[1] Primastuti R A, Prastianto R W, Alfian D, Susanto I, Fadlianto P and Syalsabila F 2023 Comparative Study of Tubular Joint Design Formulas According to API RP 2A WSD 21st Edition and 22nd Edition *IOP Conf Ser Earth Environ Sci* 1198 012018
[2] API RP 2A-WSD 22nd Ed 2014 API RP 2A-WSD 22nd Ed. Planning, Designing, and Constructing Fixed Offshore Platforms - Working Stress Design. API Recomm Pract. 2014;(November 2014)324.
[3] Tian X, Wang Q, Liu G, Liu Y, Xie Y and Deng W 2019 Topology optimization design for offshore platform jacket structure *Applied Ocean Research* 84 38–50
[4] Kurian V J, Goh S S, Wahab M M A and Liew M S 2014 Reliability assessment model for aging jacket structures in Malaysian waters *Applied Mechanics and Materials* vol 567 (Trans Tech Publications Ltd) pp 283–8
[5] Wisudawan A, Rosyid D M and Baihaqie M L 2017 Reliability analysis of APN-A offshore jacket using Monte Carlo finite element method *Applied Mechanics and Materials* 862 259–64
[6] Nava-Viveros I and Heredia-Zavoni E 2018 Assessment of statistical parameter uncertainty in the reliability analysis of jacket platforms *Ocean Engineering* 166 370–9
[7] Song Z F 2017 Reliability Evaluation of Jacket-type Offshore Platforms Subjected to Wind, Wave, and Current Loads *IOP Conference Series: Earth and Environmental Science* vol 86 (Institute of Physics Publishing)
[8] Hermanto M F and Nandalianadhira N 2022 Kehandalan Struktur Anjungan Lepas Pantai Terpancang Empat Kaki di Perairan Natuna *Rekayasa* 15 283–91
[9] Tawekal R L and Heriana R 2007 *Analisis Kehandalan Anjungan Lepas Pantai Tipe Jacket Berdasarkan Kapasitas Fatigue pada Sambungan* vol 14
[10] Karadeniz H, Toğan V and Vrouwenvelder T 2009 An integrated reliability-based design optimization of offshore towers *Reliab Eng Syst Saf* 94 1510–6
[11] Lee Y S, Choi B L, Lee J H, Kim S Y and Han S 2014 Reliability-based design optimization of monopile transition piece for offshore wind turbine system *Renew Energy* 71 729–41

Modelling a sorghum logistics as a substitution of wheat to encourage food diversification program and enhance the productivity of local crops

F. Hadi[1,2], D. V. Devintasari[1,3], A. Suryoatmojo[1,4]

[1] Marine Transportation Engineering, Sepuluh Nopember Institute of Technology, Surabaya, Indonesia

e-mail: [2] firmantohadi@seatrans.its.ac.id, [3] dika.virginia@its.ac.id, [4] abyans.2203@gmail.com

Abstract. the current diet trend in Indonesia suggests noodles made from wheat as the alternative to rice. This condition leads Indonesia to import the wheat as much as 11.5 million tonnes per year. In fact, Indonesia has become the world's second largest wheat importer and ranks among East Asia's largest staple food importers, mainly for wheat. The distribution of wheat commodity is being hampered due to the conflict between Russia and Ukraine, both of whom are major wheat suppliers in global market. As a solution, President of Indonesia, declaring a means to utilise other crops or implementing food diversification by using sorghum. Sorghum is a grain that able to thrive in tropical and dry areas, such as Indonesia and is possible to substitute that of some wheat. This final project aims to plan door to door sorghum logistic model whenever used as a substitution of wheat in a dry noodle factory located in Sidoarjo Regency by acquiring the lowest cost and a scheduling of sorghum deliveries to fulfil daily demand of the factory itself. The method applied to model the logistic planning is Vehicle Routing Problem (VRP) to determine which scenario giving the most minimum cost. The result of the analysis performed showing that the operational pattern chosen is port to port with voyage charter option obtained a freight cost of Rp17.549.076.988 or a unit cost of Rp558.571 /ton. The scheduling model shows that scenario of warehouse renting in every field is chosen and by doing the sensitivity analysis giving the total cost of Rp11.566.026.812 which includes trucking cost, warehouse cost, and stacking cost. The result shows the overall cost of the sorghum logistic model is Rp29.115.103.800 or equal to Rp926.708 /ton. Furthermore, the partial substitution of sorghum flour for wheat flour results in an 9,81% rise in production costs

1. Introduction

Wheat is one of the most produced and consumed in all around the world, leading the market alongside the rice and corn in terms of production and sales. In addition to rice, the current diet trend in Indonesia suggests noodles made from wheat as the alternative to rice. This condition leads Indonesia to import the wheat as much as 11.5 million tonnes per year [1]. In fact, Indonesia has become the world's second largest wheat importer and ranks among East Asia's largest staple food importers, mainly for wheat. Besides, Indonesia depends on imports due to domestic production of grains is close to zero (the climate does not fit the cultivation). [1]

Both Ukraine and Russia are the main producers and exporters of wheat. On February 24, 2022, there have been a conflict between Russia and Ukraine that give impact on the precarious food security situation [1]. Indonesia imports wheat from Ukraine which is utilized primary component in creating bread and noodles in a quick. Ukraine is Indonesia's largest supplier of wheat, about 30% of wheat imports from Ukraine.

According to the president, Indonesia needs food diversification and alternative food to face the food crises mainly for Wheat. In recent days, the use of sorghum can be utilized as the substitute of the wheat flour since sorghum is in the same family as wheat and has similar characteristics to wheat [3]. As a food material, sorghum has a high nutritional content because of its rich macronutrients and micronutrients, including mineral and water [3]. Sorghum is also gluten free, therefore it is safe for people who have gluten allergies. In terms of its cultivation and production, sorghum plants were cultivated without tillage farming system, no irrigation system (water required only $^1/_3$ of cane's and $^1/_2$ of corn) and used only limited fertilizer. All the parts of the sorghum are economically utilized, such as stalks and leaves used for livestock feed. The average production of the sorghum seeds is between 8 and 9 ton/ha [1].

Based on data of The Ministry of Agriculture, the area of the production of Sorghum in Indonesia spreads out to the six main provinces. There are East Nusa Tenggara that has around 3.400 hectares, West Java that has about 488 hectares, West Kalimantan 305 hectares, East Java 200 hectares, West Java 120 hectares and West Nusa Tenggara starts with 100 hectares. Therefore, the total cultivation area of Sorghum is more than 4.355 and the total production is 15.243 tons with the average production of 3,63 ton per hectare in 2022. Start in 2023, Government will expand the cultivation area in West Nusa Tenggara and set the area around 154.000 hectares in 2024. With the sorghum development and supply in Indonesia, the government expects to reduce wheat import by 11 (eleven) million tons in 2024 [4].

Nevertheless, Postharvest handling processes is quite challenging in term of the processing itself and the equipment. At the primary production level, sorghum processing has numerous steps such as threshing, polishing and grinding which all those processes are more complicated than rice and grain. All the steps are hard to handle due to the limited number of tools that do not meet the needs. However, there is a time constraint to storage the Sorghum, Sorghum seeds are easily damaged in storage for more than 3 months (due to pets, bugs, etc.), the ideal duration is 1 to 3 months. Storage of the silo model at cold temperatures can prevent the incident of damage during the sorghum storage time. As sorghum requires time consideration during the storage and the distribution time such that it can be used in an optimum quality by the factories, supply and logistics model of Sorghum is required to be built. Moreover, the distribution of sorghum is cross-island, so that the sea transportation model is urgently needed.

2. Methodology

The research methodology will be conducted as:

2.1. Problem Identification

The current issue related to the sorghum will be identified. The wheat import activity will be hampered due the impact on Russia and Ukraine conflict which lead to decreasing number of wheats in Indonesia. In addition, it is found that there is a substation of wheat, sorghum, that expedite to the scenario analysis of sorghum distribution.

2.2. The Data Collection and Process

Primary and secondary process of data collection will be conducted. Numerous data required are cultivating area, number of productions, ships/mode specifications, etc.

2.3. The Data Analysis

Data analyses are divided into supply-demand analysis, Scenario analysis, optimization, and scheduling.

2.3.1. Supply Analysis. The analysis is undertaken to obtain the amount of sorghum production in term of locations and area. There are 18 (eighteen) supply side area in the West Nusa Tenggara (NTB) and East Nusa Tenggara (NTT). There is the formulation to find the amount of sorghum production:

$$T_s = T_g.X_s.C_s.R_s \tag{1}$$

define,

S = Sorghum Productivity (ton)
A_s = Cultivation area (hectare)
P_s = Sorghum Productivity per Cultivation area (ton/hectare)

2.3.2. Demand Analysis. The Amount of sorghum required by the noodle factory in Sidoarjo. History data are required to undertake the analysis. There is the formulation to find the amount of the demand,

$$S = A_s.P_s \tag{2}$$

define,

Ts = The amount of substitution demand (ton)
Tg = The amount of the latest demand (ton)
Xs = Percentage of sorghum needed (%)
Cs = Flour composition in one-gram noodle
Rs = Ratio of sorghum required per one-gram Sorghum Flour

2.3.3. Logistics Model Analysis. Planning of the shipping sorghum using the maritime transportation. Type of vessel operated are container ship, bulk carrier, general cargo, and Ro-Ro Vessel. Below is the scenario analysed in this research:

- Scenario 1 (port to port operated undertake the time charter)

In this scenario 1, there are 5 (five) routes in where container ships carry cargo (sorghum) from five origins to a destination, Tanjung Perak. Each route will be served by a ship that has different specifications depend on the supply of the origin. A time charter as a type of vessel charter categories in this scenario.

- Scenario 2 (port to port undertake the voyage charter)

In this scenario 2, there are 5 (five) routes where container ships carry cargo (sorghum) from five origins to a destination, Tanjung Perak. Each route will be served by a ship that has different specifications depending on the supply of the origin. A voyage charter as a type of vessel charter categories in this scenario.

- Scenario 3 (Multi port undertake the time charter)

In this scenario, there will be 325 route permutation in which each port must be visited once in a one or more route. Each route will be served by a ship that has different specifications depending on the supply of the origin. A time charter as a type of vessel charter categories in this scenario.

- Scenario 4 (Multi port undertake the voyage charter)

In this scenario, there will be 325 route permutation in which each port must be visited once in a one or more route. Each route will be served by a ship that has different specifications depending on the supply of the origin. A time charter as a type of vessel charter categories in this scenario.

- Scenario 5 (Hub and spoke undertake the time charter)

In this scenario, there will be 2 types of routes, a route from origin to hub and a route from hub to destination. Each route will be served by a ship that has different specifications depending on the supply of the origin. A time charter as a type of vessel charter categories in this scenario.

- Scenario 6 (Hub and spoke undertake the voyage charter)

In this scenario, there will be 2 types of routes, a route from origin to hub and a route from hub to destination. Each route will be served by a ship that has different specifications depending on the supply of the origin. A voyage charter as a type of vessel charter categories in this scenario.

- Scenario 7 (Port to Port undertake Container Liners)

In this scenario, sorghum will be shipped by container ship. Unit cost obtained by existing data in West Nusa Tenggara and East Nusa Tenggara. Cost components calculated are ocean freight and cargo handing cost an

2.3.4. Operational Shipping Analysis. Total shipping cost will be calculated. In the time charter, the cost components are voyage cost, cargo handling cos., and time charter rate. Besides, in the voyage charter, the cost components are voyage charter rate and cargo handling cost. Ship operational consist of sea time, port time (the duration ship in the port area including the cargo handling time), roundtrip, frequency and the amount of cargo carried which will impact the load factor of each ship on each route. The formulation to calculate the total cost in terms of Voyage Charter given:

$$TC = \sum_{i=1}^{A}\sum_{j=1}^{B}\sum_{k=1}^{C} X_{ijk}(Y_{ijk}.VCH_{ijk}.f_{ijk}) \tag{3}$$

define,
T_s = Number of Substitution demand (ton)
TC = Total Transportation Cost (Rp/year)
VCH_{ijk} = Voyage charter hire rate of ship-*i* from the origin port-*j* dan to the destination port-*k* (Rp/year)
f_{ijk} = Frequency of ship-*i* from the origin port-*j* dan to the destination port-*k* (times/year)
X_{ijk} = Decision Variables of ship-*i* from the origin port-*j* dan to the destination port-*k*
Y_{ijk} = Binary Decision of Charter Scheme for ship-*i* from the origin port-*j* dan to the destination port-*k*

The formulation to calculate the total cost in terms of Time Charter given:

$$TC = \sum_{i=1}^{A}\sum_{j=1}^{B}\sum_{k=1}^{C} X_{ijk}(Y_{ijk}.TCH_{ijk}) \tag{4}$$

define,
TC = Total Transportation Cost (Rp/year)
TCH_{ijk} = Time charter hire rate of ship-*i* from the origin port-*j* dan to the destination port-*k*
X_{ijk} = Decision Variables of ship-*i* from the origin port-*j* dan to the destination port-*k*
Y_{ijk} = Binary Decision of Charter Scheme for ship-*i* from the origin port-*j* dan to the destination port-*k*

2.3.5. Optimization. Optimization run to obtain the optimum number of ships, operational planning, charter type that all leads to minimizing the total transportation cost as the objective function. There are constraints that must be satisfy:

- The amount of demand fulfilled.

$$\sum_{i=1}^{A}\sum_{j=1}^{B}\sum_{k=1}^{C} cargo_{ijk} = demand \tag{5}$$

define,
Cargo*ijk* = The amount of cargo carried by ship-*i* from the origin port-*j* dan to the destination port-*k* *(ton/year)*
Demand = Total demand (Ton/year)

- The number of origin ports visited.

$$\sum_{j=1}^{B}\sum_{k=1}^{C} X_{ijk} \leq |S| - 1 \tag{6}$$

define,
S = Total number of origins port
X_{ijk} = Decision Variables of ship-*i* from the origin port-*j* dan to the destination port-*k*

- Bulk carrier does not compatibility in the origin D

$$\sum_{j=1}^{B}\sum_{k=1}^{C} X_{jk} = 0 \tag{7}$$

define,
X_{ijk} = Decision Variables of ship-*i* from the origin port-*j* dan to the destination port-*k*

The objective function on each charter scenario (3) and (4) minimizes the total transportation cost as the multiply of ship type and number of ships chosen along the route. Constraints (5) set the number of total cargos carried must satisfy the total number of demands. As seen from constraints (6), the total number of origins port visited must be less than the number of origin ports available, and constraints (7) tell us about the ship compatibility.

2.3.6. Scheduling. Trucking and warehouse planning is conducted. Scheduling aims to regulate stock flow at the destination warehouse to meet demand in real time every day for a period of one year. In scheduling planning, it is necessary to plan land transportation in the form of trucking which is integrated with supply points and demand points. Land transportation in the form of trucks serves to carry cargo from the original warehouse to the loading port, as well as the unloading port to the factory warehouse. There are two scenarios regarding the scheduling to compare the total cost:

- Scenario 1 (warehouse at each island)

There are warehouses on each island (Figure 1), so that each rice field must send its entire load to the warehouse. The trucking used will adjust the needs of the cargo to be carried with the specified delivery duration. Trucking costs consist of rent, salaries, and fuel costs. The warehouse area is the maximum of warehouse capacity with a margin of 5%. Warehouse costs are calculated based on the area of the warehouse requirements. Stacking costs will arise as the result of sorghum being carried from the unloading port to the warehouse, where deliveries are made periodically according to the specified delivery duration.

- Scenario 2 (Warehouse at each rice field)

There are warehouses in each rice field, so that each rice field only sends sorghum to the loading port according to the optimization results obtained from sorghum which must be carried to each island. The trucking used will adjust the needs of the cargo to be carried with the specified delivery duration. Trucking costs consist of rent, salaries, and fuel costs. Warehouse area is the maximum of warehouse capacity with a margin of 5%. Warehouse costs are calculated based on the area of the warehouse requirements. Stacking costs will arise as a result of sorghum being carried from the unloading port to the factory warehouse and loading port while waiting for all the sorghum from the rice fields to be sent in its entirety.

2.3.7. Sensitivity Analysis. A sensitivity analysis is conducted on the duration of trucking delivery at the point of origin and destination based on the selected scenario in the scheduling model. In addition, there are 2 variables whose values will be changed within a certain range, namely the duration of truck delivery at the point of origin and point of destination.

3. Existing Condition

3.1. Rice Field of Sorghum

Sorghum (Sorghum bicolour) is a cereal plant originating from Africa. In Indonesia, sorghum has been cultivated for a long time as a source of food and animal feed. Indonesia itself has similarities in topography and geography with Africa. Tropical countries with temperatures that tend to be hot and high humidity which support sorghum to thrive. However, its development is still very minimal because sorghum is less attractive to local farmers. In the food industry, sorghum is also of less interest because the taste it gives is less familiar to the public.

In the context of developing sorghum, the government is seeking programs that can increase sorghum production in Indonesia, one of which is food diversification. In an effort to reduce dependence on rice as the main source of carbohydrates, the Indonesian government is encouraging agricultural diversification by promoting sorghum. Through these programs, it is hoped that they can increase the sustainability of agricultural systems and provide alternative sources of food and income for farmers.

As an alternative food: Sorghum has become an important choice as an alternative in Indonesia. Its good nutritional content, especially fibre, protein and iron, makes it a food with high nutritional value. Sorghum is also a source of carbohydrates which can be processed into flour, wheat or cereals which are used in foods such as bread, biscuits, noodles, and baby food. Following are the characteristics of sorghum:

Table 1. Sorghum characteristics

Characteristics	Number	Unit
Productivity	8	Ton/ha
Density	0,76	Ton/m^3
Harvesting	3	Month/harvest time
Ranum	2	Times/harvest time

Sorghum can be planted and harvested throughout the year regardless of the weather and available water reserves. The sorghum harvest period is once every 3-4 months and can be cultivated up to 2 times. The soil preparation period is about 2 weeks for the next planting. Productivity varies greatly in various locations in Indonesia with an average of 8 tonnes/ha and has the potential to reach 10 tonnes/ha. The density of sorghum is 0.76 ton/m3. It takes 4 times the amount of sorghum grain to produce equivalent grades. One of the sorghum growing locations with the largest land area is in the West Nusa Tenggara (NTB) and East Nusa Tenggara (NTT) regions. The supporting factor is the dry climate found in parts of NTB and NTT. However, the harvested sorghum is only consumed by local residents, this is because its mass processing, including irrigation system, addition of fertilizer, pest and disease control, requires adequate technology. Based on data from the Directorate General of Food Crops, the following is the distribution of land that can be used to grow sorghum.

Table 2. Cultivating Area for Sorghum in NTB and NTT

Code	Council	Area (ha)
A1	Bima	250
A2	Bima	250
B3	Sumba Barat	250

Code	Council	Area (ha)
B4	Sumba Timur	50
B5	Timor Tengah Utara	50
B6	Belu	350
B7	Flores Timur	100
B8	Sikka	400
B9	Ende	350
C10	Ngada	90
C11	Manggarai	150
C12	Rote - Ndao	50
C13	Manggarai Barat	300
D14	Manggarai Timur	250
E15	Timor Tengah Selatan	200
E16	Sumba Tengah	50
E17	Sumba Barat Daya	100
E18	Kupang	200

The following is an illustration of the distribution of rice fields in NTB and NTT along with loading ports on each island:

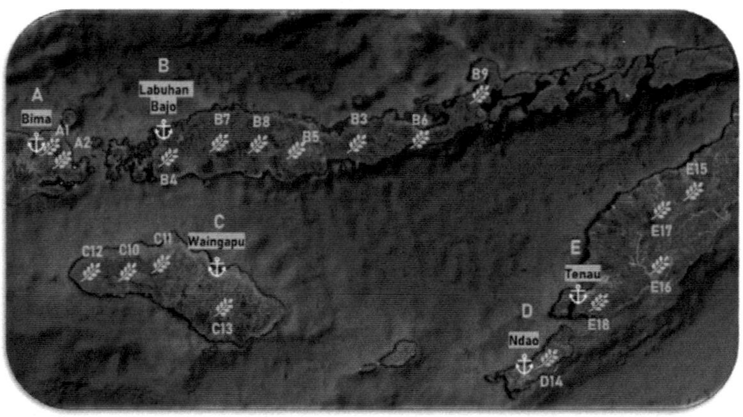

Figure 1. The Distribution of Rice Field and Ports on each Island

3.2. Factory

Sorghum can be used as a staple for food production. Its characteristics are similar to wheat, making it a multifunctional plant. Sorghum seeds can be ground into sorghum flour which is used in the manufacture of breads, cakes, cereals, and other desserts. Sorghum flour can be used as a substitute for or mixed in with wheat flour in a variety of recipes. The use of wheat can be partially substituted by sorghum flour to produce new variants with different Flavors and nutrients as well. High fiber content and low gluten have the potential to become a staple ingredient in food manufacture.

To be able to create a logistics model for sorghum, this research targets the X dry noodle factory industry, which is one of the most popular brands in the community. The factory is located in an area in East Java, namely Taman District, Sidoarjo Regency. Astronomically, the factory is located at latitude 7°21'31.3"S and 112°41'12.1"E. Factory access is an arterial road and main road that is connected to inter-provincial routes, so that large road vehicle.

Based on existing data, total dry noodle factory production was 25,696 tonnes/year or 70.4 tonnes/day in 2014. The dry noodle production process uses various adequate technology and human resource assistance. In addition, it involves several steps to produce dry noodles which include the stages of mixing, roll press, forming, steaming, frying, cooling, and packaging.

4. The Result

4.1. Supply and Demand

Based on the distribution of sorghum rice fields in the NTB and NTT regions, there is a land area of 3,440 ha. By using this approach, the production value of rice fields is 8 tons/ha. Based on these data, the amount of production for each point can be calculated using equation (1) as follows:

Table 3. The Number of Sorghum Production on Each Rice Field

Code	Kabupaten	Panen (Ton/year)
A1	Bima	6.000
A2	Bima	6.000
B3	Sumba Barat	6.000
B4	Sumba Timur	1.200
B5	Timor Tengah Utara	1.200
B6	Belu	8.400
B7	Flores Timur	2.400
B8	Sikka	9.600
B9	Ende	8.400
C10	Ngada	2.160
C11	Manggarai	3.600
C12	Rote - Ndao	1.200
C13	Manggarai Barat	7.200
D14	Manggarai Timur	6.000
E15	Timor Tengah Selatan	4.800
E16	Sumba Tengah	1.200
E17	Sumba Barat Daya	2.400
E18	Kupang	4.800

Based on the table above, harvesting is done in 3 months once as many as 3 times in one year. The total sorghum harvest at one time is 27,520 tons.

The demand is obtained by projecting existing production data on the growth rate of noodle consumption in Indonesia in the last 5 years. The following data consumption of noodles:

Table 4. Noodle Consumption in Indonesia in 2018-2022

Year	Noodle (Million-Unit)
2018	12.540
2019	12.520
2020	12.640
2021	13.270
2022	14.260

The noodle consumption growth was 3.31%. By using the projection, the demand for dry noodles in 2014 was 34,449 tons/year. Referring to equation (2), a substitute demand value of 31,418 tons/year is obtained.

4.2. Optimization on Operational Shipping

In planning sea transportation, the following is the ship data used:

Table 5. Ship Data

Code	DWT	Payload (Ton)	Vs (knot)	Daya ME (kWh)	Daya AE (kWh)
GC1	3.700	3.330	11,3	1.350	338
GC2	1.499	1.349	18	1.400	350
GC3	4.557	4.101	14	1.450	363
GC4	8.074	7.267	15,7	3.000	120
GC5	12.352	11.117	15,4	3.900	975
GC6	17.675	15.908	11,2	4.201	1.050
BC1	2.732	2.459	11	2.178	545
BC2	4.385	3.947	11	2.099	525
BC3	2.848	2.563	16,1	940	235
RO1	7.001	590	16,3	4.707	810
RO2	7.070	690	15,3	13.580	3.395
RO3	7.376	340	23,5	12.640	4.200

4.2.1. Scenario 1 (port to port undertake time charter). In scenario 1, the selected vessels are GC2 for the APA, BPB, CPC, DPD, and EPE routes with the respective frequencies of 4, 11, 4, 3, and 4 voyages. The load factors obtained for each route were 84.62%, 95.39%, 99.85%, 56.41% and 93.08%. The total cost obtained is IDR 35,883,276,615.

4.2.2. Scenario 2 (port to port undertake voyage charter). In scenario 2, the selected vessels are GC2 for the APA and EPE routes, and GC3 for the BPB, CPC and DPD routes. The frequency of each route is obtained sequentially as many as 4, 4, 2, 1, and 4 times the voyage. The load factors obtained for each route were 84.62%, 86.29%, 65.69%, 55.67% and 93.08%. The total cost obtained is IDR 17,549,076,988.

4.2.3. Scenario 3 (multi-port undertake time charter). In scenario 3, the selected ship was obtained, namely GC1 with the ABCEDPA route with a total distance of 1,349 nm. The frequency of sailing is 30 times. The load factor obtained is 79.39%, so the value of the cargo transported is 1,047 tons / RTD. The total cost obtained is IDR 61,264,155,028.

4.2.4. Scenario 4 (multi-port undertake voyage charter). In scenario 4, the selected ship is obtained, namely GC1 with the BDCAEPB route with a total distance of 1,600 nm. The frequency of sailing is 27 times. The load factor obtained is 85.12%, so the value of the cargo transported is 3,491 tons / RTD. The total cost obtained is IDR 19,768,711,668.

4.2.5. Scenario 5 (hub and spoke undertake time charter). In scenario 5, the selected vessels are GC1 for the BAB, CAC, DAD, and EAE routes with a frequency of 53, 5, 2, and 2 cruises respectively, and GC4 for the APA route with a frequency of 5 times. The load factors obtained for each route sequentially, namely 5.96%, 3.95%, 3.23%, 3.04% and 2.84%. The total cost obtained is IDR 192,454,176,564.

4.2.6. Scenario 6 (hub and spoke undertake voyage charter). In scenario 6, the selected ships are GC3 for the BAB, CAC, DAD, and EAE routes, and GC6 ships for the APA route with the respective frequencies of 4, 2, 1, 2 and 2 voyages. The load factors obtained for each route sequentially, namely 86.29%, 65.69%, 55.67%, 61.24% and 98.75%. The total cost obtained is IDR 19,524,248,234.

4.2.7. Scenario 7 (Liner Containerization). In scenario 7, the value of ocean freight is IDR 18,045,206,062 and CHC is IDR 3,931,500,000. The total cost for the container scheme is IDR 21,976,706,062 or a unit cost of IDR 699,499 / ton.

Based on the comparison of scenarios, scenario 2 is obtained, namely port to port with a voyage charter scheme is an option with a minimum cost of IDR 17,549,076,988.

4.3. Scheduling

4.3.1. Scenario 1 (Warehouse on each island). In scenario 1, the trucking value is IDR 11,142,132,024 and the total warehouse rental fee is IDR 4,170,506,298. The cost of stacking is IDR 16,672,661, so the total cost for scenario 1 is IDR 15,329,310,984.

4.3.2. Scenario 2 (Warehouse on each Rice Field Area). In scenario 2, the trucking value is IDR 7,449,939,892 and the total warehouse rental fee is IDR 4,523,883,680. The cost of stacking is IDR 28,775,536, so the total cost for scenario 2 is IDR 12,002,599,108.

Based on a comparison of scenarios, scenario 2 was chosen, namely a warehouse in each rice field is an option with a minimum cost of IDR 12,002,599,108. The following is a stock flow diagram in the factory warehouse:

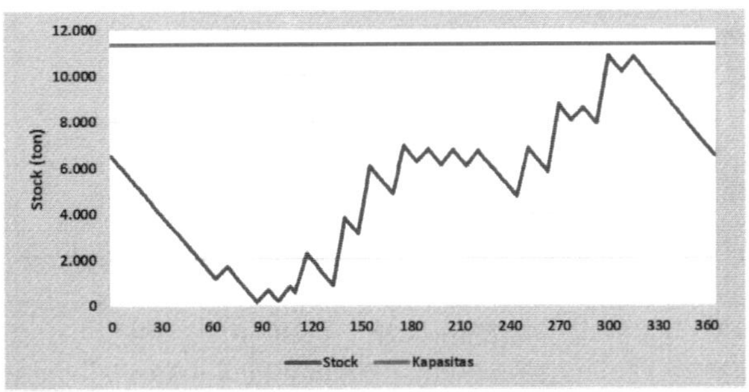

Figure 2. Warehouse stock flow

4.4. Sensitivity Analysis

Sensitivity analysis was carried out to determine the duration of truck delivery that provides the minimum cost, both for trucks at the point of origin and destination. The following is a sensitivity analysis table for trucks at the origin:

Table 6. Sensitivity Analysis Table for Trucks at the origin

Duration (Days)	Truck Cost (Million-Rp)	Warehouse Cost (Million-Rp)	Stacking Cost (Million-Rp)	Total Cost (Million-Rp)
1	Rp7.031,1	Rp4.513,2	Rp21,7	Rp11.566,0
2	Rp7.064,4	Rp4.513,2	Rp24,1	Rp11.601,6
3	Rp7.105,2	Rp4.513,2	Rp25,6	Rp11.644,0
4	Rp7.144,7	Rp4.513,2	Rp26,8	Rp11.684,7
5	Rp7.230,2	Rp4.513,2	Rp27,8	Rp11.771,2
6	Rp7.251,2	Rp4.513,2	Rp28,5	Rp11.792,9
7	Rp7.365,4	Rp4.513,2	Rp29,2	Rp11.907,8
8	Rp7.287,1	Rp4.513,2	Rp29,8	Rp11.830,1
9	Rp7.378,3	Rp4.513,2	Rp30,3	Rp11.921,8
10	Rp7.449,9	Rp4.513,2	Rp30,8	Rp11.993,9

The following is a sensitivity analysis table for trucks at the destination:

Table 7. Sensitivity Analysis of Trucks's shipping Duration form origin to Destination

Duration (Days)	Truck Cost (Million-Rp)	Warehouse Cost (Million-Rp)	Stacking Cost (Million-Rp)	Total Cost (Million-Rp)
1	Rp2.926,1	Rp4.545,3	Rp11,3	Rp7.482,7
2	Rp2.952,1	Rp4.539,9	Rp14,5	Rp7.506,6
3	Rp2.926,1	Rp4.534,6	Rp16,7	Rp7.477,4
4	Rp3.004,1	Rp4.529,2	Rp18,4	Rp7.551,7
5	Rp2.926,1	Rp4.523,9	Rp19,7	Rp7.469,7
6	Rp3.004,1	Rp4.518,5	Rp20,8	Rp7.543,4
7	Rp2.926,1	Rp4.513,2	Rp21,7	Rp7.461,0
8	Rp3.108,0	Rp4.507,8	Rp22,5	Rp7.638,4
9	Rp3.004,1	Rp4.502,5	Rp23,2	Rp7.529,8
10	Rp3.056,0	Rp4.497,1	Rp23,9	Rp7.577,1

After conducting a sensitivity analysis of the two variables, a decrease in costs arising from scheduling was obtained with a value of IDR 436,572,296 or 3.77% lower.

4.5. Sorghum Production Cost

Optimal results for planning sea transportation, land transportation, and warehousing, comparisons can be made regarding the production costs of dry noodle production. However, in this study, comparisons were conducted for the raw materials in terms of the flour used only. Related to production costs for dry noodles which include materials and tools used are not included in the discussion of research problems.

To be able to calculate the cost of production of raw materials in the form of flour, it is necessary to know the existing cost components, including raw material costs, milling costs, and packaging costs. The data used is secondary data, so an approach is used to get costs with the appropriate value. The following is the cost calculated to produce 1 kg of flour.

Table 8. Flour Production Cost (Rp/Kg)

Description Cost	Wheat Flour	Sorghum Flour
Raw Material (/kg)	Rp2.900	Rp4.087
Milling (/kg)	Rp1.487	Rp1.540
Packing (/kg)	Rp260	Rp50
Total (/kg)	Rp4.647	Rp5.677

Total Production cost can be carried out according to the demand required by the factory, which is 31,418 tons. However, the flour composition used was 57% which is the amount of flour used as the main ingredient for making dry noodles. The following is a calculation to find out the needs and costs of dry noodle production in existing conditions and when substitution is made using sorghum flour.

Table 9. Flour Production Cost According to Demand

Component	Wheat Flour	Sorghum Flour
Wheat Flour (ton)	19.636	11.782
Sorghum Flour (ton)	0	7.854
Total Cost (Jt-Rp)	**Rp91.251**	**Rp99.342**

Based on the results, there is a cost difference in the existing and replacement conditions of IDR 8,091,373,208 or 8.87% more expensive. This is due to the need for sorghum, whose production costs are more expensive and shipping costs are relatively more expensive than wheat because the number of suppliers is still limited.

If implemented, the production costs will be more expensive than the existing conditions due to the limited number of cultivation and suppliers are still limited. However, there are more advantages that can be provided by substitution with sorghum when viewed from its nutritional value. The most important nutritional content is that sorghum contains gluten-free, so it is safe for consumption for people with celiac disease who cannot consume gluten which is found in dry noodles in general.

5. Summary

The conclusion of this research is that the area of sorghum land is 3,440 hectares with a production of 82,560 tonnes per year spread over 18 rice fields in the Provinces of NTB and NTT. Meanwhile, the value of demand for dry noodle factories as a destination for substituting wheat with sorghum is 31,418 tons per year. However, there was an increase in production costs for factory X which produced dry noodles by 9.91%.

In optimizing the selection of fleets, operating routes, and charter schemes, an alternative delivery method was chosen in the form of a port to port operation pattern with a voyage charter scheme, where GC2 ships were selected on the A-P-A and E-P-E routes with 4 times the shipping frequency and GC3 ships on B-P-B, C-P-C routes and D-P-D as much as 4, 2 and 1 times the shipping frequency consecutively. From these calculations, a minimum total sea transportation cost of IDR 17,549,076,988 or a unit cost of IDR 558,571 / ton is obtained.

An alternative to renting a warehouse in each selected rice field is a warehouse fee of IDR 4,523,883,680, trucking costs IDR 7,449,939,892, and stacking fees IDR 28,775,536. After conducting a sensitivity analysis of the duration of truck delivery, a cost reduction of IDR 436,572,296 or 3.77% was obtained.

The final calculation results consisting of sea transportation costs, land transportation costs, and warehousing costs show a value of IDR 29,115,103,800 or a unit cost of IDR 926,708 / ton. There is a cost difference in the existing and replacement conditions of IDR 8,091,373,208 (8.87% higher) due to the small amount of harvested sorghum on each rice field.

References

[1] Ministry of Agriculture 2022 *Statistics of Food Consumption* (Jakarta: Center for Agricultural Data and Information System)

[2] Mottaleb, K. A., Kruseman, G., & Snapp, S. (2022). Potential impacts of Ukraine-Russia armed conflict on global wheat food security: A quantitative exploration. *Global Food Security, 35.* https://doi.org/10.1016/j.gfs.2022.100659

[3] Aryani N F, Khatimah K, and Tajuddin F N 2022 *Budidaya Tanaman Sorghum* (Makassar: State University of Makassar)

[4] Directorate General of Food Crops 2021 *Annual Report 2021* (Jakarta: Directorate General

The impact on downstream policy implementation for mineral export products in Indonesia: marine transportation point of view

S D Lazuardi[1*], F Hadi[1], D V Devintasari[1], P Wuryaningrum[1], M Riduwan[1], O S Noarista[1] and S F Alifia[1]

[1] Department of Marine Transportation Engineering, Institut Teknologi Sepuluh Nopember, Surabaya, Indonesia

*dwilaz@seatrans.its.ac.id

Abstract. Indonesia is a country with abundant natural resources, particularly for metal mineral mining. In 2020, Indonesia is recorded to have approximately 1.2 billion tons of bauxite ore reserves, which made Indonesia the 6th largest owner of bauxite reserves in the world. Apart from bauxite, around 52% or 72 million tonnes of world nickel reserves are also located in Indonesia. However, most of the bauxite and nickel mining production have been exported in the form of ore or concentrate, which has a much lower selling value compared to the processed products. Therefore, the Indonesian government imposed a ban on the export of unprocessed minerals through Law Number 3 of 2020. The export ban is intended for mineral mining products, especially bauxite and nickel, to be processed first into semi-finished goods or finished goods, to provide added value in terms of foreign exchange for Indonesia. This study aims to determine the impact on the implementation of the downstream program, especially for nickel mineral product from marine transportation point of view. The method used in this study is optimizing non-linear programming to determine the type and number of fleets needed to transport nickel from the mine site to the smelter and its processed products from the smelter to the importing country by considering the minimum unit transport costs. The output of this research is the configuration of the sea transport fleet required for the implementation of the mineral product downstream program as well as the potential added value of transporting those processed products into the Indonesia's balance of payment.

1. Introduction

Indonesia is a country with abundant natural resources, one of which is metal mineral mining. These mineral reserves include nickel and bauxite [1]. This wealth then provides great potential for Indonesia, especially to create added value for the economy if it can be processed domestically. Wealth in nickel and bauxite mining products can be seen in the growth in the production of these minerals each year, which has increased by an average of 40.3% and 196%, respectively [2].

Growth in the mining sector in Indonesia, especially bauxite itself, has not succeeded in maximizing the country's foreign exchange. This is because the results of the mining sector are directly exported without being processed into work-in-processed goods or finished goods which causes the results from these mines to be less than optimal. Therefore, in 2020, the Indonesian government plans to impose a ban on the export of unrefined minerals such as concentrate through the Law on Mineral and Coal Mining which includes bauxite. The export ban applies to Law Number 3 of 2020 where the enforcement of this law will be implemented 3 years after stipulating, which is on June 10th, 2023 [3]. The export ban is intended for mineral mining products, especially bauxite, to be processed first into semi-finished goods or finished goods, to provide added value and optimize the Indonesia's Balance of Payment, in terms of export product value.

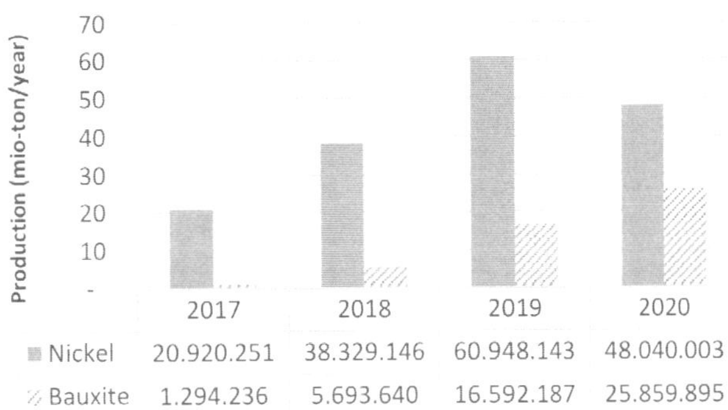

Figure 1. Nickel and Bauxite Production (2017-2020)

The existence of a ban on the export of mineral mining products has positive and negative impacts on various sectors where one of the positive impacts is the increase in the selling value of mineral mining products, especially bauxite and nickel [4]. As for the negative impact, the number of smelters in Indonesia is currently unable to meet the production of nickel ore or bauxite, so there will be potential for the construction of new smelters. Therefore, the purpose of this study is to plan the optimum marine transportation particularly for nickel from the mine site to the smelter, then continue to the importing country by considering the minimum total transportation costs. In addition, the existence of a downstream program will have an impact on the Indonesia's Balance of Payment, especially on exports of goods (minerals) related to work-in-process commodity prices which are more expensive than raw materials so that there will be potential for additional foreign exchange from exports of these goods.

2. Literature review

Downstream policy is associated with the concept of added value and product competitiveness. The farther a product is produced from industrial activities the higher the value or price. In industrial practice the concept of added value is defined as the difference or comparison between the selling value of the product produced and the costs required to create the product in equivalent units of volume or weight. [5].

There has been previous research on sea transportation planning due to the implementation of downstream policy, but it only focused on copper mining products. In 2022, Achmadi et al. showed that the results of the study include the optimum transportation scenario from the mine site to the smelter, then continue from the smelter to the importing country, considering the fleet ownership, whether the ship is new-building ship, second-hand ship, or charter ship [6]. However, this study focuses more on the magnitude of the impact of the downstream policy on Indonesia's Balance of Payment, in terms of transportation and added value of goods. Moreover, this study concentrates more on the magnitude of the impact of the downstream program on Indonesia's trade balance, in terms of transportation and added value of goods. The total amount of exports for nickel will be the main factor in determining the sea transportation model and calculating the amount of potential addition to the Indonesia's foreign exchange after being processed in Indonesian smelters.

In the Indonesia's Balance of Payment, there are two transaction records, namely: (1) current account and (2) capital and financial transactions [7]. The current account includes the recording of:

1. Goods Trade Balance

a. non-oil and gas trade balance
b. oil and gas trade balance
2. Services trade balance
3. Primary Income Balance
4. Secondary Income Balance

Meanwhile, capital, and financial transactions include recording of (1) direct investment, (2) portfolio investment and (3) other investments.

In this study, the scope of the discussion of the balance of payments is limited to measuring the impact of implementing a downstream program for export mineral products, particularly for nickel, on the amount of the current account, in terms of goods (non-oil and gas) and services, particularly sea transportation services.

3. Methodology

There are two methodology approaches to determine the optimum shipment of nickel after implementing a downstream program, as follows:

The calculation of transportation costs is carried out to find out how much transportation costs are in exporting nickel after the implementation of the downstream policy. Non-linear programming was used to define the optimum ship including type, size and number of ship required by considering the minimum transportation cost.

a. Shipment form mine site to smelter

The first analysis of fleet planning is the calculation of needs and determining the type of sea transportation to fulfil nickel shipments from mine site to ferronickel smelter. In this shipment, the various types of ship used in the calculation are four bulk carriers and four tug-barges. The combined route from the port of origin to the port of destination. By using a decision variable in the form of the number of ship frequencies in a year (nf_{abc}), the parameters set in this optimization model are as follows:

$$\boldsymbol{Min\ ZC} = \left(\sum_{a=1}^{8} \boldsymbol{FC_a}\right) + \left(\sum_{a=1}^{8} \sum_{b=1}^{3} \sum_{c=1}^{3} \boldsymbol{VC_{abc}}\right) + \boldsymbol{PL} \quad (1)$$

Subject to:

$$\sum_{b=1}^{3} \sum_{c=1}^{3} Q_{bc} \le S_b \quad (2)$$

$$\sum_{b=1}^{3} \sum_{c=1}^{3} Q_{bc} \ \ge D_c \quad (3)$$

$$\sum_{a=1}^{8} \sum_{b=1}^{3} \sum_{c=1}^{3} F_{abc} \ \le F_{abc}max \quad (4)$$

Where:
ZC: minimum transportation cost (USD/year)
a : type of ship selected
b : port of origin node
c : port of destination node
n : frequency
FC: fixed cost (USD/year)
VC: variable cost (USD/year)
K : number of ships required (unit)
Qbc: number of cargoes carried from b to c (ton/year)
Dc: total demand in c (ton/year)
Sb : total supply in b (ton/year)
Fabc : number of frequencies used of ship from b to c (roundtrip/year)
Fabcmax : maximum frequency required of ship to deliver cargo from b to c (roundtrip/year)

Equation (2) is a constraint to limit the amount of cargo transported by ships from the mine not exceeding the volume of production. For equation (3), this constraint limits the amount of cargo that enters the smelter not less or fulfils the smelter's demand. Equation (4) is the selected shipping frequency limit not to exceed the shipping frequency from the calculation results.

b. Shipment from smelter to manufacturer

In this shipping process, the various types of sea transportation used are general cargo. For shipping routes, there are two options, namely direct shipping from the ferronickel smelter to the stainless-steel factory, while the second option is to pass through the hub port first. Moreover, the ferronickel carried from the smelter is sent and collected first to the hub port using six small general cargo ships (DWT 1,000-7,000), then from the hub port it is sent to the nearest port by the stainless-steel factory using five ships of general cargo (DWT 11,000-27,000). There are two hub ports considered in this study, namely the Port of Makassar and the Port of Surabaya. The hub port in Makassar is chosen because the location is close to the ferronickel smelters, most of which are located on Sulawesi and North Maluku. On the other hand, the Port of Surabaya is selected because of the current conditions, shipments of ferronickel for export usually transhipment at the Port of Surabaya first. In addition, the Port of Surabaya is also known as a port that acts as a hub and opens the economic gate to Eastern Indonesia [8]. Then in optimizing the ferronickel shipping process from the smelter to the factory, what acts as a decision variable is the amount of cargo carried by ships in one year. The parameters set in this optimization model are as follows:

$$\text{Min } ZC = \left(\sum_{i=1}^{6} FC_i\right) + \left(\sum_{i=1}^{6} \sum_{j=1}^{7} \sum_{k=1}^{4} VC_{ijk}\right) + \left(\sum_{h=1}^{6} FC_h\right) + \left(\sum_{i=1}^{6} \sum_{j=7}^{8} \sum_{k=2}^{4} VC_{hjk}\right) + PL \quad (5)$$

Subject to:

$$\sum_{j=1}^{7} \sum_{k=2}^{4} Q_{jk} \leq S_j \quad (6)$$

$$\sum_{j=1}^{7} \sum_{k=2}^{4} Q_{jk} \geq D_k \quad (7)$$

$$\sum_{i=1}^{6} \sum_{j=1}^{7} \sum_{k=1}^{4} KT_{ijk} \leq KT_{ijk}max \quad (8)$$

$$\sum_{h=1}^{6} \sum_{j=7}^{8} \sum_{k=2}^{4} KT_{hjk} \leq KT_{hjk}max \quad (9)$$

$$\sum_{j=1}^{6} \sum_{k=1}^{4} FI_{jk} - FO_{jk} = 0 \quad (10)$$

$$\sum_{j=1}^{6} \sum_{k=2}^{4} FI_{jk} - FO_{jk} \geq D_r \quad (11)$$

Where:

ZC: Minimum transportration cost (USD/year)
i : the first group of general cargo ships
h : the second group of general cargo ships
j : port of origin node
k : port of destination node
FC: fixed cost (USD/year)
VC: variable cost (USD/year)
Qjk: number of cargoes carried from j to k (ton/year)
Dk: total demand in k (ton/year)
Sj : total supply in j (ton/year)
KTjk : number of frequencies used of ship from j to k (roundtrip/year)
KTjkmax : maximum frequency required of ship to deliver cargo from j to k (roundtrip/year)
Dr : total demand in Surabaya
FI : number of flow-in cargoes to hub port
FO: number of flow-out cargoes from hub port

Equation (6) is a constraint to limit the amount of cargo transported by ships from the mine not exceeding the volume of production. For equation (7), this constraint limits the amount of cargo that enters the smelter not less or fulfils the smelter's demand. Equations (8) and (9) are limits to the number of optimum transported cargo not exceeding the maximum transported cargo from the calculation results. Next, the limitations for hub ports. This limitation is needed to help the model regulate the rate of entry and exit of cargo at the hub port [9]. Makassar Port Constraints in equation (10) illustrates that the number of goods entering the port must be the same as the goods leaving. For the constraints at the Port of Surabaya, it is stated in equation (11), the difference between the cargo that enters and the cargo that leaves the Port must be able to meet the demand from factories in the region.

4. Results and discussion

4.1. Supply and Demand

Currently, there are a total of 296 nickel mines recorded in Minerba One Data Indonesia (MODI) belongs to the Ministry of Energy and Mineral Resources. The 296 nickel mines are a combination of Contract of Work (KK), OP (Production Operation) and IUP (Mining Business Permit) permits. On the other hand, as of 2020, there are 12 IUP OP and KK nickel mines which are suppliers of operational smelters. The smelters that were reviewed in this study were 10 ferronickel smelters which were already in operation and were still under construction as shown in **Table 1**.

Table 1. Smelters with their port locations

No	Smelter	Input Capacity (Nickel)	Output Capacity (Ferronickel)	Port
		Ton/year	Ton/year	
1	Smelter 1	4.024.581	363.655	Gebe island
2	Smelter 2	2.229.656	175.000	Obi island
3	Smelter 3	2.772.971	264.211	Obi island
4	Smelter 4	2.400.000	152.400	Bahodopi
5	Smelter 5	3.500.000	351.018	Bahodopi
6	Smelter 6	1.866.510	130.508	Bahodopi
7	Smelter 7	7.000.000	647.000	Kendari
8	Smelter 8	3.500.000	302.506	Bau-bau
9	Smelter 9	5.069.856	237.600	Pomalaa
10	Smelter 10	2.079.783	129.090	Pomalaa

Nickel ore, which is input to produce ferronickel smelters in **Table 1**, is supplied by mine sites that are still the same holding company or are affiliated with related smelters, which are shipped by land (trucks, conveyors). For several smelters, the company's supply of nickel is also bought from mines outside the island. The following is a list in **Table 2**. There are four smelters with different locations, namely Smelter 1 on Obi Island, Smelter 2 on Gebe Island, and Smelter 4 and Smelter 5 in Bahodopi.

Table 2. Mine sites of nickel in Indonesia

No	Mine site	Location	Production (ton/year)	Port
1	Mine site-1	Weda	2.210.000	Weda
2	Mine site-2	Gag island	1.800.000	Gag island
3	Mine site-3	Kendari	800.000	Kendari

The amount of supply and demand is an assumption of the demand for nickel ore from this smelter, usually around 20-30% of the total production requirement. This is because these smelters are already supplying nickel from nearby companies and supplying from outside the island to cover their shortage of nickel needs, as shown in **Table 3**.

Table 3. Smelter for shipment from mine site to smelter

No	Smelter	Location	Smelter's demand	Supply from mine site	Supply from other island
			Ton/year	Ton/year	Ton/year
1	Smelter 2	Obi Island	4.024.581	3.578.650	445.931
2	Smelter 1	Gebe Island	2.229.656	1.675.062	554.594
3	Smelter 4	Bahodopi	2.400.000	1.700.000	700.000
4	Smelter 5	Bahodopi	3.500.000	3.126.698	373.302

In the ferronickel smelter scheme to a stainless steel factory, the product focus under review is stainless steel flat products with ferronickel as a raw material. The following is a list of these factories in **Table 4**. The stainless steel factories reviewed are located in four provinces on the island of Java, namely two in East Java (Surabaya and Sidoarjo) with port locations in Tanjung Perak, three in West Java (Bekasi), one in the Special Capital Region of Jakarta (Pulogadung) with a port location in Tanjung Priok, and two in Banten (Cilegon) with a port location in Ciwandan, as well as different needs for ferronickel and stainless steel production.

Table 4. Ferronickel plant overview list

No	Company	Location	Ton/Year	Ton/Year
			Input Capacity (FeNi)	Output Capacity (SS)
1	Steel Factory 1	Surabaya	26.160	130.800
2	Steel Factory 2	Sidoarjo	100.000	500.000
3	Steel Factory 3	Bekasi	130.000	650.000
4	Steel Factory 4	Bekasi	480.000	2.400.000
5	Steel Factory 5	Pulogadung	81.400	407.000
6	Steel Factory 6	Bekasi	126.000	630.000
7	Steel Factory 7	Cilegon	600.000	3.000.000
8	Steel Factory 8	Cilegon	650.000	3.250.000

Due to limited data, the amount of ferronickel in stainless steel processing that is the demand for these factories assumes that the composition of the use of ferronickel as a raw material for stainless steel ranges from around 10-30% of the total production per tonne, depending on the type and quality of the stainless steel will be made. In this study, the assumption is that the need for ferronickel is 20% of the total production output per year. **Figure 2** is the location points of mine sites, smelters, stainless factories, and ports that are the subject of this study.

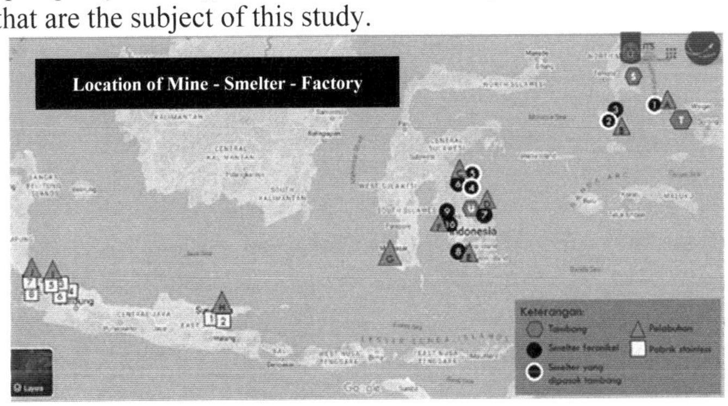

Figure 2. Location of Mine site-Smelter-Factory

4.2. Create Optimum Route

- Mine site to smelter

There are three ports of origin for nickel mines located on Gag, Weda and Kendari Islands. The same number also occurs in the smelter which also has three destination ports located on Gebe Island, Obi Island and Bahodopi. Following are the details of the port of origin and destination, as well as the mileage matrix in **Table 5**.

Table 5. Distance from mine to smelter

Route		**Obi Island**	**Gebe Island**	**Bahodopi**
Route	**Code**	**W**	**X**	**Y**
Weda	S	96	308	629
Sorong	T	52	228	517
Kendari	U	504	333	0

- Ferronickel smelter to a stainless-steel factory

From a total of 10 ferronickel smelter reviews, eight port points of origin and four port points of destination were obtained, of which two are hub ports in **Table 6.**

Table 6. Port of origin and destination

Port of Origin		**Port of Destination**	
Code	**Port**	**Code**	**Port**
A	Gebe Island	G	Makassar
B	Obi Island	H	Surabaya
C	Bahodopi	I	Jakarta
D	Kendari	J	Banten
E	Bau-bau		
F	Pomalaa		
G	Makassar		
H	Surabaya		

With these various origins and destinations, the permutation method was used to find the number of alternative routes and 2,604 route combinations (port to port and multiport) were obtained [10]. The route combination is then simplified for the optimization process by using the milk-run method [11]. Thus, obtained 368 alternative routes used in optimization.

4.3. Marine Transportation for Nickel ores

This research examines two alternative modes of nickel transportation, namely tug-barges **(Figure 3)** and dry bulk carriers **(Figure 4).** Dry bulk ships that cannot dock at jetties owned by mining companies will usually anchor at the anchorage area, and loading and unloading activities are carried out using a ship to ship system. For tug-barge specifications can be seen in **Table 7** and **Table 8.**

Source: http://www.riverswaterways.com

Figure 3. Tug-Barge

Source: http://www.tradewindsnews.com

Figure 4. Dry Bulk Carrier

Table 7. Specification of tug boat

Ship's Name		Tug 1	Tug 2	Tug 3	Tug 4
Main Dimention (meter)	LOA	27	23,53	25,51	29,75
	B	8,2	6,5	7,5	8,3
	T	3	2,75	2,9	3,8
	Vs	12	9	10	10
	GT	207	117	151	75
Engine (KW)	ME	2.646	7.943	9.480	9.731
	AE	560	441	460	653

Table 8. Specification of barge

Items	Barge 1	Barge 2	Barge 3	Barge 4
Size	270 ft	300 ft	330 ft	365 ft
Capacity	5.000 ton	8.000 ton	10.000 ton	12.500 ton
LOA	72 m	80 m	90 m	111 m
B	16 m	18 m	20 m	28 m
T	3,7	3,97	4,71	5,42
GT	2.448	2.538	5.959	5.289

While the specifications for dry bulk carriers can be seen in **Table 9.** The following ships were selected based on variations in capacity, and four alternative dry bulk carriers with different DWT with a DWT range of 4,000 to 50,000 were obtained.

Table 9. Specification of dry bulk carrier

Ship's Name		BC 1	BC 2	BC 3	BC 4
DWT		4.549	11.196	28.202	50.296
Payload (ton)		3.639	8.957	22.562	40.237
Main Dimention (m)	LOA	100	115	169	190
	B	15	20	27	32
	H	8	10	14	18
	T	5	6	10	13
	GT	2.760	7.451	17.019	31.250
Speed (Knot)	Vballast	10	12	16	15
	Vladen	9	6	10	14

4.4. Marine Transportation for Ferronickel

Marine transportation that usually transports ferronickel is general cargo **(Figure 5)**, because ferronickel is a breakbulk type of cargo. Ferronickel in the form of granules or slag will be packed in one ton sacks.

Source: http://www.shipspottersteve.com

Figure 5. General Cargo Ship

Ferronickel general cargo ships are divided into two categories, the first is small general cargo with a capacity range of 1,000 to 7,000 DWT, the port of origin is from the TUKS Smelter in **Table 10**.

Table 10. Specification of general cargo group (1)

Ship's Name		GC1	GC2	GC3	GC3	GC4	GC6
DWT		1.599	2.419	3.262	4.680	5.664	7.748
Payload (ton)		1.308	1.979	2.669	3.829	4.634	6.339
Main Dimention (m)	LOA	72	74	86	96	125	101
	B	12	12	14	15	20	19
	H	7	7	6	7	9	13
	T	3,9	5,1	4,7	5,3	5,9	7,7
	GT	498	1.448	2.003	2.996	6.361	6.155
Speed (Knot)	Vballast	12	9	8	9	10	14
	Vladen	8	8	8	8	9	11

While general cargo category two in **Table 11** is larger in size, has a capacity range of 11,000 to 27,000 DWT, and the port point of origin is from the hub port.

Table 11. Specification of general cargo group (2)

Ship's Name		GC7	GC8	GC9	GC10
DWT		11.981	17.785	20.501	24.159
Payload (ton)		9.803	14.551	16.774	19.766
Main Dimention (m)	LOA	129	162	153	159
	B	20	26	24	26
	H	11	13	14	13
	T	8,7	9,1	9,7	9
	GT	8.812	13.235	13.066	15.349
Speed (Knot)	Vballast	13	14	17	9
	Vladen	12	10	14	6

Table 12. Recapitulation of shipment optimization results from mine to smelters

Route	Carrier	Ship Technical Aspects			Operational		Financial	
		Number of Ships	Size	Utility	Frequency	Cargo Carried	Total Cost	Unit Cost
		(unit)	feet		(times /year)	(ton /year)	(USD /year)	(USD /ton)
S-Y	*Tug-barge 1*	1	270	100%	51	255.000	$2.875.235	$ 11,28
T-W	*Tug-barge 2*	1	300	100%	26	208.000	$1.235.789	$ 5,94
T-X	*Tug-barge 2*	1	300	100%	70	560.000	$4.657.024	$ 8,32
T-Y	*Tug-barge 1* *Tug-barge 4*	2	270 365	100% 100%	59 42	820.000	$8.978.059	$ 21,43

Route	Carrier	Ship Technical Aspects: Number of Ships	Ship Technical Aspects: Size	Ship Technical Aspects: Utility	Operational: Frequency	Operational: Cargo Carried	Financial: Total Cost	Financial: Unit Cost
		(unit)	feet		(times /year)	(ton /year)	(USD /year)	(USD /ton)
U-W	*Tug-barge 1*	1	270	100%	50	250.000	$2.500.485	$ 10,00
						TOTAL	$20.246.592	$ 56,97

Analysis results of Phase 1, from Mine site to Smelter is an analysis of the need for sea transportation for the delivery of nickel ore from the mine to the smelter. From three mine sites and four smelters considered in this study, the following table is optimization results based on frequency as shown in **Table 12**. Obtained, a total of five selected routes and six tug-barge units with a size of 270 feet, 300 feet, and 365 feet. The total cost is 20,246,592 USD/year and the total unit cost is 56.97 USD/ton.

Table 13. Recapitulation of shipment optimization results from smelter to factory

Route	Carrier	Ship Technical Aspects: Number of Ships		Ship Technical Aspects: Size	Ship Technical Aspects: Utility	Operational: Frequency	Operational: Cargo Carried	Financial: Total Cost	Financial: Unit Cost
				DWT		(times /year)	(ton /year)	(USD /year)	(USD /ton)
A-G	Ship 1	1	3	1.559	100%	29	218.522	$9.406.986	$139,8
	Ship 3	1		3.262		26			
	Ship 4	1		4.680		27			
B-G	Ship 3	1	2	3.262	100%	35	290.715	$9.476.002	$69,5
	Ship 5	1		5.664		41			
C-G	Ship 5	1	2	5.664	100%	50	598.557	$12.778.993	$47,4
	Ship 6	1		7.748	94%	57			
D-G	Ship 1	1	3	1.559	100%	57	418.763	$11.909.290	$92,3
	Ship 3	1		3.262		50			
	Ship 4	1		4.680		53			
E-G	Ship 4	1	1	4.680	96%	79	302.506	$5.773.198	$19,1
F-G	Ship 2	1	2	2.419	100%	75	366.690	$8.453.517	$46,2
	Ship 4	1		4.680	73%	56			
G-H	Ship 7	1	1	11.981	20%	12	126.160	$5.621.601	$44,6
G-I	Ship 10	1	2	24.159	100%	35	817.400	$21.014.940	$97,6
	Ship 11	1		27.403	9%	4			
G-J	Ship 9	1	2	20.501	36%	18	1.252.193	$28.185.880	$49,9
	Ship 11	1		27.403	100%	41			
				TOTAL				**$112.620.407**	**$606,4**

On the other hand, the results of stage two, from Smelter to Factory. This stage is an analysis of the shipment from the ferronickel smelter to the stainless steel factory. The shipping option to meet the demand for 2.19 million domestic ferronickel is through transhipment at Makassar Port as a hub port and requires 18 general cargo ships and nine selected routes. The shipping method used is port to port. By selecting the transshipment pattern, the delivery system is divided into two, namely from the smelter to the hub port, then from the hub port to the stainless-steel factory. Summary or recapitulation of optimization results in stage two is in **Table 13**.

5. Conclusion

The need of sea transportation after downstream policy implementation can be divided into two phases: (a) from the nickel mine site as a provider of raw materials to the ferronickel smelter as a place to purify and process nickel ore, and (b) from the ferronickel smelter as a provider of semi-finished commodities to the stainless-steel factory, which will later process these commodities into finished goods.

- From nickel mine site to smelter, the optimum ship selected is tug barge.
- The shipping option to meet the demand for 2.19 million domestic ferronickel is through transshipment with Makassar Port as a hub port and a total of 18 general cargo ships are required. By selecting the transshipment pattern, the delivery system is divided into two, namely from the smelter to the hub port, then from the hub port to the stainless-steel factory.

6. Further research development

This study requires improvement on determining the impact of mineral export products to the Indonesia's Balance of Payment (BOP) particularly for bauxite and copper.

Acknowledgments

Authors would like to thank Department of Marine Transportation Engineering, Faculty of Marine Technology, Institut Teknologi Sepuluh Nopember for fully supporting the financial of this study under the departmental funding scheme with contract number: 2279/PKS/ITS/2023 and researcher contract number: 2457/PKS/ITS/2023 dated 26 June 2023.

References

[1] Kementerian ESDM. Realisasi Produksi Ferro Nickel. Retrieved from Kementerian ESDM: https://modi.esdm.go.id/fero_nikel; 2020.
[2] Kementerian Perindustrian, Buku Rencana Induk Pengembangan Industri Nasional 2015-2035, Jakarta: Komunikasi Publik Kementerian Perindustrian; 2015.
[3] Kementerian ESDM. Undang-Undang No.3 Tahun 2020. Retrieved from Kementerian ESDM RI: https://jdih.esdm.go.id; 2020.
[4] Kementerian ESDM. Realisasi Produksi Ferro Nickel. [Online] Available at: https://modi.esdm.go.id/fero_nikel Kementerian Perindustrian; 2015.
[5] I. Arif. Perkembangan Hilirisasi Nikel Indonesia. Jakarta; 2020.
[6] T Achmadi et al. IOP Conf. Ser.: Earth Environ. Sci. 1198 012024; 2023.
[7] Bank Indonesia. Laporan Neraca Pembayaran Indonesia. Jakarta: Bank Indonesia; 2022.
[8] PT Aneka Tambang. Laporan Tahunan 2021. Aneka Tambang. Jakarta; 2022.
[9] Lazuardi, S. D., van Riessen, B., Achmadi, T., Hadi, I., & Mustakim, A. *Analyzing the national logistics system through integrated and efficient logistics networks: A case study of container shipping connectivity in Indonesia*. In Applied Mechanics and Materials (Vol. 862, pp. 238-243). Trans Tech Publications Ltd; 2017.
[10] Stopford, M. Maritime Economics 3rd Edition. London: Routledge Taylor & Francis Ltd.; 2008.
[11] Yunianto, I. T., Lazuardi, S. D., & Hadi, F. *Freight calculation model: a case study of coal distribution*. In IOP Conference Series: Earth and Environmental Science (Vol. 135, No. 1, p. 012013). IOP Publishing; 2018.

IOP Conf. Series: Earth and Environmental Science 1298 (2024) 012019 doi:10.1088/1755-1315/1298/1/012019

A protection method for the coast of Phan Thiet City, Vietnam

T T Nguyen[1]* and T D T Vo[2]

[1] Department of Civil Engineering, Thuyloi University, Vietnam
[2] Binh Thuan Irrigation Project Exploitation Company, Vietnam

* Email: nguyentrongtu@tlu.edu.vn

Abstract. Vietnam is located in monsoon tropical region. That is received a lot of typhoons and tropical depresion annual. Coastal area has been erosion seriously from north to south of Vietnam recently. For developing social economic in the area, some coastal protection solutions have been applied away from natural and hurman influence. Phan Thiet belonging to Binh Thuan province is one of active cities in the South Central Coast. There has strengths in developing aquaculture, fishing and tourism. With a long coastline and beautiful beachs, Phan Thiet attracts a large number of domestic and foreign tourists any time of the year. Some coastal areas were constructed concrete wall or revetment to protect from erosion, however, that are damaged, depleaced or sinked by wave and natural impact. At present, government is supporting in recovering or rebuilt the new protection construction for creating lanscape at the coastal area. The research shows a solution revetment to protect the coastline that could be suiltable for local requirement of stability construction, serving tourism and community activities. That could reduce wave run up and foundation erosion, adapt to working condition of construction.

1. Introduction

The coast of Phan Thiet City is along 42km with sand dune, villages and coastal streets. The topography of the coastal seabed is steeply, the 2m deepth isoline close to the coast. In particular, the area has two large estuaries: Ca Ty River and Phu Hai River, as shown in figure 1 (Google map 2023). The coastline in the construction area has the main direction of northeast - southwest, almost parallel to the northeast monsoon direction. At the northeast area, there are jetties to protect Phan Thiet fishing port and Phan Thiet seafood industrial zone. These jetties act as a barrier against the northeast waves, but also reduce sediment movement along the coast.

Before the jetties were built at the estuaries, in the dry season, the left bank of the estuaries appeared sand spit extending from the bank to the south, obstructing the access to the estuary. In the rainy season, large river flows can erode these spits and reopen the rivermouth. This evolution of estuary morphology is a characteristic of the estuaries in Binh Thuan province with effects of longshore current mainly from the Northeast to the Southwest. Actual developments also show that this area is strongly eroded in the dry season, with East or Northeast winds. The study area was a fairly flat topography, dense residential areas of Thanh Hai Ward, located along the coast. Previously, there was a widely sandy beach with the fisherman activities. In recent years, the coast has been carved deep into the residential area, creating an almost vertical slope, lot of houses have been destroyed, losing residential land. The purpose of study is choosing an appropriate solution to build a work that can be stable, protect the coast from erosion, contribute to the development of economic, tourism and community.

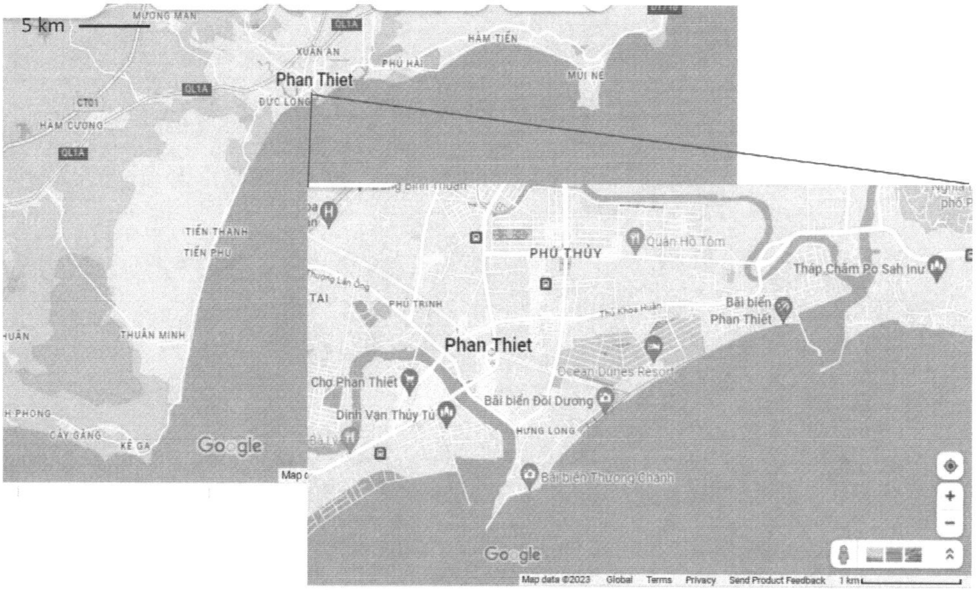

Figure 1. Location of study area

2. Natural condition

According to the survey results, the topography of the region is in the ranges from 0.0 m ÷ -1.0m elevation, the seabed slope is about 0.2%. The geological structure along the coast in order from top to bottom includes the following geological layers:

Layer 1: Fine sand to medium sand, at the deepth from 2.5m to 5.5m, the thickness around of 4m.

Layer 2: Gravel sand with seashell, the medium thickness of 2.25m.

Layer 3: Mix clay, hard plastic, the thickness around 1.75m.

The highest temprature is 37.6^{0}c, the lowest temprature is 12.2^{0}c.

Rainy season is from May to December. The highest rainfall is 467mm.

Table 1. Prevailing wind dierection at Phan Thiet

Time	Direction	Frequency (%)	Everage monthly wind speed (m/s)
January	E	35.2	3.9
February	E	37.2	3.9
March	E	34.0	3.9
April	E	21.2	3.3
May	W	15.9	2.8
June	W	37.5	3.0
Junly	W	40.4	3.0
August	W	44.7	3.3
September	W	31.0	2.6
October	E	12.0	2.4
November	E	21.5	3.0
December	E	26.7	3.2

Phan Thiet area in particular and Binh Thuan province in general are influenced by the monsoon regime. In the dry season, from November to April of the following year there is a northeast monsoon with prevailing wind directions being the East and Northeast, with a frequency of 20-40%, the speed of average wind 6 - 9m/s. The maximum wind speed could be 31m/s. [1]
In the rainy season, there is a southwesterly wind with the prevailing wind direction being West and Southwest, with a frequency of 30-40%, with an average wind speed of 4-6m/s.
The tides in Phan Thiet bay belong to the mixed tidal regime because it is located in the transition zone between the uniform diurnal tide regime in the north (Qui Nhon) and the irregular semi-diurnal tide in the south (Vung Tau), but natural about the irregular diurnal regime. The high tidal range is 1.5m to 2.0m, usually occurs in the months of October, November and December. According to monitoring data and reference documents, some basic characteristics of water level and tidal amplitude in this area are shown as follows:
- Average tide level: - 0.11m
- Maximum high tide level: + 1.12m
- Minimum low tide level: - 1.42m
- Average tide range: 122cm
- Maximum tidal amplitude: 222cm

3. Revetment design

The coast in the center of a tourism city as Phan Thiet is one of the places that citizen and managers pay the most attention to. A construction being built to protect the coast should be adapted a combinative requirement suitable with the community activities. The concrete block revetment was choosen for the coast at this area. The natural topography is low, the altitude of 2.2m to 2.5m. It is necessary a high crest of the revetment to avoid wave run-up or overtopping on the coast. The construction could break the lanscape and enviromental of coastal zone in the city. The revetment with wave overtoping on the crest is a suitable solution for protecting coast in the area.
The discharge of wave overtopping allowed to overflow on the embankment depends on the quality of the structure protecting the crest, protecting the slope on the field side and the area that allows flooding on the residential side. This embankment is combined with a pedestrian traffic road with a self-inserting brick structure placed on a cement-sand base and a compacted buffer layer. Thus, the top of the embankment is considered as a solid protection with drainage ditches. Therefore, it is possible to choose the allowable overflow flow as q=20-50 (l/s/m), by the National Standard of Vietnam TCVN 9901-2014. The project area is characterized as a beach resort, behind the embankment are resorts and entertainment activities. It is recommended to choose the allowable overflow discharge of 30 l/s/m to limit the amount of water overflowing the crest to facilitate tourism exploitation.
So, [q]= 30l/s/m=0.03 m3/s/m.

3.1 Embankment crest elevation

Crest altitude of the embankment is determined as equation 1 [2].

$$Z_c = Z_{SWL} + R_{cp} \quad (1)$$

In which: Z_c is the crest altitude. Z_{SWL} is sea water level with designing frequency. R_{cp} is the height on the sea water level basing on wave overtopping.
Sea water level is determined basing on the location of contruction and designing frequency. With the type of embankment, the designing frequency is choosen 3.33%. As shown in table B.4 of National Standard of Vietnam [2], the sea water level is interpolated Z_{SWL} = +1.36m.
The height on sea water level R_{cp} is calculated aproximately.
The surf similarity parameter ξ_0 is defined as [3]:

$$\xi_0 = \frac{\tan\alpha}{\sqrt{s_0}} \quad (2)$$

in which: α is angle of revetment slope. $s_0 = \frac{2.\pi.H_{sp}}{g.T_{m-1.0}^2}$
H_{sp} indicates wave height at the construction.

$T_{m-1.0}$ is frequency spectra.

The height of requirement on the sea water level is determined as equation 3.

$$2.0 < \gamma_b.\xi_o \leq 7.0: \quad \frac{q}{\sqrt{g.H_s^3}} = 0.2.\exp\left(-2.3\frac{R_{cp}}{H_s}.\frac{1}{\gamma_f \gamma_\beta}\right) \tag{3}$$

Table 2. Results of crest altitude calculation

Factors		Values	Units
Revetment slope	α	0.245	-
Wave height at the construction	H_s	1.75	m
Wave period	T_p	12	s
Frequency spectrum	$T_{m-1,0}$	10.4	s
Slope coefficient	m	4.0	-
Wave steepness	S_0	0.010	-
The surf similarity parameter	ξ_0	2.39	-
Wave overtopping discharge	q	0.03	$m^3/s/m$
Coefficient incident wave with $\beta = 10^0$	γ_β	0.967	-
Coefficient effection berm	γ_b	1.0	-
Coefficient effection of friction	γ_f	0.85	-
The height of requirement on the sea water level	R_{cp}	1.64	m
Sea water level	SWL	1.36	m
Crest altitude	Z_C	3.00	m

The embankment crest elevation is defined of +3.0m, it is ensured with the wave overtopping discharge of q = 30l/s/m. That is suitable to the characteristics, nature and objectives of the area. project; therefore, it is recommended to be used as the design elevation for the embankment crest.

3.2 Concrete block calculation

The thickness of concrete block for revetment is defined as following [4].

$$D \geq \frac{H_s}{\Psi u.\Phi.\Delta_m.\cos a}\xi_p^b \tag{4}$$

In which:

Ψu indicates stability coefficient of the embankment, Concrete block Tsc178 => $\Psi u = 2.5$

Φ is critical coefficient of the block being movement. That is determined as following:

$$\Phi = 6.2P_b^{0.18}.\left(\frac{S_b^2}{N}\right)^{0,1} \quad \text{in case } (\xi) < 3 \tag{5}$$

P_b is coefficient effecting of drainage of revetment and foundation. Pb = 0.1

S_b is the Initial failure parameter. Using from 0.5 to 2.0.

N indicates number of incident wave in a storm.

Δ_m is the rate of block material and structural style. $\Delta_m = \Delta = \frac{\gamma_b}{\gamma_n} - 1$

With b = 0.5; $\Delta_m = 1.33$; $\xi_p = 1.36$; $H_s = 1.75$m

Thus, the thickness of the blocks is determined larger than 27.6cm. Choosen D = 28 cm.

3.3 Stability calculation for revertment toe

The critical erosion depth of the revetment toe depends on wave energy and the geotechnical conditions of the embankment site, which is determined by formula 6.

$$\frac{S_{max}}{H_o} = \sqrt{\frac{22{,}72h}{L_o} + 0{,}25} \quad (6)$$

In there:

S_{max} is the equilibrium scour depth. H_o is the deep-water wave height. L_o is the deep water wave length. h is the depth of water in front of the structure.

Using two rows of concrete pipe at the revetment toe, with 2m depth, that could protect the toe from erosion.

The weight of the embankment foot stone must be stable under the effection of currents generated by waves at the foot. The stable weight Gd of the rock at the foot of embankment is not less than the value specified in The National Standard of Vietnam, defined by local velocity at the foot. The maximum velocity of the flow at the embankment foot is determined by the formula:

$$V_{max} = \frac{\pi . H_{sp}}{\sqrt{\frac{\pi . L_{sp}}{g} sin\frac{4\pi . h}{L_{sp}}}} \quad (7)$$

Where, V_{max} is the maximum velocity of current at the embankment foot.

L_{sp} is the design wave length, L_{sp} = 52m;

H_{sp} is the design wave height, H_s = 1.86 m;

h is the water depth in front of the dike, h=2.9m;

g is the gravity acceleration, 9.81 m/s2.

Thus, the velocity V_{max} is detemined as V_{max} = 1,78 m/s, and the weight of rock should be more than 40kg.

3.4 Stability calculation for the revetment

The embankment cross-sections at the locations and the geological condition along the construction area are relatively similar, that a representative cross-section is calculated. The revetment stability is evaluated the slip failures. The method for calculating slope stability is Slope/W in GeoStudio software program. Geostudio Office is a geotechnical software from GEO-SLOPE International Canada that is used to analyze the stability of proven rock-soil slope.

Some key features of Slope/W software are:

+ Slope/W is a software that applies limitation balance theory to determine the safety factor of earth and rock slope. In Slope/W includes many different calculation methods to calculate the factor of safety such as: Bishop method, Janbu, Spencer, Mogor-Price, Crop of Engineers, GLE method and finite element method. Therefore, the operator is free to choose the method of calculating the factor of safety.

+ Slope/W has options to calculate the sliding block including many types of soil and rock, submerged in water or not and according to different sliding blocks such as cylindrical slide, broken form in case of soft soil layer, with rock foundation, sliding along the assumed cross-sections like the slope of the foundation pit.

+ Slope/W allows integration with Seep/W so that results from Seep/W can be used.

Calculation case:

Revetment cross section with slope of m=4.0.

Case 1: Loading on crest of revetment, lowest sea water level at the toe.

Case 2: Revetment working with highest sea water level.

Case 3: Revetment working with wave rundown from highest level to the toe.

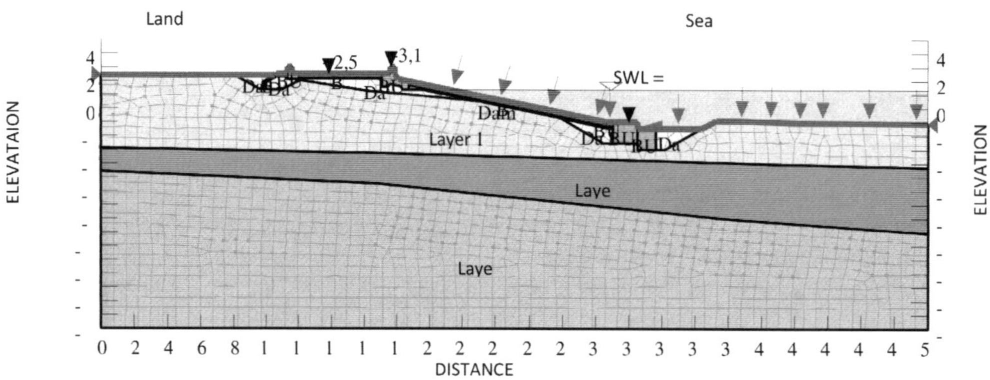

Figure 2. Calculation model for revetment stability

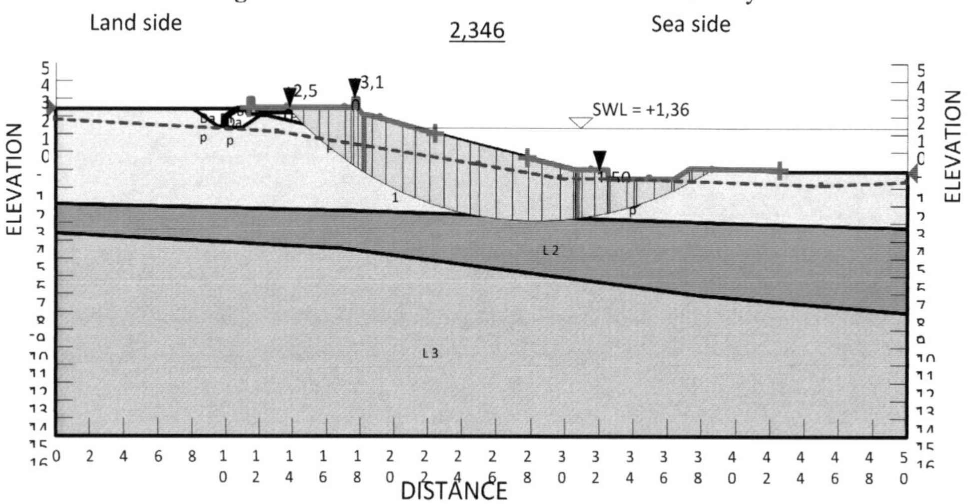

Figure 3. Results of revetment stability – Case 1

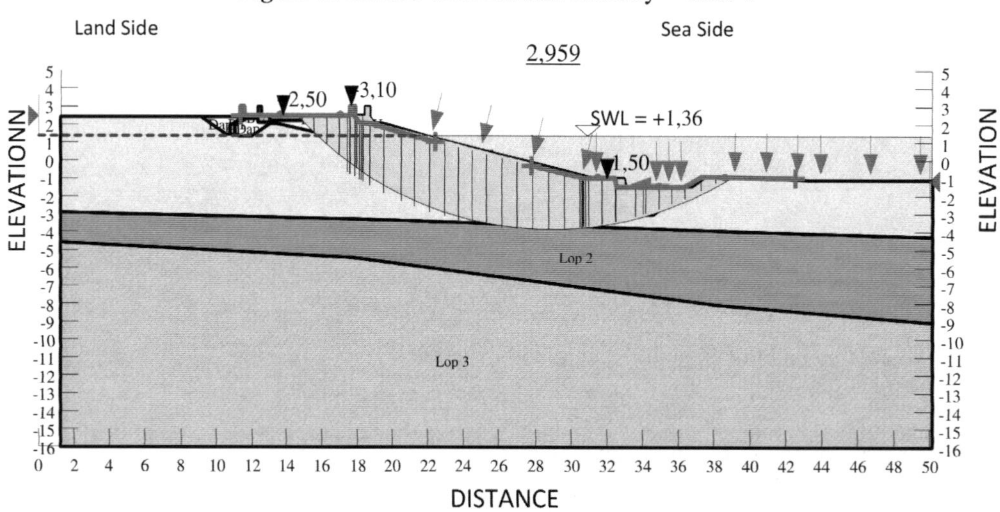

Figure 4. Results of revetment stability – Case 2

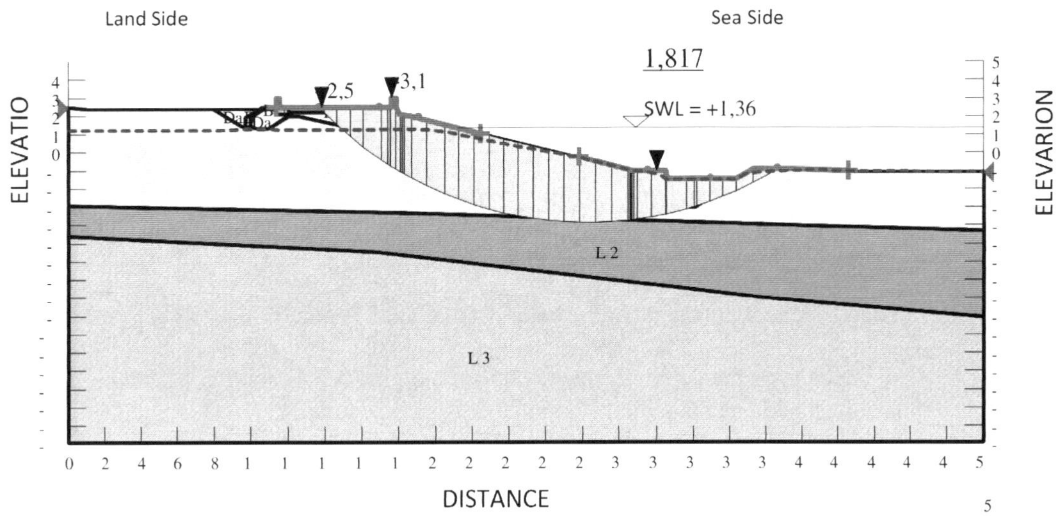

Figure 5. Results of revetment stability – Case 3

The result in figure 3 shows the stability coefficient of Case 1 is Kmin = 2.346 larger than the requirement [K] = 1.25 by National Standard of Vietnam. This is a dangerous case in loading on embankment, the structure is stable of slipping.

The situation in case 2, revetment working with highest sea water level, the struction submerges under water. The core of embankment could be seepaged leading to failure of construction stability. However, the stability coefficient in slipping slope Kmin = 2.959 is much higher than requirement from standard of 1.25. The embankment is stable in high rank of this case, as results shown in figure 4.

Result of stability calculation of Case 3, revetment working with wave rundown from highest level to the toe is presented in figure 5. In this case, the seepage water draining to surface of the slope combinate with wave rundown, the slope could be slipped. The calculation result shows the stability coefficient in slipping of the slope is 1.817, larger than standard requirement [K]= 1.25. Thus, the embankment is completely stable.

4. Conclusion

Recently, effecting of sea water level rising and climate change, the coastal area is severely affected. The coastline that is undergoing eroded due to sea wave is likely more gravely erosion in coming decades, especialy in the South Central Coast of Vietnam, include of Binh Thuan province [5]. The coast in Phan Thiet City, Binh Thuan, Vietnam is damaged seriously. Some coastal areas have been protected by embankments but also destroyed after period. To save the community activities and residential areas in the central coast of the city, a protection method from wave erosion was introduced. The solution could be stable in adverse situations. Calculation results have shown that the structure is able to stable with a high safety coefficient.

5. References

[1] Pham Trung and Nguyen Duc Vuong 2016 *Computaion of wind field and pressure, which creates by storm in the sea "Dong" and establish research scenarios in Binh Thuan province* (Proceeding of science and technology results 2015, Southern Institute of Water Resources, Vietnam)

[2] National Standard of Vietnam TCVN 9901:2014

[3] Sawaragi T 1995 *Coastal Engineering – Waves, Beaches, Wave-Structure Interactions* (Elsevier)

[4] Pilarczyk K 1998 *Dikes and Revetments: Design, Maintenance and Safety Assessment* (A.A. Balkema)

[5] Pham Trung 2019 *Mophology change of coastal estuaries in the South Central region under the impact of sea level rise* (Journal of Water Resources Science and Technology, Vietnam Academy for Water Resources)

IOP Conf. Series: Earth and Environmental Science 1298 (2024) 012020 doi:10.1088/1755-1315/1298/1/012020

Intact stability analysis of crane barge due to loading orientation effect during heavy lifting operation

M Faiz[1], Murdjito[1], R W Prastianto[1*], E B Djatmiko[1], S Nugroho[2], and Y B Hadasa[1]

[1] Ocean Engineering Department, Institut Teknologi Sepuluh Nopember, Surabaya 60111
[2] Marine Transportation Engineering Department, Institut Teknologi Sepuluh Nopember, Surabaya 60111

*Email: rudiwp@oe.its.ac.id

Abstract. Changes in the position of the lifting load on the crane barge during heavy lifting operation significantly affect the stability of the vessel. The changes in stability during this heavy lifting operation must be analysed to prevent operational failures. This study conducted an intact stability analysis of the crane barge during lifting of topside structures in the installation phase. The analysis was performed with variations in the position of the lifted load from starboard to bow for every 15° change in the crane angle. The stability results of the Crane Barge for all variations, using both analytical and numerical methods, have met the IMO requirements. The extreme condition occurred in load case 8, with a maximum GZ value of 2.904 m at a 22° angle and an area of 103,894 m.deg. The comparison of the stability results of the crane barge during the structural lifting operation using analytical calculation and numerical methods indicates that the overall validation error values have met the ±5% requirement according to the ABS Rules. However, when considering the corrections for changes in the lifting load moment in the analytical calculation, stability error values increase significantly as the heeling angle of the vessel increases.

1. Introduction

Floating crane is a floating structure equipped with accommodation facilities and one or more cranes used for lifting operations at sea. Various activities in offshore areas, such as transportation, structure assembly, installation, and decommissioning, typically involve heavy lifting and are commonly performed using floating crane vessels [1].

Heavy lifting is generally defined as lifting operations with a load weight ranging from 60 to 60,000 metric tons, primarily for transportation purposes [2]. Due to the significant weight and lifting moment of the load on the vessel, heavy lifting operations by a crane barge can significantly impact the stability of the vessel. According to the previous research, the mass load percentage to displacement ratio during lifting operations of 0.66 (% Δ) above the lifting boom of a crane vessel can reduce the GZ value by approximately 40% in calm water conditions [3].

In a more detailed research study by Australian Marine Safety Authority (AMSA), calculations were conducted on the transverse stability of fishing vessels under the combination of fishing gear, wind, and waves. The moments caused by the fishing gear and load, in many cases, are more critical than those generated by the prevailing sea conditions. With the catch lifted above the lifting boom, the ship's center of gravity shifts to the upper and lateral sides, resulting in a smaller metacentric height (GM). Additionally, when the lifting boom extends outward, the load leverage induces a torque that tends to cause the vessel to heel to one side [4]. This condition can be observed in Figure 1.

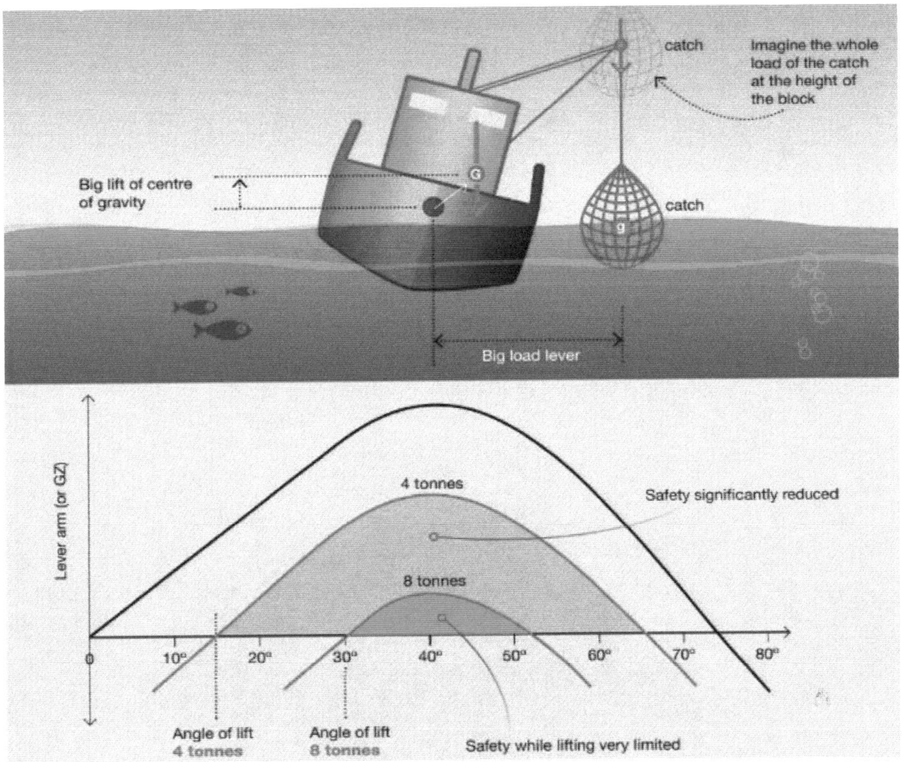

Figure 1. Lifting condition stability loss [4].

Therefore, the safe and effective operation of a floating crane in the open sea becomes a crucial requirement, especially with the current increase in demand for offshore structure lifting operations using crane barges [5]. The International Maritime Organization (IMO) mandates that specific vessels during operation must comply with the applicable stability requirements [6]. This study aims to analyze the intact stability of a floating crane barge in a free-floating condition during offshore structure lifting operations. The research's objective is to determine the vessel's stability during the lifting of offshore structures for each loading direction variation using both analytical and numerical methods. In the end, the study aims to understand the stability conditions of the crane barge when subjected to various loadings of the lifting structures using both analytical and numerical approaches.

2. Method

2.1. Topside Structure Lifting Installation

The operation conducted involves lifting the installation of the topside structure from an offshore platform. The lifting operation is carried out from the side of the ship, and subsequently, the crane is rotated until the structural load is positioned at the front side of the crane barge. Load case variations are determined based on the lifting operations of the topside structure. The following topside structure weight data presented in Table 1 and lifting configuration model as shown in Figure 2 is used in the study.

Table 1. Summary weight data of the topside structure.

Description	Net Weight (kN)	Cont. Factor	Factor Weight (kN)
Structural Self Weight	479.01	1.07	512.54
Stubbing Guide	18.00	1.07	19.26
Topside Non Generated Dead Load	66.56	1.10	73.22
Instrument Load	10.99	1.10	12.09
Electrical Load	67.16	1.10	73.88
Telecom Load	6.06	1.10	6.67
Safety Load	1.15	1.10	1.27
Mechanical Load Pre-Service	103.79	1.10	114.17
Piping Load Pre-Service	25.28	1.10	27.81
Davit Crane Selfweight	16.48	1.10	18.13
Total Lifting Load Condition Weight			859.02

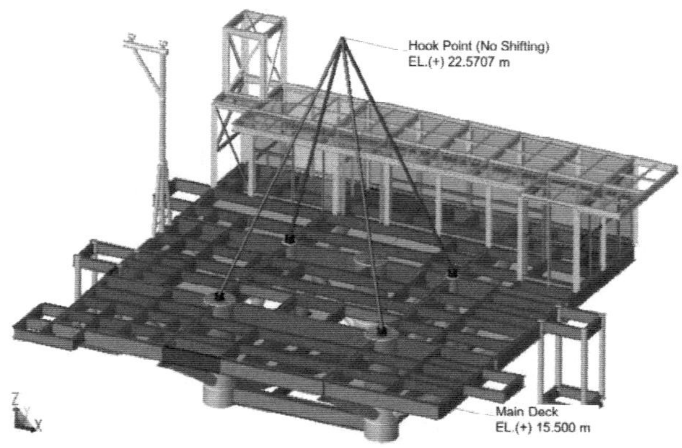

Figure 2. Topside structure model and lifting configuration.

The lifting operation's configuration involves an intact stability analysis for every 15 degrees of crane rotation, starting from the starboard position to the bow area of the crane barge. This operation loadcase condition can be seen in Figure 3. The percentage of the load is determined based on the booklet, while ballasting is determined based on the lifting operation's load variations, so the final equilibrium is achieved with trim and heel at approximately close to 0 degrees. The ballasting configuration is as follows: 97% in ballast tank No.3 S.W.B TK(S), 94% in No.3 S.W.B TK(P), 80% in tank No.4 S.W.B TK(S), and 43% in tanks No.8 S.W.B TK portside and starboard for all of loadcase configuration.

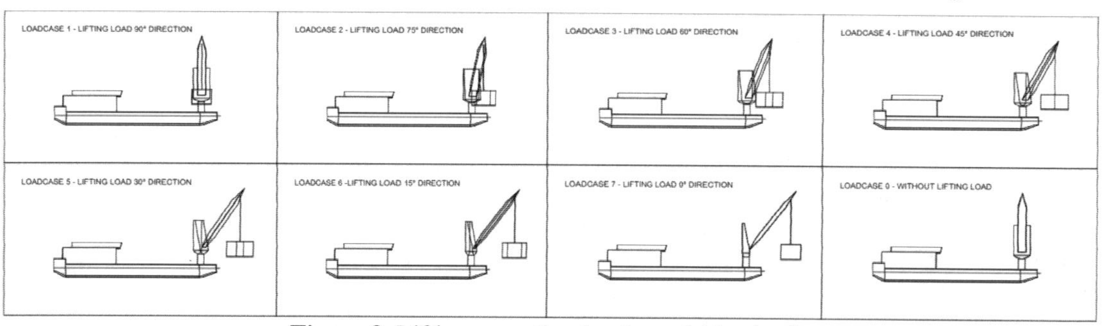

Figure 3. Lifting operation loadcase (side view).

2.2. *Crane Barge Intact Stability Criteria*

In operational conditions, all loading scenarios must be calculated, and stability conditions must be verified according to the applicable regulations. The vessel must meet stability requirements to ensure safe operations. In this vessel's stability booklet, the IMO International Code on Intact Stability (2008) criteria is used, referring to MODU Code (2009) subsections 3.2 and 3.3, with stability requirements as follows:

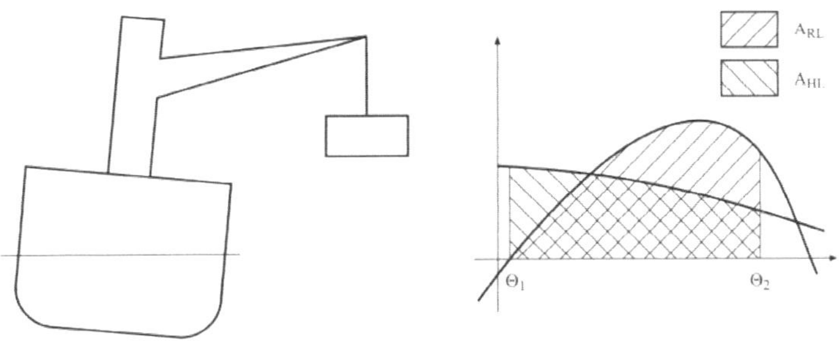

Figure 4. Lifting stability criteria.

- For the surface and self-elevating unit categories, the area under the righting moment curve (A_{RL}) to the second intercept or downflooding, whichever is reached first, must not be less than 40% of the area under the wind heeling moment curve (A_{HL}).
- For column-stabilized units, the area under the righting moment curve (A_{RL}) to the second intercept or downflooding, whichever is reached first, must not be less than 30% of the area under the wind heeling moment curve (A_{HL}).
- The righting moment curve (GZ) must be positive throughout the range of angles from the upright condition to the second intercept.

The righting moment and wind heeling moment curves as shown in Figure 4, along with their supporting calculations, must cover the entire draft range of operational conditions, including transit conditions, while considering maximum loading at the most unfavorable position that can be applied. Additionally, the free surface effect of fluids in tanks must be taken into account. The wind heeling moment curve should be depicted for wind forces calculated using the formula as follows:

$$F = 0.5 \times C_s \times C_H \times r \times V_2 \times A \quad (1)$$

Furthermore, wind forces should be considered from all directions relative to the unit, and the wind speed values should be :

- Generally, a minimum wind speed of 36 m/s (70 knots) for offshore service should be used for normal operating conditions, and a minimum wind speed of 51.5 m/s (100 knots) should be used for extreme conditions.
- If a unit will be restricted to operations in sheltered locations (inland waters such as lakes, bays, swamps, rivers, etc.), consideration should be given to reducing the wind speed to not less than 25.8 m/s (50 knots) for normal operating conditions.

2.3. *Numerical Stability Analysis*

The numerical simulations were carried out with Maxsurf Stability software. The surface model of the crane barge is modeled using Maxsurf Modeller software first, where the model is created based on the vessel's principal dimension presented in Table 2 and the general arrangement. Then, modeling of the tanks and compartments of the crane barge in Maxsurf Stability software is done regarding the General Arrangement (GA) and tank data from the Stability Booklet.

IOP Conf. Series: Earth and Environmental Science 1298 (2024) 012020 doi:10.1088/1755-1315/1298/1/012020

Table 2. Crane barge principal dimension.

Principle Dimensions		
Description	**Value**	**Unit**
LOA	110.57	m
LPP	106.15	m
B	31.70	m
H	7.30	m
T (min)	3.60	m
T (max)	5.08	m
DWT	11,355.69	MT
Displacement	17,474.18	MT

The numerical model created using Maxsurf software has been validated against the hydrostatic data in the vessel's stability booklet to ensure alignment with the actual conditions. The error value limits are determined according to the ABS Rules for Building and Classing Mobile Offshore Units [7]. The final numerical model for the vessel's hull, tanks, and compartments of the crane barge is presented in Figure 5 and subsequently utilized as input for further analysis of equilibrium and stability during operational conditions using Maxsurf Stability software.

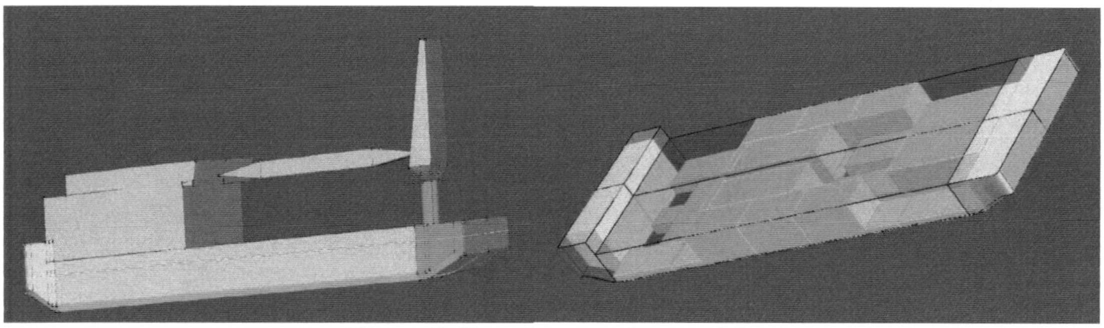

Figure 5. Crane barge surface model.

2.4. Analytical Stability Calculation

2.4.1. Equilibrium and Hydrostatic

The total displacement of the crane barge, the load, and the total CoG are calculated according to equations (1)-(4).

$$Disp = LWT + DWT \tag{2}$$

$$LCG = \frac{M_{LCG}}{Disp} \tag{3}$$

$$KG = \frac{M_{KG}}{Disp} \tag{4}$$

$$TCG = \frac{M_{TCG}}{Disp} \tag{5}$$

Where displacement ($Disp$) is the total weight of the ship, which is the sum of the lightship weight (LWT) and deadweight (DWT). The values for the ship's center of gravity (LCG, KG, TCG) are obtained by dividing the total longitudinal, vertical, and transversal moment by the total displacement value. Next, from the known displacement and total moment, the trim and draft values are calculated using the following formulas:

$$BG = LCB - LCG \tag{6}$$

$$Trim = \frac{BG \times Disp}{MTC \times 100} \tag{7}$$

$$D_{AP/FP} = D_{eq} \pm \frac{Trim * LCF}{LBP} \tag{8}$$

For the trim calculation, the values of LCB (Longitudinal Center of Buoyancy) and MTC (Midship Transverse Coefficient) are obtained by interpolating data from the stability booklet's even keel condition. Then, the KG (Vertical Center of Gravity) correction due to the free surface is calculated as follows:

$$FSM = IT \times \rho \tag{9}$$

$$GG_0 = \frac{FSM}{Disp} \tag{10}$$

The value of the free surface moment (FSM) is the product of the tank's inertia (IT), obtained from the stability booklet, and the density of the fluid contained (ρ). From this free surface moment, a correction for the vertical Center of Gravity (CoG) value is derived. Using the obtained CoG value along with the free surface, the ship's metacenter height (GM) can be calculated using equations (10) and (11).

$$GM = KMT - KG \tag{11}$$

$$GMt = GM - GG_0 \tag{12}$$

2.4.2. Stability Analysis
With the initial stability data obtained from the stability booklet in the form of KN (Righting Arm Curve), the calculation of the restoring arm (GZ) from the KN data is done as follows:

$$GZ = KN - KG \times \sin\theta - TCG \times \cos\theta \tag{13}$$

In the ideal condition, the lifted load is connected to the structure (crane pedestal) through a rope, making the position of the lifted load dynamic as shown in Figure 6, following the vessel's inclination. During heel conditions, the shifting of the load can cause a shift in the total CoG, and the TCG (Transverse Center of Gravity) position must be corrected as follows:

$$TCG_{corrected} = TCG + \left(\frac{W_L \times L_L \times \sin\theta}{Displacement}\right) \tag{14}$$

The corrected TCG value is the sum of the initial TCG value and a correction factor, which is the product of the lifting load mass (W_L) and the transverse arm shift ($L_L \times \sin\theta$) divided by the total displacement value.

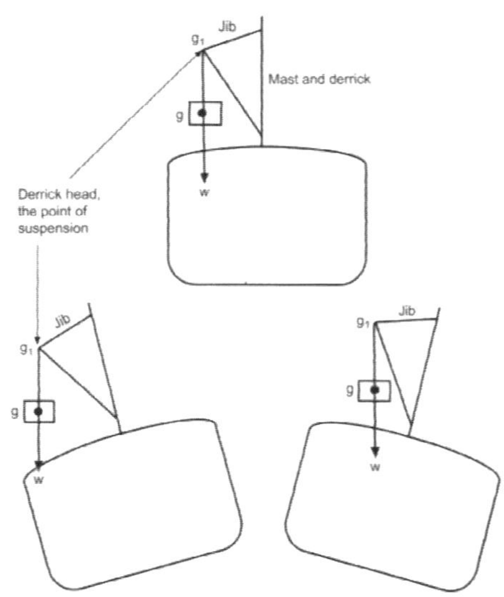

Figure 6. Lifting load on heel condition [8]

This CoG shifting factor affects the final GZ value, so the calculation of the GZ value after the correction becomes:

$$GZ_{corrected} = KN - KG \times \sin\theta - \left(TCG + \left(\frac{W_L \times L_L \times \sin\theta}{Displacement}\right)\right) \times \cos\theta \quad (15)$$

$GZ_{corrected}$ represents the final righting arm value with the correction for the change in lifting load moment. The value of KN is the length from the keel to the vertical line when the ship heels, KG is the vertical centre of gravity of the ship, and TCG is the transverse centre of gravity of the ship that has been corrected.

3. Result & Discussion

3.1. Equilibrium Condition

The equilibrium results for each loadcase condition with all loading directions using analytical and numerical methods have been obtained. Overall, the values of each hydrostatic property meet the planned operational conditions, as specified in the numerical and analytical equilibrium calculations. The designed equilibrium conditions are considered satisfactory, as the trim values range from 0 to 0.12 m, and the heeling values remain below 1 degree.

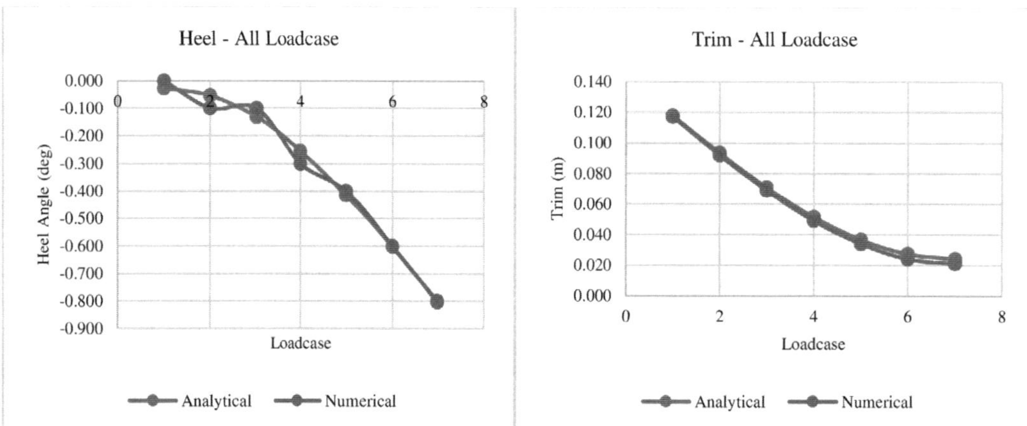

Figure 7. Heel and trim condition.

From the analysis conducted, a graph illustrating the changes in trim and heel conditions during operational circumstances is presented in Figure 7. This finding is consistent with the research conducted by T. V. Tao [9], where the trim and heel angle of the vessel increase as the lifting load shifts to the transverse side of the vessel. Throughout the operation, with constant ballasting conditions and a load weighing 88 tons continuously shifting from the ship's starboard side to the bow will result in dynamic changes in trim and heel conditions during the operation. This needs to be specifically attended in order to prevent failures during the course of the operation.

3.2. Stability Analysis

The stability analysis results conducted using numerical and analytical methods with and without corrections are obtained. The results include the GZ values, wind heeling moment, and downflooding point (DF) of the vessel for heeling angles ranging from 0 to 70 degrees under each loadcase condition, and all loading directions are shown in Figure 8.

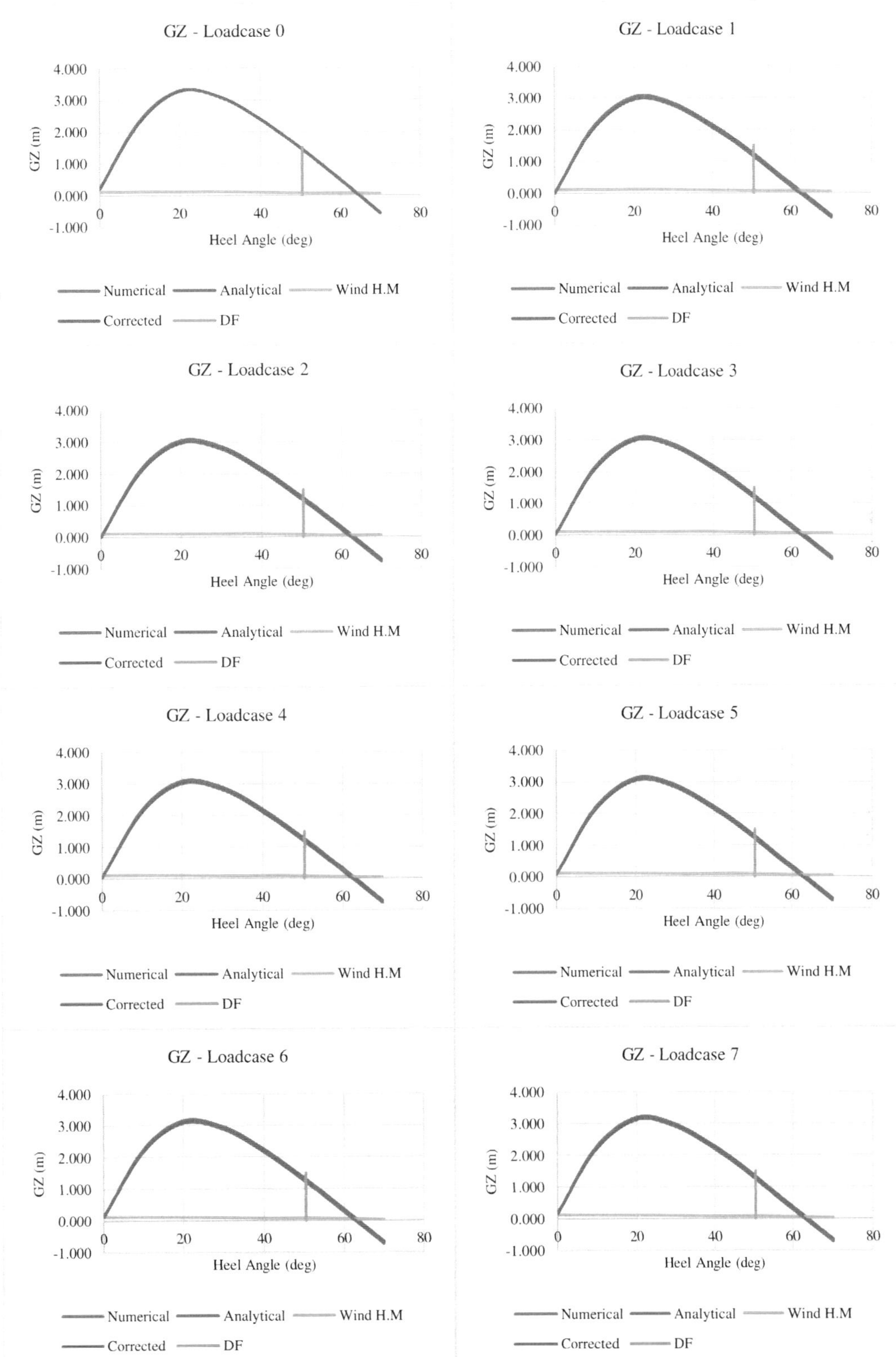

Figure 8. Stability curve (GZ) for all loadcases.

IOP Conf. Series: Earth and Environmental Science 1298 (2024) 012020 doi:10.1088/1755-1315/1298/1/012020

The stability curve for all the loadcases in both the analytical and numerical methods is presented in Figure 9. From that graph, we can observe that the stability curve's value and area will decrease further as the load shifts to the ship's lateral side and the transverse arm of the lifting load increases. When the crane barge has the lowest stability value, the extreme condition occurs when the lifting load is positioned on the ship's side (portside). The extreme condition occurs in loadcase 8, with a maximum GZ value of 2.904 m at a 22° angle and an area of 103,894 m.deg using the numerical method.

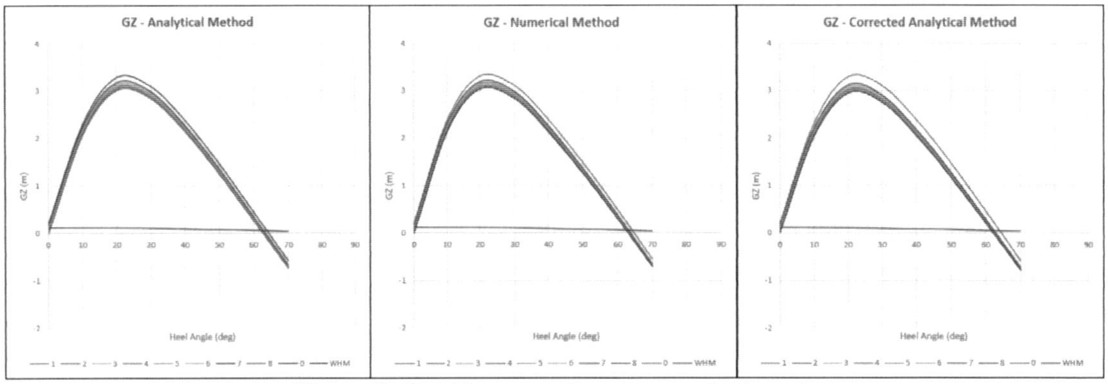

Figure 9. Stability curve for every method used.

3.3. IMO Criteria

The stability analysis must be verified against the stability parameters required by IMO, as specified in the booklet. The IMO criteria, specifically for the specific unit MODU category [6], apply to the overall results and are presented in Table 3. This table shows the value of the wind heeling moment area, the comparison of the righting moment and heeling moment area, and the GZ value compared to the IMO criteria.

Table 3. Loadcase stability criteria.

Criteria	Loadcase 1			Loadcase 5		
	Analytic	Numeric	Corrected	Analytic	Numeric	Corrected
A_{HL} **(m.deg)**	5.214	5.214	5.214	5.214	5.214	5.214
A_{RL}/A_{HL}	19.577	19.735	19.025	20.260	20.386	19.707
GZ Value	positive	positive	positive	positive	positive	positive
Criteria	Loadcase 2			Loadcase 6		
	Analytic	Numeric	Corrected	Analytic	Numeric	Corrected
A_{HL} **(m.deg)**	5.214	5.214	5.214	5.214	5.214	5.214
A_{RL}/A_{HL}	19.624	19.772	19.072	20.587	20.710	20.035
GZ Value	positive	positive	positive	positive	positive	positive
Criteria	Loadcase 3			Loadcase 7		
	Analytic	Numeric	Corrected	Analytic	Numeric	Corrected
A_{HL} **(m.deg)**	5.214	5.214	5.214	5.214	5.214	5.214
A_{RL}/A_{HL}	19.761	19.901	19.209	20.939	21.056	20.386
GZ Value	positive	positive	positive	positive	positive	positive
Criteria	Loadcase 4			Loadcase 0		
	Analytic	Numeric	Corrected	Analytic	Numeric	Corrected
A_{HL} **(m.deg)**	5.214	5.214	5.214	5.214	5.214	5.214
A_{RL}/A_{HL}	19.978	20.107	19.426	19.480	19.638	18.928
GZ Value	positive	positive	positive	positive	positive	positive

3.4. Analytical and Numerical Method Comparison

From the comparison of stability results of the crane barge during structural lifting operation using analytical and numerical methods, as shown in Table 4, it is known that the overall validation error values for analytical and numerical stability calculations are below 5%, meeting the requirement of ±5% according to the ABS Rules for Building and Classing Mobile Offshore Units [7]. However, when considering the lifting moment correction in the analytical stability calculation, the error values increase significantly as the heeling angle of the vessel increases. It is because the correction of CoG and moment calculations for the lifted load led to a reduction in GZ value that cannot be validated using the Maxsurf numerical method. It can be observed in Table 5.

Table 4. Analytical method stability error validation.

Heel Angle (deg)	Loadcase								
	1	2	3	4	5	6	7	8	0
0°	3.6%	-2.4%	2.1%	-0.2%	0.5%	0.4%	-0.2%	4.1%	0.3%
10°	2.2%	2.2%	2.2%	2.2%	2.1%	2.1%	2.0%	2.2%	1.9%
20°	1.0%	0.9%	0.8%	0.7%	0.6%	0.6%	0.5%	1.0%	0.4%
30°	0.0%	0.0%	0.0%	0.0%	0.0%	0.0%	-0.1%	0.0%	0.0%
40°	0.1%	0.0%	0.0%	-0.1%	-0.1%	-0.1%	-0.1%	0.1%	0.1%
50°	0.5%	0.4%	0.4%	0.3%	0.3%	0.3%	0.3%	0.5%	0.6%
60°	4.6%	4.5%	4.3%	4.0%	3.7%	3.4%	3.3%	4.6%	3.4%
70°	-3.0%	-2.9%	-2.8%	-2.9%	-2.8%	-2.9%	-3.0%	-3.0%	-4%

Table 5. Analytical method stability with correction error validation.

Heel Angle (deg)	Loadcase								
	1	2	3	4	5	6	7	8	0
0°	3.6%	-2.4%	2.1%	-0.2%	0.5%	0.4%	-0.2%	4.1%	0.3%
10°	3.6%	3.6%	3.6%	3.5%	3.5%	3.4%	3.3%	3.6%	1.9%
20°	2.8%	2.7%	2.6%	2.5%	2.4%	2.3%	2.3%	2.9%	0.4%
30°	2.7%	2.6%	2.6%	2.5%	2.5%	2.5%	2.5%	2.7%	0.0%
40°	4.0%	4.0%	3.9%	3.8%	3.8%	3.7%	3.7%	4.0%	0.1%
50°	7.2%	7.1%	7.1%	6.9%	6.8%	6.7%	6.5%	7.3%	0.6%
60°	29.7%	29.4%	28.6%	27.4%	26.0%	24.5%	23.2%	30.2%	3.4%
70°	-10.9%	-10.8%	-10.7%	-11.0%	-11.0%	-11.2%	-11.5%	-10.8%	-4%

From the research findings obtained, during heavy lifting operational conditions, the stability calculation results with correction for changes in lifting moment will yield a significant error compared to the results from analytical calculation using a fixed lifting load and the Maxsurf stability analysis. It can be observed in Table 5 for angles 50 degrees or above. It shows the potential for errors and the ineffectiveness of the Maxsurf Stability Software when used for heavy lifting analysis on ships or in conditions involving significant variable of lifted loads on the ship. The Maxsurf Stability Software cannot model conditions of variable loads that can dynamically change positions and input them as loads with fixed positions.

4. Conclusion

Based on the intact stability research on crane barge during the installation phase of lifting the topside structure, it is concluded that the stability of the Crane Barge for all variations of the rotational angle in the structural lifting operation, using both analytical and numerical methods, complies with IMO A.749 Intact Stability Code (IS Code) 4.6 MODU with extreme condition occurs in loadcase 8, with a maximum GZ value of 2.904 m at a 22° angle and an area of 103,894 m.deg. For stability calculation validation, the results obtained from the analytical method have an error value below 5% and have met the requirement of ABS Rules at ±5%. However, when considering a change of lifting weight moment correction in the analytical method, the error value will increase significantly as the heeling angle of the vessel increases for all loadcases and cannot be validated using Maxsurf Stability numerical methods.

5. References

[1] Firdaus N, Djatmiko E B, Prastianto R W and Muryadin 2022 *Experimental Study on Coupled Motion of Floating Crane Barge and Lifted Module in Irregular Waves* IOP Conf. Series: Earth and Environmental Science Indonesia: Surabaya 972, p 1–9.

[2] Van der Heiden A J 2019 *Vessel and Cargo Motions: A Frequency Domain Method to Study Combined Vessel and Cargo Responses* Thesis Delft University of Technology Netherlands: Delft.

[3] Yu Y R and Johnson N A 2022 Lifting-induced Stability Loss of Vessels in Shallow Water *Ocean Engineering* 253.

[4] AMSA 2017 *A Guide to Fishing Vessel Stability* Australian Maritime Safety Authority Australia: Northbourne Avenue.

[5] Firdaus N, Ali B, Hidayat A, Nantabah Z K, Djatmiko E B, and Prastianto R W 2023 *Preliminary Modeling of Coupled Motion Response for Floating Crane and Suspended Module by Using Boundary Element Approach and Empirical Model* EVERGREEN Joint Journal of Novel Carbon Resource Sciences & Green Asia Strategy Vol. 10

[6] IMO 2008 *International Code on Intact Stability* International Maritime Organization United Kingdom: London.

[7] ABS 2023 *Rules for Building and Classing Mobile Offshore Units - Part 3, Hull Construction and Equipment (MOU-Part 3)* American Bureau of Shipping USA: Houston.

[8] Barrass C B and Derrett D R 2013 *Ship Stability for Masters and Mates* Butterworth-Heinemann Unite Kingdom: Oxford.

[9] Tao T V, Luy N C 2020 *The Effect of the Change in the Position of Load on the Inclination Angle of Floating Crane* International Journal of Transportation Engineering and Technology Vol. 6, pp. 118-122

IOP Conf. Series: Earth and Environmental Science 1298 (2024) 012021 doi:10.1088/1755-1315/1298/1/012021

Study on the effects of lifting loads on the longitudinal strength of a crane barge during heavy lifting operations due to various angle of crane rotation

Y B Hadasa[1], Murdjito[1], R W Prastianto[1*], E B Djatmiko[1], Wahyudi[1], and M Faiz[1]

[1] Ocean Engineering Department, Institut Teknologi Sepuluh Nopember, Surabaya 60111

*Email: rudiwp@oe.its.ac.id

Abstract. The different positions of lifting loads on a crane barge during Heavy Lifting operations have a significant impact on longitudinal strength. This study focuses on the analysis of longitudinal strength on a crane barge with a configuration of Topside structure loading direction during the installation. The variations used the crane rotation angle from starboard to bow at 15° increments. The results showed that the load distribution for all variations followed similar trendlines, using analytical and numerical methods. Significant changes occurred at frame 3–19 as the location of the superstructure, frame 13–40 as the ballast tank loads, and frame 45–55 as the lifting loads. Regarding the maximum shear force and bending moment, occurred when the net load changes from positive to negative and the position of the crane. The maximum shear force occurred at frame 40, while the maximum bending moment occurred at frame 47, 48, and 49. The comparison between analytical and numerical methods for shear force and bending moment showed that the error values are below 5% of the allowable value according to the DNV standards.

1. Introduction

In the construction of offshore platforms, there is a phase called installation, a series of activities to place structures at predetermined locations offshore; a commonly used method for this installation is the lift method [1]. This lifting method is referred to as a conventional method that uses a crane barge to lift and position the structures installed onto the jacket, and one of the operational categories is Heavy Lifting [2]. This crane barge provides flexible and efficient maritime transportation in Heavy Lifting operations, including installation and decommissioning activities [3]. Heavy Lifting operations on a crane barge typically handle the structure with weights ranging from 60 to 60,000 tons, with any Structure lighter or heavier falling outside the scope of Heavy Lifting [4]. As infrastructure supports activities, the use of crane barges for Heavy Lifting operations is steadily increasing within the maritime business sector. Therefore, ensuring the safe operation of crane barges in the open sea is essential [5].

A crane barge is a type of barge equipped with a crane and accommodation facilities. It is a support vessel for offshore platform activities, particularly for lifting operations, where structures are lifted and moved from one place to another [6]. During Heavy Lifting operations, the loads the ship receives affect the crane barge, such as the load of the lifted structure. It can affect the longitudinal strength of the crane barge because of the load distribution that occurs during Heavy Lifting operations [7]. Analyzing the longitudinal strength of the crane barge during Heavy Lifting operations is necessary to find out how much the loading effects occur and determine the minimum required strength for safe operation. The longitudinal strength analysis is carried out in free-floating conditions without considering environmental factors. The structure lifted in this Heavy Lifting operation is the Topside of the Meliwis Wellhead Platform, where various angle variations of the lifting load conditions are used in this Heavy Lifting operation scenario, as shown in Figure 1. The resulting longitudinal strength of the crane barge is checked against the safe limits according to Det Norske Veritas (DNV) Rules (DNV RU-SHIP Pt. 3 Ch. 5).

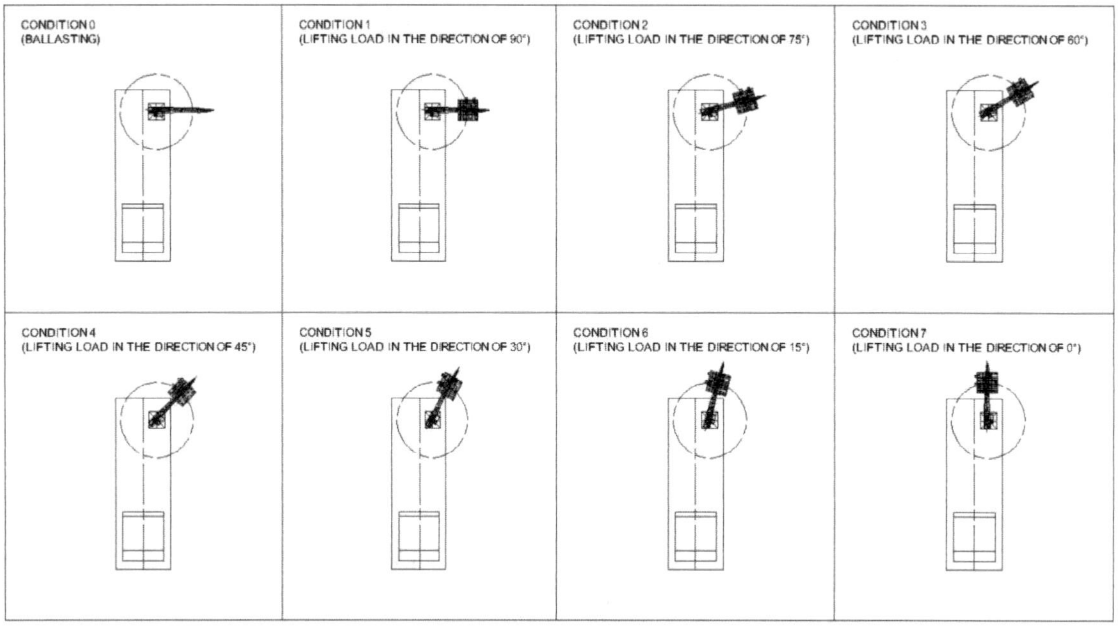

Figure 1. XYZ crane barge lifting operation loadcase (top view).

The longitudinal strength calculation is performed using analytical and numerical methods for each loading angle, where the weight data of the lifted load is shown in Table 1, categorizing the load as a Heavy Lifting operation. The utilized operational configuration involves ballasting conditions and variations in the lifting angle of the Topside structure from the ship's starboard side to the ship's bow in 15° crane rotations, as shown in Figure 1. The load distribution data is determined from the Stability Booklet, and the ballasting configuration for this operation is determined through trim and heel degree parameters at equilibrium conditions. Thus, the ballasting configuration used involves tank No.3 W.B TANK 1 (S) & (P); No.4 W.B TANK 1 (S) & (P); No.5 W.B TANK 1 (S) & (P); No.6 W.B TANK 1 (S); and No.6 W.B TANK 2 (P), all set at 100% for all loading conditions.

Table 1. Summary weight data of the Topside structure.

Description	Net Weight (kN)	Cont. Factor	Factor Weight (kN)
Structural Self Weight	479.01	1.07	512.54
Stubbing Guide	18.00	1.07	19.26
Topside Non-Generated Dead Load	66.56	1.10	73.22
Instrument Load	10.99	1.10	12.09
Electrical Load	67.16	1.10	73.88
Telecom Load	6.06	1.10	6.67
Safety Load	1.15	1.10	1.27

Mechanical Load Pre-Service	103.79	1.10	114.17
Piping Load Pre-Service	25.28	1.10	27.81
Davit Crane Selfweight	16.48	1.10	18.13
	Total Lifting Load Condition Weight		859.02

2. Methodology

2.1. Data of the Crane Barge

The XYZ crane barge data was designed based on the X barge with the operating location area in the Meliwis Gas Field. It showed the following data used in this research in Table 2–3.

Table 2. Principal dimension of XYZ crane barge.

Description	Unit	Value
Length Overall (LOA)	m	100.584
Length Between Perpendicular (LPP)	m	100.584
Breadth (moulded)	m	33.528
Depth (moulded)	m	6.096
Draft (moulded)	m	4.692
Displacement	ton	14,422.11

Table 3. Crane data.

Description	Information
Model	SL6000
Type	Heavy Duty Crane Boom Lifting
Radius of Main Hook	36.00 m
Main Hook Lifting Capacity	340 ton @8.3 m

2.2. Modelling of Crane Barge

The numerical modeling of the XYZ crane barge consisted of the barge hull, crane, and superstructure, using MAXSURF Modeller regarding the data in Table 1–2 and the general arrangement (GA) in the Stability Booklet. Meanwhile, the numerical modeling of the tanks referred to the data in the Stability Booklet using MAXSURF Stability—the numerical modeling of the XYZ crane barge, as shown in Figure 2. Then, the numerical modeling was validated based on the data hydrostatic and volume tanks from the existing ship regarding the validation tolerance limits according to American Bureau of Shipping (ABS) regulations [8]. The result of the numerical modeling shows that the numerical modeling has satisfied the tolerance limits of the ABS regulations. Thus, the model can be used for numerical longitudinal strength calculations using MAXSURF Stability.

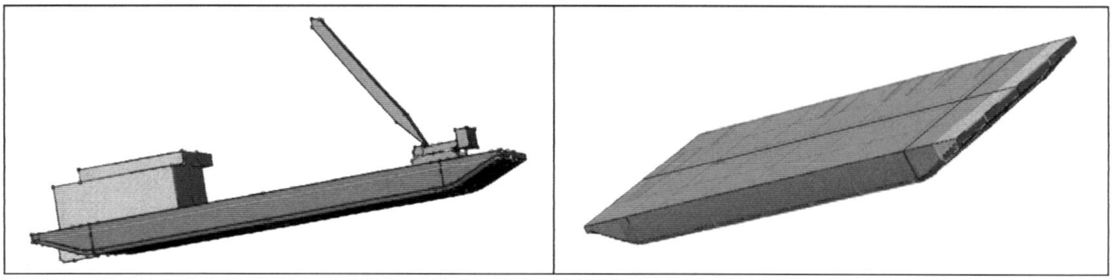

Figure 2. Model of XYZ crane barge.

2.3. Calculate the Longitudinal Strength

The analysis is conducted first by analyzing the buoyancy distribution, considering the effect of trim (x) using the equation (1)–(3):

- Fore draft (dF) > Aft draft (dA)

$$x = rs \cdot \frac{(dF-dA)}{LPP} \tag{1}$$

- Fore draft (dF) < Aft draft (dA)

$$x = (LPP - rs) \cdot \frac{(dF-dA)}{LPP} \tag{2}$$

- For station/frame 0

$$x = (LPP + rs) \cdot \frac{(dF-dA)}{LPP} \tag{3}$$

with:
rs = distance of station/frame

After obtaining the draft values at each station/frame, the buoyancy values can be determined using interpolation. Then, for the calculation of load distribution in the form of LWT and DWT, the equation (4) is used [9]:

$$Q_{x1} = \frac{\frac{(Q_{x0,n}+Q_{x0,n+1})}{2} \cdot d}{distance\ of\ station/frame} \tag{4}$$

with:
Q_{x0} = first mass distribution at each station/frame
d = distance of distribution

The shear force and bending moment can be determined from the obtained load distribution and buoyancy distribution, where shear force is the integration of the difference between the load distribution and buoyancy distribution (net load), and bending moment is the integration of the shear force. The equations for shear force and bending moment used shown in equation (5)–(6) [10]:

$$Shear\ Force = \int_{Aft\ St./Fr.}^{x} Net\ Load(x)dx \tag{5}$$

$$Bending\ Moment = \int_{Aft\ St./Fr.}^{x} Shear\ Force(x)dx \tag{6}$$

The steps for conducting a numerical analysis of longitudinal strength are:

- Modeling the barge hull, crane, and superstructure.
- Modeling the tanks.
- Modeling the operational conditions.

The rules or standards in this case are DNV RU-SHIP Pt. 3 Ch. 5 [11] for the allowable shear force and bending moment. The equations used for the still water conditions are shown in equation (7)–(10):

- Shear Force

$$Q_{sw-pos-min} = +\frac{5f_{qs}M_{sw}}{L} \quad (7)$$

$$Q_{sw-neg-min} = -\frac{5f_{qs}M_{sw}}{L} \quad (8)$$

- Bending Moment

$$M_{sw-h-min} = +0.1f_{sw}[C_W L^2 B(1.197 - 0.19C_B)] \quad (9)$$

$$M_{sw-s-min} = -0.05186f_{sw}[C_W L^2 B(C_B + 0.7)] \quad (10)$$

with:
f_{qs} = distribution factor along the ship
M_{sw} = absolute maximum of M_{sw-h} dan M_{sw-s} with
$f_{sw} = 1.0$
L = length overall (m)
f_{sw} = distribution factor along the ship
C_W = wave coefficient
B = breadth
C_B = block coefficient

For the comparison of shear force and bending moment results based on the analytical method against the numerical method with an error percentage below 5%, where the numerical method results are used as a reference due to the lack of existing ship data, the equation (11) is used [12]:

$$Deviation\ (\%) = \frac{(Base\ Value - Applicant's\ Value)}{Permissible\ Value} \cdot 100 \quad (11)$$

with:
$Base\ Value$ = numerical method data
$Applicant's\ Value$ = analytical method data
$Permissible\ Value$ = allowable value

3. Result and Discussion

The analysis of longitudinal strength on XYZ crane barge is reviewed based on 55 frames, under ballast conditions (condition 0), and with lifting load directions ranging from the starboard side to the bow of the ship at every 15 degrees (conditions 1–7), as shown in Figure 3.

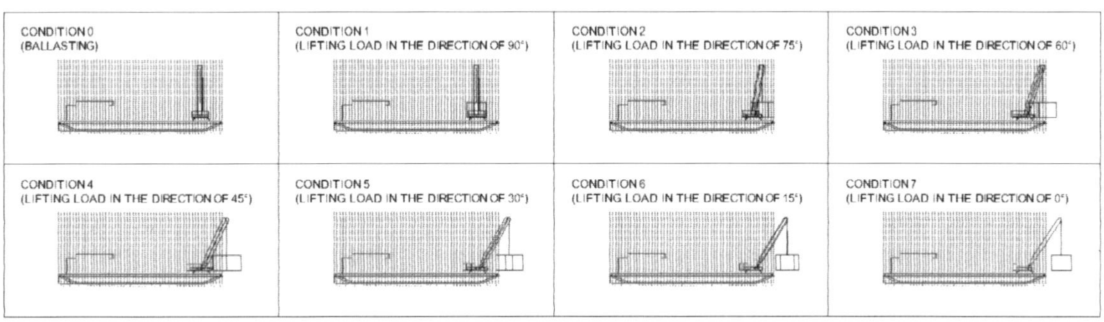

Figure 3. Division of XYZ crane barge based on 55 frames in operational condition (side view).

The results from the conducted analysis using analytical and numerical methods include the load distribution, shear force, bending moment that occurs in response to the load distribution, and the comparison between the analytical and numerical results for shear force and bending moment. For the load distribution on XYZ crane barge under each lifting load direction condition shown in Figure 4, the shear force and bending moment values occurring on XYZ crane barge under each lifting load direction condition indicated in Figure 5 and Figure 6, as well as the comparison of longitudinal strength results on XYZ crane barge during operations, are presented in Table 4.

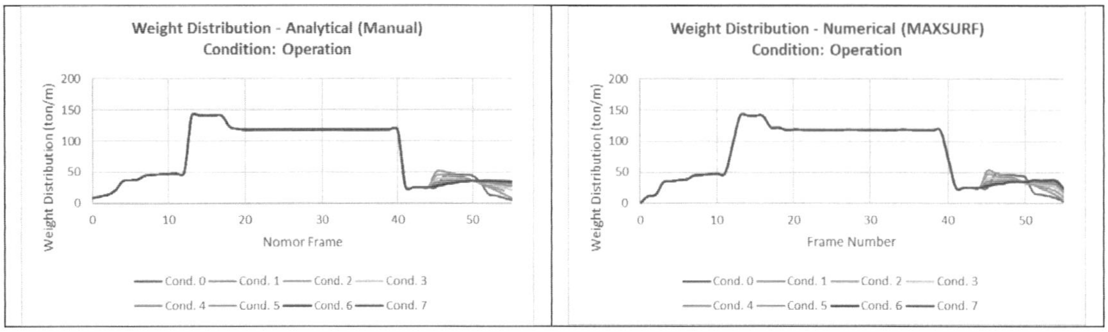

Figure 4. Load distribution graphics in each condition of lifting load direction using analytical and numerical methods.

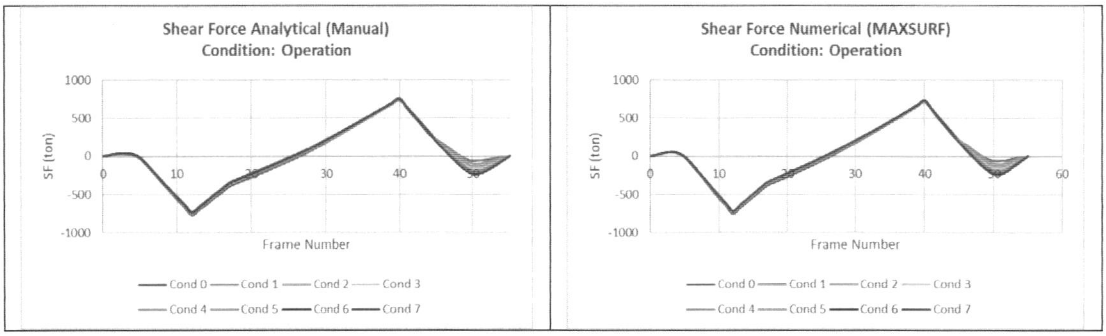

Figure 5. Shear force graph in each condition of lifting load direction using analytical and numerical methods.

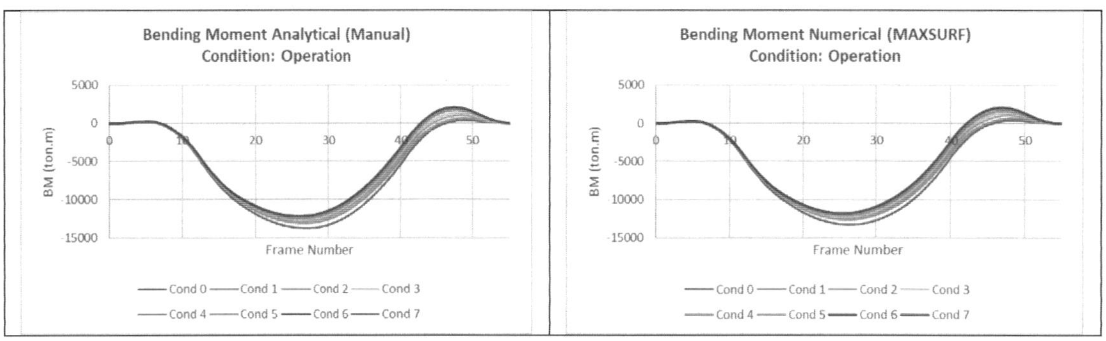

Figure 6. Bending moment graph in each condition of lifting load direction using analytical and numerical methods.

It can be observed that the trendline of the load distribution occurring on the XYZ crane barge, using both the analytical and numerical methods, exhibits similarities. The positioning of each load along the vessel's length caused significant changes in the spreading pattern along the XYZ crane barge. The increases are noted in the position from frame 3 to frame 19, which corresponds to the location of the superstructure, from frame 13 to frame 40, where the water ballast tanks are located, and from frame 45 to frame 55, which corresponds to the lifting load area with variations in the crane rotation angle. The shear force and bending moment results on the XYZ crane barge, using both the analytical and numerical methods, show similar trendline shapes. The maximum shear force occurs at frames with significant net load values transitioning from positive to negative, and the maximum bending moment occurs at frames where the crane is positioned. Specifically, the shear force occurs at frame 40, while the bending moment occurs at frame 47, frame 48, and frame 49, as tabulated in Table 4.

IOP Conf. Series: Earth and Environmental Science 1298 (2024) 012021 doi:10.1088/1755-1315/1298/1/012021

Based on the results in Table 4, the analytical and numerical methods exhibit increased shear force and bending moment values with each change in the lifting load rotation angle. The highest shear force and bending moment values occur at condition 7 using analytical and numerical methods. Specifically, for condition 7, the shear force values are 755 tons and 736 tons, while the bending moment values are 2107 tons and 2126 tons.m. Table 5 shows that comparing shear force and bending moment results between the analytical and numerical methods on the XYZ crane barge complies with the tolerance limits. The percentage errors obtained are below 5% when using the standard DNV RU-SHIP Pt. 3 Ch. 5 as allowable. The maximum percentage errors occur at frame 12, frame 40, and frame 51. The most significant percentage errors for shear force and bending moment among all conditions are observed in condition 7, where the percentage errors are 1.52% and 4.82%.

Table 4. Maximum shear force and bending moment values.

Maximum Shear Force and Bending Moment									
Longitudinal Strength	**Unit**	**Cond 0**	**Cond 1**	**Cond 2**	**Cond 3**	**Cond 4**	**Cond 5**	**Cond 6**	**Cond 7**
Analytical Method									
Shear Force	ton	758 (*fr*. 40)	732 (*fr*.40)	738 (*fr*. 40)	743 (*fr*. 40)	748 (*fr*. 40)	752 (*fr*. 40)	754 (*fr*. 40)	755 (*fr*. 40)
Bending Moment	ton.m	400 (*fr*. 49)	680 (*fr*. 49)	1011 (*fr*. 48)	1364 (*fr*. 48)	1666 (*fr*. 48)	1899 (*fr*. 48)	2051 (*fr*. 47)	2107 (*fr*. 47)
Numerical Method									
Shear Force	ton	739 (*fr*. 40)	713 (*fr*.40)	719 (*fr*. 40)	724 (*fr*. 40)	729 (*fr*. 40)	733 (*fr*. 40)	735 (*fr*. 40)	736 (*fr*. 40)
Bending Moment	ton.m	399 (*fr*. 49)	690 (*fr*. 48)	1034 (*fr*. 48)	1358 (*fr*. 47)	1676 (*fr*. 47)	1920 (*fr*. 47)	2073 (*fr*. 47)	2126 (*fr*. 47)

Note: fr. = frame

Table 5. Percentage of maximum longitudinal strength error.

Error Maximum Longitudinal Strength (LS) [%]								
LS	**Cond 0**	**Cond 1**	**Cond 2**	**Cond 3**	**Cond 4**	**Cond 5**	**Cond 6**	**Cond 7**
Shear Force	1.52% (*fr*. 40)	1.47% (*fr*. 40)	1.47% (*fr*. 40)	1.52% (*fr*. 40)	1.51% (*fr*. 40)	1.49% (*fr*. 40)	1.51% (*fr*. 40)	1.50% (*fr*. 40)
Bending Moment	3.19% (*fr*. 40)	3.12% (*fr*. 40)	3.11% (*fr*. 40)	3.35% (*fr*. 51)	3.96% (*fr*. 51)	4.45% (*fr*. 51)	4.74% (*fr*. 51)	4.82% (*fr*. 51)

Note: fr. = frame

4. Conclusions

It can be concluded that the trendline of the load distribution on the crane barge shows changes in the shape of the load distribution for each rotation angle of the lifting load. These changes occur in the lifting load region, specifically between frame 45 and frame 55. The maximum shear force occurs at frame 40 for each rotation angle of the lifting load. It corresponds to a significant change in the net load value from positive to negative, where the analytically and numerically calculated shear force values are 775 tons and 736 tons, respectively. The maximum bending moment occurs at frame 47, corresponding to the crane's position, where the analytically and numerically calculated bending moment values are 2107 ton.m and 2126 ton.m, respectively. With these results, the percentage error for both shear force and bending moment, compared between analytical and numerical calculations, remains below 5% of the allowable limits based on DNV Rules, where the maximum percentage errors are 1.52% and 4.82%.

5. References

[1] ESDP 2023 ESDEP Course http://fgg-web.fgg.uni-lj.si/~/pmoze/ESDEP/master/toc.htm

[2] Xu X, Yang J, Li X and Xiao L 2012 *A Study of Floatover Installation onto Jacket with Two Barges* Proceedings of the 17th Offshore Symposium USA: Houston, Texas p D3–D14.

[3] Firdaus N, Djatmiko E B, Prastianto R W and Muryadin 2022 *Experimental Study on Coupled Motion of Floating Crane Barge and Lifted Module in Irregular* IOP Conf. Series: Earth and Environmental Science Indonesia: Surabaya 972, p 1–9.

[4] Van der Heiden A J 2019 *Vessel and Cargo Motions: A Frequency Domain Method to Study Combined Vessel and Cargo Responses* Thesis Delft University of Technology Netherlands: Delft.

[5] Firdaus N, Ali B, Hidayat A, Nantabah Z K, Djatmiko E B and Prastianto R W 2023 Preliminary Modeling of Coupled Motion Response for Floating Crane and Suspended Module by Using Boundary Element Approach and Empirical Model *EVERGREEN Joint Journal of Novel Carbon Resource Sciences & Green Asia Strategy* **10** p 00-00.

[6] Sun Y, Li W, Dong D, Mei X and Qiang H 2015 Dynamics Analysis and Active Control of a Floating Crane *Technical Gazette 22* **22** p 1383–1391.

[7] Utomo B and Santoso B 2019 S*trength Analysis of Vessel Ship Type PBL Conversion from Barge* International Conference on Maritime and Archipelago (ICoMA 2018) 167, p 364–367.

[8] ABS 2023 *Rules for Building and Classing Mobile Offshore Units - Part 3 Hull Construction and Equipment. MOU-Part 3* American Bureau of Shipping USA: Houston.

[9] Selby, B A 1967 *Computer Analysis of Longitudinal Strength of Ships* Masters Thesis Durham: Durham University.

[10] Henschke W 1957 *Schiffbautechnisches Handbuch Band 1* Veb Verlag Technik Berlin Berlin.

[11] DNV 2023 *Ships - Part 3: Hull, Ch 4: Loads. DNV-RU-SHIP Pt. 3 Chp. 4* Det Norske Veritas Norway: Høvik.

[12] BKI 2021 *Guidelines for Certification of Loading Computer System* Biro Klasifikasi Indonesia Indonesia: Jakarta.

Real time AIS tracking system for fishermen using microcontroller-based approach

A Rohman[1], J Rulianto[1], A Wahid[1]

[1]Department of Ship Manufacture, College Mechanical Engineering, Politeknik Negeri Banyuwangi, East Java, Indonesia.

*Corresponding author: jangka.rulianto@poliwangi.ac.id

Abstract. Fishing activities in the waters of Banyuwangi are relatively intense. Local fishermen often venture 50 nautical miles from the coastline using small boats. These boats are lacking adequate communication and navigation equipment, causing difficulties in obtaining assistance during emergencies. This study aims to develop an Automatic Identification System (AIS) device to be installed on fishing boats. The AIS device developed for this research incorporates the STM32F4 microcontroller, CMX7032 Evaluation Board, VHF Antenna, and Global Positioning System (GPS) as its main components. The integration and configuration of each component will be elaborated in this paper. Performance testing has yielded promising results, as the AIS device was capable of transmitting data up to 10 nautical miles from the Base Station. Utilization of AIS technology would enhance safety and communication among fishermen. Through careful integration and configuration of these components, the AIS device effectively ensures that the fishermen's positions are accurately tracked and communicated. The successful development and testing of this device underscore its potential to bridge the communication gap, offering timely assistance and enhancing the overall safety of these seafarers.

1. Introduction

This research aims to enhance the security and efficiency of small-scale fishing activities through the development of an enhanced Automatic Identification System (AIS) [1]. This is crucial as small-scale fishing often occurs in remote and high-risk areas, necessitating a system that can help reduce risks and increase productivity [2]. Within this study, an assessment of the needs and challenges faced by fishermen in small-scale fishing activities is conducted. Additionally, an analysis of existing AIS systems is performed to understand their strengths and weaknesses. Based on the analysis results, an enhanced AIS system with various additional features that better cater to fishermen's needs is developed. Subsequently, this upgraded AIS system is tested at several small-scale fishing locations in Indonesia [3]. It is anticipated that the testing outcomes will demonstrate the capability of the enhanced AIS system to improve safety and efficiency in small-scale fishing activities [4]. Furthermore, the enhanced AIS system can aid in the supervision and management of fisheries resources more effectively. With the implementation of this upgraded AIS system, fishermen engaged in small-scale fishing can mitigate safety risks at sea and elevate their fishing productivity [5]. Moreover, the enhanced AIS system can also assist the government in more effective and sustainable supervision and management of fisheries resources [6]. Overall, the development of this enhanced AIS system can bring positive impacts to small-scale fishing activities, especially in hard-to-reach and high-risk areas [7]. This research is expected to serve as a foundation for the future optimization of AIS systems, thereby contributing to the well-being of fishermen and the sustainability of fisheries resources [8]. The utilization of AIS on fishing boats is expected to aid fishermen in reporting their emergency situations and facilitate Search and Rescue (SAR) operations by relevant parties in case of maritime accidents [8]. However, this solution still presents challenges. The issue that arises is that the cost of commercial AIS devices available in the market is not affordable for the fishing community [6]. This is the driving force behind conducting research to develop an inexpensive AIS Transmitter that can be customized to meet the needs of the Banyuwangi fishermen [9]. The prototype of the Automatic Identification System (AIS) Transmitter developed in this research is equivalent to AIS Class B. This device is capable of periodically transmitting the fishermen's position information, which will be monitored by VTS LPTK and other vessels equipped with AIS transponders [10]. Additionally, a distress function is embedded as an additional feature in this device. In emergency situations, fishermen can activate this distress function to send messages to nearby VTS and AIS transponders [10].

The Automatic Identification System (AIS) is a system designed to transmit information between vessels and maritime traffic control centers (Vessel Traffic Control Centers, VTCC) using VHF radio signals [11]. This system was first introduced in 2002 and has since become an international standard required for vessels with a gross tonnage above 300 and passenger vessels with a capacity exceeding 12 passengers. AIS is used to assist vessel traffic services (VTS) and enhance maritime navigation safety. AIS has undergone several technological developments since its introduction in 2002[12]. Recent research has focused on improving the accuracy and reliability of the AIS system and expanding its capabilities for uses beyond navigation and vessel traffic monitoring [13].

The Automatic Identification System (AIS) technology is a digital device used as a navigation instrument in maritime navigation. This device integrates Global Positioning System (GPS) technology, Very High-Frequency (VHF) radio technology, and information technology into one system. AIS enables the exchange of maritime data between ships, base stations, and other vessels [14]. The exchanged data includes ship identity (MMSI, Call Sign, and ship name), ship position, direction, size, cargo type, and other important information [15]. The initiation of AIS technology was by the International Association of Maritime Aids to Navigation and Lighthouse Authorities (IALA), followed by development and standardization by the International Maritime Organization (IMO), International Telecommunication Union (ITU), and International Electrotechnical Commission (IEC) in the early 2000s [16]. Over time, AIS became mandatory for all passenger ships, tanker ships, and cargo ships with a Gross Tonnage (GT) 300 for international routes [17]. Since 2008, AIS has also been mandatory for all domestic ships with a GT 500. Koto, J. and Amran, N.A (2016) classified AIS devices into six types: Class A, Class B, Base Station, Aids to Navigation (AtoN), Search and Rescue Transponder (SART), and Specialist AIS Transponder. Each device can operate on the maritime VHF frequencies of 161.975 MHz (AIS channel 1 or channel 87B) and 162.025 MHz (AIS channel 2 or channel 88B). Communication among AIS devices use the TDMA (Time Division Multiple Access) protocol established by ITU-R (2014). TDMA protocol regulates the usage of VHF channels by dividing each channel into 2250 time slots per minute. Each time slot can be used for 25.6 milliseconds to transmit AIS messages [18].

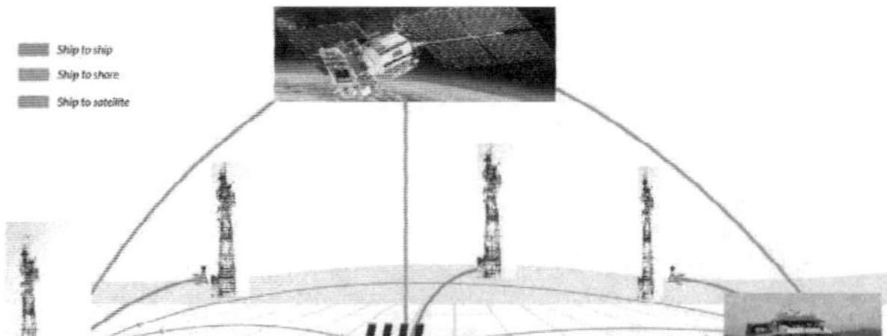

Figure 1. Marine Automatic Identification System

IOP Conf. Series: Earth and Environmental Science 1298 (2024) 012022 doi:10.1088/1755-1315/1298/1/012022

Importance of using AIS as a navigation instrument has led to research efforts to develop affordable devices. Some research explored the development of low-cost AIS receivers using a fast time simulator based on the AIS CMX7042 chip [19]. Similar research was conducted using the Radiometrix RX1 chip and the Arduino Uno R3 microcontroller [20]. Referring to these research findings, the technical specifications for the developed AIS Transmitter are outlined as follows:

Table 1. Technical Specifications of AIS

Parameter	Value
AIS channel 1	161.975 MHz
AIS channel 2	162.025 MHz
Channel spacing	25 kHz
Channel bandwidth	25 kHz
Modulation mode	GMSK
Bit rate	9.6 Kbps
Transmit power	2 Watts

2. Method

The method employed in this research is the design and construction method with an engineering approach. The AIS Transmitter is developed by integrating several modules into a single unit. These modules consist of the STM32F4 microcontroller, AIS CMX7032 Evaluation Board (DE70321T), GPS Receiver, and Shakespeare VHF Antenna.

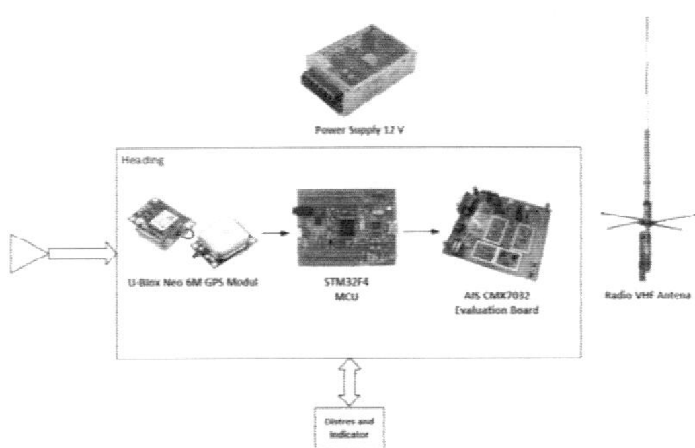

Figure 2. Wiring Diagram of AIS

2.1. GPS Module

The GPS Receiver (Figure 4) is used to obtain position information, speed, and other dynamic information from GNSS (Global Navigation Satellite System) satellites. The Neo-6M GPS Receiver from U-blox is used in this research. This GPS Receiver has three advantages that support the design requirements of the AIS Transmitter: low power consumption, compact size, and affordability.

Figure 3. GPS Mono Module

IOP Conf. Series: Earth and Environmental Science 1298 (2024) 012022 doi:10.1088/1755-1315/1298/1/012022

2.2. STM32F4 Microcontroller

The STM32F4 microcontroller (Figure 5) is one of the products from STMicroelectronics that utilizes the ARM Cortex-M4 32-bit architecture. It operates within a voltage range of1.8 to 3.6 volts. The STM32F4 microcontroller processes and controls data flow in the AIS system. Position data from the GPS receiver is compiled along with additional information such as MMSI (Maritime Mobile Service Identity) and ship name, then transmitted to the AIS data processor (CMX7032 Evaluation Board) via the C-Bus communication protocol.

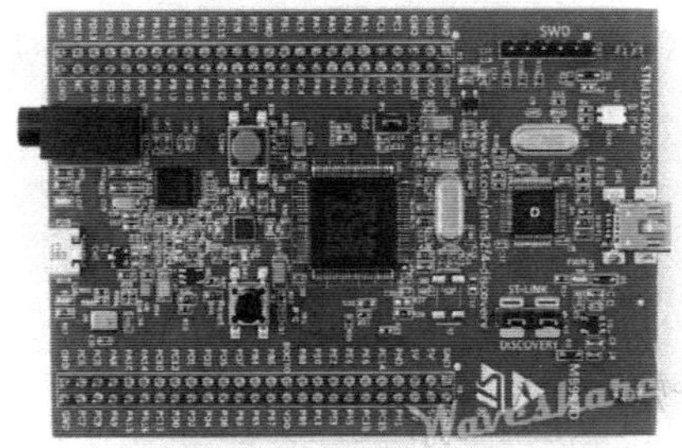

Figure 4. STM32f4 Micro *Controller*

2.3. CMX7032 Evaluation Board (DE70321T)

The DE70321T module (Figure 6) integrates the CMX7032 Chip with a 2-Watt Power Amplifier, Voltage Controlled Oscillator (VCO), Harmonic Filter, and Antenna Switching. In the AIS system, this module encodes position data, speed, MMSI, ship name, and other data into AIS Messages.

Figure 5. CMX7032 Evaluation Board (DE70321T)

2.4. VHF Antenna

The developed AIS device uses a Shakespeare VHF Antenna. This antenna is 1.2 meters long with a 3dB gain and a maximum input power of 50 Watts. An example of the Shakespeare VHF Antenna is shown in Figure 7.

Figure 6. VHF *Antenna*

3. Result and discussion

The implementation of the AIS Transmitter device in this study refers to the specifications outlined in Table 1. The configuration begins by assembling all modules and connecting them to the power supply unit. The STM32F4 microcontroller is set up to acquire position data from the GPS receiver and embed additional information such as the MMSI (MMSI 525150296 was used for testing purposes). The STM32F4 is then connected to the CMX7032 Evaluation Board using the C-Bus communication protocol. The next step involves uploading the function image to the CMX7032 Chip, containing the message encoding settings (AIS Message).

Figure 7. Electronic Modul

The types of AIS Messages utilized in this device are AIS Message 14 and AIS Message18. Following the ITU-R regulations (2014), AIS Message 14 is designated for AIS Safety Related Broadcast Messages, while AIS Message 18 is used for position reporting in AIS Class B devices. In this AIS Transmitter device, AIS Message 14 is employed to transmit emergency messages (distress). When activated (by pressing the red button on the device), the emergency message will be broadcasted on VHF channels 87B and 88B (161.975 MHz and 162.025 MHz). AIS Message 18 is configured to periodically transmit ship position information at a 30-second interval.

The device's performance displayed in Table 2 is obtained through two testing scenarios: periodic data transmission testing and emergency data transmission testing. Periodic data transmission testing ensures that position information and MMSI can be received by the VTS Ketapang Banyuwangi. This testing involves activating the device by pressing the power button. If the periodic data transmission function works properly, the position information and MMSI will be recorded by the VTS Ketapang Banyuwangi. The results shown in Figure 9 demonstrate that AIS Message 18 transmission (MMSI and ship position) can be received by VTS Ketapang Banyuwangi up to a distance of 10 nautical miles.

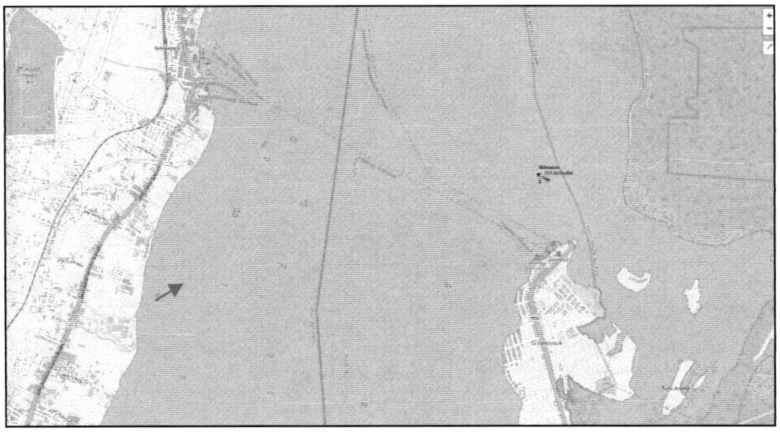

Figure 8. AIS transmitter's position and MMSI information recorded by VTS Ketapang Banyuwangi.

4. Conclusion

It can be concluded that the prototype of the AIS Transmitter based on the STM32F4 microcontroller functions effectively. The modular concept employed proves to save time and costs in the development of this device. The packaging used only provides protection against seawater splashes, therefore further research is needed to design packaging that offers maximum protection for the device.

Acknowledgement

This work was supported by Research Pusat Penelitian dan Pengabdian Masyarakat Politeknik Negeri Banyuwangi. with entitled "Analisa Mini Automatic Identification System Untuk Nelayan". The Support is gratefully recognized by the authors.

Reference

[1] A. F. , T. S. R. and H. D. A. Molland, *Ship resistance and propulsion : practical estimation of ship propulsive power.* Cambridge. GB. Cambridge University Press, 544pp., 2011.
[2] Sv. A. Harvald, *Tahanan dan Propulsi Kapal.* Airlangga University Press: Surabaya, 1992.
[3] E. V. Lewis and Society of Naval Architects and Marine Engineers (U.S.), *Principles of naval architecture. Volume II, Resistance, propulsion and vibration.* Society of Naval Architects and Marine Engineers, 1988.
[4] "Ship Construction - D.J. Eyres".
[5] "Principles Of Naval Architecture Vol III - Motion in Waves & Controllability".
[6] D. G. M. Watson, "ELSEVIER OCEAN ENGINEERING BOOK SERIES VOLUME 1 PRACTICAL SHIP DESIGN."
[7] Y.-S. Wu, W.-C. Cui, and G.-J. Zhou, "Practical Design of Ships and Other Floating Structures Volume I Edited by."
[8] "Basic Ship Theory."
[9] M. G. Parsons, "CHAPTER 11 PARAMETRIC DESIGN 11.1 NOMENCLATURE."
[10] H. (Herbert) Schneekluth and Volker. Bertram, *Ship design for efficiency and economy.* Butterworth-Heinemann, 1998.
[11] J. M. J. Journée, "Prediction of Speed and Behaviour of a Ship in a Seaway," 1976. [Online]. Available: www.shipmotions.nl
[12] P. Kyropoulos, "MARINE DIESEL POWER PLANTS and Ship Propulsion ENGINEERING AND SCIENCE MONTHLY."
[13] "Bulk Carrier Safety _19 January 2009".
[14] "Introduction to Marine Engineering."
[15] "Marine engineering. Vol.3. 1899".
[16] "Principles of Naval Architecture Vol I - Stability And Strength".

[17] H. (Herbert) Schneekluth and Volker. Bertram, *Ship design for efficiency and economy*. Butterworth-Heinemann, 1998.

[18] H. Schneekluth and V. Bertram, "Chapter 6 - Ship propulsion," in *Ship Design for Efficiency and Economy (Second edition)*, Second edition., H. Schneekluth and V. Bertram, Eds., Oxford: Butterworth-Heinemann, 1998, pp. 180–205. doi: https://doi.org/10.1016/B978-075064133-3/50006-2.

[19] K. J. Rawson and E. C. Tupper, "10 - Powering of ships: general principles," in *Basic Ship Theory (Fifth Edition)*, Fifth Edition., K. J. Rawson and E. C. Tupper, Eds., Oxford: Butterworth-Heinemann, 2001, pp. 365–410. doi: https://doi.org/10.1016/B978-075065398-5/50013-3.

[20] T. Greenhalgh, R. Stones, and D. Swinglehurst, "Choose and Book: A sociological analysis of 'resistance' to an expert system," *Soc Sci Med*, vol. 104, pp. 210–219, 2014, doi: https://doi.org/10.1016/j.socscimed.2013.12.014.

A comprehensive analysis of hydrogen fuel cell integration in fishing vessels

I G N A Satria[1*], J Rulianto[1], Kanom[1]

[1]Department of Ship Manufacture, College Mechanical Engineering, Politechnic Negeri Banyuwangi, Esat Java, Indonesia, Indonesia.

*Corresponding e-mail : jangka.rulianto@poliwangi.ac.id

Abstract. HHO gas is a fuel supplement produced through the electrolysis process of water in an HHO generator, where HHO gas falls under the category of alternative energy and is classified as a new energy source. The fundamental principle of internal combustion engines is generally to convert the chemical energy of liquid fuel into mechanical energy. The production of hydrogen through the electrolysis process is continuously evolving to facilitate the use of environmentally friendly fuels, with potential impacts on engine performance and exhaust gas emissions. This study examines the effect of adding HHO gas to electronic fuel injection (EFI) type-L gasoline engines on exhaust gas emissions and fuel consumption. It has been proven that the addition of HHO gas to the combustion process of gasoline engines, particularly EFI type-D, results in reduced emission levels (CO by 15.31% and HC by 16.27%). Moreover, the introduction of 0.1 Mol of HHO gas into EFI type-L engines through the air filter leads to a significant impact on exhaust gas emissions and a 2.8% increase in fuel consumption compared to not using HHO gas. This could be attributed to the distinct characteristics of EFI systems compared to conventional engines, as they are not accustomed to fuel supplements being injected into the engine.

1. Introduction

HHO stands for Hydrogen-Hydrogen-Oxygen, also known as Brown's Gas. HHO is a mixture of hydrogen and oxygen gases generated through the electrolysis of water using electric current3. It is believed that HHO can assist in reducing fuel consumption in fishing vessels by enriching the air-fuel mixture entering the engine [1]. HHO is injected into the engine's fuel system, resulting in an air-fuel mixture that is richer in hydrogen and oxygen [2]. This is claimed to enhance fuel combustion efficiency, thereby reducing fuel consumption and leading to cleaner exhaust emissions [3]. However, it should be noted that several studies and field tests have not provided sufficient evidence to conclusively demonstrate that HHO usage significantly reduces fuel consumption in fishing vessels. Moreover, the use of HHO also entails risks, such as the potential for explosions if the produced HHO gas is not stored or utilized properly [4]. Therefore, prior to implementing HHO for fuel consumption reduction in fishing vessels, thorough evaluation and testing are necessary to ensure its safety and effectiveness [5].

The utilization of HHO for fishing vessels can have varying economic impacts contingent upon numerous factors, including the cost and effectiveness of HHO usage, as well as fluctuating fuel prices [6]. On one hand, employing HHO in fishing vessels can contribute to operational cost reduction by lowering fuel consumption [7]. This can aid fishermen in augmenting their income by diminishing operational expenses. Furthermore, HHO usage can assist in mitigating the environmental impact of fossil fuel consumption, thereby enhancing the image and reputation of fishermen in the eyes of the public and consumers. Conversely, the acquisition and installation costs of HHO devices on fishing vessels can be quite high and require significant investment4. Additionally, the efficacy of HHO usage has not been fully proven and remains a subject of debate among experts [8]. If HHO usage proves ineffective, it can lead to losses for fishermen and diminish their income [9]. Moreover, fluctuating fuel prices can also influence the economic implications of HHO usage in fishing vessels5. If fuel prices drop significantly, the benefits of HHO usage may become insignificant and not justify the investment costs [10].

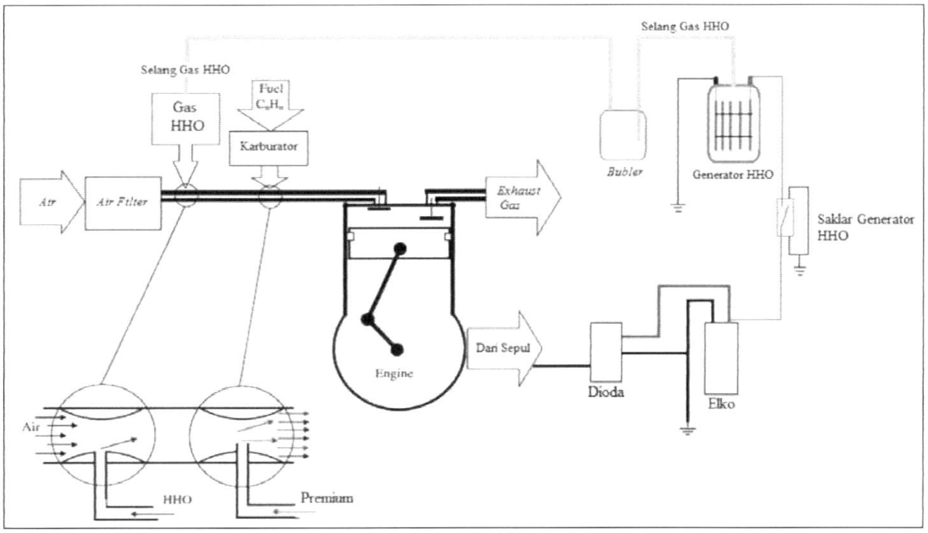

Figure 1. General HHO Schematic Diagram

1.1. HHO Gas

Hydrogen is one of the most abundant elements, forming the basis of all matter in the universe. Its atoms consist of a single electron orbiting around its nucleus. Hydrogen transitions into a liquid state at a temperature of 20 K (-423 ° F; -253 ° C) [11]. When reacted fully with oxygen, fuel generates a considerable amount of energy. The energy content of various fuels is presented in Table 1.

Table 1. Fuel Oil Calories

Fuel Oil	**Top Heat (25°C and 1 atm)**	**Lower Heat (25°C dan 1 atm)**
Hidtogen	121.86 MJ/kG	119.93 MJ/kG
Metana	55.53 MJ/kG	50.02 MJ/kG
Propana	50.36 MJ/kG	45.6 MJ/kG
Bensin	47.5 MJ/kG	44.5 MJ/kG
Diesel	44.8 MJ/kG	42.5 MJ/kG
Metanol	19.96 MJ/kG	18.05 MJ/kG

HHO gas or Brown's gas is a gas produced by the electrolysis of water, which is a combination of two gases, hydrogen (H2) and oxygen (O2). The simple equation for converting water into HHO can be expressed as follows:

$$H2O\ (l) \rightarrow HHO\ (g) \quad (1)$$

Electrolysis cells connected to an electric current will separate water into hydrogen and oxygen:

$$2H2O + energy \rightarrow 2H2 + O2 \quad (2)$$

The electrolysis process can be accelerated with the aid of a catalyst, and during the process, the catalyst does not affect the form of the electrolytic outcome. The catalyst aims to expedite electrolysis reaction and reduce the amount of activation energy required in a process:

$$2H_2O + \text{catalyst} + \text{energy} \rightarrow 2H_2 + O_2 + \text{catalyst} \quad (3)$$

Thus, the chemical reaction occurring in electrolysis is [13]:

$$2H_2O \rightarrow 2H_2 + O_2 \quad (4)$$

Electrolysis is a chemical process that converts electrical energy (electric current) into chemical energy (redox reactions) [11]. It involves the decomposition of the constituent elements of water. By applying an electric current, two water molecules react by capturing two electrons at the cathode, reducing into H2 gas and hydroxide ions (OH-) are formed [12]. At the anode pole, two other water molecules will be broken down into oxygen gas (O2), releasing 4 H+ ions and flowing electrons to the cathode. As a result of these reactions, H+ ions and OH- ions neutralize each other and form water molecules again. The hydrogen and oxygen gases produced by this reaction form bubbles and accumulate around the electrode. This principle is then utilized to produce hydrogen and hydrogen peroxide (H2O2) [13].

The key components of this electrolysis process are the electrodes and the electrolyte solution. In the electrolysis process, two electrodes are required: the cathode as the negative pole and the anode as the positive pole [14]. The electrolysis process occurs in a specially designed chamber, called an HHO generator, which is a gas-producing unit. HHO generators are classified into 2 types: dry cell HHO generators and wet cell HHO generators. In the dry cell type, the electrodes are not immersed in the electrolyte, and the electrolyte only fills the gaps between the electrodes themselves. The area of the electrode plates submerged in water is where electrolysis occurs to generate HHO gas, while the remaining area is not submerged in water and the plates are in a dry state. In the wet cell type, all electrodes are immersed in an electrolyte solution inside a water container [15].

1.2. Catalyst

A catalyst is defined as a substance that accelerates the rate of a reaction, yet the catalyst itself undergoes no chemical change. The catalyst is not included in the stoichiometric reaction equation, and its concentration in the reaction mixture remains unchanged [14], [16]. This is only possible if, in one stage of the reaction, the catalyst participates and is regenerated in subsequent stages [17]. Hence, a catalyst is absent from the reaction equation, and its concentration does not appear in the equilibrium expression. Thus, a catalyst doesn't affect the position of the reaction equilibrium. Generally, a catalyst will lower the activation energy [18].

In the process of electrolysis, the catalyst is a dissolved substance in water (H2O) and forms a solution as an electrolyte. Commonly used catalysts in electrolysis include vinegar (H3C-COOH), baking soda (NaHCO3), and potassium hydroxide (KOH). Using a smaller amount of catalyst reduces the rate of HHO gas production. Conversely, excessive use of catalyst raises the process temperature [19]. Typically, the electrolysis process to generate HHO gas employs KOH and NaHCO3 catalysts. The use of KOH catalyst in electrolysis is preferable to NaHCO3 catalyst because it produces more significant amounts of HHO gas and consumes less electrical power. Calculating catalyst concentration with a solvent is done using the molarity formula as follows [20]:

$$Jk = Mr \times M \times 1 \text{ mol/liter (grams)} \quad (5)$$

Where:
Jk = Catalyst amount (grams)
M = Molarity
Mr = Relative atomic mass

1.3. Design and Fabrication of an HHO Generator

An electrolysis cell with a generator casing measuring 25 cm × 30 cm is employed. The plates used are made of stainless steel 316L. The plate configuration involves 2 anode and cathode plates measuring 9 cm × 14.5 cm, along with a cover and neutral plates measuring 3 cm × 14.5 cm. This design is tailored to meet the current requirements of a four-wheeled vehicle's automotive battery, ensuring that the current does not exceed the total capacity of 15 Amps. Within the reactor cell, the neutral plates are arranged vertically, consisting of a total of 3 plates. These plates are separated by dividers, each spaced 1 mm apart for every cell. This arrangement facilitates a separation space during the electrolysis process, as depicted in Figure 2.

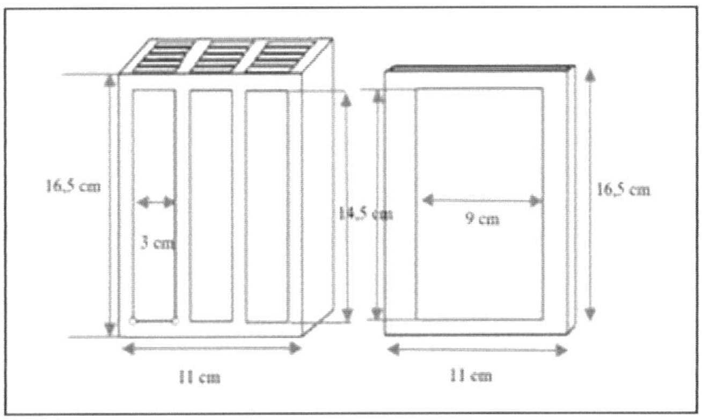

Figure 2. Hidrogen Fuel Cell Generator

2. Methode

2.1. Fungtion Test

Functional testing of the HHO generator is conducted to determine whether the constructed generator operates as expected and to ensure that there are no gas leaks. The functional test involves passing an electric current through the HHO gas generator using an automotive battery (accu) and utilizing KOH catalyst with a solution concentration of 0.1 mol.

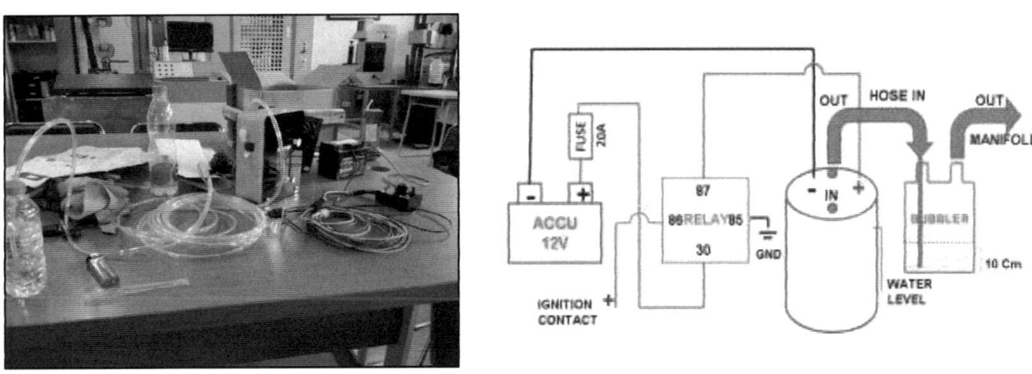

Figure 3. Schematic Diagram Of HHO

The generator is installed using the installation scheme illustrated in Figure 3. The output hose for HHO gas and the positive cable connected to the relay and automotive battery are covered with cable protectors to shield them from engine heat.

2.2. Exhaust Gas Emission and Fuel Consumption Testing

The testing process involves two stages: one without the addition of HHO gas and the other with the addition of HHO gas. The testing is carried out at idle RPM, 1500 RPM, and 2500 RPM. Fuel consumption testing consists of two phases: the first phase involves testing with gasoline fuel alone, and the second phase involves testing with the addition of HHO gas. The tests are conducted with a full tank of gasoline, and the vehicle covers a specific distance while collecting data three times.

3. Results and Discusion

This section presents the research findings and discussions, which include carbon monoxide, hydrocarbons, carbon dioxide, oxygen, and air-fuel ratio from the combustion process in internal combustion engines (IC engines).

3.1. Emision

Carbon monoxide is a colorless, odorless, and tasteless gas that is produced from incomplete combustion of carbon compounds, a common occurrence in internal combustion engines. Carbon monoxide forms when there is insufficient oxygen during the combustion process. It is highly flammable and produces a blue flame, resulting in the formation of carbon dioxide.

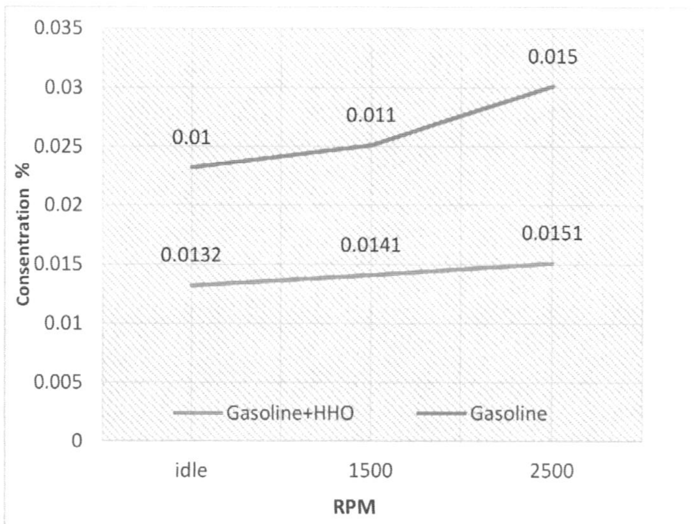

Figure 4. Graph Carbon Monoxide Consentration

The graph illustrates the emission levels of CO without using gasoline fuel at idle engine speed, which is 0.00%, 1500 RPM engine speed, which is 0.00%, and 2500 RPM engine speed, which is 0.00%. With the addition of 0.1 Mol of HHO gas, the data obtained at idle engine speed is 0.01%, 1500 RPM engine speed is 0.01%, and 2500 RPM engine speed is 0.01%. Elevated CO levels are generally caused by an imbalanced mixture of gasoline fuel and air, where the gasoline fuel content is excessively high. In injection engines, especially Type L, gasoline fuel is injected into the combustion chamber based on the airflow sensor that detects the amount of air in the air filter. The addition of HHO gas results in more fuel being injected, leading to incomplete combustion in the combustion chamber.

Hydrocarbons are compounds consisting of carbon (C) and hydrogen (H) elements. All hydrocarbons have carbon chains and hydrogen atoms bonded to those chains. The term is also used to define aliphatic hydrocarbons. Methane gas (swamp gas) is a hydrocarbon with one carbon atom and four hydrogen atoms (CH_4). Ethane is a hydrocarbon composed of two carbon atoms joined by a single bond, each binding three hydrogen atoms: C2H6. Propane contains three C atoms (C3H8), and so forth (CnH2·n+2).

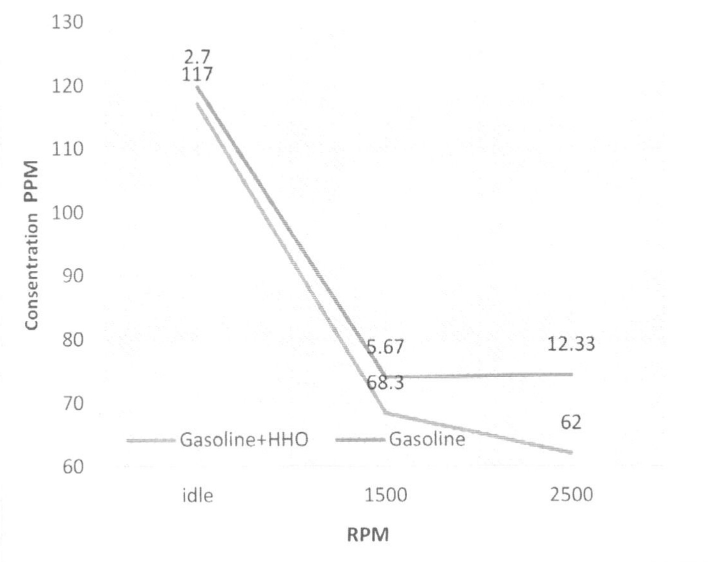

Figure 5. Hydrocarbon Concentration

The results show hydrocarbon gas emissions with gasoline fuel at idle engine speed, which is 2.7 ppm, 1500 RPM engine speed, which is 5.67 ppm, and 2500 RPM engine speed, which is 12.33 ppm. With the addition of 0.1 Mol of HHO gas, the data obtained at idle engine speed is 111 ppm, 1500 RPM engine speed is 68.3 ppm, and 2500 RPM engine speed is 62 ppm. The elevated HC emissions after adding 0.1 Mol of HHO gas indicate three possible causes: a malfunctioning catalytic converter (CC), incorrect air-fuel ratio (AFR), or incomplete combustion of gasoline in the combustion chamber. If the catalytic converter is functioning normally but HC remains high, it suggests symptoms of an incorrect AFR or misfire. For vehicles equipped without a catalytic converter (CC), the tolerable HC emission level is 500 ppm, and for those equipped with a CC, the tolerable HC emission level is 50 ppm. The test engine with code 1 NZ-FE has been equipped with a catalytic converter (CC). Before adding 0.1 Mol of HHO gas, the average HC results were low and below the tolerance limit. However, after adding 0.1 Mol of HHO gas through the air filter, the average HC emissions exceeded the tolerance limit. According to the Ministry of Environment regulations, these emission levels are still considered acceptable because the CO level is below 1.5% and HC is below 200 ppm for vehicles manufactured after 2007 [14].

Carbon dioxide, also known as carbonic acid, is a chemical compound consisting of two oxygen atoms and one carbon atom. It exists as a gas at standard temperature and pressure and is an important greenhouse gas due to its strong infrared absorption. Carbon dioxide plays a crucial role in the carbon cycle and is produced as a byproduct of fossil fuel combustion.

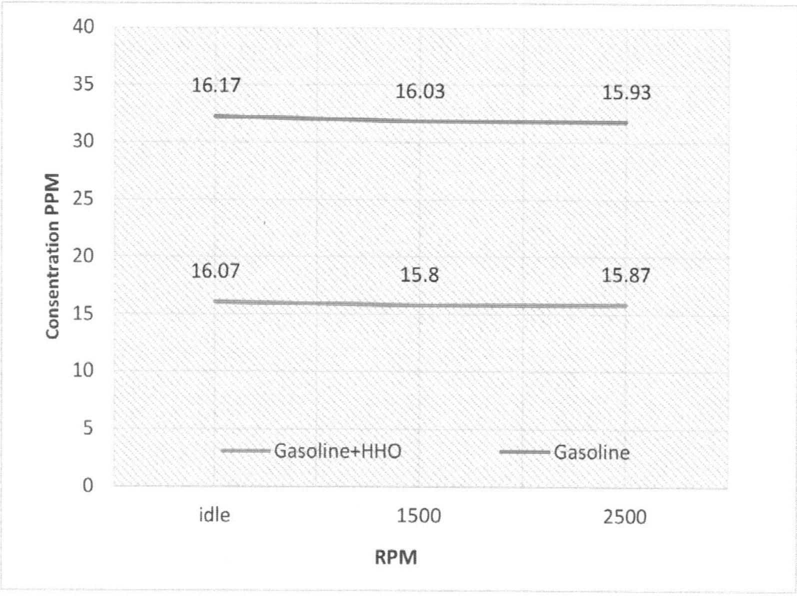

Figure 6. Carbon Dioxide Concentration

In Figure 6, the emission results for carbon dioxide (CO2) with gasoline fuel show that at idle engine speed, it is 16.17%, at 1500 RPM engine speed, it is 16.03%, and at 2500 RPM engine speed, it is 15.93%. With the addition of 0.1 Mol of HHO gas, the data obtained at idle engine speed is 16.33%, at 1500 RPM engine speed, it is 16.03%, and at 2500 RPM engine speed, it is 15.87%.

CO2 concentration directly indicates the combustion process status in the combustion chamber. Higher levels are generally better. When the air-fuel ratio (AFR) is ideal, CO2 emissions typically range between 12% and 15%. CO2 emissions at idle engine speed without HHO gas and with the addition of HHO gas at idle and 1500 RPM engine speeds show non-ideal concentrations. The ideal CO2 emission concentrations occur without HHO gas at 1500 and 2500 RPM engine speeds and with the addition of HHO gas at 2500 RPM engine speed. If the AFR provides too little or too much fuel, CO2 emissions will drop dramatically. If CO2 is below 12%, other emissions should be considered to determine whether the AFR is fuel-rich or fuel-lean. It's important to note that the source of CO2 is solely within the combustion chamber. If CO2 is too low but CO and HC are normal, it indicates exhaust pipe leaks [7].

Oxygen, represented by the symbol O and atomic number 8, is a chemical element. In the periodic table, oxygen is a non-metal in Group VIA (chalcogen) and can readily react with almost all other elements (typically forming oxides). At standard temperature and pressure, two oxygen atoms bond to form O2, a colorless, tasteless, and odorless gas. Oxygen is the third most abundant element in the universe by mass, following hydrogen and helium.

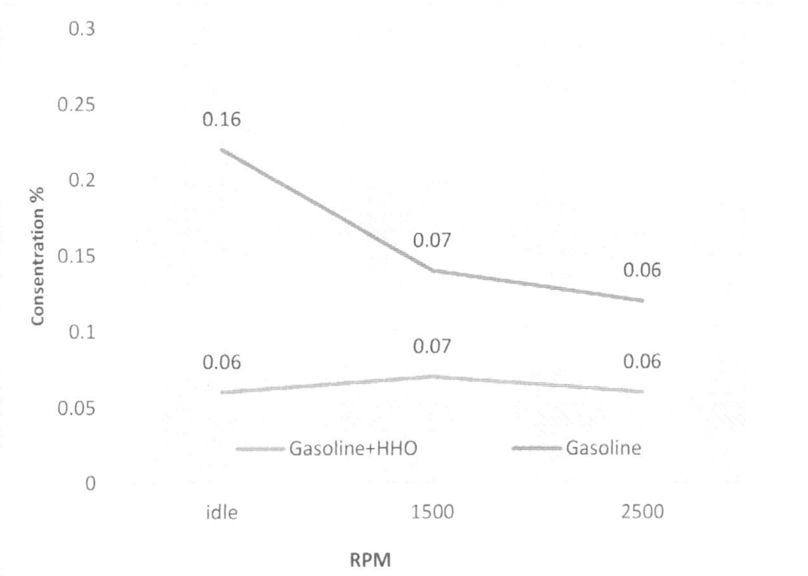

Figure 7. Oxygen Concentration

The emission results for oxygen (O2) with gasoline fuel show that at idle engine speed, it is 0.16%, at 1500 RPM engine speed, it is 0.07%, and at 2500 RPM engine speed, it is 0.06%. With the addition of 0.1 Mol of HHO gas, the data obtained at idle engine speed is 0.08%, at 1500 RPM engine speed, it is 0.07%, and at 2500 RPM engine speed, it is 0.06%. The oxygen concentration in the exhaust gas emissions before and after adding 0.1 Mol of HHO gas indicates low values. Low oxygen concentration suggests that all oxygen is utilized in the combustion process, which can indicate a tendency towards a fuel-rich mixture. In such conditions, low oxygen concentration would be accompanied by high CO emissions. In the case of CO emissions from an EFI Type L gasoline engine without HHO gas, the values are low, indicating complete combustion. After adding HHO gas, CO concentration rises, indicating an excessive fuel mixture in the combustion process. High oxygen concentration can suggest a lean fuel mixture condition, but it can also indicate other factors and high CO and HC emissions when oxygen is excessively high [2].

3.2. Air-Fuel Ratio Relative (λ)

To achieve optimal power generation through the combustion process in an engine, three main conditions must be met: high compression pressure, accurate ignition timing, and strong spark plug ignition with the proper air-fuel mixture.

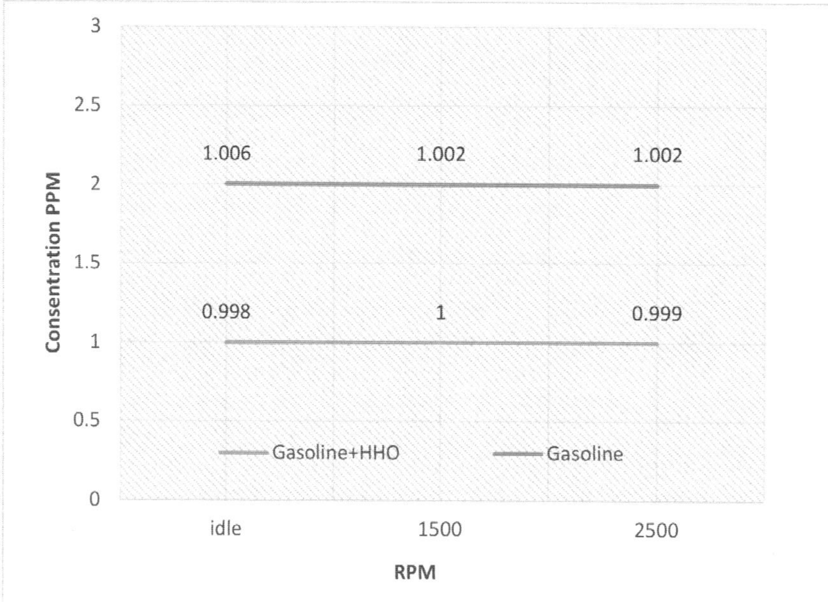

Figure 8. Air Fuel Ratio

The results of the air fuel ratio relative (λ) for gasoline fuel are shown. At engine idle speed, the ratio is 1.006, at 1500 rpm it is 1.002, and at 2500 rpm it is 1.002. Upon adding 0.1 M of HHO gas, the data obtained at engine idle speed is 0.998, at 1500 rpm it is 1.000, and at 2500 rpm it is 0.999. The air fuel ratio relative (λ) values without using HHO gas, at idle speed, 1500 rpm, and 2500 rpm, indicate stoichiometric mixtures. Adding 0.1 Mol of HHO gas demonstrates a stoichiometric mixture at 1500 rpm and a fuel-rich mixture at idle and 2500 rpm.

Combustion conditions can also exist in fuel-lean mixtures, stoichiometric mixtures, and fuel-rich mixtures. Fuel-lean mixtures occur when the ratio of air to fuel is imbalanced due to a small amount of fuel, fuel-rich mixtures arise from an imbalanced air-fuel mixture with an excess of fuel, and stoichiometric mixtures occur when the air-fuel ratio is in perfect or precise condition [10].

3.3. Fuel Consumption

Internal combustion engines used in motor vehicles have varying cylinder volumes and cylinder counts according to the purpose of the vehicle. The cylinder volume determines the amount of fuel consumed for the engine to operate effectively.

In this study, an analysis was conducted on fuel consumption when the engine is operated with gasoline fuel both without and with the addition of HHO gas. The data obtained is presented in the form of data tables and graphs, as shown in Figure 9. The average fuel consumption with gasoline fuel alone can cover a distance of 9.823 km/l. With the addition of 0.1 Mol of HHO gas, the mileage is reduced to 9.556 km/l, which is lower than the engine without HHO gas. This is related to the air-fuel ratio system, which might not be compatible with the fuel-air mixture controlled by the injector system in the engine. As a result, the process of achieving a complete fuel-air mixture is not attained. In this research, using an L-type injection engine, it's evident that fuel consumption becomes 2.8% less efficient after adding 0.1 Mol of HHO gas compared to not using HHO gas. This is because the airflow sensor detects an increase in the amount of air due to the addition of HHO gas to the air filter, causing the air-fuel ratio (AFR) to read too much air sucked in, resulting in a fuel-rich mixture. Gasoline fuel is injected into the combustion chamber based on the airflow sensor detecting the amount of air in the air filter.

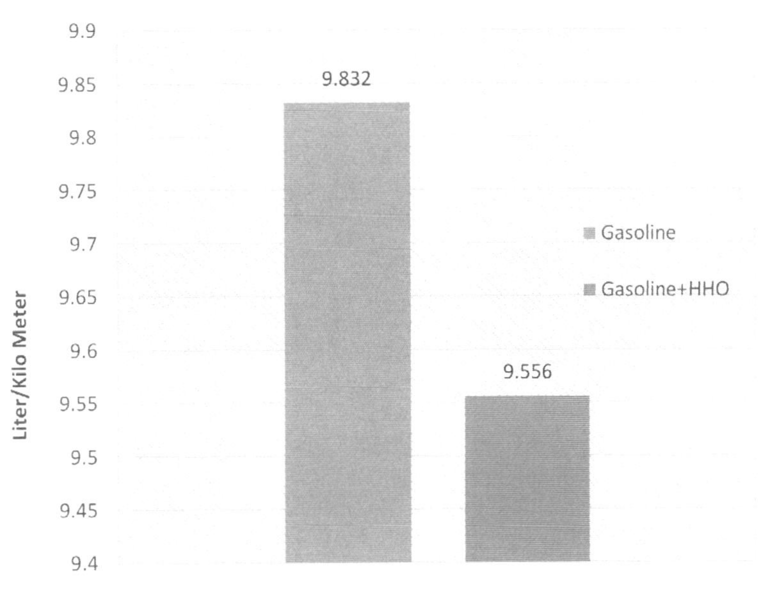

Figure 9. Fuel Oil Consuption

3.4. Comparative Exhaust Gas Emission Analysis

This research was conducted with the addition of 0.1 Mol of HHO gas using an L-type EFI engine. As there is no direct comparison available, a comparative analysis is performed by referring to a study conducted with a D-type EFI engine. the addition of HHO gas to a 1200 cc cylinder capacity EFI engine of the D type showed a reduction in HC emissions by 5.5% and CO emissions by 3.57%, as indicated in Table 4. The decrease in exhaust gas emissions from the gasoline EFI engine of the D type is due to the engine's operational principle, which measures the vacuum pressure in the intake manifold rather than the amount of air. This causes the injection of gasoline fuel to remain unaffected by the addition of HHO gas, and the HHO gas contributes to more complete combustion in the combustion chamber.

In contrast, the gasoline EFI engine of the L type experiences an increase in exhaust gas emissions. This is because the engine operates based on a sensor that measures the amount of air entering the combustion chamber through the air filter, determining the quantity of fuel to be injected. After adding HHO gas to the gasoline engine of the L type, the sensor detects an increase in both air and fuel injected into the combustion chamber. This results in the HHO gas not leading to complete combustion in the combustion chamber. By comparing the measurement results with the standards for a gasoline engine with a 1300 cc cylinder capacity (Table 3) and a gasoline engine with a 1200 cc cylinder capacity (Table 4), the exhaust gas emissions are evident.

4. Conclusion

The addition of 0.1 Mol of HHO gas to the gasoline EFI engine of the L type through the air filter during the combustion process leads to an increase in the emissions of carbon dioxide, hydrocarbons, carbon monoxide, oxygen, and the air-fuel ratio number, indicating an excessive fuel mixture. The increased exhaust gas emissions occur due to the operational principle of the EFI engine of the L type. After adding 0.1 Mol of HHO gas, the EFI system of the L type experiences a malfunction, resulting in an excessive amount of fuel in the combustion chamber and causing the mixture of HHO gas and gasoline fuel to not burn completely during the combustion process of the internal combustion engine.

The average fuel consumption is more wasteful by 2.8% due to the addition of HHO gas, as the air-fuel ratio (AFR) is in a fuel-rich mixture state. This condition arises because the EFI engine of the L type experiences a malfunction after adding 0.1 Mol of HHO gas through the air filter. This malfunction results in excessive fuel being injected into the combustion chamber, leading to a fuel-rich mixture state.

IOP Conf. Series: Earth and Environmental Science 1298 (2024) 012023 doi:10.1088/1755-1315/1298/1/012023

5. Reference

[1] A. F. , T. S. R. and H. D. A. Molland, *Ship resistance and propulsion : practical estimation of ship propulsive power*. Cambridge. GB. Cambridge University Press, 544pp., 2011.
[2] Sv. A. Harvald, *Tahanan dan Propulsi Kapal*. Airlangga University Press: Surabaya, 1992.
[3] E. V. Lewis and Society of Naval Architects and Marine Engineers (U.S.), *Principles of naval architecture. Volume II, Resistance, propulsion and vibration*. Society of Naval Architects and Marine Engineers, 1988.
[4] "Ship Construction - D.J. Eyres".
[5] "Principles Of Naval Architecture Vol III - Motion in Waves & Controllability".
[6] Y.-S. Wu, W.-C. Cui, and G.-J. Zhou, "Practical Design of Ships and Other Floating Structures Volume I Edited by."
[7] "Basic Ship Theory."
[8] J. Carlton, "Marine Propellers and Propulsion," 2012.
[9] H. (Herbert) Schneekluth and Volker. Bertram, *Ship design for efficiency and economy*. Butterworth-Heinemann, 1998.
[10] J. M. J. Journée, "Prediction of Speed and Behaviour of a Ship in a Seaway," 1976. [Online]. Available: www.shipmotions.nl
[11] "Introduction to Marine Engineering."
[12] P. Kyropoulos, "MARINE DIESEL POWER PLANTS and Ship Propulsion ENGINEERING AND SCIENCE MONTHLY."
[13] "Bulk Carrier Safety _19 January 2009".
[14] "Marine engineering. Vol.3. 1899".
[15] "Principles Of Naval Architecture Vol I - Stability And Strength".
[16] H. (Herbert) Schneekluth and Volker. Bertram, *Ship design for efficiency and economy*. Butterworth-Heinemann, 1998.
[17] H. Schneekluth and V. Bertram, "Chapter 6 - Ship propulsion," in *Ship Design for Efficiency and Economy (Second edition)*, Second edition., H. Schneekluth and V. Bertram, Eds., Oxford: Butterworth-Heinemann, 1998, pp. 180–205. doi: https://doi.org/10.1016/B978-075064133-3/50006-2.
[18] K. J. Rawson and E. C. Tupper, "10 - Powering of ships: general principles," in *Basic Ship Theory (Fifth Edition)*, Fifth Edition., K. J. Rawson and E. C. Tupper, Eds., Oxford: Butterworth-Heinemann, 2001, pp. 365–410. doi: https://doi.org/10.1016/B978-075065398-5/50013-3.
[19] T. Greenhalgh, R. Stones, and D. Swinglehurst, "Choose and Book: A sociological analysis of 'resistance' to an expert system," *Soc Sci Med*, vol. 104, pp. 210–219, 2014, doi: https://doi.org/10.1016/j.socscimed.2013.12.014.
[20] G. Li, C. Zhang, and Y. Li, "Effects of diesel injection parameters on the rapid combustion and emissions of an HD common-rail diesel engine fueled with," *Appl Therm Eng*, vol. 108, pp. 1214–1225, 2016, doi: 10.1016/j.applthermaleng.2016.08.029.

Pesisir Barat Lampung Offshore Wind Characteristic in Local Renewable Energy Development Studies

M F Hermanto[1*], N Nandalianadhira[1], E R Kencana[1], M H I Khaldun[1], I D Hutajulu[2], D Epipanus[2], and J D Samosir[2]

[1] Ocean Engineering Department, Institut Teknologi Sumatera, South Lampung, Indonesia
[2] Ocean Engineering Department Student, Institut Teknologi Sumatera, South Lampung, Indonesia

[*]Email: mochammad.hermanto@kl.itera.ac.id

Abstract. Renewable energy is one of the most researched topics. One renewable energy source is wind energy. As with some other energy sources, there are 2 locations for wind energy conversion, onshore and offshore. Using offshore wind energy requires more accurate wind data due to the higher risk. The characteristics of the wind at sea are one of these data. This study used data sets from ECMWF and BMKG from 2001 to 2020 (20 years). This research aims to examine the wind characteristics in the offshore of Pesisir Barat Lampung. The wind characteristics being examined are wind speed, wind direction, wind blowing duration, and wind speed distribution. The maximum hourly mean wind speed in these waters is 16.35 m/s. The dominant wind direction is blowing from the South East (SE). The maximum average wind-blowing duration to SE direction is 11 hours based on ECMWF and 14 hours based on BMKG. Based on the Chi-Square and Kolmogorov-Smirnov distribution fitting methods, the wind speed distribution in these waters is best fitted by the Weibull distribution for the ECMWF and the Lognormal distribution for BMKG. Using these two data sources, on average, ECMWF provides wind speeds 5% to 20% higher than BMKG. In analyzing wind potential, BMKG data will certainly provide more critical results than ECMWF data. In the analysis of offshore support structures, ECMWF data will certainly produce more critical structures. While the calculation of potential is better using BMKG data. Wind power potential in Pesisir Barat offshore is in the range of 44.6 to 82.9 W/m^2.

1. Introduction

The Indonesian government is committed to reducing the use of fossil power plants to participate in reducing greenhouse gas emissions, which is targeted to reduce the greenhouse effect to 314 million tons by 2030 [1]. Indonesia is currently starting to replace the use of fossil energy with renewable energy, the Indonesian government has released several guidelines for the development of eco-friendly renewable energy [2]. With Indonesia's commitment and potential, it is expected to achieve 17% by 2025 [3] and 50% by 2050 [4] in replacing fossil fuels with renewable energy.

Wind, as one renewable energy source, is clean and always available in nature [5]. Indonesia's estimated wind power potential is about 60.6 GW, with wind speeds ranging from 3-6 m/s and installed facility for only 0.15 GW in 2020 [6], [7]. With other methodologies, the Energy Sector Management Assistance Program (ESMAP) released offshore wind technical potential in Indonesia is

277 GW [8]. With those potentials, Indonesia has only 2 facilities commercially, Sidrap and Jeneponto, and none of them are located offshore.

The offshore wind turbine is very popular in many countries and increasingly developed because the industry is very promising. For many years Northern Europe has been developing offshore wind turbines for various reasons. The greater wind speeds at sea and no land use are the top reasons [9].

In the case of Indonesia, offshore wind energy research has not made significant progress because the majority is still in the form of potential studies in several areas, but those will make good progress for the development [10]–[16]. Although this industry is very promising, many challenges must be resolved such as the large costs incurred for its study, development, and construction [17]. Primary data become a challenge in offshore wind energy development because of its high costs for maritime countries like Indonesia. But now, secondary wind data can be provided from satellites that conduct measurements of wind [18]. Due to renewable energy's high risk [19], secondary data could only be used for initial research. Research has already done for Southeast Brazil, China, Korea, North Europe, West Mediterranean are using that kind of data [20]–[24]. This implies that the secondary data could be valuable in evaluating the offshore wind energy potential from wind characteristics.

The western and southern waters of Lampung are two potential offshore wind energy areas in Indonesia with an average wind speed range of 5-7 m/s [8]. Western Lampung has an average wind speed of about 6 m/s which was obtained from the MTSAT and Lognormal Distribution model [25]. With the knowledge of the wind characteristics and potential, it can provide good information for the development of offshore wind energy. Each region has specific and different wind characteristics. Some wind characteristics are wind speed, wind direction, wind blowing duration, and wind speed distribution.

2. Statistical Distribution

Wind data statistic is being analyzed in this research. The primary statistical parameter being used in determining the potential of wind energy is the data distribution, in this case, wind distribution. Wind distribution is fitted by theoretical distribution. The fitting process uses using probability density function of 3 theoretical distributions, Normal, Lognormal, and Weibull. The best-fit distribution can be used for further analyses in the wind energy sector for Pesisir Barat offshore. Another result of that analysis is the extreme value of wind speed. That can be an input for wind turbine structural analyses. Maximum and average wind duration blowing from a direction is summarized in this research. That can be an input for wind energy analyses.

2.1. Probability Density Function

The probability density function is shown in equations (1) for Normal distribution, (2) for Lognormal distribution, and (3) for Weibull distribution. Each parameter for Lognormal and Weibull is described in equation (4) to (7).

$$f_{normal}(x) = \frac{1}{\sigma_X\sqrt{2\pi}} \exp\left[-\frac{1}{2}\left(\frac{x-\mu_X}{\sigma_X}\right)^2\right] \quad (1)$$

$$f_{lognormal}(x) = \frac{1}{\sqrt{2\pi}\zeta_X x} \exp\left[-\frac{1}{2}\left(\frac{\ln x-\lambda_X}{\zeta_X}\right)^2\right] \quad (2)$$

$$f_{weibull}(y_1) = \frac{k}{w_1}\left(\frac{y_1}{w_1}\right)^{k-1} \exp\left[-\left(\frac{y_1}{w_1}\right)^k\right] \quad (3)$$

$$\lambda_X = \ln \mu_x - \frac{1}{2}\zeta_X{}^2 \quad (4)$$

$$\zeta_X{}^2 = \ln\left(1 + \delta_x{}^2\right) \quad (5)$$

$$k = \delta_{Y_1}^{-1.08} \quad (6)$$

$$w_1 = \frac{\mu_{Y_1}}{\Gamma\left(1+\frac{1}{k}\right)} \quad (7)$$

2.2. Distribution Fitting

The distribution fitting methods used in this research are Kolmogorov-Smirnov and Chi-Square. The Kolmogorov-Smirnov method uses the cumulative density function. Whereas, the Chi-Square method uses the probability density function described above. Kolmogorov-Smirnov result value is calculated using equation (8). Chi-Square result value is calculated using equation (9).

$$D_n = \max|F_X(x_i) - S_n(x_i)| \quad (8)$$

$$\sum_{i=1}^{m} \frac{(n_i - e_i)^2}{e_i} \quad (9)$$

2.3. Extreme Value

Extreme value is calculated depending on the best-fit distribution to the data. Equation (10) gives the general equation that determines the extreme value.

$$P(X > Ws_n) = \frac{1}{RP} \quad (10)$$

3. Pesisir Barat Offshore Wind Characteristic

The characteristics of wind in Pesisir Barat Lampung offshore are shown by the average wind speed map, windrose by the season, and occurrence table. This research analyses the wind data from 2 reanalysis satellite data, the European Centre for Medium-Range Weather Forecasts (ECMWF) and Badan Meteorologi, Klimatologi, dan Geofisika (BMKG). Much research has been done using one of those data sources for another specific area [24], [26], [27]. The data points are shown in Table 1. Every point is being analyzed in this research for a duration of 20 years of hourly 10 m wind data (2001 to 2020).

Table 1. Coordinate points of interest.

No	Longitude (°)	Latitude (°)
1	103.25	-5.25
2	103.50	-5.50
3	103.75	-5.75
4	104.00	-6.00
5	104.25	-6.25
6	104.50	-6.50

The average wind speed map for BMKG and ECMWF data is shown in Figure 1. BMKG average wind speed data has ranged from 3.5 to 5.5 m/s. BMKG's average wind speed is greater the further southeast. ECMWF average wind speed data has ranged from 3.5 to 6.0 m/s. This data has the same pattern as BMKG, but there is an anomaly in point 4 that it is the smallest among all points. Secondary reanalysis data does tend to have quite large values when compared to onshore field data [28]. Turbines have a minimum cut-in speed to operate. Referring to the example of the small-sized turbine with rotor diameter of 9.8 m, the cut-in speed is 2 m/s [29]. The average wind speed as obtained by both BMKG and ECMWF in this site is sufficient to rotate the reference turbine.

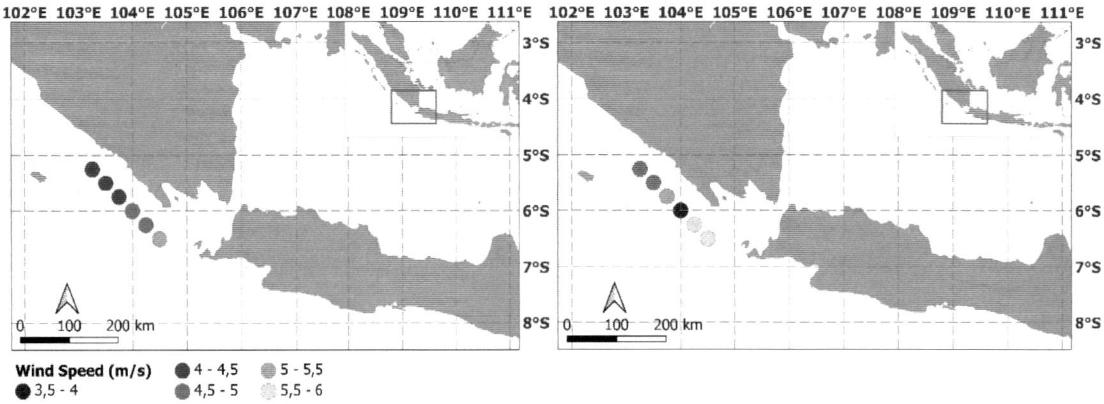

Figure 1. Average wind speed data BMKG (left) and ECMWF (right).

The windrose for every season in Indonesia is shown in Figure 2 and Figure 3. The wind blows from SE in Transition 1, Southeast Monsoon, and Transition 2 seasons while the wind blows from NW in Northwest Monsoon season, both BMKG and ECMWF data.

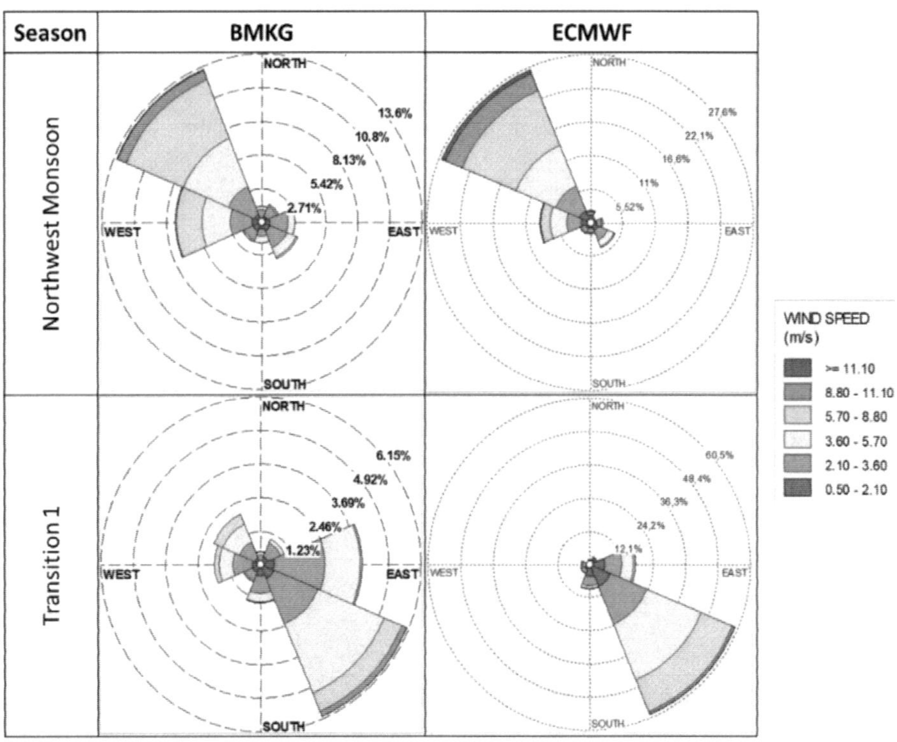

Figure 2. Windrose for Northwest Monsoon and Transition 1 season.

The incidence percentage of every wind speed class is quite similar for both BMKG and ECMWF data. This implies that both data source has a fairly high similarity in hourly wind speed and direction. The northwest monsoon season covers the months of December to March. The transition 1 season covers the months of April to May. The southeast monsoon season covers the months of June to September. The transition 2 season covers the months of October to November.

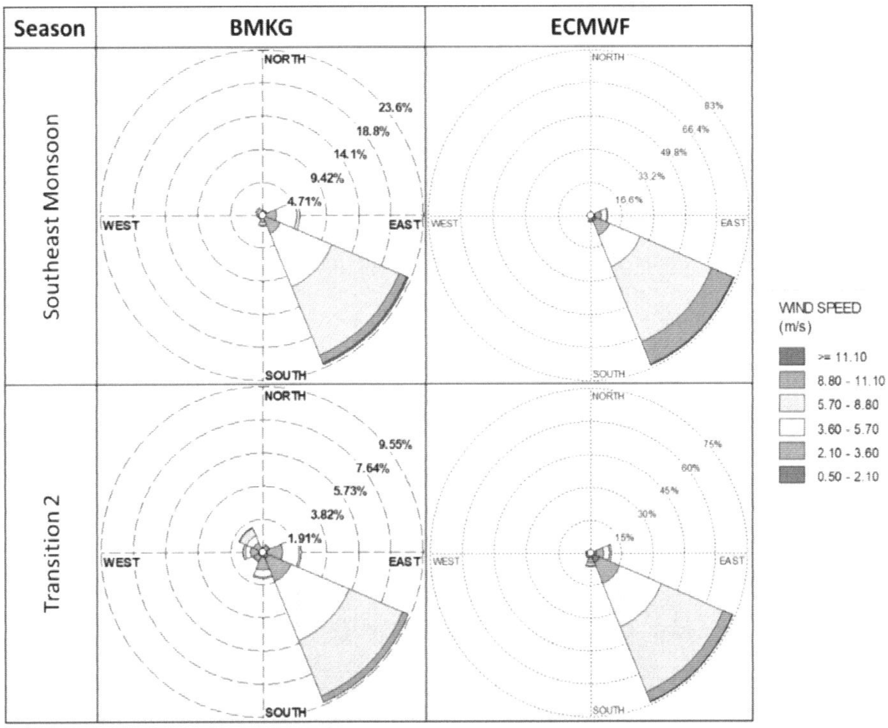

Figure 3. Windrose for Southeast Monsoon and Transition 2 season.

Figure 4 and Figure 5 show the average and maximum duration of wind blowing in the same direction for ECMWF and BMKG data for representing Point 6. All the direction data shows the direction of wind blowing from, in 16 directions.

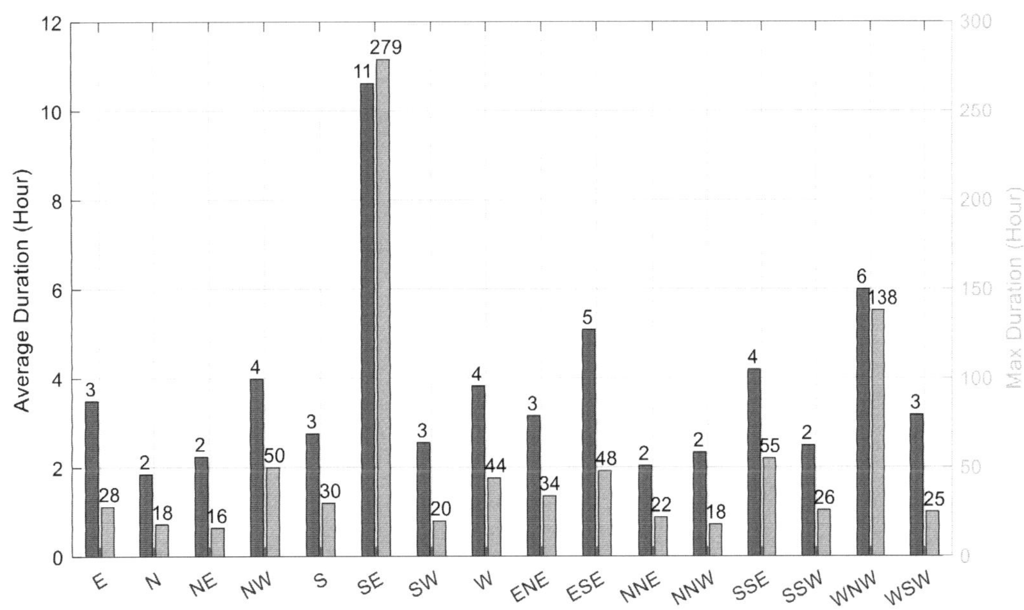

Figure 4. The average and maximum duration of wind-blowing ECMWF data.

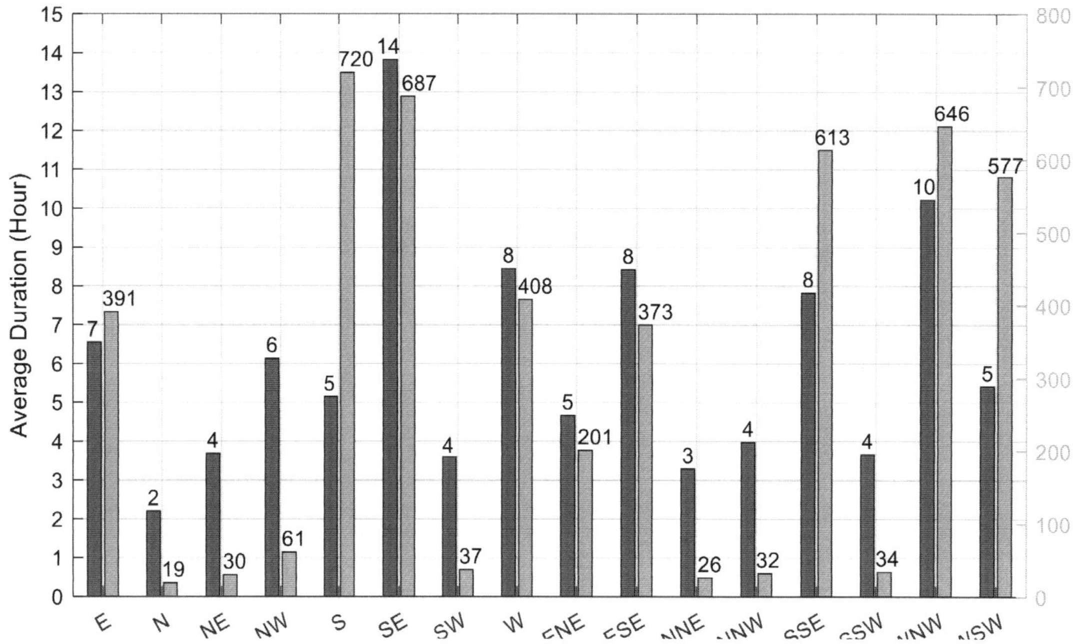

Figure 5. The average and maximum duration of wind-blowing BMKG data.

It shows that in ECMWF data, the wind blows dominantly from the SE direction with an average duration of 11 hours and a maximum duration of 279 hours. However, the BMKG data is more diffuse than the ECMWF data. According to BMKG, the maximum wind-blowing duration is 720 hours from the S direction. The average wind duration is 14 hours from the SE direction, while this direction also has a huge max duration of 687 hours of wind blowing. BMKG data also gives more direction with a huge wind-blowing duration (more than 500 hours), SSE, WNW, and WSW. We can't see that from ECMWF data that dominantly gives SE the center of interest.

4. Pesisir Barat Offshore Wind Distribution

Maximum wind speed in a year or annual maximum wind speed of all points of interest in this research is shown in Figure 6 for ECMWF data and Figure 7 for BMKG data, while the right side of the figure shows the Weibull distribution of annual maximum wind speed data of all points of interest. From those figures, the annual maximum BMKG data looks more uniform for all points. That is because BMKG gives the data in units of knots. In ECMWF data, point 4 has the smallest annual maximum data among all points. That makes the distribution, in the right figure, of points 4 plotted very left among all points in ECMWF data. In the distribution of BMKG data, all points are in a similar position. The only similarity between BMKG and ECMWF annual maximum wind speed data is that they range from 9.0 to 16.0 m/s. Maximum wind speed for ECMWF data is 16.35 m/s in point 6 in 2018. The maximum wind speed for BMKG data is 15.43 m/s in point 1 in 2001.

For comparison, this offshore maximum wind speed data is far bigger than onshore data in South Lampung, Pontianak, and Banyuwangi, 2.5-3.5 m/s [30]–[32]. Of course, offshore winds have more potential than onshore winds. For further study, data validation with filed data is a must. Data validation could be done by measuring the wind speed in the field at a certain period. As done by Novrita [27], the difference is that for offshore wind potential, of course, it needs to be done offshore. That data validation will compare and then correct the satellite wind speed data for all periods.

IOP Conf. Series: Earth and Environmental Science 1298 (2024) 012024 doi:10.1088/1755-1315/1298/1/012024

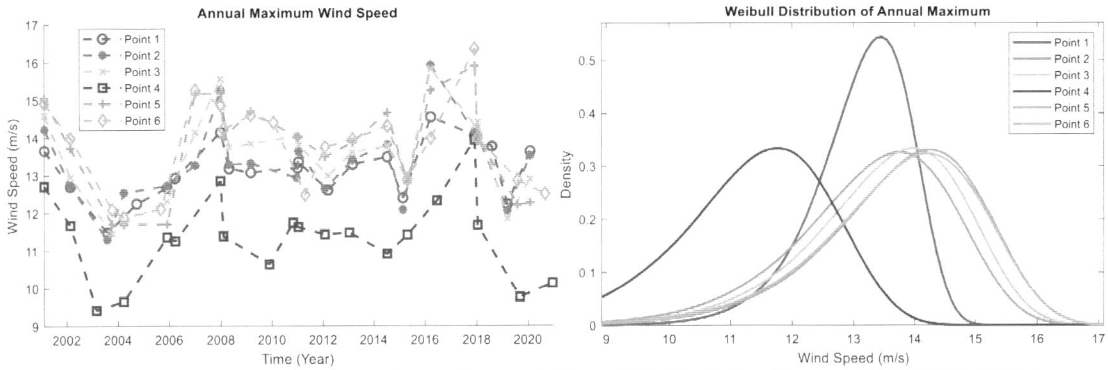

Figure 6. Annual-maximum wind speed data (left) and Weibull distribution (right) ECMWF data.

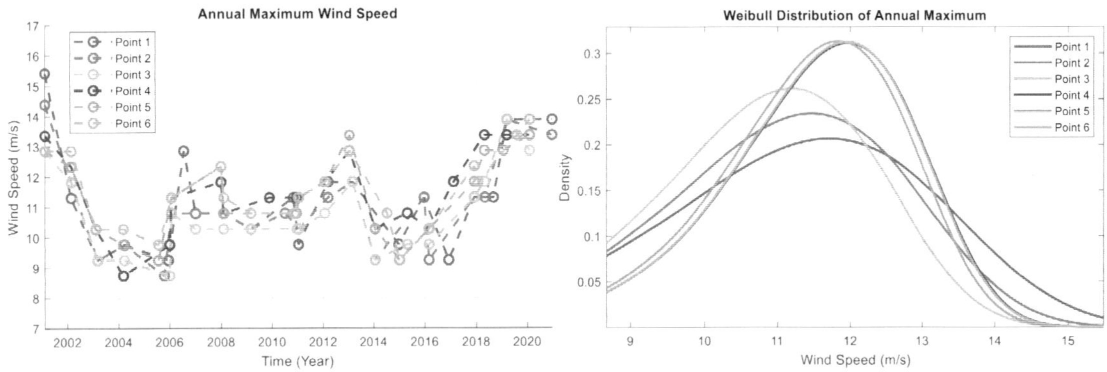

Figure 7. Annual-maximum wind speed data (left) and Weibull distribution (right) BMKG data.

Nonetheless, the trend of those data is different. Annual maximum wind speed ECMWF data was decreased from 2003 to 2006, 2015, and from 2019 to 2020. While BMKG data decreased from 2003 to 2006, and from 2014 to 2017. The annual maximum wind speed increased in 2001, 2007 to 2008, and 2016 to 2018 for some points for ECMWF data. While BMKG data increased in 2001, and from 2019 to 2020. Based on the things that had been learned, the 2 data sources provide different things when analyzed annually, but not when analyzed at the overall statistics.

To get the best distribution fit to the data distribution, goodness of fit analyses were conducted on both data. The result of the analyses is shown in Table 2 for Kolmogorov-Smirnov analyses and Table 3 for Chi-Square analyses.

Table 2. Kolmogorov-Smirnov Analyses Score.

Data	Distribution	Kolmogorov-Smirnov Value per Point					
		1	2	3	4	5	6
BMKG	Normal	0.188	0.176	0.188	0.120	0.123	0.116
	Lognormal	0.159	0.150	0.172	0.110	0.121	0.103
	Weibull	0.209	0.209	0.193	0.152	0.152	0.149
ECMWF	Normal	0.075	0.106	0.124	0.139	0.135	0.150
	Lognormal	0.085	0.100	0.141	0.155	0.153	0.165
	Weibull	0.086	0.148	0.106	0.100	0.125	0.108

Table 3. Chi-Square Analyses Score.

Data	Distribution	Chi-Square Value per Point					
		1	2	3	4	5	6
BMKG	Norm	0.015	0.257	0.188	0.200	0.431	0.091
	Lognormal	0.127	0.077	0.159	0.064	0.210	0.014
	Weibull	0.000	0.511	0.209	0.703	0.004	0.449
ECMWF	Norm	0.112	0.071	0.589	0.766	0.940	0.827
	Lognormal	0.187	0.151	0.834	0.121	1.281	1.100
	Weibull	0.205	0.710	0.147	0.146	0.249	0.134

The result varies depending on the point being analyzed. Weibull distribution has the best fitting to the ECMWF data distribution with all points average value of 0.112 for Kolmogorov-Smirnov and 0.265 for Chi-Square. Whereas, the Lognormal distribution has the best fitting to the BMKG data distribution with all points average value of 0.109 for Chi-Square and 0.136 for Kolmogorov-Smirnov. With the value limit of Kolmogorov-Smirnov value of 0.304 and Chi-Square value of 5.991, all the results, don't exceed the limit. The other research with various locations has the same result that wind data is best fit for Weibull distribution [10], [18], [34], [35]. A study using BMKG data in South Lampung before has the same result, that the wind data distribution is best fit for Weibull [30].

Based on the theoretical distribution best fit to the data distribution above, the wind speed extreme value is calculated. The calculation for BMKG data uses Lognormal distribution. Whereas, ECMWF data uses Weibull distribution. Extreme value wind speed is calculated for 1-year and 100-year return periods for each data. The results are shown in Figure 8 for 1 year return period and Figure 9 for 100 years return period.

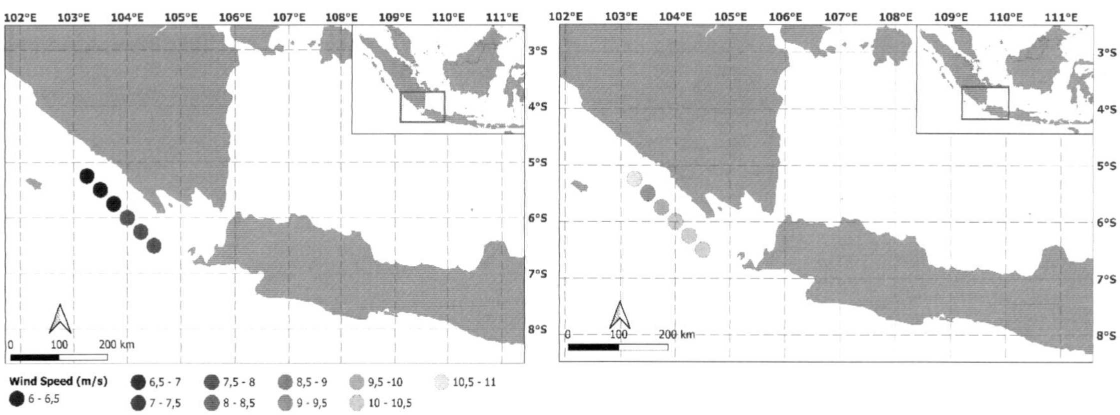

Figure 8. 1-year extreme wind speed for BMKG (left) and ECMWF (right).

The result showed no difference from the average wind speed data before. ECMWF result has a greater extreme wind speed value than BMKG result, but 1 year and 100 years return period. ECMWF results for 100 years return period have ranged from 14.5 to 16.1 m/s for all points. Whereas, BMKG results for 100 years return period only range from 14.3 to 15.6 m/s for all points. ECMWF result is 5.2% greater than BMKG result for 100 years return period. ECMWF results for 1 year return period have ranged from 9.5 to 10.7 m/s for all points. Whereas, BMKG results for 1 year return period only ranged from 7.9 to 8.9 m/s for all points. ECMWF result is 20.0% greater than BMKG result for 1

IOP Conf. Series: Earth and Environmental Science 1298 (2024) 012024 doi:10.1088/1755-1315/1298/1/012024

year return period. This extreme wind speed value may not be used in the calculation of wind energy potential. But it will be used in the design process of wind turbine support structures.

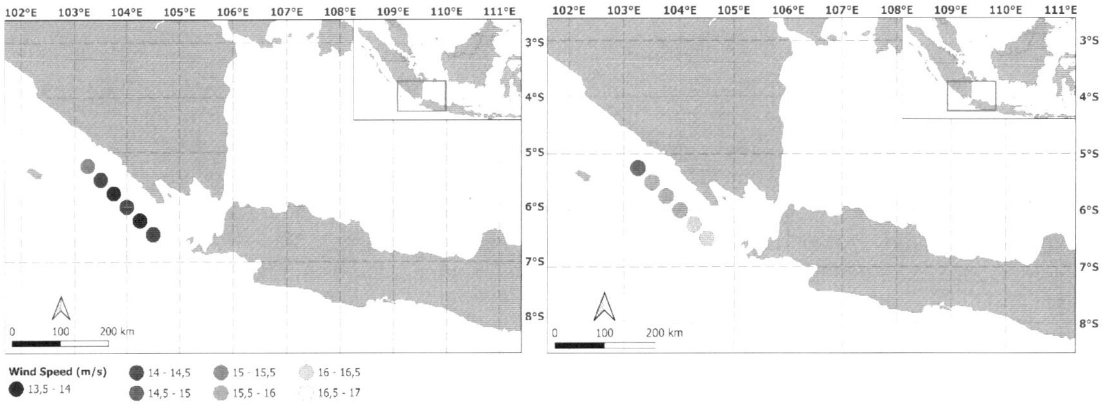

Figure 9. 100-year extreme wind speed for BMKG (left) and ECMWF (right).

5. Wind Potential

The wind potential or energy is often measured and calculated by its power. Wind power per swept area can be calculated using the equation (11). $\boldsymbol{U}$ denotes wind speed, $\boldsymbol{\rho}$ denotes air density with value of 1.23 kg/m^3, and $\boldsymbol{A}$ denotes swept area. Based on that equation, when the wind blows faster, the greater the power generated quadratically. Furthermore, the swept area only has a linear effect on wind power. Therefore, the development of wind energy relies heavily on finding sites with a strong wind speed.

$$\boldsymbol{P/A = \frac{1}{2}\rho U^3} \tag{11}$$

It is crucial to use the more reliable wind speed for a more realistic potential estimation, as overestimation is not desirable in this potential calculation. Therefore, the average wind speed from BMKG is selected. BMKG data is a little bit lower than ECMWF for annual average data. The wind potential is shown in Figure 10.

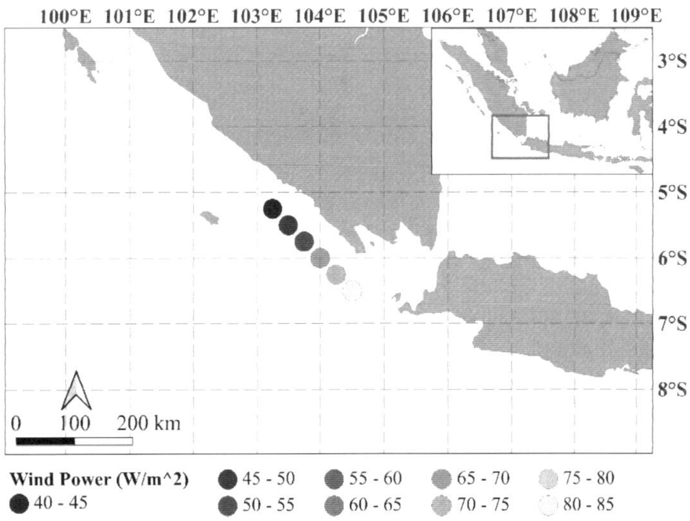

Figure 10. Wind Power Potential per Swept Area in Pesisir Barat Lampung Offshore.

From the result, wind power potential in Pesisir Barat offshore is in the range of 44.6 to 82.9 W/m^2. The potential of this site for small-medium scale utilization is promising. A similar result from a study in Sidoarjo with wind speed more than 5 m/s that produced 89 W/m^2 [33].

Note that the calculated potential concerns only the wind potential at a height reference of 10 m above ground. The wind speed increases with a higher elevation of wind measurement. With a higher wind turbine altitude, the potential or energy produced would exceed that produced in this study. For instance, a study carried out in Jember, which is situated at an elevation of 43.2 m, yielded a power of 277.03 W/m^2 at wind speed of 6.5 – 7.5 m/s [10]. In the study conducted in Banyuwangi, the same methodology yielded an average of 1.34 W/m2, based on wind speed with average of 2 m/s over 10 days [36]. In onshore measurements, Abdullah [26] obtained a potential value of 0.23 - 0.80 W/m^2 for the coastal area of North Sumatra with an average wind speed of 1.4 m/s sourced from BMKG.

In line with Burke et al [28], Indonesia has wind potential, especially in Pesisir Barat offshore. This location is just an example of wind potential analysis in the rural area. With many small and outer islands, Indonesia has the potential to develop it even more with an off-grid system. This will help people living on small and outer islands could get green energy not far from where they live.

6. Conclusion

From all the calculations and analyses done in this research, some conclusions have been obtained are:

1. Wind blows from SE in Transition 1, Southeast Monsoon, and Transition 2 seasons while the wind blows from NW in Northwest Monsoon season for both of the data sources.
2. Average wind speed has ranged from 3.5 to 5.5 m/s for BMKG data. Whereas, average wind speed has ranged from 3.5 to 6.0 m/s for ECMWF data.
3. From all the data source analyses, the wind blows dominantly from the SE direction with an average wind duration of 11 hours and a maximum duration of 279 hours for ECMWF data and the average wind duration is 14 hours from the SE direction for BMKG data, while this direction also has a huge max duration of 687 hours of wind blowing from.
4. The Weibull distribution has the best fitting to the ECMWF data distribution while the Lognormal distribution has the best fitting to the BMKG data.
5. ECMWF results for 100 years return period have ranged from 14.5 to 16.1 m/s, 5.2% greater than the BMKG result. Besides that, ECMWF results for 1 year return period have ranged from 9.5 to 10.7 m/s, 20.0% greater than the BMKG result.
6. Using BMKG average wind speed data, wind power potential in Pesisir Barat offshore is in the range of 44.6 to 82.9 W/m^2.

References

[1] Kementerian ESDM, 'Turunkan Emisi 314 juta ton CO2 Tahun 2030, Indonesia Perlu Investasi Rp 3.500 Triliun'. KEMENTERIAN ENERGI DAN SUMBER DAYA MINERAL REPUBLIK INDONESIA, Aug. 19, 2020. Accessed: Aug. 19, 2023. [Online]. Available: https://ebtke.esdm.go.id/post/2020/08/24/2613/turunkan.emisi.314.juta.ton.co2.tahun.2030.indonesia.perlu.investasi.rp.3.500.triliun

[2] D. P. Sari and W. B. Kusumaningrum, 'A Technical Review of Building Integrated Wind Turbine System and a Sample Simulation Model in Central Java, Indonesia', *Energy Procedia*, vol. 47, pp. 29–36, 2014, doi: 10.1016/j.egypro.2014.01.193.

[3] N. P. Purba *et al.*, 'Preliminary Research of Using Ocean Currents and Wind Energy to Support Lighthouse in Small Island, Indonesia', *Energy Procedia*, vol. 47, pp. 204–210, 2014, doi: 10.1016/j.egypro.2014.01.215.

[4] Erdiwansyah *et al.*, 'Biomass and wind energy as sources of renewable energy for a more sustainable environment in Indonesia: A review', *Archives of Environmental Protection*, vol. 48, no. 3, pp. 57–69, Aug. 2023, doi: 10.24425/aep.2022.142690.

[5] S. Martosaputro and N. Murti, 'Blowing the Wind Energy in Indonesia', *Energy Procedia*, vol. 47, pp. 273–282, 2014, doi: 10.1016/j.egypro.2014.01.225.

[6] T. Hidayat *et al.*, 'Wind Power in Indonesia: Potential, Challenges, and Current Technology Overview', in *Indonesia Post-Pandemic Outlook: Strategy towards Net-Zero Emissions by 2060 from the Renewables and Carbon-Neutral Energy Perspectives*, H. Ardiansyah and P. Ekadewi, Eds., Penerbit BRIN, 2022. doi: 10.55981/brin.562.

[7] Pusat Pengkajian Industri Proses dan Energi, *Outlook Energi Indonesia 2021: Perspektif Teknologi Energi Indonesia: Tenaga Surya untuk Penyediaan Energi Charging Station*. Jakarta: Badan Pengkajian dan Penerapan Teknologi, 2021. [Online]. Available: https://www.bppt.go.id/dokumen/outlook-energi

[8] Energy Sector Management Assistance Program, 'Technical Potential for Offshore Wind in Indonesia'. The World Bank, Mar. 2020. [Online]. Available: https://documents1.worldbank.org/curated/en/439761586851827372/pdf/Technical-Potential-for-Offshore-Wind-in-Indonesia-Map.pdf#:~:text=Offshore%20WindTechicl%20Potenial%20RISE%20score%3A%2054%20Fixed%3A%20198,megawatts%20%28MW%29%20within%20200%20kilometers%20of%20the%20shoreline.

[9] J. F. Manwell, 'Wind Energy Explained: Theory, Design and Application'.

[10] T. Hardianto, B. Supeno, A. Saleh, D. K. Setiawan, Gunawan, and S. Indra, 'Potential of Wind Energy and Design Configuration of Wind Farm on Puger Beach at Jember Indonesia', *Energy Procedia*, vol. 143, pp. 579–584, Dec. 2017, doi: 10.1016/j.egypro.2017.12.730.

[11] W. Hayyatul and A. Bachtiar, 'Analisis Potensi Pembangkit Listrik Tenaga Angin PT. Lentera Angin Nusantara (LAN) Ciheras', *JTE*, vol. 7, no. 1, pp. 35–45, Jan. 2018, doi: 10.21063/JTE.2018.3133706.

[12] N. W. Hesty, T. W. Hadi, and P. Ketenagalistrikan, 'Reanalisis Weather Research and Forecast - Four Dimension Data Assimiliation (WRF-FDDA) untuk Meningkatkan Akurasi Estimasi Potensi Energi Angin di Daerah Lepas Pantai (Studi Kasus: Pantai Selatan Jawa Barat)', vol. 13, no. 1, 2015.

[13] Mukhrijal, T. Zulfadli, and A. Mulkan, 'Analisis Potensi Energi Angin Sebagai Sumber Energi Listrik Untuk Menggerakkan Pompa Air Di Areal Persawahan Desa Blang Krueng Kecamatan Baitussalam', *JITU*, vol. 2, no. 1, pp. 1–10, Jun. 2021, doi: 10.55616/jitu.v2i1.167.

[14] B. S. Premono, D. D. D. P. Tjahjana, and S. Hadi, 'Wind energy potential assessment to estimate performance of selected wind turbine in northern coastal region of Semarang-Indonesia', presented at the International Conference on Engineering, Science and Nanotechnology 2016 (ICESNANO 2016), Solo, Indonesia, 2017, p. 030026. doi: 10.1063/1.4968279.

[15] W. S. Putro, R. A. Prahmana, H. T. Yudistira, M. Y. Darmawan, D. Triyono, and W. Birastri, 'Estimation Wind Energy Potential Using Artificial Neural Network Model in West Lampung Area', *IOP Conf. Ser.: Earth Environ. Sci.*, vol. 258, p. 012006, May 2019, doi: 10.1088/1755-1315/258/1/012006.

[16] N. A. Satwika, R. Hantoro, E. Septyaningrum, and A. W. Mahmashani, 'Analysis of wind energy potential and wind energy development to evaluate performance of wind turbine installation in Bali, Indonesia', *JMES*, vol. 13, no. 1, pp. 4461–4476, Mar. 2019, doi: 10.15282/jmes.13.1.2019.09.0379.

[17] M. F. Umam, S. Selia, A. F. Sunaryo, and M. R. Al Asy'ari, 'Energy Storage Applications to Address the Challenges of Solar PV and Wind Penetration in Indonesia: A Preliminary Study', *IJE*, vol. 5, no. 1, Feb. 2022, doi: 10.33116/ije.v5i1.110.

[18] F. Mahmuddin, 'Analysis of Wind Energy Potential With a Mobile Floating Structure Around Sulawesi and Maluku Islands of Indonesia', in *Volume 9: Ocean Renewable Energy*, St. John's, Newfoundland, Canada: American Society of Mechanical Engineers, May 2015, p. V009T09A064. doi: 10.1115/OMAE2015-41588.

[19] D. L. Pristiandaru and N. A. Pambudi, 'Wind Energy in Indonesia: Current Status, Potential, Challenge, Opportunities, and Future Policy', *Indonesian Journal of Energy*, vol. 2, no. 2, pp. 65–73, 2019.

[20] F. Pimenta, W. Kempton, and R. Garvine, 'Combining meteorological stations and satellite data to evaluate the offshore wind power resource of Southeastern Brazil', *Renewable Energy*, vol. 33, no. 11, pp. 2375–2387, Nov. 2008, doi: 10.1016/j.renene.2008.01.012.

[21] D. Jiang, D. Zhuang, Y. Huang, J. Wang, and J. Fu, 'Evaluating the spatio-temporal variation of China's offshore wind resources based on remotely sensed wind field data', *Renewable and Sustainable Energy Reviews*, vol. 24, pp. 142–148, Aug. 2013, doi: 10.1016/j.rser.2013.03.058.

[22] I. Karagali, M. Badger, A. N. Hahmann, A. Peña, C. B. Hasager, and A. M. Sempreviva, 'Spatial and temporal variability of winds in the Northern European Seas', *Renewable Energy*, vol. 57, pp. 200–210, Sep. 2013, doi: 10.1016/j.renene.2013.01.017.

[23] H.-G. Kim, H.-J. Hwang, S.-H. Lee, and H.-W. Lee, 'Evaluation of SAR wind retrieval algorithms in offshore areas of the Korean Peninsula', *Renewable Energy*, vol. 65, pp. 161–168, May 2014, doi: 10.1016/j.renene.2013.08.013.

[24] F. Ardhuin *et al.*, 'Comparison of wind and wave measurements and models in the Western Mediterranean Sea', *Ocean Engineering*, vol. 34, no. 3–4, pp. 526–541, Mar. 2007, doi: 10.1016/j.oceaneng.2006.02.008.

[25] W. S. Putro, R. A. Prahmana, M. Y. Darmawan, and H. T. Yudistira, 'Investigation Wind Speed Characterization Over West Lampung Area For Wind Power Potential Application', vol. 5, no. 2, 2018.

[26] I. Abdullah, J. Nurdin, and Hasanudding, 'Kajian Potensi Energi Angin di Daerah Kawasan Pesisir Pantai Serdang Bedagai Untuk Menghasilkan Energi Listrik', *Jurnal Ilmiah "MEKANIK" Teknik Mesin ITM*, vol. 2, no. 1, pp. 31–38, 2016.

[27] A. Prasetio and D. Pratiwi, 'Studi Hidro Oseanografi Pantai Sebalang Kecamatan Katibung Kabupaten Lampung Selatan', *Jurnal of Infrastructural in Civil Engineering*, vol. 02, no. 02, pp. 57–64, Jul. 2021.

[28] F. Abrianto and L. M. Jaelani, 'Evaluasi Pengukuran Angin dan Arus Laut Pada Data Sentinel-1, Data BMKG Dan Data In-Situ (Studi Kasus: Perairan Tenggara Sumenep)', *JTITS*, vol. 5, no. 2, pp. G153–G158, Dec. 2016, doi: 10.12962/j23373539.v5i2.17145.

[29] 'Ryse Energy. E-20 HAWT Datasheet.' Accessed: Sep. 12, 2023. [Online]. Available: https://www.ryse.energy/wp-content/uploads/2020/07/Ryse-Energy-Data-Sheet-E-20.pdf

[30] R. A. Prahmana, M. Y. Darmawan, H. T. Yudistira, and M. S. Anrokhi, 'Wind Characteristics Analysis in South Lampung Indonesia', *Journal of Science and Applicative Technology*, vol. 2, no. 1, pp. 10–13, Jun. 2018, doi: doi.org/10.35472/281439.

[31] A. Ihwan and I. Sota, 'Kajian Potensi Energi Angin untuk Perencanaan Sistem Konversi Energi Angin (SKEA) di Kota Pontianak', *Jurnal Fisika FLUX*, vol. 7, no. 2, pp. 130–140, 2010.

[32] Sudarti and F. A. Dani, 'Potensi Pembangkit Listrik Tenaga Angin di Pantai Blimbingsari Kabupaten Banyuwangi', *CRC*, vol. 5, no. 2, pp. 93–102, Sep. 2021, doi: 10.22373/crc.v5i2.9565.

[33] R. R. Novrita, Sudarti, and Yushardi, 'Analisis Potensi Energi Angin di Tambak Untuk Menghasilkan Energi Listrik', *Journal of Research and Education Chemistry*, vol. 3, no. 2, pp. 96–112, 2021, doi: 10.25299/jrec.2021.vol3(2).7165.

[34] R. J. Barthelmie and S. C. Pryor, 'Can Satellite Sampling of Offshore Wind Speeds Realistically Represent Wind Speed Distributions?', *J. Appl. Meteor.*, vol. 42, no. 1, pp. 83–94, Jan. 2003, doi: 10.1175/1520-0450(2003)042<0083:CSSOOW>2.0.CO;2.

[35] M. Soeripno and M. Ibrochim, 'Analisa Potensi Energi Angin dan Estimasi Energi Output Turbin Angin di Lebak Banten', vol. 7, no. 1, pp. 51–59, 2009.

[36] M. E. Murniati and Sudarti, 'Analisis Potensi Energi Angin Sebagai Pembangkit Enegi Listrik Tenaga Angin Di Daerah Banyuwangi Kota Menggunakan Database Online-BMKG', *JSE*, vol. 6, no. 1, pp. 9–16, Sep. 2021, doi: 10.32502/jse.v6i1.3364.

[37] P. J. Burke, J. Widnyana, Z. Anjum, E. Aisbett, B. Resosudarmo, and K. G. H. Baldwin, 'Overcoming barriers to solar and wind energy adoption in two Asian giants: India and Indonesia', *Energy Policy*, vol. 132, pp. 1216–1228, Sep. 2019, doi: 10.1016/j.enpol.2019.05.055.

IOP Conf. Series: Earth and Environmental Science 1298 (2024) 012025 doi:10.1088/1755-1315/1298/1/012025

Assessment of coastal vulnerability index (CVI) and its application along the Labuhan Maringgai coast, east Lampung Indonesia

T A Tarigan[1*], M A R Fauzi[1], U J Wisha[2,3] and Y Martin[1]

[1] Department of Ocean Engineering, Institut Teknologi Sumatera, Indonesia
[2] Coastal Processes Research Group, Research Center for Oceanography, National Research and Innovation Agency (BRIN), Jakarta 14430, Indonesia
[3] Physical Oceanography Laboratory, Department of Physics and Earth Sciences, Graduate School of Engineering and Science, University of the Ryukyus, Nishihara 903-0213, Japan

*Email: trika.tarigan@kl.itera.ac.id

Abstract. The Coastal Vulnerability Index (CVI) is one of the simplest and commonly used methods to assess coastal vulnerability. In this way, it is a common tool contributing to the decision-making process in long-term coastal planning and management. This study aims to assess the level of coastal vulnerability according to coastal typology of Labuhan Maringgai Coast section. The variable formulation of CVI is divided into hydrodynamic factors (mean significant wave heights, mean tide range, and mean sea level rise) and morpho dynamics factors, i.e. geomorphology, coastal slope, and average width of emerged beach. The morpho dynamics data were obtained through observation, direct measurements in the field by systematic sampling, extraction from aerial photographs and digital surface model (DSM). The results of modelling and predictions by several agencies are derived to obtain the hydrodynamics data. The coast of Labuhan Maringgai was divided into 12 transects and the coastal vulnerability index were classified into three categories such as low, moderate, and high. The results of CVI method showed that the most vulnerable transect is located at Desa Bandar Negeri characterized by mud flats and damaged mangroves.

1. Introduction

The coastal region is a very complex and dynamic area because it is a meeting place between land and ocean. This condition causes coastal areas to have a high degree of susceptibility to environmental change. This makes coastal environment potentially highly vulnerable and environmentally unstable in settlement areas along the coastal, tourist, and fishing sectors so that they cannot be exploited properly and sustainably over a long period of time. The vulnerability of coastal areas is a description of conditions that are easily influenced by natural factors and human activity factors [1]. Labuhan Maringgai district located in Lampung East district is a coastal area that has the potential of coastal areas to be developed from the mining, catch fishing, and tourism sectors. The potential exploitation can be sustainable if supported by the balance and sustainability of coastal regions by identifying the level of vulnerability in coastal areas by mapping the area vulnerability using the Coastal Vulnerability Index (CVI) method.

This research aims to identify the physical parameters of vulnerabilities that are located along the coastal areas of Labuhan Maringgai with a variety of data and processing will be subsequently included in the scoring system of the CVI method. Variables used as calculation weights include coastal geomorphology, coastal inclination, waves, change of coastline, and sea level rise. The results of all data processing with variables – such variables will be categorized according to the range of values on the method of CVI. The result of this study is a map of coastal vulnerability level that can be used as a material for the consideration of the local government to develop dam preserving coastal regions.

2. Material and methods

This section provides details on the CVI formulation describing the variables used, the ranking score and the mathematical formulation applied to construct the synthetic index. This section also describes the sensitivity analysis carried out for the preparation of this work, and the application and result of CVI formulations present for the case study area.

The area has been divided into 11 transects which were the area considered to Bandar Negeri Village, Karya Makmur Village, Karya Tani Village, Margasari Village, Muara Gading Mas Village, dan Sriminosari Village (Figure 1).

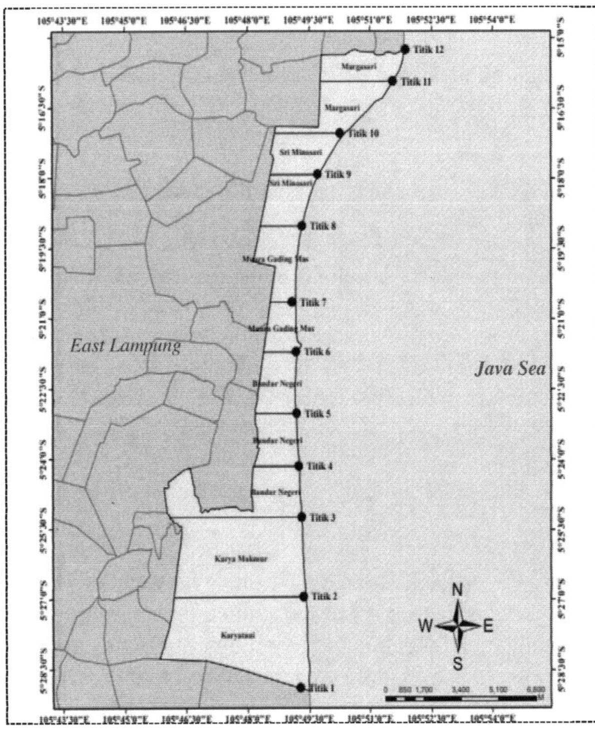

Figure 1. Case study area and related transects at Labuhan Maringgai, East Lampung.

2.1. Data Sources

The data collection was carried out along the coastline of Labuhan Maringgai district, the eastern district of Lampung which comprises 6 villages that are located along coastal areas with a coastal line length of ±26 km by purposive sampling method with 12 station.

2.1.1. Geomorphology

The geomorphological observations of the coast were conducted directly at the 12 points of the research station visually to determine the geomorfological conditions of the shore. Coastal geomorphology relates to the sighting of coastal areas [2]. At 12 points of stations reached using GPS Hand Garmin 64s. The Geomorphological observation aims to know the characteristics of the coastal area of each village.

2.1.2. Mean Significant Wave Height

Significant high-wave data came from 1 January 2011 – 31 December 2020 wind data obtained from the official website of the European Centre for Medium Range-Weather Forecast (ECMWF) or the site https://www.ecmwf.int/, the location of the wind data collection at the coordinates 5°28'48.00"S, 105°46'12.00"E located on land. The wind data was used for Hindcasting Wave with the SPM 1984 (Shore Protection Manual) method aimed at obtaining significant high wave values.

2.1.3. Tide Range

Recession data is obtained from the MIKE21 prediction with the time used from January 1, 2005, to June 30, 2022, the location of the recession prediction at the coordinates of 5°21'42.62"S, 105°49'11.06"E with a time interval of 1 hour. The Mike 21 prediction data is verified against field recession measurement data from February 7, 2022, to February 21, 2022, to determine the accuracy of the data obtained from the forecast of MIKE21.

2.1.4. Coastal Slope

Collecting coastal slope data using DEMNAS secondary data obtained from the official INA-Geoportal website or https://tanahair.indonesia.go.id/. Elevator data used in Labuhan Maringgai district, East Lampung district. The elevation data is grouped using a reclassify tool, then a buffer is carried out at the point of the research station (land) to create a new point (sea) with varying distances, so that the reduction between the elevation on land and at sea obtains a difference in elevation at the station point. The obtained coastal angle is interpolated to find out the area of the spread of the value of the coastal tilt, to obtain the angle of coastal inclination using the equation below [3].

2.1.5. Shoreline

Data on coastline changes are derived from Landsat Image 8 data for 2011 and 2021 on Labuhan Maringgai district obtained from the official United States Geological Survey (USGS) website or https://earthexplorer.usgs.gov/, Landsat image data 8 is used to obtain values of the rate of coastal changes at each point of the research station. Image data processing is performed there are several stages such as performing image pre-processing aimed at fixing image data to do the extraction. The threshold approach is the way to separate land and ocean in the image, the band ratio approach is a way to cover the weaknesses resulting from the thresold. The stages were performed using ENVI 5.3 software, then an image overlay was performed to determine the speed of coastline change at the research site using the DSAS method. (Digital Shoreline Analysis System).

2.1.6. Sea Level Rise

Data on sea level rise using secondary data obtained from from National Oceanic and Atmospheric Administration, NOAA (https:// www. star. nesdis. noaa. gov/ socd/ lsa/ SeaLe velRi se/).

2.2. The CVI and Its Components

The basic physical-geological parameters (1) geomorphology, (2) mean significant wave height, (3) mean tide range, (4) coastal slope, (5) shoreline erosion/accretion rates, and (6) relative sea-level rise are mostly adapted from Gornitz [4] and the basic of CVI studies addressing geophysical vulnerabilities.

Geomorphological landforms represent erosion resistance, and erosion rates are considered markers of susceptibility to coastal processes. Wave energy is connected to erosion capacity, and relief and vertical land motions are seen as indications of flooding danger. Variables are classified on a scale of 1 to 5 (low risk to high risk), with the range divided into five equal parts (Table 1). Apart from the geospatial comparison of various risk classes, the distributions of vulnerability classes are analyzed using statistical analysis (skewness and mean). The cell distribution around moderate vulnerability classes (mean) and their peaks or extreme vulnerability categories are examined in this context.

Table 1. Coastal Vulnerability Index (CVI) Rangking [5].

Variabel	Very Low 1	Low 2	Moderate 3	High 4	Very Hight 5
Geomorphology	Rocky cliffed coasts, Fiords, Fiards.	Medium Cliffs Indented coasts.	Low cliffs, glacial drift, alluvial plains.	Cobble beaches, estuary lagoon.	Barrier beaches, sand beaches, salt marsh, mud flats, deltas, mangroves, coral reefs.
Mean Significant Wave Height	0 – 2.9 m	3.0 – 4.9 m	5.0 – 5.9 m	6.0 – 6.9 m	> 6.9 m
Mean Tide Range	< 1.0 m	1.0 - 1.9 m	2.0 – 4.0 m	4.1 – 6.0 m	> 6.0 m
Coastal Slope	> 10^0	$6^0 – 9.9^0$	$4^0 – 5.9^0$	$2^0 – 3.9^0$	< 2^0
Shoreline Erosion/Accretion (m/yr)	> +0.1	0.0	-0.1 - -0.5	-0.6 - - 1.0	> - 1.0
Sea Level Rise	< -1.0	-1.0 – 0.9	1.0 – 2.0	2.1 – 4.0	> 4.0

The Coastal Vulnerability Index (CVI) of each cell is determined by combining several risk parameters to produce an indicator. The calculation of the index vulnerability score value is based on the originality of the concept of calculating the index of vulnerability value in the CVI method, i.e. is the root of the multiplication of each variable weight value divided by the number of variables as follows [5].

$$CVI = \sqrt{\frac{axbxcxdxexf}{6}} \qquad (1)$$

where CVI = value (score) Coastal Vulnerability Index, a, b, c, d, e and f are sequential variable weights; geomorphology, coastline change, coastal bottom tilt, significant wave height, rise range cycle, and sea level change. The determination of the category of the CVI method is classified into 3 category index such as low, moderate, and high.

3. Results

3.1. Geomorphology

Geomorphological observations were conducted directly at 12 research stations located in 6 villages in Labuhan Maringgai district of East Lampung. The results of the observations show that at Labuhan Maringgai district has a type of sandy and muddy beach as well as has a beach building at some point of the station and mangrove that grows at certain points of the Station and at some points it is used as a beach resort and tourist area by the surrounding community. The average of the conditions of the coastal area on Labuhan Maringgai which is beaches with sandy and mud that can be classified vulnerability is very high with the value obtained ranging from 4 – 5.

3.2. Mean Significant Wave Height

Significant high-wave data generated by wind came from the ECMWF in 2011 until 2020 obtained from the 1984 SPM method. In the coastal waters of Labuhan Maringgai district produced a significant wave height of 0.32 m – 0.49 m and a wave period of 3.07 s – 3.74 s, so that the significant high value of the wave produced was 0.5 m and the significant wavelength resulted was 3.7 s.

3.3. Tide Range

MIKE21 prediction data was used to generate tide data over 18.6 years, which was then confirmed using field data. The findings of the field verification of tidal measurement data attained an accuracy of 84.97% with a Mean Realative Error value of 15.03%.

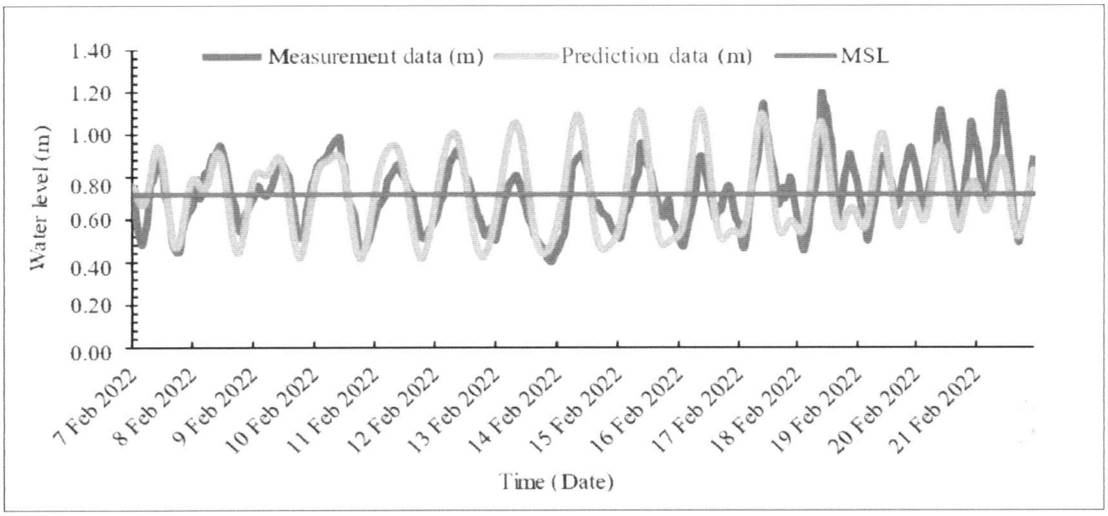

Figure 2. Verification of predicted and field tidal data results.

This elevation data is gathered and utilized to determine the Formzhal number. The mixed tidal type tends to be single daily, with one high tide and one low tide in one day, although there are occasionally two high tides and two low tides with varying heights and periods. The greatest tidal value (HHWL), 1.22 m, was attained, with the lowest ebb value (LLWL), 0.30 m, giving a tidal rise value of 0.92 meters.

3.4. Coastal Slope

Dem data processing for coastal angle was carried out at 12 research stations located at 6 villages in Labuhan Maringgai district of Lampung East. Results of data analysis by means of interpolation showed that at Labuhan maringgai has a coastal tilt angle ranging from 0.92o – 5,05o, the percentage of coastal inclination at 12 states of the research range from 2.05% – 11.21%.

3.5. Shoreline

The average rate of coastline change in Labuhan Maringgai district in 2017 – 2021 was obtained with the highest accretion at the point of research station 9 of 239.80 m/year and highest abrasion at research station 4 point of -3.67 m/ year. Labuhan Maringgai has a variety of vulnerability classification values, namely highly vulnerable, moderate and non-sensitive with values 1, 3, 4 and 5.

3.6. Sea Level Rise

Secondary data analysis from National Oceanic and Atmospheric Administration, NOAA (https:// www. star. nesdis. noaa. gov/ socd/ lsa/ SeaLe velRi se/) shows that Indonesia's waters have an average trend of sea level rise of 6.8 mm/year.

3.7. Coastal Vulnerability Index

Based on the calculation of the Coastal Vulnerability Index of each parameter, the coastal vulnerability class of every parameter (i.e. A = geomorphology, B = mean significant wave height, C = tide range, D = coastal slope, E = shoreline, and F = sea level rise) are obtained, as shown in Table 2.

Table 2. Coastal Vulnerability Index (CVI) Results

Segments	Village/Location	A	B	C	D	E	F	CVI	Category
1	Karya Tani	4	1	2	3	4	5	8,9	High
2	Karya Makmur	4	1	2	4	1	5	5,2	Moderate
3	Bandar Negeri	5	1	2	3	4	5	10,0	High
4	Bandar Negeri	5	1	2	4	5	5	12,9	High
5	Bandar Negeri	5	1	2	4	3	5	10,0	High
6	Muara Gading Mas	4	1	2	4	1	5	5,2	Moderate
7	Muara Gading Mas	4	1	2	5	1	5	5,8	Moderate
8	Sri Minosari	5	1	2	4	1	5	5,8	Moderate
9	Sri Minosari	5	1	2	5	1	5	6,5	Moderate
10	Sri Minosari	5	1	2	3	1	5	5,0	Moderate
11	Marga Sari	5	1	2	4	1	5	5,8	Moderate
12	Marga Sari	4	1	2	3	1	5	4,5	Moderate

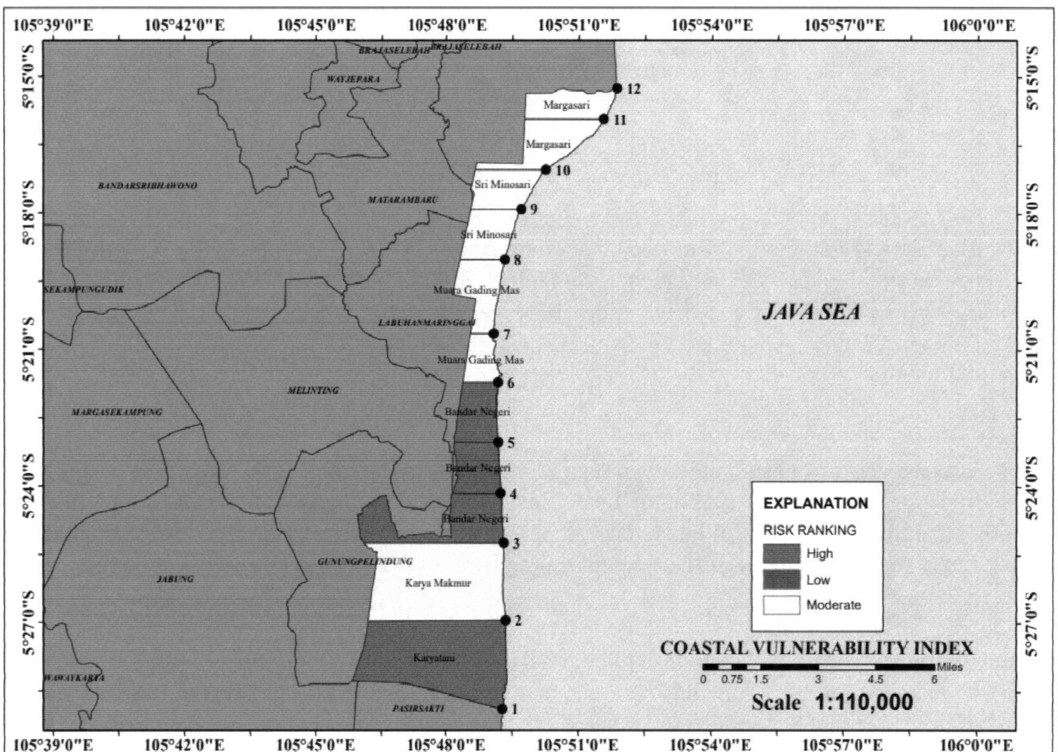

Figure 3. The CVI Map of Labuhan Maringgai, East Lampung.

4. Conclusions

The CVI results obtained at 12 research stations located in 6 villages, covering 6 physical parameters, range from 4.5 to 12.9, categorizing Labuhan Maringgai District as a moderate and vulnerable area. Table 1 shows that the vulnerable category includes 4 research stations, while the moderate category comprises 8 research stations. Villages classified as vulnerable are Karya Tani and Bandar Negeri, while those in the moderate category include Karya Makmur, Muara Gading Mas, Sri Minosari, and Marga Sari. The vulnerability in Labuhan Maringgai District is attributed to the high classification values of vulnerability in each physical parameter, such as coastal geomorphology, coastal slope, changes in coastline, and trends in sea level rise, excluding parameters like significant wave height and tidal range. The moderate category in Labuhan Maringgai District is due to the high classification values of vulnerability in each physical parameter, such as coastal geomorphology, coastal slope, and trends in sea level rise, excluding physical parameters like changes in coastline, significant wave height, and tidal range.

Based on the results obtained from the six physical parameters, the most dominant physical parameters causing very high vulnerability are coastal geomorphology, coastal slope, and the rise in sea level, which tends to result in tidal flooding. Therefore, an analysis is needed for the design of coastal structures at the research location and the planting of mangroves in areas where there is currently no mangrove vegetation as a form of reducing the potential for disasters.

5. References

[1] Kaiser, G 2007 *Coastal Vulnerability to Climate Change and Natural Hazards. Forum DKKV/CEDIM:* Disaster Reduction in Climate Change. Karlsruhe University. Karlsruhe

[2] Suhana, M. P., Nurjaya, I. W., & Natih, N. M. N 2016 *Analisis Kerentanan Pantai Timur Pulau Bintan Menggunakan Digital Shoreline Analysis dan Coastal Vulnerability Index Tahun 2016. Jurnal Teknologi Perikanan Dan Kelautan*, **7 (1),** 21–38.

[3] Kalay DE, Lopullissa VF, Noya YA 2018 *Analysis of The Slope of The Coastal Slope and Distribution of Coastal Sediments In Waai State Waters, Salahutu District, Maluku Province. Jurnal Triton.* **Vol. 14 (1)**: 10 – 18.

[4] Gornitz V, Kanciruk P 1989 *Assessment Of Global Hazards From Sea Level Rise. Presented In Proceedings Of 6th Symposium On Coastal And Ocean Management ASCE. Charleston, SC.* 11-14.

[5] Gornitz, V 1991 *Global coastal hazards from future sea level rise. Palaeogeography, Palaeoclimatology, Palaeoecology (Global and Planetary Change Section).* **89,** 379–398.

IOP Conf. Series: Earth and Environmental Science 1298 (2024) 012026 doi:10.1088/1755-1315/1298/1/012026

Assessment of the Quality of Raw Water for Salt Production in Pangarengan, Sampang, East Java

Maulinna Kusumo Wardhani*, Dyah Ayu Sulistiyorini

Department of Marine Science and Fisheries, Trunojoyo University of Madura

Email: maulinnakusumo@trunojoyo.ac.id

Abstract. Seawater is the primary raw material for the production of sea salt. Seawater quality dramatically affects to salt production. The purpose of this study was to evaluate the seawater's quality as a raw material for salt production in Pangarengan, Sampang, East Java, Indonesia. The assessment of water quality in this study includes physical and chemical characteristics. The location of this research was in the salt fields that close to settlements with high anthropogenic activity. Sampling points were the seawater inlet (P1), reservoir plot/bozem (P2), and evaporation plot (P3). Assessment of temperature, pH, salinity and water saturation using in-situ measurements. Determination of heavy metal parameters (Mercury, Cadmium, Lead, and Arsenic) using UV spectrometry. The physical characteristics of the highest water temperatures were in evaporated plot (40°C), while the water temperatures in the seawater entrance and reservoir ponds were respectively 33.5 °C and 36 °C. pH, one of the chemicals parameter, at the seawater entrance, reservoir plot, and evaporated plot were 8.2, 7.7, and 9. The salinity of the three points is more than 33 ‰, and the raw water saturation in the evaporated pool is 24° Be. Analysis of heavy metals in this study showed levels below seawater quality standards for consumption purposes.

1. Introduction

Salt fields are generally located in coastal areas to facilitate the flow of water as a raw material into the plots. However, coastal areas are often places where various types of waste are disposed of, making them vulnerable to pollution [1]. Therefore, salt farms must be located on the coast with clean, unpolluted sea waters, free of garbage, and slightly suspended solids, far from ports, industrial zones, agriculture, settlements, and major cities. Thus, pre-production of salt needs to pay attention to salt land management so that the salt produced is free of pollutants [2].

The quality of the sea salt will depend on the quality of the seawater, the manufacturing process, and the technology used [3]. The salt produced from sea water, which is pumped into salt production plots, is highly prone to heavy metal contamination. Salt containing heavy metals such as Lead (Pb), Cadmium, Mercury, and Arsenic will pose a risk to human health if ingested [4]. The presence of heavy metals in marine water can be directly ingested by marine organisms and subsequently build-up in their tissues. This build-up can spread throughout the food chain and pose a risk to marine life and human health. [5]. In addition, other studies have shown that low-level concentrations of Pb and Cd, as well as Hg and As, can have adverse effects on human health. The presence of heavy metals in food can lead to a variety of health issues, including damage to the central nervous system, brain injury, paralysis, delayed growth, renal impairment, bone weakness, DNA alteration, and cancer [6].

Salt field samples from Sandpit in Karachi, Pakistan that receive untreated effluent through Lyari River show health risk assessment (HRA) index 2 to 6 times higher than non-carcinogenically relevant THQ (Target Hazard Quotient) (TQ) (As), (Pb), (Al), (Hg), and (Cr). The hazard index similarly indicates that salt is at least 20 times more hazardous than other materials, with a hazard index value of 20.29. The carcinogenic rate (CR) index of heavy metals Cd, As, Cr and Ni is also higher than the CR reference value of 1×10^{-4} [7].

Pangarengan Village, Sampang Regency is one of the salt centers on Madura Island which uses the evaporation method in producing salt. The salt fields in Pangarengan Village are currently up to 5 km from the shoreline and are in a residential area. Land adjacent to human activities allows for a decrease in the quality of seawater as a source of raw material for the production of consumption salt. Based on the condition of the salt field in Pangarengan Village, Sampang Regency, determining the quality of salt requires testing the quality of raw water that is free from heavy metal contamination. The purpose of this study was to evaluate the seawater's quality as a raw material for salt production in Pangarengan, Sampang, East Java, Indonesia. This study examined the physical and chemical properties of water, including its temperature, pH, salinity and water saturation, as well as the presence of heavy metals such as mercury, cadmium, lead, and arsenic. Water chemistry generally influences the toxicity of environmental pollutants, particularly metals.

2. Method

2.1. Time and place of research

This study conducted purposive sampling by taking samples from the salt fields farthest from the seashore and located around the settlements of Pangarengan Village, Sampang Regency, East Java, Indonesia (Figure 1). Sampling points were seawater inlet (P1), reservoir plot (P2), and evaporated plot (P3).

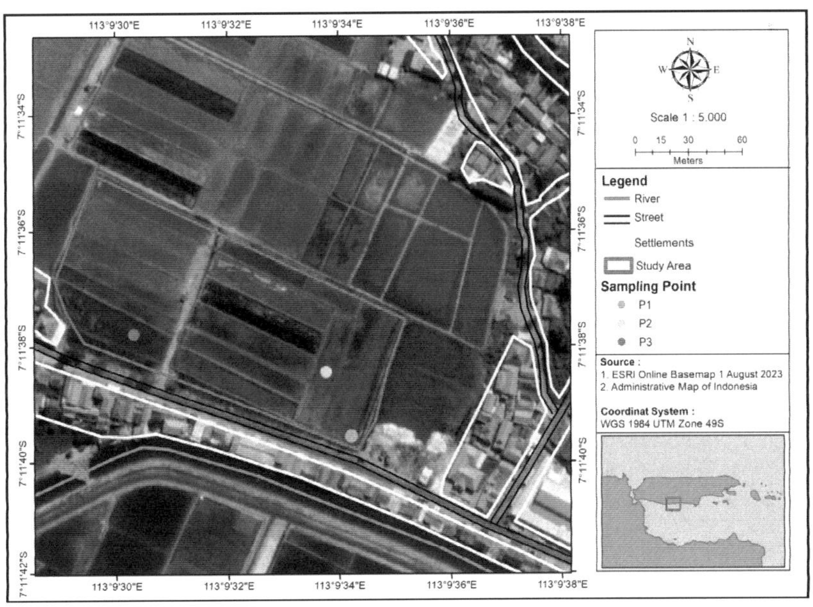

Figure 1. Study area.

2.2. Research methods

2.2.1. Tools and materials

This research uses the tools and materials listed in Table 1.

Table 1. Tools and materials

No	Tools and materials	Function
	Tools	
1	Thermometer	Measure temperature
2	Refractometer	Measure salinity
3	pH meter	Measure pH
4	Baume meter	Measure the degree of saturation of a water sample
5	Dark glass bottle	Take water samples
6	Cooling box	Storing water samples
7	Fume hood	Heavy metal analysis
8	Beaker glass	Heavy metal analysis
9	Volumetrik flask	Heavy metal analysis
10	Graduated pipettes	Heavy metal analysis
11	Hot plate	Heavy metal analysis
12	Spectrophotometry UV Vis	Heavy metal analysis
13	Book	Recording
	Materials	
1	Ice	Water sample preservation
2	HNO_3	Heavy metal sample preservation and heavy metal analysis
3	H_2SO_4	Heavy metal analysis
4	H_2O_2	Heavy metal analysis
5	$HgCl_2$	Heavy metal analysis
6	NaOH	Heavy metal analysis
7	Cd	Heavy metal analysis
8	Silver nanoparticles	Heavy metal analysis
9	Ditizon 0,005%	Heavy metal analysis
10	KCN 10%	Heavy metal analysis
11	NaOBr	Heavy metal analysis
12	$NH^{4-}H_2SO_4$	Heavy metal analysis
13	Aquadest	Heavy metal analysis

2.2.2. Temperature, salinity, pH and saturation

Assessment of temperature, salinity, pH and the degree of water saturation using in situ measurements. Assessment of temperature using a thermometer, measuring salinity using a refractometer, pH data using a pH meter, and the degree of saturation of a water sample using baume meter (Figure 2).

Figure 2. In situ data collection

2.2.3. Heavy metal

Determination of mercury levels in samples was carried out by means of sample measurements through destruction methods, standard solutions, and blank measurements.

Sample preparation

Destruction aims to remove/separate other ion content, with the initial treatment it is expected that errors during analysis can be minimized. The initial treatment method used is the destruction method, namely by breaking the bonds of metal elements with other components in the matrix so that these elements are in a free state [8]. Destruction of water samples was carried out by taking 5 ml of seawater samples then adding 250 ml aquadest, 30 ml of nitric acid (HNO_3) and 10 ml of sulfuric acid (H_2SO_4), and stirring. Heating for 3 hours in a fume cupboard until the solution becomes clear and cooled. Add 10 ml of 30% H_2O_2 (hydrogen peroxide), and filter using Whatmen No. 42.

Standard solution

Mercury (Hg)

Take 0.068 g of Mercury II Chloride solution ($HgCl_2$) each as much as 0; 0.3; 0.6; 0.9; 1.2 ml and add distilled water at each concentration up to 10 ml. Each sample was taken as much as 1 ml and added 2 ml of silver nanoparticles, then stirring was carried out. Make observations using UV-Vis Spectrophotometry with waves of 280-700 nm and make a calibration curve [9].

Cadmium (Cd)

Take each Cd solution as much as 0.1; 0.3; 0.5; 0.7; 0.9 ml and add distilled water at each concentration up to 10 ml. Each sample was taken as much as 1 ml and added 0.1 N Sodium Hydroxide (NaOH) 5 ml, stirred for 15 minutes. Make observations using UV-Vis Spectrophotometry with waves of 250-300 nm and create a calibration curve [10].

Lead (Pb)

Take each Pb solution 0; 0.5; 1; 1.5; 2; 2.5; 3 ml and add 5 ml of 0.05 M Sodium Hydroxide (NaOH), 5 ml of 0.005% Ditizon and 5 ml of 10% KCN, then add distilled water up to 50 ml and let stand for 5 minutes. Each sample was taken as much as 1 ml and made observations using UV-Vis Spectrophotometry with waves of 460-570 nm and made a calibration curve [11].

Arsenic (As)

Take each As solution as much as 1; 2; 3; 5; 6 ml and add aquadest to each sample up to 25 ml. Take 1 ml of each sample and add 3 ml of Sodium Hyprobromite (NaOBr), 15 ml of Aquadest, 0.5 ml of NH_4-H_2SO_4, and 1 ml of hydrozine sulfate. Stir for 75 minutes prior to observation using UV-Vis Spectrophotometry with 845 waves and create a calibration curve [12]

Observation of water samples

Mercury (Hg)

Take 1 ml of water sample and add distilled water at each concentration up to 10 ml. Each sample was taken as much as 1 ml and added 2 ml of silver nanoparticles, then stirring was carried out. Make observations using UV-Vis Spectrophotometry with a wave of 280-700 nm [9].

Cadmium (Cd)

Take 1 ml of water sample and add distilled water at each concentration up to 10 ml. Each sample was taken as much as 1 ml and added 0.1 N Sodium Hydroxide (NaOH) 5 ml, stirred for 15 minutes. Make observations using UV-Vis Spectrophotometry with a wave of 250-300 nm [10].

Lead (Pb)
Take 1 ml of water sample and add 5 ml of 0.05 M Sodium Hydroxide (NaOH), 5 ml of 0.005% Ditizon and 5 ml of 10% KCN, then add distilled water up to 50 ml and let stand for 5 minutes. Each sample was taken as much as 1 ml and observed using UV-Vis Spectrophotometry with a wave of 460-570 nm [11].

Arsenic (As)
Take 1 ml of water sample and add distilled water to each sample up to 25 ml. Take 1 ml of each sample and add 3 ml of Sodium Hyprobromite (NaOBr), 15 ml of Aquadest, 0.5 ml of NH_4 - H_2SO_4, and 1 ml of hydrozine sulfate. Stir for 75 minutes before observing using UV-Vis Spectrophotometry with a wave of 845 [12].

2.3. Data analysis
Analysis of metal content in this study uses the following formula [13]:

$$Metal\ content\ (mg/l) = \frac{C}{W}\ x\ V\ xfp$$

C : Metal concentration from the calibration curve (µg/ml)
V : volume of final solution (l)
W : the weight of the test sample (g)
fp : dilution factor

3. Result and Discussion

3.1. Temperature, salinity, pH and saturation of sea water

Assessment of temperature, salinity and pH at the sampling point is presented in Table 2.

Tabel 2. The results of the assessment of air temperature, salinity, pH and seawater saturation

No.	Parameters	Quality standards	P1	P2	P3
1.	Air temperature (°C) [14]	>32	33.6	33.6	33.6
2.	Water temperature (°C) [14]	>32	34	36	40
3.	Salinity (‰) [15]	≥30	33	39	>100
4.	pH [16]	>7	8.2	7.7	9.9
5.	Seawater saturation (°Be) [17]	>2	0	3.8	24

The physical characteristics of the highest water temperatures were in evaporated plot (40°C), while the water temperatures in the seawater entrance and reservoir ponds were respectively 33.5 °C and 36 °C. pH, one of the chemicals parameter, at the seawater entrance, reservoir plot, and evaporated plot were 8.2, 7.7, and 9. The salinity of the three points is more than 33 ‰, and the raw water saturation in the evaporated pool is 24° Be. The parameters of water as a raw material show the results of water quality suitable for salt production based on parameters of land suitability for salt ponds [14-17] and seawater quality standards based on Government Regulation of the Republic of Indonesia No 22 of 2021 about Implementation and Protection of Environmental Management Attachment VIII concerning quality standards sea water. A different thing is shown in the water saturation parameter at the water inlet of less than 2°Be which shows a low concentration. The degree of water saturation increases from the holding pond to the evaporation pond (P3) and is ready to enter the crystallization compartment.

Salt ions have the potential to alter the mobility and toxic effects of heavy metals within estuarine wetland ecosystems [18]. Environmental parameters such as temperature, pH, DO, and salinity also influence metal concentrations and behavior along estuaries [19]. In addition, a rise in temperature will impede the absorption of heavy metal compounds by particulates, causing them to settle to the bottom. The rise in salinity results in a decrease in the concentration of toxic metals as a result of desalination. Salinity has a big role in the process of decaying heavy metals Cd and Pb when mixing in estuaries. The increase in salinity causes the concentration of Pb and Cd to decrease. The presence of high levels of carbonate in water can lead to a decrease in its solubility due to the formation of carbonate-hydroxide bonds with particles in the water due to an increase in pH [20]. pH is one of the essential chemical determinants of water quality. By knowing the degree of acidity (pH) of the water, the type and rate of reaction of some materials in the water can be controlled. The high or low pH of water is influenced by the compounds contained in the water. Pond water has an ideal pH between 7.5–8.5. Generally, changes in water pH are influenced by soil properties [15].

This study utilizes a calibration curve in heavy metal analysis to compare the concentration of an unknown substance in a sample to a collection of standard samples with known concentrations. The standard curve uses the absorbance spectrum data of a standard solution that has been prepared with certain concentrations. The absorbance spectrum for the standard solution is presented in Table 3 and in Figure 3 below.

Table 3. The absorbance spectrum of the standard solution

Mercury (Hg)		Cadmium (Cd)		Lead (Pb)		Arsenic (As)	
Wavelength (nm)	416.5		219		541		845
Consentration (ppm)	Absorbance	Consentration (ppm)	Absorbance	Consentration (ppm)	Absorbance	Consentration (ppm)	Absorbance
0	0,1085	1	1,0573	0	0,0145	0,4	0,0038
0,3	0,1105	3	1,2084	1	0,0668	0,8	0,0084
0,6	0,1193	5	1,5206	2	0,1035	1,2	0,0122
0,9	0,1252	7	1,6255	3	0,1122	2	0,02
1,2	0,1289	9	1,6368	4	0,1418	2,4	0,0236
				5	0,1558		
				6	0,1694		

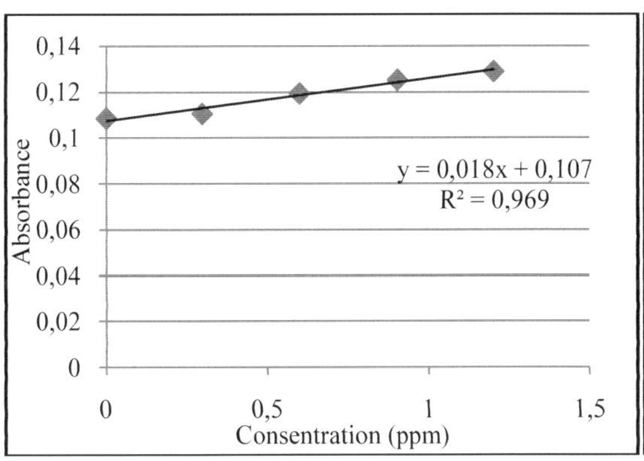

Mercury (Hg)

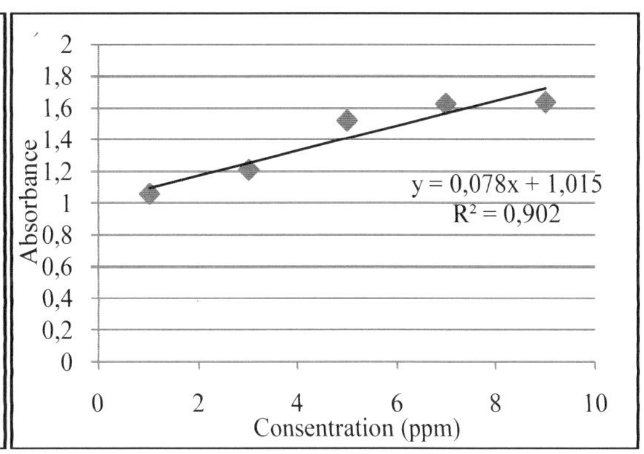

Cadmium (Cd)

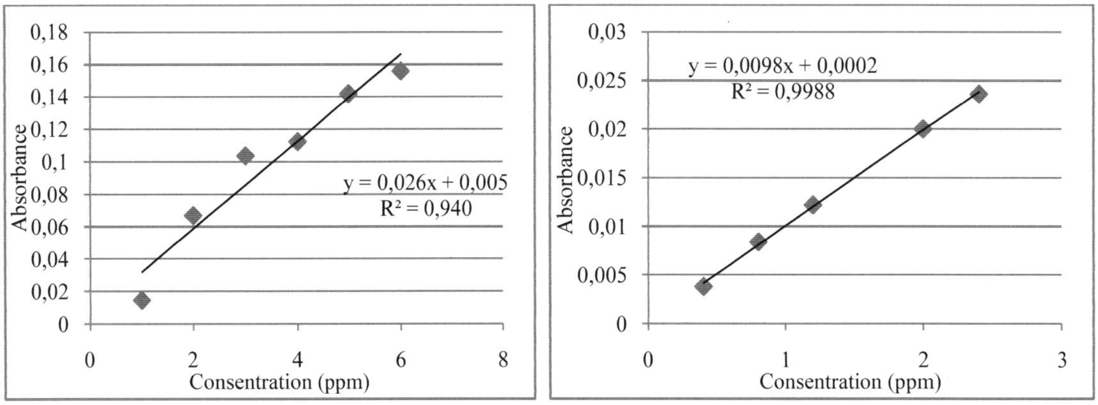

Lead (Pb) Arsenic (As)

Figure 3. Calibration Curve.

The calibration curves for the four heavy metals above show a value of $R^2 > 0.9$, so the equation for each calibration curve can be used in the analysis of heavy metal content. The results of the analysis of heavy metal content in raw water for salt production in Pangarengan Village, Sampang Regency are shown in Table 4.

Table 4. The result of analysis of heavy metal content

		Heavy metal content (mg/l)							
		Mercury (Hg)		Cadmium (Cd)		Lead (Pb)		Arsenic (As)	
Sampling point	Equation	y = 0,018x + 0,107	Content	y = 0.078x + 1.015	Content	y = 0.026x + 0.005	Content		Content
P1		-1,935185185	0	1,4128	0,0057	0,500	0,0020	1,680272	0,006721
P2		-3,761111111	0	1,3224	0,0053	0,333	0,0013	1,656463	0,006626
P3		-4,107407407	0	1,5462	0,0062	0,223	0,0009	1,880952	0,007524

Sea waters close to land generally have high metal content. This condition indicates that anthropogenic activity contributes to metal richness in the waters [20]. Heavy metals can contaminate at every stage of sea salt production. The quality standard for heavy metal contamination in raw water for salt production uses 1st class quality standards, namely water whose designation can be used as raw water for drinking water, and/or other uses that require the quality of water to be the same as that used. Based on the national water quality standard PPRI No 22 of 2021, it shows that raw water for salt production in Pangarengan Village, Sampang Regency is still below the set quality standards, namely 0.001 mg/l (Hg), 0.1 mg/l (Cd), 0.03 mg/l (Pb), and 0.05 mg/l (As). The results of the analysis showed that the raw water at the research location was free of mercury with a negative value of the metal concentration indicated on the calibration curve.

Even though the sampling location is in an anthropogenic area, it has not been polluted by heavy metals. This condition is due to the location of this salt land which is not adjacent to mining or industry which can increase the concentration of heavy metal pollutants [10]. Industrialization and rapid economic development have created serious problems regarding heavy metal contamination in the soil and large amounts of these pollutants are discharged through streams and other routes into the coastal and estuarine environment [5]. The disposal of waste from tin mining, battery industry waste disposal, shipping, and port operations can lead to an increase in the presence of heavy metals, such as copper, chromium, and manganese, in marine waters. This wastewater is discharged into river waters and transported to sea waters [2].

4. Conclusion

1. The physical characteristics of the highest water temperatures were in evaporated plot (40°C), while the water temperatures in the seawater entrance and reservoir ponds were respectively 33.5 °C and 36 °C. pH, one of the chemicals parameter, at the seawater entrance, reservoir plot, and evaporated plot were 8.2, 7.7, and 9. The salinity of the three points is more than 33 ‰, and the raw water saturation in the evaporated pool is 24° Be. The parameters of water as a raw material show the results of water quality suitable for salt production based on parameters of land suitability for salt fields.
2. Even though the sampling location is in an anthropogenic area, it has not been contaminated with any heavy metals. The results of the analysis of heavy metals show that all sampling locations are still below the PPRI 1st class standard water quality standard No 22 of 2021.
3. It is necessary to compare the same research using Atomic Absorption Spectrophotometry (AAS).

Acknowledgement

This research was funded by DIPA Trunojoyo University Madura in 2023.

References

[1] Suhelmi I R, Zainuri M, and Hafiluddin 2013 *Garam Madura Tradisi dan Potensi Usaha Garam Rakyat* (Jakarta: Pusat Penelitian dan pengembangan Sumberdaya Laut dan Pesisir, Badan Penelitian dan Pengembangan Kelautan dan Perikanan Kementerian Kelautan dan Perikanan)

[2] Samsiyah N, Moelyaningrum A D, Ningrum P T 2019 *Jurnal Ilmiah Perikanan dan Kelautan* **11** (1) 43-48 (DOI: 10.20473/jipk.v11i1.11058)

[3] Sumada K et al 2018 *IOP Conf. Series: Journal of Physics: Conf. Series 954*

[4] Salsabiela M 2019 *Journal of Green Science and Technology* **5** (2) 39-44 (DOI:10.33603/jgst.v5i2.5678)

[5] *Xia Sun, Bao-Shi L, Xuan-Li L,and Cheng-Xuan L 2020 Journal of Chemistry* **2** 1-13 (DOI:10.1155/2020/7971294)

[6] Agustina T 2014 *Teknobuga* **1** (1) 53-65 (DOI: 10.15294/teknobuga.v1i1.6405)

[7] Chan M W H, Hasan K A, Balthazar-Silva D, Mirani Z A, and Asghar M 2021 *Marine Pollution Bulletin* **166** 112215 (DOI: 10.1016/j.marpolbul.2021.112215)

[8] Asmorowati D S, Sumarti S S, and Kristanti I I 2020 *Indo. J. Chem. Sci.* **9** (3) 169-173

[9] Nordan H, Firdaus M L and Elvia R 2020 *Jurnal Pendidikan dan Ilmu Kimia* **4** (1) 8-15 (DOI: 10.33369/atp.v4i1.13695)

[10] Ariska N, Supriyanto R, Rinawati, Kiswandono A A 2018 *Analit: Analytical and Environmental Chemistry* **3** (2) 79-90 (DOI: 10.23960%2Faec.v3i2.2018.p)

[11] Marlina A 2019 *10th Industrial Workshop and Research Seminar*

[12] Maryati Sri 2012 *Jurnal Standardisasi* **14** (3) 228 (DOI: 10.31153/js.v14i3.87)

[13] Standar Nasional Indonesia 8207:2016 *Garam Industri Aneka Pangan* (Jakarta: Badan Standarisasi Nasional Indonesia)

[14] Muga T A, Perwira I Y, Kartika I W D 2021 *Current Trends in Aquatic Science* **4** (1) 170-177

[15] Hartati R, Widianingsih, Broto R T D W, Puspa M B, Supriyo E 2022 *Journal of Marine Research* **11** (4) 657-666 (DOI: 10.14710/jmr.v11i4.35353)

[16] Government Regulation of the Republic of Indonesia No 22 of 2021 about Implementation and Protection of Environmental Management Attachment VIII

[17] Kurniawan A, Jaziri A A, Amin A A, Salamah L N 2019 *JFMR* **3** (2) 236-244 (DOI: 10.21776/ub.jfmr.2019.003.02.14)

[18] Bai J, Zhao Q, Wang W, Wang X, Jia J, Cui B, and Liu X 2019 *Ecological Indicators* **96** (1) 91-98 (DOI: 10.1016/j.ecolind.2018.08.026)

[19] Jia Z, Li S, Liu Q, Jiang F, and Hu J 2021 *Environmental Pollution* **287** 117277 (DOI: 10.1016/j.envpol.2021.117277)

[20] Setyaningrum E W, Dewi A T K, Yuniartik M, and Masithah E D 2019 *J Life Sci Biomed* **9** 10-18

[21] Harmesa, Lestari dan Budiyanto F 2020 *Oseanografi dan Limnologi di Indonesia* **5** (1) 19-32 (DOI: 10.14203/oldi.2020.v5i1.310)

IOP Conf. Series: Earth and Environmental Science 1298 (2024) 012027 doi:10.1088/1755-1315/1298/1/012027

Analysis of Environmental Factors on Maritime Security Operations Using the Bayesian Network Method

David Artanto[1], Trika Pitana[2,3*], Udisubakti Ciptomulyono[4], and Athaya Gusti Adisha [5].

[1] Department of Marine Engineering, Institut Teknologi Sepuluh Nopember, Surabaya, Indonesia.
[2,3] Department of Marine Engineering, Institut Teknologi Sepuluh Nopember, Surabaya, Indonesia.
[4] Industrial Engineering Department of ITS, Institut Teknologi Sepuluh Nopember, Surabaya, Indonesia.
[5] Department of Marine Engineering, Institut Teknologi Sepuluh Nopember, Surabaya, Indonesia.

Abstract. The navy as law enforcement with a main duty of maintaining security in the maritime space of national jurisdiction in accordance with the provisions of national law and ratified international law certainly has warships as a support system to achieve such goal. Certain maritime security operations not only require the readiness of the equipment or ships used, but also the supporting environmental factors. In fact, optimal maritime security operation needs a method for an assessment to determine the implementation of maritime security operation. In this discussion, the researcher found a probability as a result of influential environmental factors by using the Bayesian network to determine whether the area to be passed by operating ships is feasible.

1. Introduction

A country that has a water zone must face its own challenges regarding the security of its territorial waters. Since Indonesia is an archipelagic country, about 70% of its territory consists of waters [1]. A country's navy has a military role, a diplomatic role and also acts as a security guard for a country's territory [2]. The diplomatic role aims to carry out politics, both foreign and domestic politics of a country. According to Marsetyo, in carrying out a diplomatic role, a country's Navy must have excellence characters to be displayed consistently from time to time which include human resources (HR) excellence, organizational excellence, and operational capability excellence [1]. Associated with the role as a security guard for territorial waters from various types of law violations at sea, in general, a /0/country's navy has a duty to manage security disturbances that occur within the scope of maritime security, namely non-traditional threats such as maritime terrorism, separatism, radicalism that lead to communal conflicts at sea, social unrest among sea users, piracy and piracy at sea, illegal immigrants, illegal fishing and logging, as well as smuggling and marine pollution [3].

This section aims to determine a probability of safety at sea as a result of environmental factor to provide input to decision makers so that the maritime zone security operations can be carried out as optimally as possible. The probability is obtained based on analysis of the Bayesian Direct Acyclic

Graph (DAG) network with influencing nodes in the form of environmental factors, operational factors of a warship and violations that occur.

2. Factors which Influence Law Enforcement at Sea

Law enforcement at sea refers to all activities related to carrying out the law in the waters in accordance with its authority (constabulary function) which applies globally and in accordance with applicable legal provisions to overcome risk of acts of violence, navigation threats, as well as law violations in maritime zone within national jurisdiction [4]. One of the steps in carrying out law enforcement at sea is to increase strict supervision by local governments through strengthening security in their maritime zone and developing relevant regulations to prevent violations of the law. Weakness in monitoring the maritime zone may have an impact on the loss of rich natural resource potential in national waters [5].

Warships have a duty as maritime security guards against crimes such as piracy, acts of terrorism, various forms of smuggling, human trafficking, illegal fishing, and damage to the marine environment. Such duty is influenced by various factors, including internal operations and environmental conditions within the area of operation. The operational implementation of warships in an effort to maintain the security of a country's maritime territory is closely related to environmental conditions which play an important role, including the ship's endurance in journeys with limited logistical supplies, operational mileage, and maintenance schedules of the ship itself. Operational logistics include liquid logistics (fuel, slip oil, fresh water, etc.), foods and beverages, as well as logistics for ship maintenance. Ship operations are also influenced by environmental factors in certain areas. Ships sailing against the winds and ocean waves may experience significant obstacles, resulting in the need for more powerful propulsion engines. An increase in the use of engine power will have an impact on the limited operational distance of the ship as well as the greater use of logistics in consumption.

2.1 Lawlessness at sea. "Lawlessness at sea" refers to a situation where the sea zone loses the control and effective rule of law, including piracy, illegal fishing, human trafficking and other illegal activities. The existence of irregular sea zone threatens international trade, disrupts navigation, and has a negative impact on global stability. Such phenomena often occur due to a lack of supervision, weak international cooperation, and low law enforcement capabilities in the region as well as unsupportive local government regulations. Therefore, cross-country cooperation is important to overcome "lawlessness at sea. States should cooperate in the exchange of intelligence, joint patrols, and the exercise of strong law enforcement in the affected sea zones. This cooperation will have a positive impact in reducing threats to maritime security and global stability. Based on the report of the UN Secretary General in 2008 concerning Ocean and the Law of the Sea, there are 7 types of threats at sea [6]:

- Piracy and armed robbery;
- Terrorist acts;
- Illicit tracking in arms and weapons of mass destruction;
- Illicit trafficking in narcotics;
- Smugling and trafficking of person by sea;
- IUU Fishing;
- International and unlawful damage to the marine environment.

Of the various types of violations within the scope of maritime security that have been mentioned above, the navy in a country certainly has an important role to protect its maritime territory.

2.2 Warship Operations. Steps in law enforcement at sea include strict supervision and patrols, imposition of strict punishments for offenders, as well as development of appropriate regulations to deal with various forms of maritime crime. In addition, advanced technologies such as satellite monitoring system and automatic identification system (AIS) can be applied to track and observe the movements of vessels at sea. In this context, violations or crimes at sea are not only considered a national problem, but also a global concern. Success in dealing with violations at sea certainly requires cooperation between countries, cross-sectoral coordination, and a commitment to maintaining the security and sustainability of the sea as a shared asset [7].

Operational factors play an important role in determining the level of maritime security [8], which involve some elements as follows:

- Sea Crimes and Violations of Jurisdiction. Maritime crimes such as piracy and illegal fishing can undermine security. In addition, violation of jurisdiction is also a serious threat. States must have the ability to enforce the law in their maritime space and prosecute violations that occur.
- Illegal Fishing. Regulation and supervision of fishing activities is very important. Sustainable legal fishing can help maintain the balance of marine ecosystems and ensure sustainable fish resources.
- Ship endurance and Ship Movement. Ships that can withstand harsh environmental conditions and have good maneuverability can support law enforcement and patrol operations in the maritime zone.
- Environmental conditions. Marine environmental conditions, such as adverse weather and pollution, can affect the operational capability of ships and law enforcement personnel. Efforts to protect the environment certainly have a positive impact on the security of the maritime zone.

2.3 Environment as a Dominant Factor during Maritime Security Operations. The marine environment has a dominant effect on ship movements. Unfavorable sea conditions, such as high waves, strong winds, rain and storms, can interfere with the duties of law enforcement officials in carrying out security and law enforcement operations in the maritime zone under their jurisdiction. Certain natural factors become the main obstacles for law enforcement officers operating at sea [7]. Such constraints affect almost all stages of law enforcement activities, ranging from monitoring to prosecution and detention of perpetrators of illegal activities [9].

2.4 Effect of Rain on Maritime Security Operations. Rain is one of the important environmental factors that can have a significant impact on maritime security operations. The presence of rain in the maritime environment can affect various aspects of ship operations, law enforcement activities, and territorial waters security. Analysis of environmental factors for marine security operations using the Bayesian Network method should consider the effect of rain on the efficiency and effectiveness of maritime security operations [10]. On the other hand, high rainfall in the mainland may also bring sufficient nutrients and cause sea water to be warm, which is suitable for fish growth in the local area. Thus, the fishing area around the area will increase [8]. Of course, this condition will certainly trigger violations of the law regarding fish theft at sea.

Rain has physical and technical impacts that can affect ship operations and law enforcement activities at sea. High rainfall during maritime security operations can cause low visibility problems, making it difficult to detect other vessels or suspicious activity. Disruption to visibility can hamper patrol and monitoring operations, especially ship navigation. Bad weather conditions due to rain can also cause interference with electronic equipment and ship communications, reducing the efficiency of communication between ships or with the control center [10]. Furthermore, rain can affect the condition of the ship's deck, making it more slippery and dangerous for the crew to move on it. Such condition can increase the risk of injury and accidents on board, which in turn can affect operational readiness, crew safety and security as well as law enforcement activities [11].

Rain can also affect ship maneuvers in maritime security operations. Heavy rain can cause stronger waves and ocean currents, resulting in unstable ship movements. Ships sailing in bad weather and rain conditions must overcome challenges in maintaining stability and navigation control [12]. Increased sea waves and currents can affect the speed and direction of ships, which in turn can affect the effectiveness of patrols, reconnaissance and law enforcement actions.

2.5 Effect of Rain on Humidity during Maritime Security Operations. Air humidity is one of the important parameters in the maritime environment that can affect various aspects of maritime security operations. Rain is one of the environmental factors that can have a direct impact on air humidity around water areas. An analysis of the effect of rain on air humidity in maritime security operations is very relevant [13], since this factor can have significant implications for the efficiency, effectiveness and safety of operations carried out by ships and law enforcement officers.

Rain has the ability to reduce air humidity in the surrounding environment. When it rains, water will evaporate from the sea surface and the earth's surface, which in turn can result in a decrease in air humidity. Water vapor in the atmosphere is converted into liquid rain, reducing the amount of water vapor present in the air. A decrease in air humidity can have an impact on micro-climatic conditions in and surrounding waters [14].

2.6 Effect of Rain on Visibility during Maritime Security Operations. Visibility plays a central role in maritime security operations. The ability to have a clear view of the maritime environment is of invaluable significance in maintaining ship integrity, identifying potential threats, and conducting law enforcement operations. Natural factors, such as rain, have a real impact on the level of visibility in waters [15]. An in-depth analysis on the effect of rain on visibility during maritime security operations is of great urgency to understand the challenges faced by law enforcement officers and vessels actively operating at sea. Rain has the potential direct impact on visibility in the maritime environment, which can be seen through several important mechanisms:

- Visual Obstacles. Rainfall may cause substantial visual disturbances. Water droplets falling from the sky may obscure the view, creating a fog-like effect or reducing visibility. Such phenomenon has the potential to make it difficult to detect objects or targets clearly, including other ships, objects above the water surface, or suspicious activities.
- Fog formation. Rain has the opportunity to trigger the formation of fog or a thin layer of fog on the surface of the sea. This kind of fog can result in very limited visibility and can cause distant objects to become completely invisible. Adjacent ships may also appear only faintly or even completely invisible.
- Effect on Lighting. Rain also has the potential to affect the intensity of light that reaches the surface of the water. The sun may be blocked or light reflections may be created from raindrops, producing a sparkling effect or light dimming which further result in blurring of vision and poor visibility.

2.7 Effect of Day and Night Periods on Environmental Temperature during Maritime Security Operations. The ambient temperature in ocean maritime operations is a critical factor which affects various aspects of safety, effectiveness and sustainability of operations. One of the factors that have a significant effect on temperature is the periods, in this case the difference between day and night conditions. Changes in temperature during periods of day and night can affect various aspects of maritime security operations, including navigation, crew working condition, and natural resource management. In this analysis, the researchers explored how the occurrence of day and night periods affected the marine environment temperature in the context of maritime operations [16].

Effect of the day and night periods on water temperature can be explained through the process of solar radiation and heat loss. During the day, sunlight radiates heat energy to the surface of the water, causing an increase in water temperature. This process is known as heating by solar radiation. On the other hand, during the night, without exposure to sunlight, water tends to lose heat to the surrounding atmosphere, causing a drop in water temperature. This condition is referred to as cooling by radiation [17].

Changes in water temperature that occur throughout the day can affect ship navigation and maritime operations. During the day, when the ambient temperature tends to increase, there is a change in the density of both water and air. The lower density of water can affect the transmission of sound waves and the use of sonar, which are important aspects of ship navigation and detection of subsurface objects. An increase in water temperature can also affect ocean currents and atmospheric conditions, which can further affect navigation plans and travel times.

Temperatures that vary between day and night can also affect marine ecosystems and the natural resources. Temperature changes can affect food cycles, migration of marine species, and other biological activities. Fishermen and maritime law enforcement officers need to pay attention to these changes in their efforts to manage marine natural resources in a sustainable manner. Higher seawater temperatures during the day can also affect the growth and distribution of plankton, which is the basis of the marine food chain [18].

2.8 Personnel as Maritime Zone Security Guards. In maritime security operations, the welfare, safety and performance of the crew are very important. Changes in temperature between day and night can affect the working conditions of the crew on deck and inside the ship [19]. Higher temperatures during the day can lead to hot and humid conditions on board, which can affect crew comfort and well-being. On the other hand, lower ambient temperatures during the night can create cold weather which potentially endangers the health of the crew. Temperature management in the ship's working environment is important to ensure the performance and welfare of the crew. In addition, rain conditions can affect the condition of the ship's deck, which may increase the risk of injury and accidents on board. In turn, such condition can affect operational readiness, crew safety and security as well as law enforcement activities [11].

3. Methods

The steps in Bayesian network analysis included compiling/identifying the factors that influence maritime security, compiling a Directed Acyclic Graph (DAG) with the nodes regarding the influencing factors, analyzing the relationship between the influencing factors by compiling a Conditional Probability Table (CPT), to analyzing the probability of security in the maritime area using the academy version of Genie software. Identification of influencing factors and the relationship between these influencing factors was obtained from the literacy process from various sources as well as opinions from several experts in certain fields. This process is one of the important steps that influence risk assessment in data collection and influce the next steps [20]. Factors that influnce the security of the maritime area were compiled in a DAG using simple mind software as shown in Figure 1 below.

The process of analyzing relationship between these influencing factors applied scores derived from questionnaire data that had been weighted through data normalization. A CPT was further compiled based on rules in Bayes' theorem as follows:

$$P(A|B) = \frac{P(A \cap B)}{P(B)} \quad (1)$$

As an illustration of figure 1 below, a probability relationship between ship operations and the factors that affect the operation of these ships can be compiled, so that a CPT table can be compiled as in Table 1 below If it is assumed that the vessel's operational conditions (Y2) with the influence factors of competent crew (A8), ideal environment (A9), ideal vessel motion (A10), and normal vessel endurance (A11), so that a van diagram of the relationship between these factors can be arranged as in Figure 3 below and a probability is arranged as in formula (2) below:

$$P(Y_2) = P(Y_2 \cap A_8) + P(Y_2 \cap A_9) + P(Y_2 \cap A_{10}) + P(Y_2 \cap A_{11})$$

$$P(Y_2) = P(Y_2|A_8)P(A_8) + P(Y_2|A_9)P(A_9) + P(Y_2|A_{10})P(A_{10}) + P(Y_2|A_{11})P(A_{11}) \quad (2)$$

The conditions of the 4 influencing factors did not occur simultaneously and did not influence each other. It was indicated that the system worked as a mutually exclusive event. Probabilities $P(A_8)$, $P(A_9)$, $P(A_{10})$, and $P(A_{11})$ showed the actual conditions obtained at the time of observation and analysis. On the other hand, Conditions $P(Y_2|A_8)$, $P(Y_2|A_9)$, $P(Y_2|A_{10})$, and $P(Y_2|A_{11})$ showed the effect size of factors A_8, A_9, A_{10} and A_{11} which influenced ship operations (R_3). Such data indicated fixed variables obtained from various sources that can be accounted for, in order to create a CPT table regarding the relationship between the 4 influencing factors and operations, as shown in table 1 below.

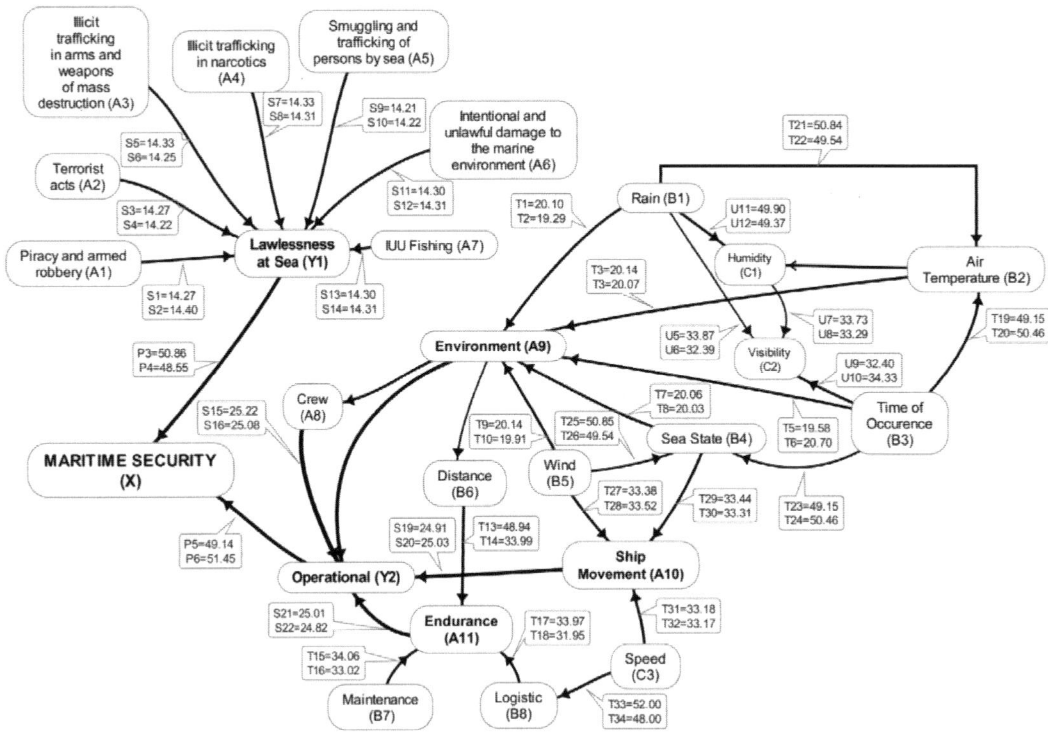

Figure 1. DAG Maritime Security.

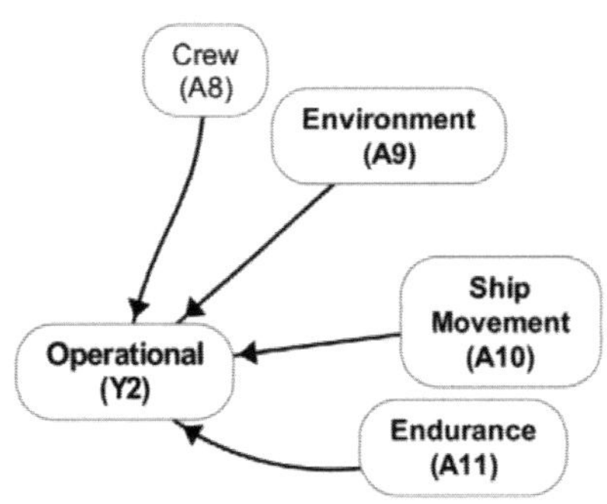

Figure 2. DAG Relationship between ship operations and influencing factors.

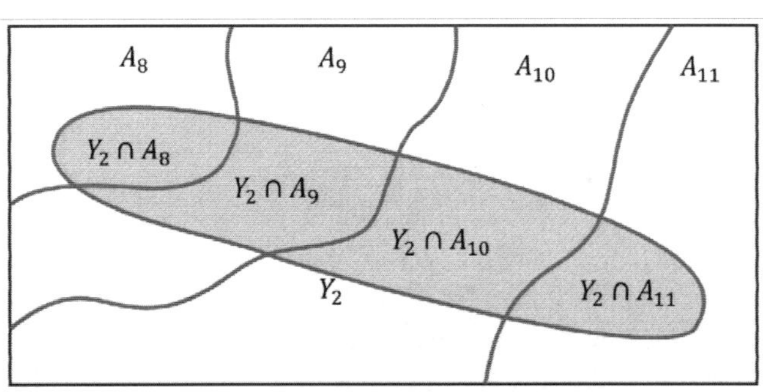

Figure 3. Van diagram Vessel operation Vs influence factors

IOP Conf. Series: Earth and Environmental Science 1298 (2024) 012027 doi:10.1088/1755-1315/1298/1/012027

Table 1. Conditional table probability (CPT) Operational Vs influence factors

A_8		S_{15}								S_{16}							
A_9		S_{17}				S_{18}				S_{17}				S_{18}			
A_{10}		S_{21}		S_{22}		S_{21}		S_{22}		S_{21}		S_{22}		S_{21}		S_{22}	
A_{11}		S_{19}	S_{20}	S_{19}	S_{20}	S_{19}	S_{20}	S_{19}	S_{20}	S_{19}	S_{20}	S_{19}	S_{20}	S_{19}	S_{20}	S_{19}	S_{20}
Ops	P_5	0.53	0.53	0.53	0.53	0.53	0.53	0.53	0.53	0.52	0.53	0.53	0.53	0.53	0.53	0.53	0.39
	P_6	0.47	0.47	0.47	0.47	0.47	0.47	0.47	0.47	0.48	0.47	0.47	0.47	0.47	0.47	0.47	0.61

DAG analysis was performed thoroughly at each node regarding the Directed Acyclic Graph (DAG) of maritime security as shown in figure 1.

4. Discussion and Analysis

In this discussion where researchers get a *Directed Acyclic Graph* (DAG) from the literacy process of discussing the probability analysis of maritime regional security using the *Bayesian network* method. After the DAG is obtained, then look for the relationship between these influencing factors, which is then carried out a sensitivity analysis to validate the DAG model and the variables that have been obtained previously. Parameter settings -1 for initial test conditions are given as in table 2 below.

Table 2. Set parameter-1 maritime security.

Variable	Code	Probability	
Rain	B1	rain= 0	Doesn't rain = 1
Time of Occurance	B3	day = 1	night = 0
Speed Ops	C3	high = 0	low = 1
Wind	B5	low = 0	high = 1
Maintenance	B7	Normal= 1	abnormal = 0
Lawlessness	Y1	Violations occurs	No violation occurs
Piracy and armed robbery (A1)		0	1
Terrorist acts (A2)		0	1
Illicit trafficking in arms and weapons of mass destruction (A3)		0	1
Illicit trafficking in narcotics (A4)		0	1
Smuggling and trafficking of persons by sea (A5)		0	1
IUU Fishing (A7)		0	1
Intentional and unlawful damage to the marine environment (A6)		0	1
Crew	A8	Competent=1	No = 0
Distance	B6	Normal=1	Abnormal=0
Logistic		Full =1	not full = 0

Furthermore, using the academy version of Genie software, with the influencing factors that have been obtained from the results of the analysis, simulation tests are carried out as with parameter set-1 as in table 2 above. The simulation results are shown in Figures 4 and 5 below.

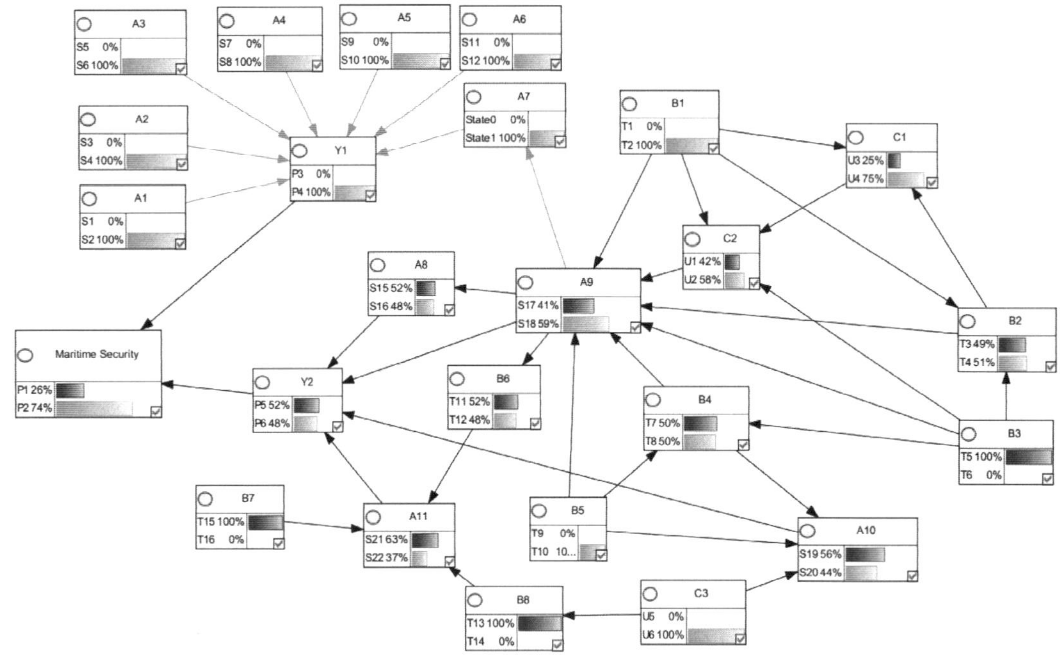

Figure 4. Maritime security in genie software – set parameter 1.

From the simulation results with parameter set-1 as shown in Figure 5 below, 25 nodes are obtained, with the type of Chance-general, 534 parameters, 267 independents, and 8 factors that have no effect between parent-child nodes from 38 factors that have been created, which are A1-Y1, A2-Y1, A3- Y1, A4-Y1, A5-Y1, A6-Y1, A7-Y1, and A9-A7. While the maritime security probability obtained is the maritime area as shown in Figure 4 above, at the level of vulnerability to violations at sea (P1) of 26%, then it means that it has a level of security in the maritime area (P2) of 74%. This is in accordance with the simulation results which show 8 factors between parent-child nodes that do not have a significant influence due to the absence of territorial violations at sea at node Y1.

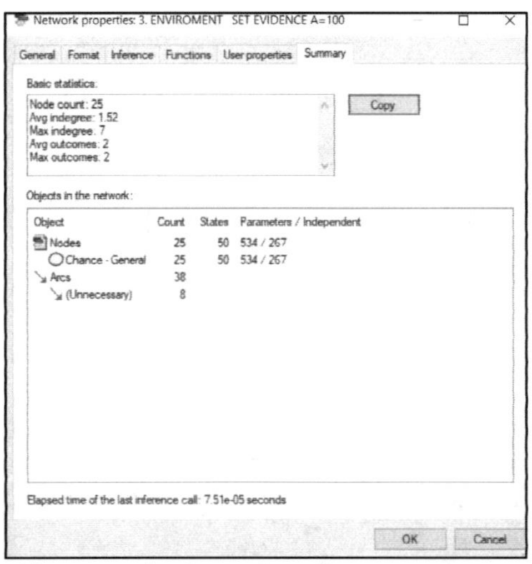

Figure 5. Maritime security in genie software report – set parameter 1.

IOP Conf. Series: Earth and Environmental Science 1298 (2024) 012027 doi:10.1088/1755-1315/1298/1/012027

From the simulation test results shown above, the DAG built with the given influence factors can be tested in the case with the parameter set-2 below.

Table 3. *Set parameter-2* maritime security.

Variable	Code	Probability	
Rain	B1	rain= 0.5	Doesn't rain = 0.5
Time of Occurance	B3	day = 0	night = 1
Speed Ops	C3	high = 0.5	low = 0.5
Wind	B5	low = 0.2	high = 0.8
Maintenance	B7	normal= 0.9	abnormal = 0.1
Lawlessness	Y1	Violations occurs	No violation occurs
Piracy and armed robbery (A1)		0.4	0.6
Terrorist acts (A2)		0.4	0.6
Illicit trafficking in arms and weapons of mass destruction (A3)		0.4	0.6
Illicit trafficking in narcotics (A4)		0.5	0.5
Smuggling and trafficking of persons by sea (A5)		0.7	0.3
IUU Fishing (A7)		0.9	0.1
Intentional and unlawful damage to the marine environment (A6)		0.7	0.3
Crew	A8	Competent=0.8	No = 0.2
Distance	B6	Normal=1	Abnormal=0
Logistic		Full =0.75	not full = 0.25

From table 3 about the maritime security parameter-2 set above, the simulation results are obtainedusing the genie academy software version as shown in figures 6 and 7 below.

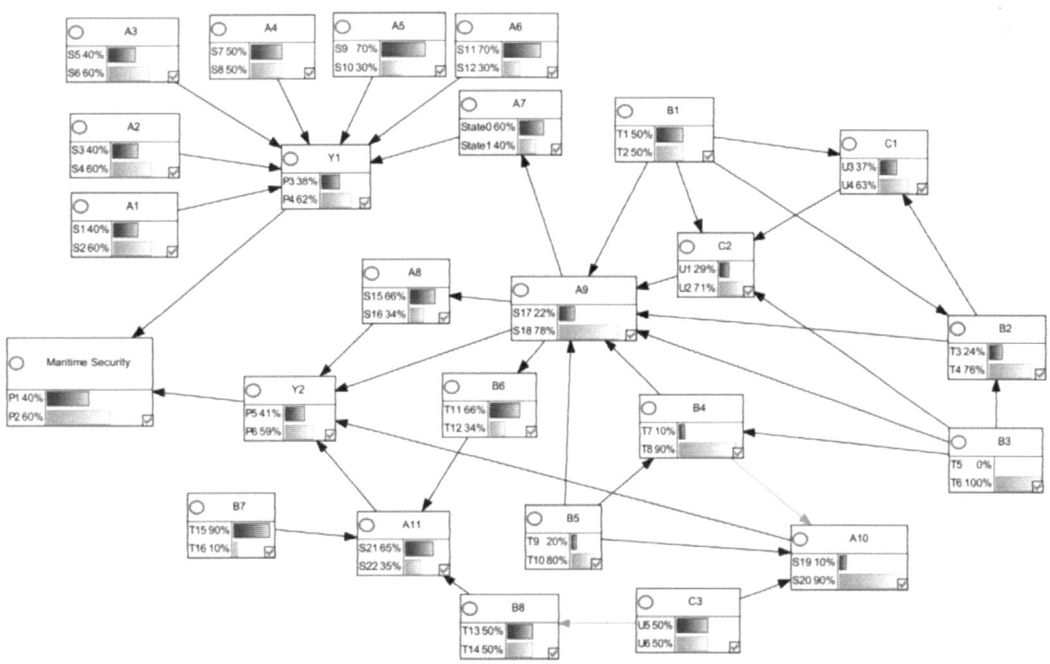

Figure 6. Maritime security in genie software – set parameter-2.

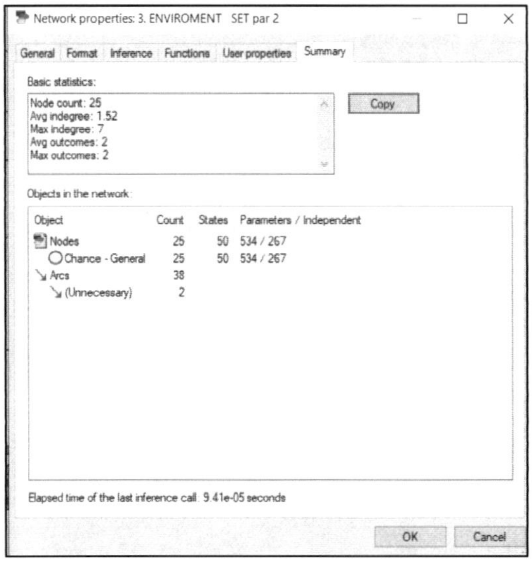

Figure 7. Maritime security in genie software report – set parameter-2.

From the simulation results with parameter set-2, 25 nodes are formed, with the type of Chance-general, 534 parameters, 267 independents, there are 2 factors that do not affect the parent-child nodes of the 38 factors/nodes that have been created, the 2 factors are C3-B8, and B4-A10. While the probability of maritime security obtained is the maritime area at the level of vulnerability to violations at sea (P1) of 40%, which means that it has a level of security in the maritime area (P2) of 60%. This is because there are still violations of the law at sea, while there are 2 factors between the parent-child nodes that do not have a significant effect. This is because in the use of normal ship speed, the use of logistics will also be at a normal level, so that the ship's departure and arrival schedule when carrying out operations is in accordance with the plan. While in the relationship between sea state-ship movement, low sea state is supported by normal ship speed, so ship movements due to sea waves (6DOF) are not affected.

5. Conclusions

This section revealed several important points related to the environmental impact on maritime security operations. Environmental factors such as bad weather, high waves and strong ocean currents can pose a serious challenge to maritime security operations. The Bayesian Network method enables adaptability and responsiveness to environmental changes, as well as adjustments to operational strategies when necessary. In addition, such analysis highlighted the importance of resource optimization by understanding the causal relationship between environmental factors and maritime security operations. Intelligent resource allocation capabilities can cope with deteriorating environmental conditions, maintain operational effectiveness and mitigate greater risks. The use of Bayesian Networks in environmental factor analysis provides a potential to increase the effectiveness of maritime security operations. Strategies can be improved for better results by understanding the effect of environmental variables on operations. The importance of cross-sectoral cooperation was also highlighted in this analysis, by recognizing that environmental factors affect not only security operations, but also other sectors such as fisheries, transport, and the environment. Cross-sectoral cooperation was recognized as a key step towards addressing common challenges and achieving sustainable solutions. In conclusion, the analysis of environmental factors using the Bayesian Network method opened up new opportunities in understanding complex interactions in the maritime zone. Such approach provides hope for a safer and more stable future in global waters through smarter decision-making, sustainable environmental protection, and close cross-sectoral cooperation. Within challenges that are constantly evolving, this scientific and responsive approach may become a foundation for maintaining the maritime zone security and the balance of the maritime ecosystem effectively.

Reference

[1] Marsetyo A S 2013 *Jurnal Penelitian Politik.* **10(2)** 93-111
[2] Booth K 2014 *Navies and Foreign Policy* (Taylor & Francis: Routledge Revivals)
[3] Setiawan I 2017 *Jurnal Perencanaan dan Pembangunan Wilayah* **1(1)** 33-48
[4] Simanjuntak M 2018 *Jurnal Ilmiah Hukum Dirgantara* **8** 2 https://doi.org/10.35968/jh.v8i2.256
[5] Rizki M J 2018 *Jurnal Ilmiah Platax* **5(3)** 45-58
[6] Juliawati T M 2018 *Jurnal Ilmu Hubungan Internasional* **6(3)** 1389-1404
[7] Supriadi D, and Alimuddin S 2011 *Jurnal Dinamika Hukum* **11(1)** 82-97
[8] Putra I N, and Hakim A 2016 *Asro Jurnal-STTAL* **6** 22
[9] Wulansari E M 2014 *Media Pembinaan Hukum Nasional* **1** 6
[10] Aryani C 2021 *Jurnal Pembangunan Hukum Indonesia* **3(2)** 155–173 https://doi.org/10.14710/jphi.v3i2.155-173
[11] Dirgantara OB, Tatit EW, and Suharto D 2020 *Jurnal Maritim Indonesia* **8** 14–23
[12] Wirgantara D R, and Sunendar I 2021 *Prosiding Ilmu Hukum* **7** 1
[13] Sadikin I 2018 E-Proceedings of Art & Design **5** 3
[14] Hutasuhut MA 2020 *Diktat Ekologi Tumbuhan*. (Medan:Fakultas sains dan teknologi Universitas Islam Negeri Sumatera Utara)
[15] Syahputra OH, Nugroho B, Kartodihardjo H, and Santoso N 2019 *Journal of Natural Resources and Environmental Management* **9 (2)**380-393 http://dx.doi.org/10.29244/jpsl.9.2.380-393
[16] Pangemanan AE 2019 *Lex Et Societatis*. **7** No. 12
[17] Ali I M, Yudo L, and Sianturi D 2021 *Jurnal Prodi Strategi Pertahanan Laut* **6(2)** 169–188
[18] Lilis, Nurgayah W, and Irawati N 2019 *Jurnal Sapa Laut (Jurnal Ilmu Kelautan)*, **4(4)**, 205. https://doi.org/10.33772/jsl.v4i4.10767.
[19] Hendrawan A 2019 *Jurnal Saintara*. **3(2)** 53–59
[20] Baroroh I 2023 *International Journal of Marine Engineering Innovation and Research* submitted

IOP Conf. Series: Earth and Environmental Science 1298 (2024) 012028 doi:10.1088/1755-1315/1298/1/012028

Design of wave spectrum in the Java Sea

W L Dhanistha[1*], Suntoyo[1], D M. Rosyid[1], R Akbar[2]
[1]Department of Ocean Engineering, Institut Teknologi Sepuluh Nopember, Surabaya, Indonesia
[2]Department of Industrial Engineering, Muhammadiyah University, Surabaya, Indonesia

*Corresponding author: wimala.danistha@gmail.com

Abstract. Information about wave characteristics plays a crucial role in the design of coastal and offshore infrastructure. This information is necessary to establish safety parameters for structures located offshore. Key parameters of interest include significant wave height and wave period, which are typically described through a wave spectrum. The creation of this wave spectrum involves using the fast Fourier transform, which converts the time domain into the frequency domain. The JONSWAP method describes how wave energy is distributed in the wave spectrum based on specific parameters, such as wind speed, duration of the wind, and the prevailing wave conditions. This method attempts to account for changes in the wave spectrum that occur when the wind blows at different intensities. As a result, the JONSWAP method allows for the generation of more realistic wave spectra under various weather conditions. From this energy spectrum, the JONSWAP shape will be sought to closely match the energy spectrum.

1. Introduction

Indonesia is an archipelagic nation with approximately 70% of its territory covered by water bodies, supporting the Sustainable Development Goal (SDG) number 9, which focuses on industry, innovation, and infrastructure. Infrastructure in this context encompasses both coastal and offshore structures. There are five stages involved in the planning process of coastal and offshore structures, which include defining operational criteria, establishing environmental criteria, foundation planning, structural planning, and construction and installation. In the stage of determining environmental criteria, environmental data such as wind, currents, and most significantly, waves, are essential. This information is crucial to establish safety standards for offshore structures. Therefore, it's imperative to acquire information about wave characteristics in the intended areas for designing these marine structures. For instance, infrastructure projects like ports, coastal and offshore structures, as well as mooring systems as depicted in Figure 1, are closely tied to hydrogeology factors such as erosion, sediment transport, and ocean wave energy [1;2].

Ocean waves exhibit nonlinear behavior in accordance with wave generation and environmental characteristics [3]. Two crucial parameters for monitoring ocean waves are the Significant Wave Height and the time period [4]. The Significant Wave Height represents the wave height that characterizes sea conditions, while the time period signifies the duration of the most dominant wave energy. This information about wave height is of paramount importance, particularly for offshore construction, maintenance, and the safety of ships navigating the sea. Therefore, the data accuracy should closely mirror real-world conditions.

The characteristics of ocean waves can be depicted or observed through wave spectra. Creating a wave spectrum can be achieved using conventional methods that estimate wave energy distribution based on measurements and analysis of actual waves [5;6]. Some commonly used methods for deriving spectrum formulas include the ISSC (Bretschneider or modified Pierson-Moskowitz), International Towing Tank Conference (ITTC), and the Joint North Sea Wave Observation Project (JONSWAP). However, in accordance with statements by Djatmiko (2012) [7] and Mahjoobi (2008) [8] that waves are inherently nonlinear, deterministic equations fall short of accurately describing wave spectra.

Djatmiko & Adrianto (2013) [9] conducted wave height measurements in Indonesian waters, specifically in the western part of the Java Sea. Wave height was measured every second for a year, but data was collected only once a day for each month, resulting in 48 wave data points per month and 576 data points for the entire year. However, based on this data, researchers felt that the periodic monthly observations were insufficient to adequately characterize the wave conditions at that specific location. It is hoped that this study will build upon previous research and yield a new formula that better suits the wave conditions in the Java Sea waters.

2. Materials and methods

2.1. FFT generation

The Fast Fourier Transform (FFT) is a mathematical computation technique used to transform analog signals into digital signals based on frequency. The Fast Fourier Transform (FFT) divides a signal into different frequencies using complex exponential functions. It's an algorithm designed for efficiently and rapidly calculating discrete Fourier transforms. Since signals in communication systems are continuous in nature, the results can be applied to Fourier transformation.

$$X[k] = \sum_{n=0}^{N-1} x[n].e^{-i2\pi kn/N} \qquad (1)$$

Where:

- X[k] is the frequency representation at index k in the frequency domain.
- x[n] is the signal sample at index n in the time domain.
- i is the imaginary unit (usually $\sqrt{-1}$).
- k is the frequency index.
- N is the total number of samples in the signal.

2.2. JONSWAP Spectrum

The Joint North Sea Wave Project spectrum (JONSWAP) is a spectral model used to describe the distribution of wave energy in a specific area of the sea. This model was initially developed by a team of researchers involved in the JONSWAP project in 1973 to study wave characteristics in the North Sea. The JONSWAP spectrum illustrates how wave energy is distributed across various wave frequencies. Typically, this model is expressed in the form of mathematical equations that describe the spectrum of wave energy in terms of wave frequency. The equation's form may vary depending on the weather conditions and the ocean wave parameters present at a specific location. The JONSWAP model finds utility in various applications, particularly in marine engineering, offshore structure planning, ship development, and diverse purposes related to ocean dynamics. This model assists experts in comprehending and predicting the characteristics of ocean waves under various weather and geographical conditions.

3. Results and discussion

In this study, the observed wave heights are located in the Jakarta Bay, indicated by the yellow point in Figure 1. The data collection is based on the European Centre for Medium-range Weather Forecast

(ECMWF) data over a span of 9 years, from 2013 to 2021, with hourly intervals, resulting in a total of 78,887 data points for wind speed and wave heights.

Figure 1. Location map of wave observation

Figure 2 represents the wind speed, and Figure 3 represents the wave height, comprising a total of 78,887 data points but in figure 2 and 3 only 8.760 data points wind speed and wave height for one year. The wave height exhibits a similar pattern to the wind speed, which is consistent with the generation of wave height based on the high or low intensity of the blowing wind. In other words, waves are generated by the wind. Within a year, there are peaks and troughs, with peaks occurring during the rainy season, while troughs occur during the dry season.

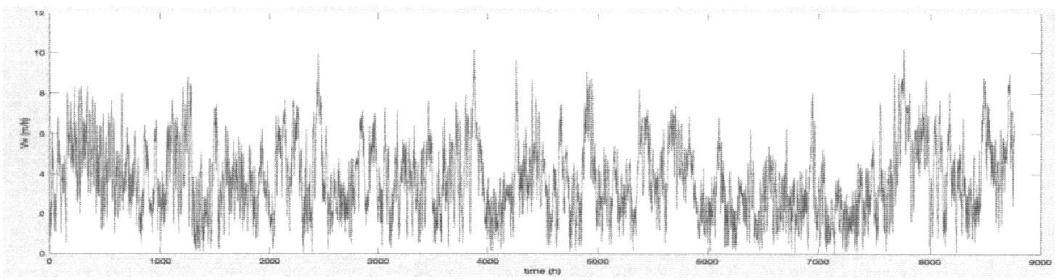

Figure 2. Wind Velocity

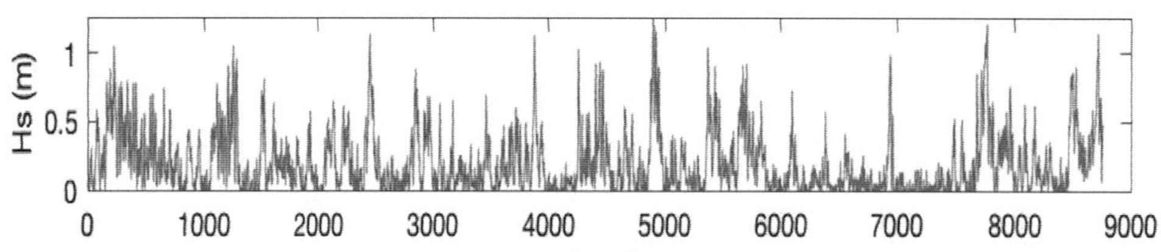

Figure 3. Wave Height

In the creation of the wave spectrum, only wave height data is utilized. The wave height data in the time domain is transformed into the frequency domain using the fast Fourier transform (FFT). The outcome of the FFT will subsequently undergo validation by inversely applying the FFT and comparing it with the original data. The FFT method operates recursively by dividing the original vector into two parts, computing the FFT of each part, and then merging them back together.

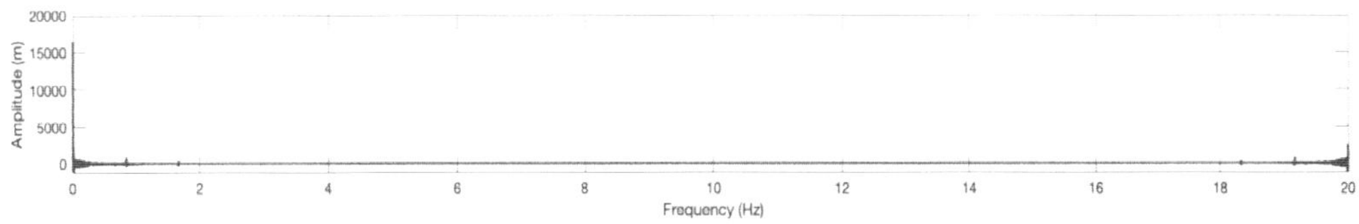

Figure 4. FFT Wave Height

Figure 4 depicts a wave height graph obtained by performing FFT on a frequency signal. In this process, the wave height information in the time domain from Figure 3 is transformed into the frequency domain in Figure 4 using Equation 1. The result of this FFT is then subjected to an inverse FFT to validate whether the FFT transformation aligns accurately with the wave height data based on ECMWF (shown in Figure 3). The comparison between ECMWF wave heights and the inverse FFT is illustrated in Figure 5. The blue line represents the wave heights derived from ECMWF data, while the green line represents the wave heights obtained from the inverse FFT. The results indicate that the inverse FFT effectively matches the ECMWF wave height data. This alignment suggests that the generated FFT is correct, validating its accuracy and allowing for progression to the energy spectrum generation phase.

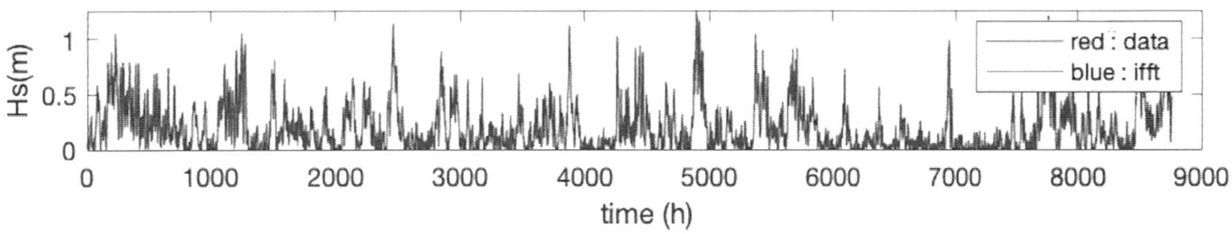

Figure 5. Data vs iFFT

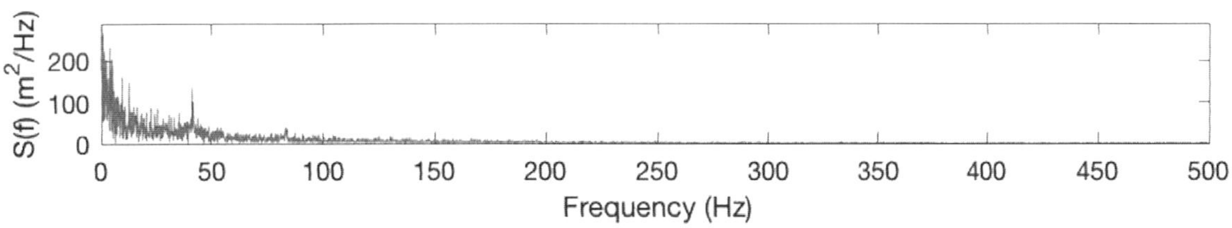

Figure 6. Wave Spectrum

In the context of wave height energy spectrum, 'wave height' refers to the amplitude of waves or the difference between wave crests and troughs. This spectrum provides insights into how energy is distributed across different wave frequencies. Within this spectrum, waves with higher frequencies possess greater energy. The wave energy spectrum aids scientists and engineers in comprehending and measuring the characteristics of oceanic or atmospheric waves, which significantly impact navigation, coastal engineering, maritime transportation, and various other aspects. This information is also crucial for understanding climate changes and ocean-atmosphere dynamics. In practical terms, wave energy spectra are often measured using instruments like radar altimeters on satellites, wave buoys, and other equipment. Figure 6 illustrates the energy spectrum formed from the previous FFT process, where the spectrum showcases a single peak at the beginning or 0 Hz. This energy spectrum serves as a reference for JONSWAP analysis and machine learning applications.

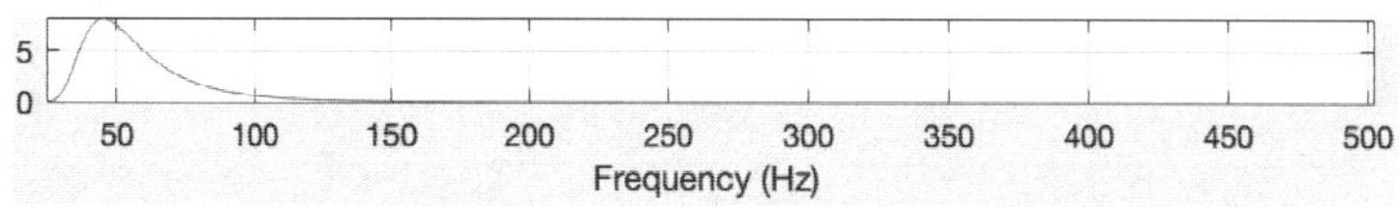

Figure 7. JONSWAP Spectrum

The Joint North Sea Wave Project (JONSWAP) spectrum is one of the wave energy spectrum models used in oceanography and marine science to describe the distribution of wave energy on the sea surface. Originally developed to depict waves in the North Sea, this model has since been widely adopted for various water bodies. JONSWAP was chosen due to its consideration of higher-frequency components. The JONSWAP spectrum incorporates contributions from higher-frequency components to describe waves with higher and sharper amplitudes that can occur under specific conditions, such as storms. Figure 7 represents the JONSWAP spectrum, where the spectrum pattern aligns with the energy spectrum but still exhibits differences in amplitude.

Acknowledgements

The authors are deeply indebted to the Department of Ocean Engineering and DRPM ITS which has sincerely made the data available. Gratitude and great appreciation without the supports from those this study might not be successfully carried out.

References

1. P V Guimaraes, L Farina, E E Toldo Jr 2014 Analysis of Extreme wave events on the southern coast of Brazil *Nat. Hazards Earth Syst. Sci.* **14**(12) 3195-3205
2. A C Eckert-Gallup, C J Sallaberry, A R Dallman, V S Neary 2016 Application of principal component analysis (PCA) and improved joint probability distributions to the inverse first-order reliability method (I-FORM) for predicting extreme sea states *Ocean Eng.* **112** (2016) 307-319
3. M R U Mustafa, S Z A S Ahmad, M K A Husain, N I M Zaki, M H Mohd, G Najafian, I B Othman, M Latheef, D B Endrayana, N Z Bt Yusof 2018 Comparison of various spectral models for the prediction of the 100- year design wave height *MATEC Web Conf.* **203** 01020
4. NOAA 2008 Peak wave period and direction forecasts
5. G Muraleedharan, M Sinha, A D Rao, N U Nair, P G Kurup 2008 Estimation of wave period statistics using numerical coastal wave model *Nat. Hazards* **49** (2) 165-186
6. D I Gopinath, G S Dwarakish 2015 Wave prediction using neural networks at new mangalore port along west coast of India *Aquatic Procedia* **4** 143-150
7. W Sheng, H Li 2017 A method for energy and resource assessment of waves in finite water depths *Energies* **10** (4) 460
8. E B Djatmiko 2012 Perilaku dan Operabilitas Bangunan Laut di atas Gelombang Acak
9. J Mahjoobi, A Etemad-Shahidi 2008 An alternative approach for the prediction of significant wave heights based on classification and regression trees *Appl. Ocean Res.* **30**(3) 172-177
10. D Adrianto 2020 Studi Karakteristik Gelombang Laut Jawa Bagian Barat

The Advantages and Challenges of Carbon Fiber Reinforced Polymers for Tidal Current Turbine Systems – An Overview

D Satrio[1], S Musabikha[2*], S Junianto[3], S Prifiharni[2], R Kusumastuti[2], A Nikitasari[2], and G Priyotomo[2]

[1] Institut Teknologi Sepuluh Nopember, Surabaya 60111, Indonesia
[2] National Research and Innovation Agency (BRIN), South Tangerang 15314, Indonesia
[3] Politeknik Elektronika Negeri Surabaya, Surabaya 60111, Indonesia

*Email: siti089@brin.go.id

Abstract. A significant step in reducing the effects of greenhouse gases is obtaining electric energy from renewable sources. Electricity from tidal currents using underwater turbines is one of the most promising and well-liked technologies. The turbine systems are the key element in the tidal current energy. They are built using hydrodynamic principles to extract the most power possible from tidal ocean currents and are designed to last for extended periods in a maritime environment. The performance of tidal turbines is also significantly influenced by their materials, i.e., carbon fiber reinforced polymers (CFRP) used in them. This paper also reviews the CFRP materials used in tidal current turbine systems. Besides, an analysis of their advantages and challenges regarding CFRP materials that can impact tidal current turbine efficiency is further explored.

1. Introduction

Tidal current energy is one of the renewable energies that uses a turbine system to capture the tidal currents' kinetic energy and convert it into electricity [1–6]. It is the most predictable [7–9], has reached the early stages of technological maturity [8,10], and is a promising source of energy [11–15]. Even without abandoning research on conventional energy sources [16–18], research on ocean renewable energy is important. Tidal current energy is produced using a similar process to how wind energy is produced, except it is entirely immersed underwater. Tidal current energy systems widely used are horizontal and vertical axis turbines [19], as seen in Figure 1.

Figure 1. Tidal current energy: Oceade (GE/Alstom) (left), Openhydro (mid) [20], Kobold (right) [21]

Tidal current turbines must be installed in areas with challenging weather and sea environment (strong currents, turbulence, waves, and storms) [22], and it has complex stresses such as corrosion, mechanical, and biological stress [23–25]. Moreover, this ocean current energy device is operated for a long time with minimal maintenance intensity and is submerged in turbulent seawater [7,26]. On the other hand, the tidal turbine's blade is its most important component to generate tidal energy. Hence, tidal turbine blades must be made of a suitable material to lower the chances of failure. Due to their superior specific strength (strength/weight ratio) and specific strength/stiffness ratio as compared to glass fibers, carbon fiber reinforced polymers (CFRP) have become the favored material for making blades that are thinner, stiffer, and lighter [22,27,28]. Therefore, this paper's goal is to study CFRP and its application to tidal current energy. The advantages and challenges of CFRP utilization are then emphasized to further the tidal current's development.

2. Carbon Fiber Reinforced Polymers (CFRP)

A composite material that is not metallic and is made of carbon fiber (the reinforcement material) and polymer resin (the fiber-holding matrix) is known as CFRP [29]. In a CFRP composite, the carbon fiber contributes to the composite's strength and stiffness, while the matrix typically controls the chemical-heat resistance and load balancing [30]. Figure 2 illustrates the standard CFRP structure.

As the first component of CFRP, carbon fiber (CF) is one of the most widely utilized reinforcement materials due to its outstanding physical qualities, like high stiffness and strength, high heat resistance, high conductivity, self-lubrication, and anti-corrosion [30]. Carbon fiber is generated from substance precursors for polymers like cellulose, polyvinyl chloride, and polyacrylonitrile (PAN). These precursors are treated with heating and tensioning processes to create carbon fibers. The diameter of carbon fibers, which ranges from 5 to 10 μm, is extremely small [29], as shown in Figure 2.

Polymer resins have two primary types: thermoplastic resin and thermosetting resin [31]. Thermoplastic resin is a polymer resin connected by intermolecular interactions. This creates a molecular structure that is either in the direction of a straight line or have branches, which is only capable of providing a very slight limitation on how far molecules can move. As a result, after curing, the thermoplastic resin is remeltable and tractable when heat and pressure are applied. Thermosetting resin, on the other hand, is a polymer resin linked with chemical connections to create a strongly cross-linked molecule. The mobility of molecular chains can be severely constrained by this cross-linked structure, which renders thermosets unmeltable and intractable when heated after curing [29]. For typical thermosets, for instance, an epoxy matrix is preferred because it offers strong adhesion [31].

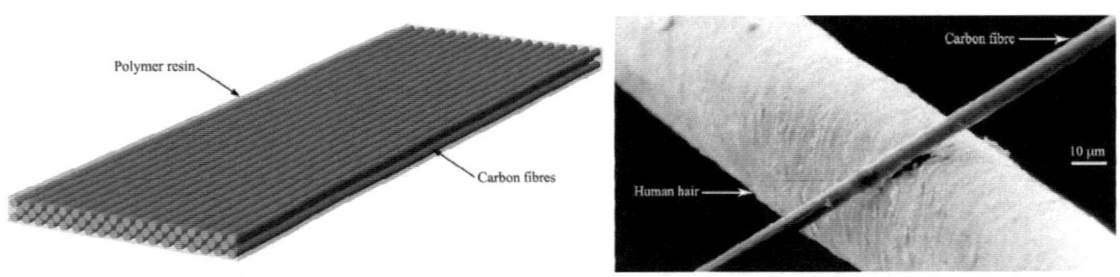

Figure 2. The CFRP's structure (left) and the comparison human hair to carbon fiber (right) [29]

The mechanical features of steel, CFRP, carbon fiber, and polymer resin are shown in Table 1. Between carbon fiber and S355 steel, it shows that despite having far lower densities and weights than steel, carbon fiber has greater tensile strengths [29]. Also, polymer resins have slightly lower densities than carbon fibers, while carbon fibers have far higher strengths than polymer resins. Carbon fibers, however, cannot be used as stand-alone engineering materials because of their filamentary character. Therefore, when carbon fiber is combined with a polymer resin in a synergistic way, it produces CFRP, a typical orthotropic material with outstanding mechanical capabilities.

Table 1. Mechanical characteristics of steel, carbon fiber, polymer resin, CFRP [8,27,29,32]

Materials	Category	Density ρ (kg/m^3)	Tensile strength σ_u (GPa)
Steel	355	7850	0.50
Carbon fiber	Standard	1760	3.53
Carbon fiber	High strength	1820	7.06
Thermoplastic resin	Polyetherether ketone	1310	0.07
Thermoplastic resin	Polyetherimide	1270	0.11
Thermoset resin	Vinylester	1250	0.08
Thermoset resin	Epoxy	1250	0.12
CFRP	Carbon fiber-epoxy	1580	2.30

There are three ways to make composite materials that can be distributed into four steps (layup, resin, joining, and finishing) [33]:

- Vacuum-assisted resin transfer molding (VARTM) is a popular technique for making composite blades. The blade's top and bottom halves, as well as the two shear webs, are manufactured in separate pieces and bonded together to create the blade's final shape. Mold is used to create the parts, which are then filled with fiber materials and reinforcing root sections before being sealed with a vacuum bag. A vacuum is then employed to pull the resin into the fiber by pumping the resin between the bag and the fiber. Curing took place at 80°C for 8 hours.
- Monocoque is a single-piece blade manufacturing method where fibers are placed out in a mold, and then the polymer resin is sucked into the fibers using vacuum. Curing took place at 80°C for 8 hours. Shear webs are not required since this single-piece process enables the production of blades that are stiffer than those produced using the VARTM process.
- The heated mold method utilizes epoxy powders or thermally induced polyester, substituting the resin solution utilized in VARTM processes. The powder and fiber are heated in a 200°C ceramic mold. As a result, the powder melts, and composites are created. The blade's upper and lower parts, as well as a pair of shear webs is partly dried before being combined and cured in one piece within 70 minutes at 170°C.

Some tidal current turbines comprise CFRP, i.e., Seaflow and Seagen (Figure 3). The two-blade rotor Seaflow has an 11-meter diameter. It is constructed of a carbon fiber-reinforced spar joined to fiberglass ribs and then covered in a pre-impregnated carbon fiber/epoxy resin matrix. The prepreg resin dried at a low temperature of 75°C in an ordinary industrial-grade oven [34,35].

Figure 3. Seaflow (left), a 300-kW trial installation from Marine Current Turbines, has been running off the coast of Lynmouth, UK [35], and SeaGen (right), a 1200-kW tidal current turbine system from Simec Atlantis Energy, located in Northern Ireland [20]

The SeaGen (Marine Current Turbines Ltd.) rotor blades have a 16-meter diameter. It was made up of a hollow intermediate-modulus unidirectional carbon fiber mixture box spar that serves as the primary load-balancing component, coupled with carbon ribs and a glass fiber composite envelope that is joined to the carbon fiber ribs. For both glass and carbon components, prepreg was employed and cured at 80°C [34,36,37].

3. Advantages

In comparison to traditional materials, CFRP is viewed as an appealing option due to its outstanding strength to weight ratio and high strength to stiffness ratio, flexibility, greater corrosion resistance, prolonged service life, and their anisotropy may be altered to allow 3-D blade deformation modification [28,30–32,38,39]. Structures with complex curves can also be produced using CFRP with ease and simplicity to maintain [22,36].

3.1. High strength and lightweight

CFRP material usage has significantly grown in recent years because of its strong specific strength and light weight. Table 2 displays the physical and mechanical characteristics of steel and composite. Comparing CFRP to steel and other composite materials, it is found to be lighter while yet having more strength. This causes a significant weight reduction and span extension for prestressed parts. However, composites usually have a lower Young's modulus than steel with the exception of CFRPs having a high elastic modulus [40,41]. Also, compared to GFRP, CFRP provides superior performance while being 13% lighter (for the blade) [39].

Table 2. Physical and mechanical characteristics of steel and composite [41]

Material	Density ρ (g/cm^3)	Tensile strength σ_u (MPa)	Young's Modulus (GPa)
Steel	7.85	483-690	200
GFRP	1.25-2.10	483-1600	35-51
CFRP	1.50-1.60	600-3690	120-680

Also, CFRP has a lower mass compared to GFRP. Jaksic et al. [42] investigated the feasibility of using CFRP instead of GFRP in the design of tidal turbine blades to make the blades lighter and less expensive. The findings demonstrate that as blade length increases, blade mass also increases. When comparing the same lengths of CFRP and GFRP blades, Figure 4 also indicates a reduction in the blade mass distribution along the blade for CFRP blades. This difference reaches up to 24 kg/m at the blade's widest point, while the mass of the blade overall is reduced by 14.5%.

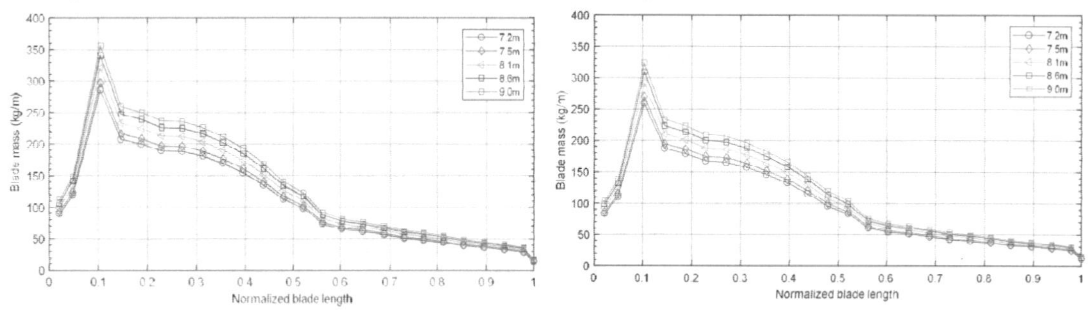

Figure 4. Mass distribution over increasing rotor diameter with blade length: GFRP (left) and CFRP (right) [42]

3.2. Corrosion and fatigue resistances

CFRP has non-metallic and non-corrosive qualities that make it less susceptible to chloride-induced corrosion than steel, which can greatly increase the structure's corrosion resistance [41]. Also, CFRP provides higher resistance to corrosion and fatigue, and it is an important design consideration for tidal turbines. Non-corrosive CFRPs are crucial in harsh corrosive environments since chloride ions and water molecules are corrosion-promoting elements [43]. Also, carbon fibers have far better seawater corrosion resistance than glass fiber [27]. Carbon fibers have the advantage over glass fibers, which lose tensile strength (by around 8% weekly) when subjected to 2% sodium chloride solutions. Glass fibers can also be corroded due to microbial attack when exposed for an extended period of time to natural polymers released by bacteria into their environment [43].

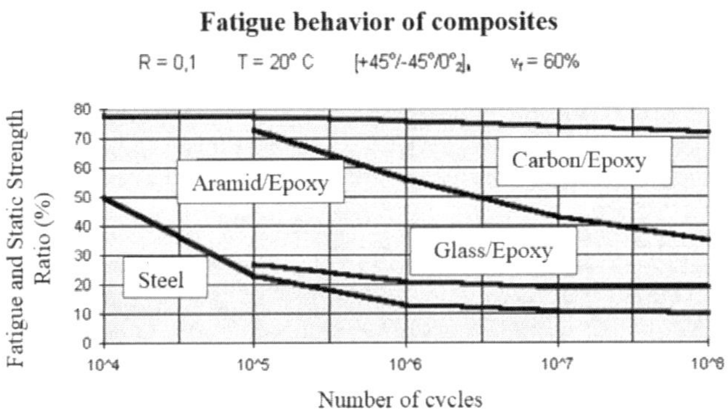

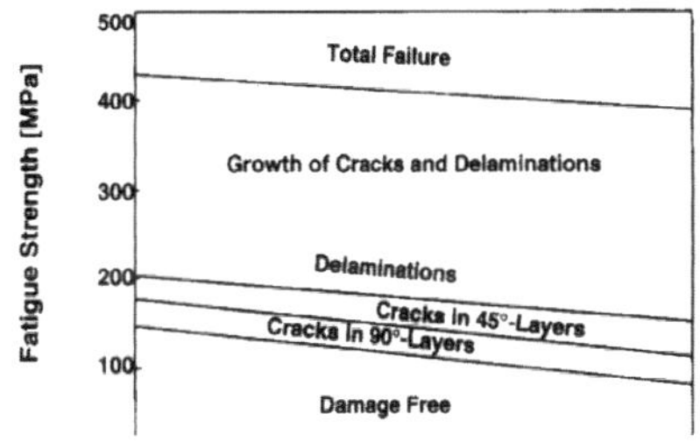

Figure 5. Comparison of the fatigue behavior of composites (left) [44] and failure diagram of carbon fiber-epoxy with regard to water content (right) [45]

Light weight, great strength, strong shock absorption capabilities, excellent durability, and favorable fatigue characteristics are the key benefits of a CFRP. Figure 5 compares the fatigue behavior of composites and other materials, and it shows CFRP has the highest fatigue and static strength. Additionally, the function of moisture content affects the fatigue strength of epoxy matrix CFRP. Moisture content can also influence how CFRP might fail in various of expected failure types, as shown in Figure 5.

4. Challenges

The challenges to tidal current turbine development are the hostile sea environment with its corrosive saline water, fouling growth, and other environmental concerns. These aspects must be addressed when developing tidal current turbine systems since they will impact the longevity and efficiency of tidal turbine blades.

In addition, CFRP can provide high performance, although it is 7/8 times more expensive than GFRP. However, the advantages of CFRP in terms of weight reduction, mechanical advantages, and durability would need to balance the higher cost [39]. The inability to recycle CFRPs is another issue with their use. The composite CFRP material was created by combining carbon fibers with thermoset epoxy resin, resulting in non-recyclable materials. Only incineration can be used to eliminate composite materials containing these resins [33,39].

4.1. Biofouling

The development of marine biofouling on the tidal current turbine's structure might be a major challenge. Organic molecules, algae, anemones, bacteria, barnacles, bivalve molluscs, bryozoans, diatoms, fungi, hydroides, kelps, tubeworms—all examples of marine biofouling—can accumulate on maritime objects submerged in water [22,46,47]. This marine biofouling can cause a considerable problem for tidal turbine systems [22,48] since it can change the hydrodynamic design, increase the structural weight, increase inertia and drag load, enlarge the surface roughness, reduce the overall efficiency and power production, and contribute to structural failures [20,22,49,50]. Figure 6 depicts the marine biofouling on a tidal current turbine system.

Figure 6. Biofouling in the Kobold tidal current turbine (left) [21], ducted axial-flow tidal current turbine in Canada's Race Rocks Ecological Reserve (mid) [51], OpenHydro (right) [52]

Farkas et al. [53] have researched how biofilm fouling affects the performance of marine current turbines using a computational fluid dynamics (CFD) method. Full-scale CFD simulations are run for eight tip speed ratios (TSRs) under six marine biofouling situations by modifying their surface coverage percentages and biofouling heights. The results show that biofilm has a considerable negative effect on the tidal current turbines efficiency. For the best TSR, the power coefficient drop is about -10.7% in the R1 marine biofouling scenario (with 500 μm biofilm height, 50% biofilm surface coverage, and a 195 μm roughness length scale).

Marine structures must have an antifouling coating to prevent the negative effects of marine biofouling growth. The antifouling paint on the market has a performance life of three to five years. On the other hand, tidal current turbines have a lifespan of 20 to 30 years. Tidal current turbine performance must be maintained by manual cleaning and paint reapplication once the coating lifespan has passed. Consequently, it is necessary to develop a more robust antifouling coating to reduce the frequency and expense of maintenance [22].

4.2. Seawater ageing and fatigue

The environment that tidal turbines must function in is quite hostile. Thus, it is critical to consider how seawater aging affects composites [36]. Davies et al. [34] examined how unidirectional (UD) carbon/epoxy from prepreg was affected by natural saltwater ageing. The material gained weight ranging from 1% to 5% after 400 days of immersion at 20°C and 60°C (Figure 7). They exhibited interlaminar shear failures between the blades and the hub, which significantly decreased the fatigue life. The defects (such as voids and fractures) brought on by high-stress concentrations are also directly responsible for the fatigue strength of CFRPs [32].

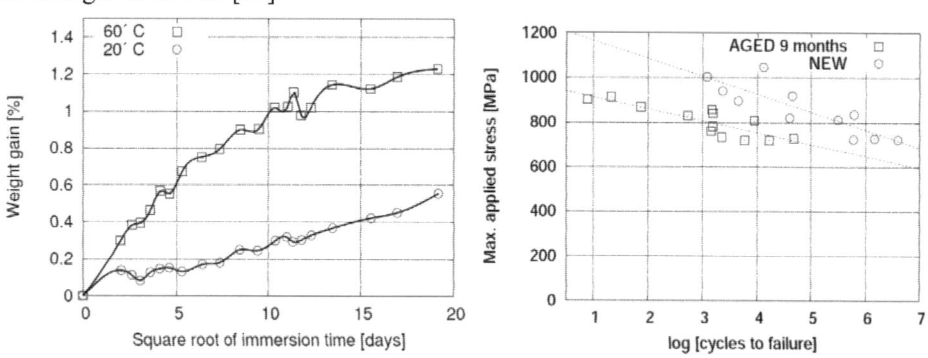

Figure 7. Left: weight increase, and right: seawater ageing effects on wet fatigue behavior, 2 Hz, R=0.1, parallel specimens, at 60°C

Additionally, Tual [28] used accelerated seawater aging studies to examine the behavior of three types of carbon epoxy composites in tidal turbines over a long period of time. The results showed that the kinetics of water diffusion were primarily unaffected by thickness and fiber orientation. Still, the water absorption kinetics and the water saturation condition were significantly influenced by the manufacturing method and the matrix resin type. The failure strength was reduced by 20% to 40%, although the elastic modulus and toughness were unaffected. The results of the interlaminar shear stress tests showed that seawater ageing impacted on the interfacial adhesion under shear loading in all three carbon/epoxy composites.

Li et al. [54] also investigated the impact of seawater ageing on the mechanical and physical properties of CFRP. The tensile strength of the CFRP samples decreased exponentially, and the moisture absorption content increased when the ageing process was prolonged. Additionally, the deterioration worsened with increasing temperature. As the ageing duration and ambient temperature increased, the interlaminar bindings were degraded. Enhancing the fiber/resin matrix interface bonds and interlaminar bonding may be a valuable strategy for increasing CFRP's resistance to deterioration in maritime environments.

4.3. Hydrodynamic loads

The hydrodynamic loads depend on inflow speed. As inflow speeds grow, the bending moments increase [55]. In addition, high seawater density provides extra difficulties, such as excessive turbine strain caused by turbulence. The load on the blades can rise noticeably with even a little increase in flow velocity [22], which might cause failure. There are two primary reasons why rotor blades fail: 1) inadequate design strength owing to underestimating or unknown loads, and 2) manufacturing errors produce a structure with significantly lower strength than intended. The list of blade fractures or failures that were recorded for various prototypes is shown below [22,55,56]:

- On December 11, 2006, Verdant Power's two tidal turbines that were deployed in the East River in New York City experienced blade failure (Figure 8).
- On June 11, 2010, blade failure occurred in the Open Hydro turbine that was installed in Canada's Bay of Fundy when there was a very high inflow speed of around 5.15 m/s, or it exceeded the designed operating inlet speed by over 2.5 times.
- In August 2010, a manufacturing flaw led to the Atlantis AR1000 turbine's blade fracture, which happened soon after it was installed and linked to the grid of the European Marine Energy Centre.

Figure 8. Blade of Verdant Power's tidal turbines that failed.

5. Discussions

The benefits of using CFRP materials were covered in the preceding section. Comparing the CFRP to other composites, it performs better and is lighter. The advantages can be useful to researchers and related stakeholders who want to develop more robust tidal current turbine blades. However, it has a higher cost. Because of this, adopting a combination or hybrid composite, with carbon mainly employed to boost stiffness, constitutes a compromise between these competitive objectives [39].

An efficient antifouling coating may stop marine biofouling problems. But issues like seawater aging and fatigue, and hydrodynamic loads that cause blade failure might be overcome with better materials. Composites can be expected to be joined using a range of methods, including adhesives and other materials. Reusable and natural-based composites can be options to unrecyclable composites, and they could be able to reduce waste through recycling and reusing [33]. However, it is vital to comprehend how the environment effects on these material qualities relating to operation and maintenance, reliability, and safety issues [57]. The impact of seawater and biofouling on these materials must therefore be thoroughly assessed in further research and work.

6. Conclusions

Tidal current turbine blades must be made of a suitable material to lower the failure chances of generating renewable energy. CFRP material has been preferred for making tidal current turbine blades that are thinner, stiffer, and lighter. The CFRP materials used in tidal current turbine systems are reviewed. CFRP offers advantages such as superior strength, weight reduction, mechanical advantages, durability, corrosion resistance, and fatigue resistance. The primary challenges that must be answered and solved in CFRP utilization are marine biofouling attached to the structures, seawater ageing and fatigue that will reduce the CFRP's physical and mechanical properties, and unpredictable hydrodynamic loads and turbulence. Consequently, further work is still needed to address the challenges to support the tidal current turbine developments.

Acknowledgments

The authors appreciate the support given by the National Research and Innovation Agency, Republic of Indonesia. The authors gratefully acknowledge financial support from the Institut Teknologi Sepuluh Nopember for this work, under the project scheme of the Publication Writing and IPR Incentive Program (PPHKI).

References

[1] Sasmita A F, Arini N R and Satrio D 2023 Numerical Analysis of Vertical Axis Tidal Turbine: A Comparison of Darrieus and Gorlov Type *2023 International Electronics Symposium (IES)* (Surabaya: IEEE) pp 26–31

[2] Junianto S, Fadilah W N, Adila A F, Tuswan T, Satrio D and Musabikha S 2022 State of the Art in Floating Tidal Current Power Plant Using Multi-Vertical-Axis-Turbines *IES 2022* (Surabaya: IEEE) pp 1–7

[3] Mubarok M A, Arini N R and Satrio D 2023 Optimization of Horizontal Axis Tidal Turbines Farming Configuration Using Particle Swarm Optimization (PSO) Algorithm *2023 International Electronics Symposium (IES)* vol 1 (Surabaya: IEEE) pp 19–25

[4] Aulia A, Athallah A G, Albasyir M A, Selvia S P, Darisman H A and Satrio D 2023 Study of Diameter-to-Draft Ratio of Pitch Point Absorber as Wave Energy Converter in Indonesia Waters Based on Boundary Element Method *IOP Conference Series: Earth and Environmental Science* (Surabaya: IOP Publishing Ltd) pp 1–13

[5] Musabikha S, Utama I K A P and Mukhtasor 2018 State of the art in protection of erosion-corrosion on vertical axis tidal current turbine *AIP Conference Proceedings* vol 1964

[6] Satrio D, Suntoyo, Albatinusa F, Junianto S, Musabikha S, Madi and Setyawan F O 2023 Effect of Positioning a Circular Flow Disturbance in Front of the Darrieus Turbine *IOP Conf. Ser. Earth Environ. Sci.* **1166** 012019

[7] Bahaj A S and Myers L E 2003 Fundamentals applicable to the utilisation of marine current turbines for energy production *Renew. Energy* **28** 2205–11

[8] Grogan D M, Leen S B, Kennedy C R and Ó Brádaigh C M 2013 Design of composite tidal turbine blades *Renew. Energy* **57** 151–62

[9] Chen L and Lam W H 2015 A review of survivability and remedial actions of tidal current

turbines *Renew. Sustain. Energy Rev.* **43** 891–900
[10] Day A H, Babarit A, Fontaine A, He Y P, Kraskowski M, Murai M, Penesis I, Salvatore F and Shin H K 2015 Hydrodynamic modelling of marine renewable energy devices: A state of the art review *Ocean Eng.* **108** 46–69
[11] Li Y and Calisal S M 2010 Three-dimensional effects and arm effects on modeling a vertical axis tidal current turbine *Renew. Energy* **35** 2325–34
[12] Satrio D, Suntoyo S, Erwandi E, Albatinusa F, Tuswan T and Hakim M L 2023 The Benefit Using a Circular Flow Disturbance on the Darrieus Turbine Performance *Int. J. Eng. Appl.* **11** 55–63
[13] Satrio D, Adityaputra K A, Suntoyo S, Silvianita S, Dhanistha W L, Gunawan T, Muharja M and Felayati F M 2023 The Influence of Deflector on the Performance of Cross-flow Savonius Turbine *Int. Rev. Model. Simulations* **16** 27–34
[14] Yahya N Y, Silvianita S and Satrio D 2023 A Review of Technology Development and Numerical Model Equation for Ocean Current Energy *IOP Conference Series: Earth and Environmental Science* (Surabaya: IOP Publishing Ltd) pp 1–13
[15] Lestari S, Mukhtasor M, Satrio D and Rahmawati S 2023 Oceanographic survey on current due to breaking wave for energy generations : a case study of Pacitan bay , East Java , Indonesia *IOP Conference Series: Earth and Environmental Science* (Surabaya: IOP Publishing Ltd) pp 1–12
[16] Kurniawati I, Pratilastiarso J and Satrio D 2020 Analysis of the Effect of Cooling Water Condenser to Power Plant Cycle Using Cycle-Tempo Software *IES 2020 - International Electronics Symposium: The Role of Autonomous and Intelligent Systems for Human Life and Comfort* (Surabaya: IEEE) pp 37–42
[17] Ariansyah M N, Diana L and Satrio D 2022 Numerical Study the Effect of Gap Ratio on Flow Characteristics and Heat Transfer in Staggered Tube Banks *IOP Conference Series: Earth and Environmental Science* vol 972 (Surabaya: IOP Publishing Ltd) p 012065
[18] Simangunsong R G, Amalia R and Satrio D 2021 Numerical Analysis of a Shell and Tube Heat Exchanger Using Computational Fluid Dynamics Software *International Electronics Symposium 2021: Wireless Technologies and Intelligent Systems for Better Human Lives, IES 2021 - Proceedings* (Surabaya: IEEE) pp 630–5
[19] Khan M J, Bhuyan G, Iqbal M T and Quaicoe J E 2009 Hydrokinetic energy conversion systems and assessment of horizontal and vertical axis turbines for river and tidal applications: A technology status review *Appl. Energy* **86** 1823–35
[20] Touimi K, Benbouzid M and Tavner P 2018 Tidal stream turbines: With or without a Gearbox? *Ocean Eng.* **170** 74–88
[21] Calcagno G and Moroso A 2007 The Kobold Marine Turbine: From The Testing Model To The Full Scale Prototype *Tidal Energy Summit* (London) pp 1–44
[22] Shiekh Elsouk M N, Santa Cruz A and Guillou S S 2018 Review on the characterization and selection of the advanced materials for tidal turbine blades *7th International Conference on Ocean Energy 2018* pp 1–7
[23] Momber A W, Plagemann P and Stenzel V 2015 Performance and integrity of protective coating systems for offshore wind power structures after three years under offshore site conditions *Renew. Energy* **74** 606–17
[24] Musabikha S, Utama I K A P, Mukhtasor, Wharton J A and Wood R J K 2020 Effects of Nickel–Aluminum Bronze Pre-Oxidized Films on the Cathodic Kinetics of Oxygen Reduction *Anal. Lett.* **53** 1218–32
[25] Nikitasari A, Astuti I M, Musabikha S, Kusumastuti R, Prifiharni S and Priyotomo G 2023 Chasew (Anacardium Occidentale) Leaves Extract as Green Corrosion Inhibitor of API 5L X52 in Acidic Media *Trends Sci.* **20** 1–12
[26] Blackmore T, Myers L E and Bahaj A S 2016 Effects of turbulence on tidal turbines: Implications to performance, blade loads, and condition monitoring *Int. J. Mar. Energy* **14** 1–

26
[27] Alam P, Mamalis D, Renewable O, Catapult E, Robert C and Bradaigh C O 2017 Mechanical properties and damage analyses of fatigue loaded CFRP for tidal turbine applications *Proceedings of the 12th European Wave and Tidal Energy Conference* pp 1–8
[28] Tual N, Carrere N, Davies P, Bonnemains T and Lolive E 2015 Characterization of sea water ageing effects on mechanical properties of carbon/epoxy composites for tidal turbine blades *Compos. Part A Appl. Sci. Manuf.* **78** 380–9
[29] Liu Y, Zwingmann B and Schlaich M 2015 Carbon fiber reinforced polymer for cable structures-a review *Polymers (Basel).* **7** 2078–99
[30] Mamalis D, Flanagan T, Doyle A and Brádaigh C M Ó 2017 A Carbon Fibre Reinforced Powder Epoxy Manufacturing Process for Tidal Turbine Blades Fatigue modelling and experimentation of CFRP composites in tidal turbine blade applications *Proceedings of the 12th European Wave and Tidal Energy Conference* pp 1–10
[31] Karaş B, Smith P J, Fairclough J P A and Mumtaz K 2022 Additive manufacturing of high density carbon fibre reinforced polymer composites *Addit. Manuf.* **58** 1–10
[32] Okayasu M and Tsuchiya Y 2019 Mechanical and fatigue properties of long carbon fiber reinforced plastics at low temperature *J. Sci. Adv. Mater. Devices* **4** 577–83
[33] Walker S R J and Thies P R 2022 A life cycle assessment comparison of materials for a tidal stream turbine blade *Appl. Energy* **309** 1–15
[34] Davies P, Germain G, Gaurier B, Boisseau A and Perreux D 2013 Evaluation of the durability of composite tidal turbine blades *Philos. Trans. R. Soc. A Math. Phys. Eng. Sci.* **371** 1–15
[35] Marsh G 2004 Tidal turbines harness the power of the sea *Reinf. Plast.* **48** 44–7
[36] Calvário M, Sutherland L S and Soares C G 2018 A review of the applications composite materials in wave and tidal energy devices *Marit. Transp. Harvest. Sea Resour.* **2** 695–701
[37] Douglas C A, Harrison G P and Chick J P 2008 Life cycle assessment of the Seagen marine current turbine *Proceedings of the Institution of Mechanical Engineers Part M: Journal of Engineering for the Maritime Environment* vol 222 pp 1–12
[38] Nicholls-Lee R F, Turnock S R and Boyd S W 2008 Performance prediction of a free stream tidal turbine with composite bend-twist coupled blades *2nd International Conference on Ocean Energy* pp 1–9
[39] Rubino F, Nisticò A, Tucci F and Carlone P 2020 Marine application of fiber reinforced composites: A review *J. Mar. Sci. Eng.* **8** 1–28
[40] Abbood I S, Odaa S A, Hasan K F and Jasim M A 2021 Properties evaluation of fiber reinforced polymers and their constituent materials used in structures - A review *Materials Today: Proceedings* vol 43 pp 1003–8
[41] Ahmed A, Guo S, Zhang Z, Shi C and Zhu D 2020 A review on durability of fiber reinforced polymer (FRP) bars reinforced seawater sea sand concrete *Constr. Build. Mater.* **256** 1–17
[42] Jaksic V, O'bradaigh C, Wallace F and Brádaigh C Ó 2017 Upscaling of Tidal Turbine Blades: Glass or Carbon Fibre Reinforced Polymers? *Proceedings of the 12th European Wave and Tidal Energy Conference* vol M pp 1–15
[43] Alam P, Robert C and Ó Brádaigh C M 2018 Tidal turbine blade composites - A review on the effects of hygrothermal aging on the properties of CFRP *Compos. Part B Eng.* **149** 248–59
[44] Zaludek M, Rusnakova S, Kubisova M, Bilek O and Karvanis K 2020 Fatigue life of thermoset composite materials *IOP Conference Series: Materials Science and Engineering* vol 726 pp 1–7
[45] Selzer R and Friedrich K 1997 Effect of Moisture on Mechanical Properties and Fracture Behavior of Carbon Fiber Reinforced Polymer Composites *Adv. Compos. Lett.* **2** 595–604
[46] Agostini V O, Macedo A J, Muxagata E, da Silva M V and Pinho G L L 2020 Non-toxic antifouling potential of Caatinga plant extracts: effective inhibition of marine initial biofouling *Hydrobiologia* **847** 45–60
[47] Musabikha S, Priyotomo G, Nikitasari A, Prifiharni S, Kusumastuti R, Satrio D and Junianto S

2023 The Advantages and Challenges of Environmentally Friendly Terrestrial Plant-Based Antifouling Coating – A Short Review *AIP Conference Proceedings*
[48] Musabikha S, Utama I K A P and Mukhtasor 2021 Characterisation and performance evaluation of marine coatings used for submerged ocean energy application *J. Eng. Sci. Technol.* **16** 4309–24
[49] Hudson J A, Phillips D C and Wilkins N J M 1980 Materials aspects of wave energy converters *J. Mater. Sci.* **15** 1337–63
[50] Priyotomo G, Prifiharni S, Nuraini L, Royani A, Sundjono and Gunawan H 2022 The Performance of Antifouling Paint for Prolonged Exposure in Madura Strait, East Java Province, Indonesia *Int. J. Adv. Sci. Eng. Inf. Technol.* **12** 732–7
[51] Stringer C C and Polagye B L 2020 Implications of biofouling on cross-flow turbine performance *SN Appl. Sci.* **2** 1–13
[52] Rashid H, Benbouzid M, Titah-Benbouzid H, Amirat Y and Mamoune A 2023 Tidal Stream Turbine Biofouling Detection and Estimation: A Review-Based Roadmap *J. Mar. Sci. Eng.* **11** 1–18
[53] Farkas A, Degiuli N, Martić I, Barbarić M and Guzović Z 2022 The impact of biofilm on marine current turbine performance *Renew. Energy* **190** 584–95
[54] Li H, Zhang K, Fan X, Cheng H, Xu G and Suo H 2019 Effect of seawater ageing with different temperatures and concentrations on static/dynamic mechanical properties of carbon fiber reinforced polymer composites *Compos. Part B Eng.* **173** 1–9
[55] Liu P and Veitch B 2012 Design and optimization for strength and integrity of tidal turbine rotor blades *Energy* **46** 393–404
[56] Li W, Zhou H, Liu H, Lin Y and Xu Q 2016 Review on the blade design technologies of tidal current turbine *Renew. Sustain. Energy Rev.* **63** 414–22
[57] Hernandez-sanchez B A, Nicholas J, Gunawan B, Miller D A, Bonheyo G T, Presuel-moreno F and Hughes S 2019 Evaluation of Composite Materials for Wave and Current Energy Technologies *Proceedings of the 13th European Wave and Tidal Energy Conference* pp 1–7

Planning on Perintis Shipping Rerouting By Utilizing Commercial Shipping: Case Study of Kupang, East Nusa Tenggara

P Wuryaningrum[1] *, I T Yunianto[1], F Hadi[1], and I G B Santosa[1]

[1] Department of Marine Transportation Engineering, Institut Teknologi Sepuluh Nopember

*Email: pratiwi@its.ac.id

Abstract. Perintis shipping is an Indonesian government program that is a domestic subsidized shipping that connects islands in Indonesia to increase the connectivity, especially for underdeveloped, remote, and border areas, or called 3TP. As Perintis shipping services have grown, with more routes and voyages, subsidies have also increased. The aim of this paper is to explore possible solutions for reducing subsidies by analyzing routes that intersect with commercial shipping. The objective is to determine the optimal route by rerouting the current Perintis route and utilizing the commercial route. This involves analyzing various factors such as operational efficiency, technical and financial considerations to identify the most feasible routes. This research focuses on the Perintis shipping route in Kupang, East Nusa Tenggara. There are five Perintis shipping routes, but only three of them intersect with commercial shipping routes. There are four potential routes available, consisting of three options for R-23 and one option for R-24. Based on the analysis, it has been determined that the most feasible option is the first option of R-24. The chosen routes have witnessed a reduction in subsidies, with a total amount of IDR 30.56 million and commercial shipping has achieved a profit increase of IDR 809.70 million.

Keywords: Subsidized Shipping, Commercial Shipping, Route

1. Introduction

Indonesia, being the largest archipelagic country in the world with over 18,000 islands and more than 80,000 km of coastline, requires efficient transportation facilities to connect different regions, especially remote areas. Transportation facilities in these remote areas are still scarce. Some islands are still not available at all, which can lead to social and economic disparities compared to regions or islands with more advanced and developed transportation infrastructure. Unequal pricing distribution in Indonesia's regions can be a concerning issue [1]. The government has launched a subsidized shipping program called Perintis Shipping to increase connectivity in Indonesia. The program has predetermined routes to serve underdeveloped, remote, and border areas called 3TP that are usually not covered by commercial transportation[2] . Remote areas have very minimal economic activity, resulting in the development of these areas being left behind compared to other areas with high accessibility. As a result, an archipelagic region has great challenges to maritime transportation and logistics[3]. In 2017, Perintis Shipping had 96 routes. As of 2023, they have 116 routes with 42 home base ports and 562 feeder ports throughout Indonesia. These routes are currently served by 116 ships[4]. With an increase in the number of Perintis shipping routes, the subsidies that the government must issue will undoubtedly increase. To support the Perintis shipping operations, the government continues to work on colonization and seek new solutions to reduce the operational costs of the Perintis shipping. One solution that can be used is to reduce the route or port of call that the Perintis ship must pass. The reduction of this route may involve commercial shipping in its implementation,ion considering that there are Perintis shipping routes that are cut off from commercial shipping routes. Route reduction can be done by analyzing routes that intersect between Perintis shipping and commercial shipping. From the results of this route intersection analysis, Perintis shipping no longer needs to go through the same route as commercial shipping. This will lead to reduced routes or ports of call that must be passed by the Perintis fleet, reducing the travel time of the Perintis fleet, which will affect the Perintis operational cost subsidies that must be spent.

Sea transportation plays a crucial role in the economic and regional development of archipelagic regions in maritime countries[5]. Sea transportation activities will cause ship operational costs consisting of fixed costs and variable costs[6]. It is essential to consider various expenses such as capital, operational, shipping, loading, and unloading costs when calculating shipping costs [7]. In sea transportation, the cost is influenced by factors such as distance, transit time, and transport conditions [8]. Indonesia, being a maritime country offers subsidies to sea transportation to promote fair development. Subsidies are provided when costs surpass income or consumer price [9]. In providing subsidies, the optimal approach is for the government to provide them without calculating the distribution of incentives [10]. The government subsidizes sea transportation in areas where commercial industries have not yet become profitable [11]. As a result, the government should provide subsidies for marine transportation in the area.

2. Research Methodology

2.1 Method

The study uses a comparative method to analyze available routes, considering cost and operational analysis. Different scenarios are created by involving commercial companies to determine the best option based on financial impact.

2.2 Rerouting Analysis

At this stage, the obtained data is processed for analyzing available route change scenarios. The following stages of analysis are carried out.

2.2.1 Intersection Routes Analysis

Analyzing the intersection of routes between Perintis and commercial fleets requires attention to compatibility, and generates route change scenarios.

2.2.2 Operational analysis

This analysis will examine Fleet Frequency, Round Trip Days (RTD), and the operational days of each merchant and Perintis fleet that has the potential to modify their routes. In order to conduct a frequency analysis, the commercial fleet must ensure that it does not surpass its maximum frequency while serving Perintis and that all Perintis cargo is serviceable. For scenarios that meet the requirements, a schedule is made to adjust the activity or time of arrival and departure of the Fleet so that it can find out the optimal time for activities or intersections at the port and also to exercise control over its operational activities.

2.2.3 Finacial Analysis

After conducting an operational analysis, a financial analysis will be carried out to compare subsidies and profits before and after route changes. Perintis Transport aims to minimize subsidies while maximizing income and cost differences, ultimately choosing the most profitable route scenarios.

3. Result and Discussion

East Nusa Tenggara Province is one of the archipelagic provinces in Indonesia, with Kupang as its capital. Based on its geographical location, the NTT region has a land area of 47,931 km2 with a coastline of 5,700 km. The province of NTT consists of many big and small islands [12]. When viewed from the population and economic conditions, East Nusa Tenggara Province had a population of 2,733,236 people in 2022 with an average Gross Regional Domestic Product (GRDP) of 118.72 trillion rupiahs in 2022 which nominally has increased in value compared to 2021, which reached IDR 110. 88 trillion rupiah [13]. There are 2 large ports in Kupang, namely Tenau Port, which is a Perintis base port serving 5 routes, and there is also Bolok Port. Here are the shipping routes for Perintis and commercial ships in Kupang.

Table 1. Perintis and Commercial Shipping Routes

Shipping Type	Shipping Route	Route	Frequency/ year	Distance (Nm)
Perintis	P-1	Kupang - Ndao - Sawu - Raijua - Ende - P,Ende - Maumbawa - Waiwole - Mamboro - Waingapu - Mamboro - Waewole - Maumbawa - P,Ende - Ende - Raijua - Sabu - Ndao - Kupang	24	894.00
	P-2	Kupang - Naikliu - Wini - Lirang - Ilwaki - Kisar - Romang - Leti - Lakor - Luang P, Kelapa - Sermata (Elo) - Tepa - Kroing - Saumlaki -Kroing - Tepa - Sermata (Elo) - Luang P, Kelapa - Lakor - Leti - Romang - Kisar - Ilwaki - Lirang - Wini - Naikliu - Kupang	24	1,364.00
	P-3	Kupang - Naikliu - Mananga - Lewoleba - Balauring -Baranusa - Kalabahai - Atapupu - Kalabahi - Baranusa - Balauring - Lewoleba - Mananga - Naikliu - Kupang	37	810.00
	P-4	Kupang - Wulandoni - Mananga - Maumere - Palue - Marapokot - Reo - Labuhan Bajo - Bima - Labuhan Bajo - Reo - Marapokot - Palue - Maumere - Mananga - Wulandoni - Kupang	28	926.00
	P-5	Kupang - Kalabahi - Lirang - Iliwaki - Kisar - Leti - Moa - Tepa - Kroing - Saumlaki - Kroing - Tepa - Moa - Leti - Kisar - Iliwaki - Lirang - Kalabahi - Kupang	25	1,468.00
Commercial	C-1	Kupang – Ende - Kupang	110	295.46
	C-2	Kupang - Naikliu -Wini -Teluk Gurita	110	350.00
	C-3	Kupang – Sabu - Kupang	165	208.10
	C-4	Kupang – Rote - Kupang	330	87.32
	C-5	Kupang – Rote - Kupang	330	87.32
	C-6	Kupang – Kalabahi - Kupang	165	198.00

Table 1, shows five routes for Perintis shipping (P-1 to P-5) and six routes for commercial shipping (C-1 to C-6). The P-5 route covers a distance of 1.468 nautical miles, which is the longest distance it covers.

3.1 Analysis of Intersection Routes

Route intersection analysis is carried out on each Perintis shipping route, which will be combined with available commercial shipping routes by taking into account several limitations, such as crossing at one transit port, being able to serve a minimum of three Perintis ports, and looking at the number of segments served. Based on the results of the analysis, it is known that there are four scenarios of route changes that can be carried out where three scenarios are on the P-1 line and one scenario is on the P-2 line. The following details the route intersections that occur in each scenario.

3.1.1 Scenario 1

Based on Figure 1, there is an intersection of the commercial fleet route C-1 route, which is drawn as a dashed line, with the Perintis route P-1 as a continuous line. By looking at the intersection of the routes, Scenario 1 can change the P-1 Perintis route involving fleet route C-1. This route change will form a new route Ende – P. Ende - Maumbawa - Waiwole - Mamboro - Waingapu for the P-1 Perintis route and Kupang - Ndao - Sabu - Raijua - Ende for the C-1 route.

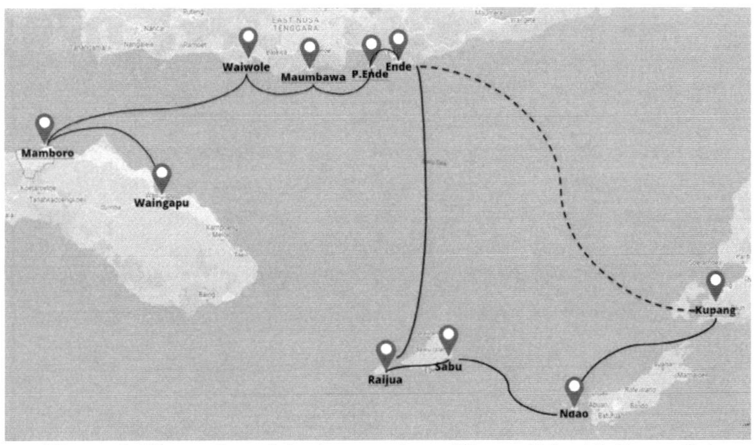

Figure 1. Illustration of the intersection of Route P-1 with Route C-1

3.1.2 Scenario 2

Based on the illustration, the C-3 commercial fleet route intersects with the P-1 Perintis fleet route. By altering the P-1 route at the intersections, a new route can be formed: Sabu - Raijua - Ende - P, Ende - Maumbawa - Waiwole - Mamboro - Waingapu for the P-1 Perintis route and Kupang - Ndao - Sabu for the K-3 route as can be seen in Figure 2.

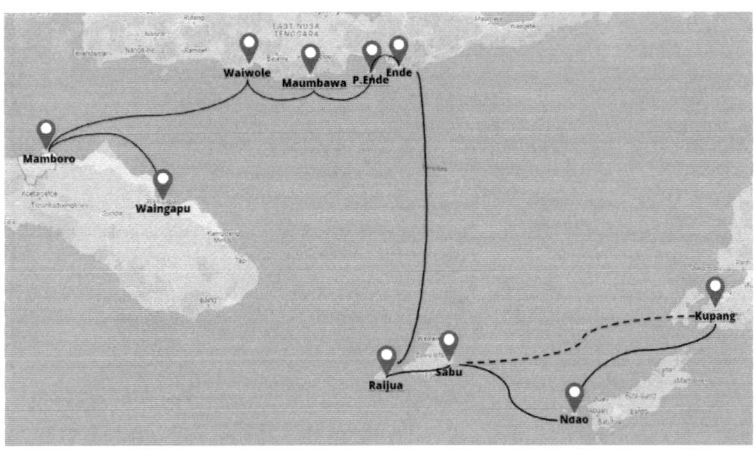

Figure 2 Illustration of the intersection of Route P-1 with Route C-3

3.1.3 Scenario 3

As can be seen in Figure 3, there is an intersection of the routes of 2 commercial fleets, namely Route C-1, which is depicted as a dashed line, and Route C-3, which is defined as a chain line with the Perintising voyage P-1, which is ddescribed as a continuous line. By looking at the project cuts that occur, Scenario 3 can be made to change the P-1 Perintis route by involving fleet routes C-1 and routes C-3. This route change will form a new route Ende - P, Ende - Maumbawa - Waiwole - Mamboro - Waingapu for the P-1 Perintis route, Kupang - Ndao - Sabu - Raijua for Route K-3 and Kupang - Ende for Fleet Route C-1.

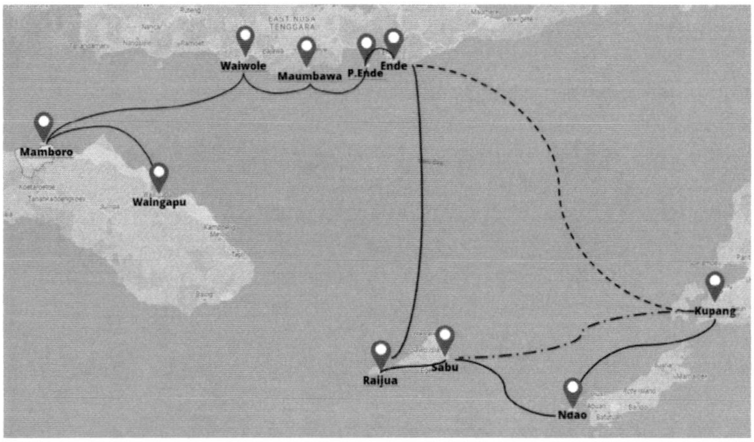

Figure 3 Illustration of the intersection of Route P-1 with Route C-1 and C-3

3.1.4 Scenario 4

Based on the illustration of Figure 4, there is a section of the commercial fleet route C-2 route with the Perintis fleet P-2 route where this route change will form a new route Wini - Lirang - Ilwaki - Kisar - Romang - Leti - Lakor - Luang P, Kelapa - Sermata (Elo) - Tepa - Kroing - Saumlaki for the Perintis route R-24 and Kupang - Naikliu - Wini - Octopus Bay for the C-2 route.

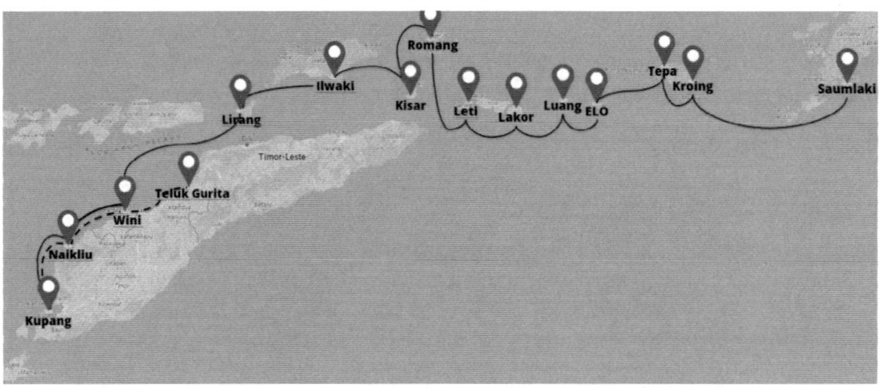

Figure 4 Illustration of the intersection of Route P-2 with Route K-2

3.2 Operational Analysis

To conduct an operational analysis, it is essential to have information on the specifications of the ships operating in the study area. Figure 5 shows an example of a Perintis ship, and Figure 6 shows an example of a Commercial Ship. The specifications of the Perintis and commercial ships that were analyzed can be seen in Table 2.

Figure 5 Perintis Ship

Figure 6 Commercial Ship

Table 2 Ship Specifications

Name	Type	LOA(m)	B (m)	T (m)	H (m)	Vs (knot)	GT	Passenger Capacity (pax)	Cargo Capacity (ton)
P-1	Perintis	68.5	14.0	2.9	6.2	7.2	2000	466	100
P-2	Perintis	68.5	14.0	2.9	6.2	6.1	2000	466	200
C-1	Commercial	56.0	14.0	2.2	3.8	12	1039	250	120
C-2	Commercial	54.0	12.0	2.6	3.8	12	1604	198	120
C-3	Commercial	52.0	9.5	2.0	3.1	18	788	500	10

Once the specifications of each ship are known, it can be continued by carrying out an operational analysis of the existing scenarios. An analysis of each scenario's operational and technical will be carried out on Fleet Frequency, RTD, and Operational Days for each available scenario. The following is the result of the operational analysis that has been carried out.

3.2.1 *Scenario 1*

Scenario 1 is an intersection on the P-1 route with the C-1 Route commercial fleet. Based on the results of the operational analysis, the following results are obtained.

Table 3 Results of Operational Changes to Perintis Shipping and Commercial Shipping for Scenario 1

Shipping Type	Data	Unit	Before	After
Perintis	Total Passengers	Pax	709	128
	Total Cargo	Ton	103	31
	RTD	Days	13.75	13.75
	Frequency	Voyage/year	24.00	24.00
	Sea Time	Days	5.17	2.11
	Port Time	Days	0.49	0.18
	Port Stay	Days	8.09	11.46
Commercial	Total Passengers	Pax	480	1,101
	Total Cargo	Ton	0	90
	RTD	Days	3.00	3.00
	Frequency	Voyage/year	110	600
	Sea Time	Days	1.03	1.84
	Port Time	Days	0.08	0.77

Shipping Type	Data	Unit	Before	After
	Port Stay	Days	1.89	0.39

According to Table 3, it is impossible to plan for rerouting as commercial shipping cannot transport all Perintis cargo. It would require 25 voyages to transport 1 Perintis cargo, which means 600 voyages would be needed to transport all Perintis cargo, exceeding the maximum frequency of commercial voyages of 110 voyages a year. Therefore, scenario 1 is not feasible.

3.2.2 Scenario 2

Scenario two is an intersection on the P-1 route with the commercial fleet on the C-3 route. Based on the results of the operational analysis, the following results are obtained.

Table 4 Results of Operational Changes to Perintis Shipping and Commercial Shipping for Scenario 2

Shipping Type	Data	Unit	Before	After
Perintis	Total Passengers	Pax	709	301
	Total Cargo	Ton	103	74
	RTD	Days	13.75	13.75
	Frequency	Voyage/year	24	24
	Sea Time	Days	5.17	3.60
	Port Time	Days	0.49	0.30
	Port Stay	Days	8.09	9.85
Commercial	Total Passengers	Pax	278	739
	Total Cargo	Ton	0	62
	RTD	Days	2.00	2.00
	Frequency	Voyage/year	165.00	165.00
	Sea Time	Days	0.48	0.63
	Port Time	Days	0.05	0.14
	Port Stay	Days	1.47	1.23

Rerouting can be planned without affecting total operational days, frequency, and number of commercial transports. Once the scenario is known, scheduling can proceed from scenario two at the intersecting ports to determine the number of simultaneous fleet arrivals.

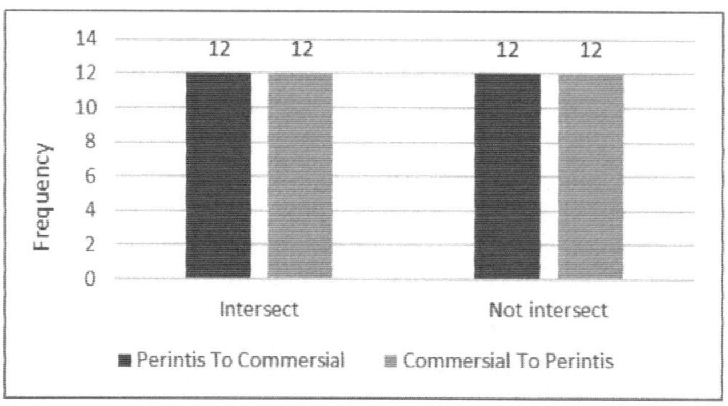

Figure 7 Frequency Intersect Scenario 2

Based on the data above, it is known that there are 12 times the frequency of transshipment of the Perintis fleet to commercial and vice versa and there is no waiting time where the schedule between Perintis ships and commercial ships is synchronized. While the remaining 12 times the frequency occurs waiting due to differences in the arrival schedule of Perintis ships and commercial ships. Furthermore, the consequences of this waiting time will be analyzed at the financial analysis stage.

3.2.3 Scenario 3

Scenario 3 is the intersection of the P-1 route with two commercial fleets for the C-1 route and the C-3 route. Based on the results of the operational analysis, the following results were obtained.

Table 5 Results of Operational changes to Perintis Shipping and Commercial Shipping for Scenario 3

Shipping Type	Data	Unit	Before	After
Perintis	Total Passangers	Pax	709	128
	Total Cargo	Ton	103	31
	RTD	Days	13.75	13.75
	Frequency	Voyage/year	24	24
	Sea Time	Days	5.17	2.11
	Port Time	Days	0.49	0.18
	Port Stay	Days	8.09	11.46
Commercial C-1	Total Passangers	Pax	480	623
	Total Cargo	Ton	0	28
	RTD	Days	3.00	3.00
	Frequency	Voyage/year	110.00	192.00
	Sea Time	Days	1.03	1.03
	Port Time	Days	0.08	0.19
	Port Stay	Days	1.89	1.78
Commercial C-3	Total Passangers	Pax	278	914
	Total Cargo	Ton	0	78
	RTD	Days	2.00	2.00
	Frequency	Voyage/year	165.00	165.00
	Sea Time	Days	0.48	0.74
	Port Time	Days	0.05	0.17
	Port Stay	Days	1.47	1.09

Planning for rerouting for Scenario 3 is not feasible as commercial transport on Route C-1 cannot carry all Perintis cargo. Eight voyages are required to carry one Perintis cargo, resulting in a need for 192 commercial voyages in a year. However, a maximum of 110 commercial voyages can be made in a year. Therefore, Scenario 3 cannot be implemented.

3.2.4 Scenario 4

Scenario 4 is an intersection on the P-2 route with the commercial fleet on the C-2 route. Based on the results of the operational analysis, the following results are obtained.

Table 6 Results of Operational changes to Perintis Shipping and Commercial Shipping for Scenario 4

Shipping Type	Data	Unit	Before	After
Perintis	Total Passangers	Pax	443	414
	Total Cargo	Ton	204	198
	RTD	Days	13.75	13.75
	Frequency	Voyage/year	24	24
	Sea Time	Days	9.32	7.75
	Port Time	Days	2.47	1.82
	Port Stay	Days	1.97	4.18
Commercial	Total Passangers	Pax	264	373
	Total Cargo	Ton	0	204
	RTD	Days	3.00	3.00
	Frequency	Voyage/year	110	110
	Sea Time	Days	1.22	1.22
	Port Time	Days	0.13	0.43
	Port Stay	Days	1.47	1.23

Planning for rerouting can be done without affecting total operational days, frequency, and number of commercial transports. Once the scenario is known, scheduling can proceed from scenario 2 at the intersecting ports to determine the number of simultaneous fleet arrivals.

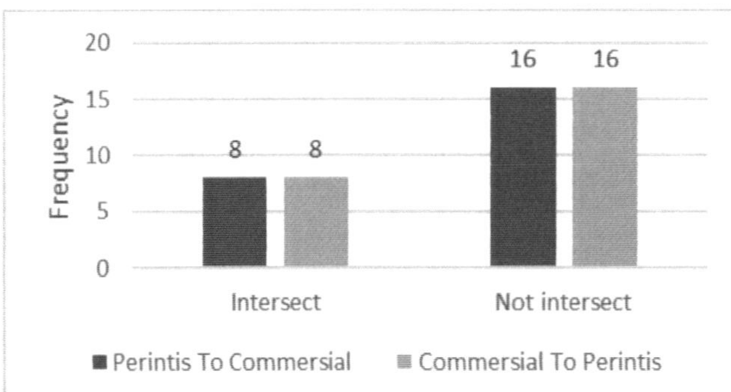

Figure 8 Frequency Intersect Scenario 4

Based on the data above, it is known that there are 8 times the frequency of transshipment of the Perintis fleet to commercial and vice versa and there is no waiting time where the schedule between Perintis ships and commercial ships is synchronized. While the remaining 16 times the frequency occurs waiting due to differences in the arrival schedule of Perintis ships and commercial ships. Furthermore, the consequences of this waiting time will be analyzed at the financial analysis stage.

3.3 *Financial Analysis*

After conducting an operational analysis, only Scenarios 2 and 4 can proceed to financial analysis. When analyzing the financial impact on Fleet operators, operational costs, Fleet revenue, and consequential costs due to route changes must be considered. The following is the result of the cost calculation of the scenario changes made. There are several cost assumptions used in making calculations for the consequences of passengers and goods. The following is the assumption of consequential costs for passengers and cargoes.

Table 7 Consequential Cost Assumptions for Passengers and Goods

Cost Component	Unit	Cost	
Accomodation	pax/day	Rp	31,500
Meal	pax/day	Rp	16,024
Trasportation	pax/km	Rp	226,89
Revenue	pax/day	Rp	77,611
Warehouse Tariff	Ton/day	Rp	30,000
Cargo Handling	Ton/day	Rp	10,470
Value of Good	Ton	Rp	6,814,333

In calculating the cost of consequences there are several equations that can be used as follows.

- Cost Consequences of Waiting for Fleet Arrival

$$(Q_a \; x \; W \; x \; C_a) + (Q_b x \; W \; x \; C_b) \quad (1)$$

Where
Q : Amount of Cargoes (Ton) or Number of Passengers Waiting (Pax)
W : Waiting time for the arrival of the ship (day)
C : Daily cost (Rp/day)
a : passengers
b : cargoes

- Cost Consequences of Using Commercial shipping

$$\sum_{k=1}^{n} Q \; x \; T_k \; x - \sum_{p=1}^{m} Q \; x \; T_p \quad (2)$$

Where
Q : Number of Passenger (Pax) or Amount of Cargoes (Ton)
T : Tariff (Rp)
p : Perintis Ship
k : Commercial Ship

- *Inventory Cost*

$$N \; x \; P \; x \; I \; x \; T \quad (3)$$

Where
N : Amount of Cargoes (Ton)
P : Tariff (Rp)
I : Interest Rate (Rp/Day)
K : Waiting Time (Day)

In order to calculate the consequences, we can group them into several categories. The cost of handling waiting cargo and passengers can be found using equation (1) ,while the cost of waiting for the fleet to arrive can be found using equation (3). We can calculate the estimated income of lost passengers while waiting to determine the lost income. Additionally, the cost of using commercial can be found using equation (2). When implementing a route change, commercial passengers will be affected due to the longer journey, resulting in an impact cost that can be found by equation (3). The value of goods waiting will decrease, which can be calculated by determining the decrease per day at the present value. Finally, when the fleet is overloaded, costs can be calculated due to the length of time the goods are waiting to be transported using equations (1) and (3).

After the financial analysis, the following results (Table 8) were obtained for Scenario 2 of the P-1 Perintis route and Scenario 4 of the P-2 Perintis route.

3.3.1 Scenario 2

Table 8 Financial Analysis for Scenario 2

Shipping Type	Financial Components	Unit	Before	After
Perintis	Shipping Cost	Million-Rp/Year	19,523.23	17,647.30
	Consequences Cost	Million-Rp/Year	0	3,317.26
	Total Cost	Million-Rp/Year	19,523.23	20,964.56
	Revenue	Million-Rp/Year	647.24	259.97
	Subsidy	Million-Rp/Year	18,875.99	20,704.58
Commercial	Shipping Cost	Million-Rp/Year	8,904.41	8,904.41
	Consequences Cost	Million-Rp/Year	8,904.41	9,132.96
	Total Cost	Million-Rp/Year	11,467.50	14,876.85
	Revenue	Million-Rp/Year	2,563.09	5,743.90

Based on the previous table, the application of changes to scenario 2 causes an increase in costs of 7.38% or IDR 1,441.33 million. In addition, it also had an impact on increasing subsidies by 9.69% or Rp. 1,828.60 million on Perintis shipping and then on commercial, causing an increase in costs of 2.57% or Rp. 228.54 million and an increase in profit of Rp. 3180.81 million from the initial profit.

3.3.2 Scenario 4

Table 9 Financial Analysis for Scenario 2

Shipping Type	Financial Components	Unit	Before	After
Perintis	Shipping Cost	Million-Rp/Year	22,710.92	21,117.32
	Consequences Cost	Million-Rp/Year	0	1,364.14
	Total Cost	Million-Rp/Year	22,710.92	22,48.46
	Revenue	Million-Rp/Year	926.71	727.81
	Subsidy	Million-Rp/Year	21,784.21	21,753.65
Commercial	Shipping Cost	Million-Rp/Year	13,782.50	13,839.75
	Consequences Cost	Million-Rp/Year	13,782.50	13,839.75
	Total Cost	Million-Rp/Year	13,896.09	14,763.04
	Revenue	Million-Rp/Year	113.60	923.29

According to the table, making changes to Scenario 4 leads to a 1.01% decrease in costs or IDR 229.46 million and a 0.14% decrease in subsidies or IDR 30.56 million. Additionally, costs for commercial shipping increase by 0.42% or IDR 57.25 million, and profit increases by IDR 809.70 million from the initial amount. The results showed that both Perintis and commercial shipping were advantageous.

After analyzing the four existing scenarios, the following results are obtained for the Perintis and commercial scenarios and can be seen in Table 10.

Table 10 Summary of The Results for All Scenarios

Scenario	Shipping Type	Operational Parameter	Financial Parameter
Scenario 1	Perintis	v	-
	Commercial	x	-
Scenario 2	Perintis	v	x
	Commercial	v	v
Scenario 3	Perintis	v	-
	Commercial C-1	x	-
	Commercial C-3	v	-
Scenario 4	Perintis	v	v
	Commercial	v	v

In the table sign (v) it can be interpreted as fulfilled where operationally, it is fulfilled if commercial shipping can serve Perintiss without affecting the day of operation. Financially it is declared fulfilled if it provides benefits to Perintiss or commercial. Sign (x) can be interpreted as non-compliance where operationally, it will not comply if commercial shipping cannot serve Perintiss and affects its operational days. Then, it is declared financially non-compliant if it does not provide benefits to Perintis or commercial. For the sign (-) it is stated that no analysis was carried out because the previous analysis did not meet the requirements to be carried out. From the table above it can be seen that Scenario 4 alone is eligible for the route change scheme because it provides benefits for both parties.

4. Summary

Based on the results of the analysis of the four calculated scenarios, the selected scenario is the fourth scenario 4 with the following specifications

a. Scenario 4 on the intersection of the route on P-2 occurs at the Port of Wini with commercial transportation Route K-2. So that route P-2 serves 11 sections, namely Wini - Lirang - Ilwaki - Kisar - Romang - Leti - Lakor - Luang P, Kelapa - Sermata (Elo) - Tepa - Kroing - Saumlaki and route changes on route K-2 to remain three the segment, namely Kupang – Naikliu – Wini – Teluk Gurita which will only serve up to the port of Wini.

b. Based on the operational analysis, it can be seen that the effect of the route change in scenario 4 is that the number of fixed frequencies is 110 times by serving Perintiss 48 times to fulfill the entire Perintis load which still does not exceed the frequency of commercial ships in a year.

c. Scenario 4 reduces Perintising shipping costs by 1.01% or IDR 229.46 million and subsidies by 0.14% or IDR 30.56 million. However, commercial transportation on Route K-2 costs increased by 0.42% or IDR 57.25 million, and an initial profit increased by IDR 809.70 million.

5. References

[1] E. Wahyono, Tangkilisan. Yuda B, and D. Marihandono, “Pelayaran Perintis Dalam Perspektif Sejarah Pembangunan Kawasan Tertinggal dan Terluar untuk Memperkuat Kedaulatan NKRI,” in *Seminar Nasional Riset Inovasi*, 2014, pp. 780–794.

[2] Kementerian Perhubungan Republik Indonesia, "PM 48 Tahun 2018 Penyelengaraan Kegiatan Pelayaran Publik Kapal Perintis ." 2018.
[3] C. Amin, H. Mulyati, E. Anggraini, and T. Kusumastanto, "Impact of maritime logistics on archipelagic economic development in eastern Indonesia," *Asian Journal of Shipping and Logistics*, vol. 37, no. 2, pp. 157–164, 2021, doi: 10.1016/j.ajsl.2021.01.004.
[4] Direktorat Jendral Perhubungan Laut, "Informasi Berkala Direktorat Jenderal Perhubungan Laut." 2023.
[5] A. Jugovic, N. Komadina, and A. Perić Hadžić, "Factors influencing the formation of freight rates on maritime shipping markets," *Scientific Journal of Maritime Research*, vol. 29, pp. 23–29, 2015.
[6] Polo Gerardo, "On Maritime Transport Costs, Evolution, and Forecast," *Ship Science and Technology*, vol. 5, pp. 19–31, 2012.
[7] T. Achmadi, P. Wuryaningrum, and A. R. Ayuningtyas, "Evaluation Models of Effectiveness and Efficiency of Pioneer Ship Services: Case Study of West Papua Region," in *Science and Technology Publications*, Portugal: Scitepress, 2022, pp. 43–52.
[8] G. Wilmsmeier and I. Martínez-Zarzoso, "Determinants of maritime transport costs - A panel data analysis for Latin American trade," *Transportation Planning and Technology*, vol. 33, no. 1, pp. 105–121, Feb. 2010, doi: 10.1080/03081060903429447.
[9] T. Acheampong and I. Ackah, "Petroleum Product Pricing, Deregulation and Subsidies in Ghana: Perspectives on Energy Security," *SSRN Electronic Journal*, Aug. 2015, doi: 10.2139/SSRN.2644561.
[10] A. Ferrari and R. Ossa, "A quantitative analysis of subsidy competition in the U.S.," *J Public Econ*, vol. 224, p. 104919, Aug. 2023, doi: 10.1016/J.JPUBECO.2023.104919.
[11] M. L. Pinedo, *Scheduling: Theory, algorithms, and systems: Fourth edition*, vol. 9781461423614. Springer US, 2012. doi: 10.1007/978-1-4614-2361-4.
[12] Pemerintah Provinsi NTT, "Geografis NTT," 2023.
[13] Badan Pusat Statistik, "Gross Regional Domestic Product of Provinces in Indonesia by Industry 2018 - 2022," 2023.

IOP Conf. Series: Earth and Environmental Science 1298 (2024) 012031 doi:10.1088/1755-1315/1298/1/012031

Operation failure of risk analysis on floating storage and offloading

Silvianita[1*], A A Harahap[1], M F Khamidi[2]

[1] Department of Ocean Engineering, Institut Teknologi Sepuluh Nopember, Surabaya, Indonesia
[2] Department of Architecture & Urban Planning, Qatar University
*Corresponding author: silvianita@oe.its.ac.id

Abstract. The need for petroleum in daily life is increasing. The oil comes from the seabed, so it is necessary to have adequate facilities during the exploration and exploitation of crude oil. The process of distributing crude oil to FSO is assisted by the Product Transfer System pipeline. In this process there is a possibility of failure that occurs form several factors, namely tools, weather and sea conditions, and human. In this study, the analysis of the risk of operational failure during loading-unloading of crude oil is discussed. The analysis was carried out using the HAZOP Analysis and Bow-Tie Analysis methods. HAZOP Analysis for hazard identification, and Bow-Tie Analysis for finding the appropriate mitigation. From this method, an overview of the risks than can occur can be obtained so that the risk control required by FSO can be implemented. From the results of this study, it was found that there are 41 potential hazards on the process of loading-unloading crude oil FSO, with the highest risk is inadequate quality of the transfer equipment components with the likelihood rank is 4 and severity rank is 4. The results of Bow-Tie visualization of dominant risk found five causes, namely corrosion, inadequate material quality, eroded material, service life, and material degradation, and five consequences, namely property damage, delayed operational activities, oil spills, environmental damage due to oil spill, and corrosion occurs on the product transfer equipment.

Keywords: Floating Storage and Offloading (FSO), crude oil, operation failure, HAZOP Analysis, Bow-Tie Analysis

1. Introduction

In daily life, the need for petroleum has many benefits in various fields, such as household needs, construction, vehicle fuel, and others. This makes the need for oil very profitable. Petroleum is a strategic natural resource, has high economic value, and is a non-renewable natural resource. Petroleum is found under land and the seabed. So, in the process of exploration and exploitation of oil off the coast, of course, it is necessary to have adequate facilities for the extraction process. Of course, the facilities needed are those that can survive the harsh marine environment. There are many types of offshore structures, this paper will discuss about Floating Storage and Offloading (FSO). Floating Storage and Offloading (FSO) is one of the most profitable structures in the petroleum industry [1]. FSO itself is a floating offshore platform designed for offshore hydrocarbon exploration activities, one of which is crude oil. FSO has several functions, which include loading-unloading crude oil. Loading-unloading is an activity where the FSO receives the processed oil, stores it, and distributes it to a carrier or shuttle tanker. While offshore, the FSO is moored with a floating structure, namely Single Point Mooring (SPM), so that its balance is not disturbed due to environmental loads during the loading unloading process of crude oil [2]. Due to its large size, the FSO significantly influences wave, wind, and current loads. SPM is connected to the product transfer system pipe, which transfers products from the seabed to the ship. At this time, the author conducted research on Y Floating Storage and Offloading System (FSO).

In distributing oil to FSO, it is undeniable that there are possibilities of failure. Operational failure during loading-unloading of crude oil can result in losses from small to large scale for several parties involved. Failure, in this case, can occur from 2 factors, namely internal and external factors, both from tools and weather and sea conditions. This can be detrimental to many parties and also harms many aspects, such as the environment, people, assets and companies, and others.

Failure can occur anytime and anywhere, so within the company, it is necessary to carry out risk management to avoid or minimize the occurrence of the risk. Risk management needs to be managed qualitatively and quantitatively to predict or control risks in a project [3]. Thus, risk management can make it easier for companies to determine the actions taken when these risks occur. The objectives of this paper is to determine the most potential hazard failure of loading-unloading crude oil operation of FSO and also their threats and consequences using HAZOP and Bow Tie Analysis.

There are several cases regarding failures in offshore structures, which then cause several major impacts, be it on reputation, assets, environmental pollution, or fatalities in the form of fatalities [4]. Cases that have occurred are like the Oil Rig owned by Transocean, which exploded, caught fire, and sank and then caused environmental pollution due to oil leaks and also fatality in the form of fatalities of as many as 11 workers [5]. This results from missed warning signals and failures in monitoring and disseminating information. Another case is the FPSO Cidade do Rio de Janeiro, which experienced an oil leak due to a cracked hull on the FPSO [6]. There have been several risk analyses regarding FPSOs and other offshore structures. The author carried out research at North West Java East, especially research on operation failures on loading-unloading crude oil at the FSO. Thus, early prevention can be done for each potential hazard and appropriate mitigation if the hazard occurs.

2. Literature review

Table 1. Literature review summary

References	Scope/Topics	Methods	Objectives
[7]	Drilling platform	BOW TIE	to represent the potential accident scenarios, their causes, and the associated consequences
[8]	Anchor Handling Operation	BOW TIE	to reduce potential risks during AHO operations and increase maritime safety.
[9]	Port Berth Construction	Hazard Identification	To identify the hazards and risks associated with the construction of a dock at Visakhapatnam Port and to evaluate existing preventive controls to avoid incidents and accidents.
[10]	Board Ships	BOW TIE	to conduct a risk analysis for confined space accidents on board ships using fuzzy bow-tie methodology. And to identify the hazards and risks associated with confined spaces on board ships and to evaluate the effectiveness of existing preventive controls to avoid accidents and incidents
[11]	Small LNG-Fueled Fishing Ship	Hazard Identification	To conduct a preliminary risk assessment on the development of the fuel gas supply system of a small LNG-fueled fishing ship.

[12]	Structural and Marine Facilities	BOW TIE	Application of process safety bow-ties into failure risks for structural and marine offshore facilities
[13]	Lima-Compresor Platform	BOW TIE	To identify potential hazards and risks associated with the decommissioning process and to evaluate the effectiveness of existing preventive controls to avoid accidents and incidents.
[14]	Ship Grounding Accident	Hazard Identification	to identify the hazard of ship grounding; where a ship runs on a rock with a forward speed, and to select a set of credible scenarios with a limited number that can still represent all possible situations of the accidents
[15]	Mobile Mooring System	HAZOP	to explain the potential causes and the possible consequences of mooring system failures using HAZOP as preliminary analysis

3. Failure risk

In this research, the case study used in Northwest Java, one of the floating offshore platforms in Indonesia. Table 2 shows the likelihood ranking used in this paper.

Table 2. Likelihood Ranking [16]

Likelihood		
1	*Almost Impossible*	Almost never happens in the oil and gas industry
2	*Very low*	Ever happened in the oil and gas industry
3	*Low*	Has happened once in 100 years
4	*Medium*	Happens at least once per year
5	*High*	Happens more than once per year

The severity rank is on a scale of 1 to 6, divided into four aspects: health and safety, environmental, equipment damage and business value, and business reputation as shown in Table 3 [16]

Table 3. Severity Ranking

Severity					
No		Health and Safety	Environmental	Equipment Damage and Business Value	Business Reputation
1	Slight	Minor injury or health effects but no effect on performance	Slight damage (oil spill <1 bbls)	Cost USD 50.000 and partial shutdown < 1 day	No media attention
2	Minor	Impact on health, medical treatment is required and impact on limited activities	Damage can still be controlled and repaired on site, has minor impact on the environment (1-15 bbls oil spil)	Costs > USD 50.000 – 25.000 and total shutdown < 1 day	Local press and regulatory requests

Table 3. Severity Ranking

3	Moderate	Unable to work for > 3 days	Moderate damage requiring cleanup or removal from company (15-100 bbls oil spill)	Costs > USD 250.000 – 1 million and total shutdown > 1 day – 1 week	Local press and potential fines by regulators
4	Serious	Accidents that cause permanent disability and require hospitalization	Oil spill and severe environmental damage (100 – 250 bbls oil spill)	Costs > USD 1 million – 10 million and total shutdown > 1 – 2 weeks	National media and potential regional coverage and demands by regulators
5	Major	Causing 2 fatalities due to accidents or illness, substances that cause death	Oil spill is out of control and causes severe environmental damage beyond site limits (>250 – 500 bbls oil spill)	Costs that reach > USD 10 million – 50 million and total production shutdowns for > 2 – 4 weeks	Regional media and potential international coverage, demands by regulators
6	Catastrophic	More than 3 fatalities due to accidents or diseases or materials with the potential to cause death	Oil spill are not controlled and cause severe environmental damage that continues to exceed site limits (>500 bbls)	Costs reached > USD 50 million and total shutdown for more than 1 month	Major impact on business reputation internationally

So, from the questionnaire results, it can be calculated for the likelihood and severity rank of each potential hazard. Each variable has different likelihood and severity category values, so the likelihood and severity ranking is calculated using the formula below:

$$Severity\ Average = \frac{\sum_{i=1}^{6} a_i n_i}{5} \quad (1)$$

$$Likelihood\ Average = \frac{\sum_{i=1}^{5} a_i n_i}{5} \quad (2)$$

Description:

a_i = rating constant (1 until 5 for likelihood, and 1 until 6 for severity)
n_i = respondent probability
I = 0,1,2,3,4,... n
N = total numbers of respondent

4.Risk matrix

From the calculation results of the average likelihood rank and severity rank, the next step is to classify the level of risk. The risk matrix used follows ISO 31000:2009 [16], divided into three levels : low risk, medium risk, and high risk. To determine the level of the risk matrix, the likelihood ranking, and severity ranking is multiplied for each potential hazard. From the multiplication results, it can be determined the level of risk for each potential hazard variable.

After assessing the causes and impacts of each existing risk is carried out, the determination of risk control or what can be called operational controls is carried out. Determination of risk control is an action to reduce or prevent the risk that has been predicted in advance. Risk control is carried out by distributing questionnaires back to stakeholders. Next step can be continued by calculating the likelihood and severity again on operational controls. This calculation is carried out again to know whether the controls that have been made are effective enough to minimize the impact of the potential hazards that exist. From the questionnaires that have been distributed, the likelihood and severity results are obtained from the five respondents for operational controls for each potential hazard. The risk level classification table can be seen below in the Risk Matrix table 4.

Table 4. Risk Matrix

		Likelihood				
		1	2	3	4	5
Severity	1	1	2	3	4	5
	2	2	4	6	8	10
	3	3	6	9	12	15
	4	4	8	12	16	20
	5	5	10	15	20	25
	6	6	12	18	24	30

High	High Risk – The activity should not be carried out until the risk has been reduced. The value for this high risk is from 15 to 30.
M	Moderate Risk – Action is needed to reduce the risk, but the cost of necessary prevention must be carefully calculated and limited. The value of this moderate risk is from 5 to 12.
L	Low Risk – Acceptable risk additional controls are not required. The value of this low risk is from 1 to 4.

The likelihood and severity rank can be calculated from the questionnaire results above. In the same way, as in the previous likelihood and severity calculations, the likelihood and severity ranking results are obtained using operational controls.

5. Risk matrix after operational controls

From the likelihood and severity ranking results above, the risk matrix can then be calculated again with the presence of operational controls. The calculation of the risk matrix is carried out the same as before, namely the multiplication between the likelihood ranking and the severity ranking. Thus, the results of the risk matrix are obtained as shown in the following table:

From the calculation of the risk matrix, it is found that the operational controls for each potential hazard are categorized in green, which is low risk. Thus, it can be concluded that the operational controls that have been made are efficient enough to be used to reduce the impact of the potential hazards that exist in the loading-unloading process of crude oil at the Y FSO.

6. HAZOP worksheet

Furthermore, the analysis results above can be entered into the HAZOP Worksheet table. HAZOP Worksheets are columns containing main activities, tasks, activity codes, activity descriptions, potential hazards, possible causes, possible impacts, ranking for likelihood and severity for each potential hazard, operational controls, and ranking for likelihood and severity for each operational control. Thus, a HAZOP Worksheet was formed on the crude oil transfer operation at FSO Y, as shown in the table 5 below.

Table 5. HAZOP Worksheet Result

Task 1: Floating and subsea hose loading crude oil from Platform to FSO										
Activity Description	Potential Hazard	Possible Causes	Hazard Effect	Likelihood	Severity	Risk	Operational Controls	Likelihood	Severity	Risk
Floating and subsea hose operation	1.1.1. Hose leaked	- Corrosion - Inadequate material quality - High pressure	- Crude oil transfer failure - Property damage - Environmental damage (oil spill)	3	4	12	- Perform fatigue life assessment - Periodic inspection and maintenance on each component	2	2	4
	1.1.2. Twisted hose	- Sea conditions - Human error	Property damage	3	4	12	Periodic inspection and maintenance	2	2	4
	1.1.3. Inadequate quality of the components on the hose	- Corrosion - Inadequate material quality - Eroded material	- Property damage - Potential hose leakage - Environmental damage (oil spill)	3	4	12	Determination of quality standards of the type of hose material	2	2	4
	1.1.4. Inadequate quality of the transfer equipment components	- Corrosion - Inadequate material quality - Eroded material	Property damage	4	3	12	Determination of quality standards of the type of transfer equipment materials	2	2	4

7. Bow-tie analysis

Bow-Tie Analysis can be continued after calculating the likelihood, severity, and risk matrix, which is then summarized in a HAZOP Worksheet table. Based on the HAZOP Analysis above, the critical risk during the loading-unloading process of crude oil at the Y FSO occurs in the floating and subsea hose operations during offloading of crude oil from the Y FSO to the Tanker, with the potential danger being that the quality of the transfer equipment components is inadequate.

Risk analysis using the Bow-Tie Analysis method was performed using BowtieXP software. This method can then clearly describe the threats and consequences. In this Bow-Tie diagram, you can find the causal and preventive variables on the left side of the diagram, the consequences and mitigation variables on the right side of the diagram, the escalation factor, and the mitigation escalation. The following is Bow-Tie Analysis Modelling for activity code inadequate quality of transfer equipment components.

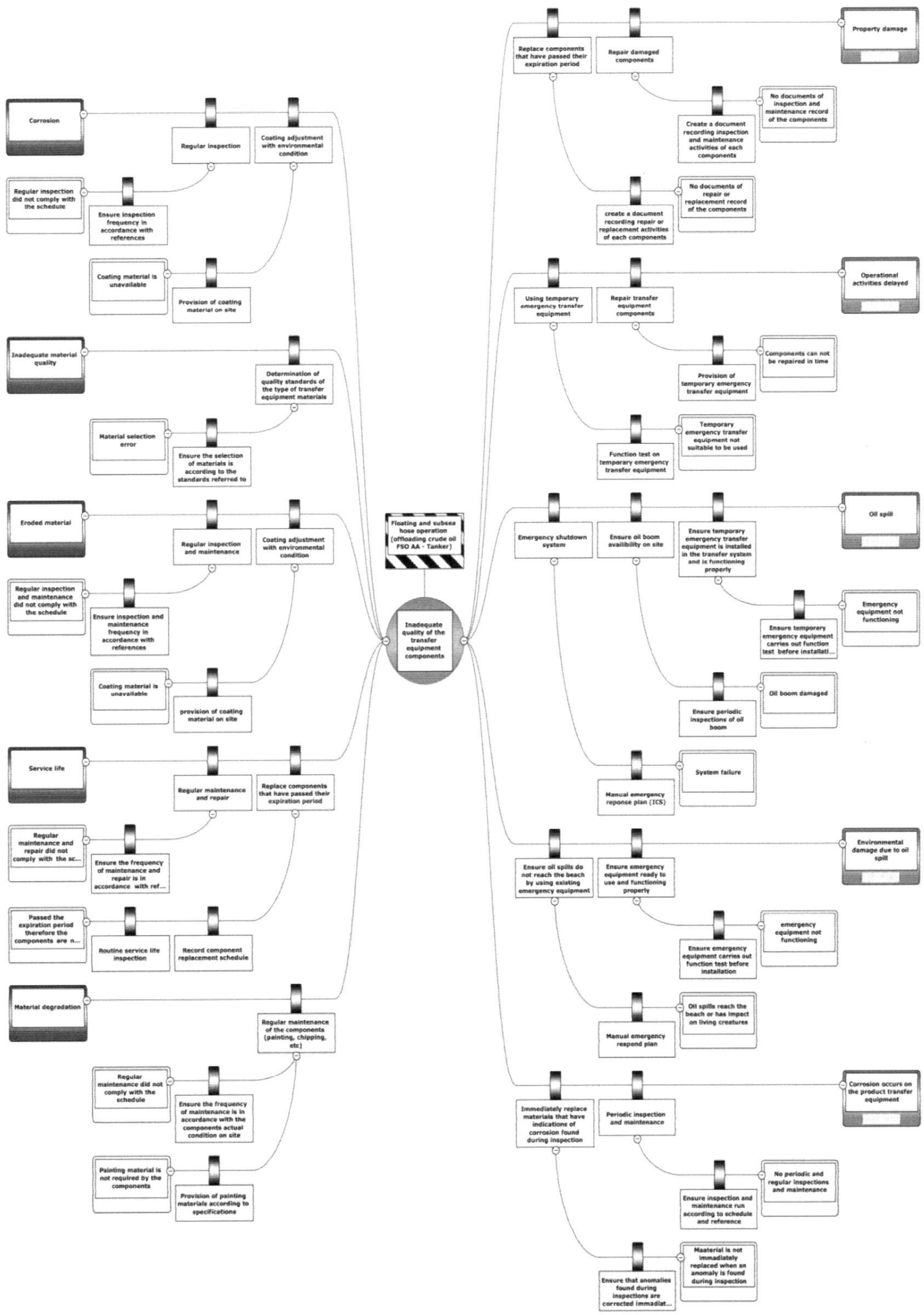

Figure 1. Bow-Tie Analysis Modeling for Indequate Quality of the Transfer Equipment Components

The description of the threat of inadequate quality of the transfer equipment components is shown in Table 6.

Table 6. Threat of Inadequate Quality of The Transfer Equipment

Risk	Threat			
	Causes	Preventive barrier	Escalation factor	Mitigation escalation
Inadequate quality of the transfer equipment components	- Corrosion	- Regular inspection - Coating adjustment with environmental condition	- Regular inspection did not comply with the schedule - Coating material is unavailable	- Ensure inspection frequency in accordance with references - Provision of coating material on site
	Inadequate material quality	Determination of quality standards of the type of transfer equipment materials	Material selection error	Ensure the selection of materials according to the standards
	Eroded material	- Regular inspection and maintenance - Coating adjustment with environmental condition	- Regular inspection and maintenance did not comply with the schedule - Coating material is unavailable	- Ensure inspection and maintenance frequency in accordance with references - Provision of coating material on site
	Service life	- Regular maintenance and repair - Replace components that have passed their expiration period.	- *Regular* maintenance and repair did not comply with the schedule - Passed the expiration period therefore the components are not adequate	- Ensure the frequency of maintenance and repair is in accordance with reference - Routine inspection on service life inspections - Record component replacement schedule
	Material degradation	Regular maintenance of the components (painting, chipping, etc)	- *Regular* maintenance did not comply with the schedule - Painting material is not as required by the components	- Ensure the frequency of maintenance is in accordance with the components actual condition on site - Provision of paint materials according to specifications

The description of the consequences of inadequate quality of the transfer equipment components is shown in Table 7.

IOP Conf. Series: Earth and Environmental Science 1298 (2024) 012031 doi:10.1088/1755-1315/1298/1/012031

Table 7. Consequences of Inadequate Quality of The Transfer Equipment

Risk	Consequences		Escalation Factor	Mitigation Escalation
	Consequences	**Mitigation**		
Inadequate quality of the transfer equipment components	Property damage	- Repair damaged components - Replace components that have passed the expiration period	- No documents of inspection and maintenance record of the components - No documents of repair or replacement record of the components	- Create a document recording inspection and maintenance activities of each components - Create a document recording repair or replacement activities of each components
	Operational activities delayed	- Repair transfer equipment components - Using temporary emergency transfer equipment	- Components cannot be repaired in time - Temporary emergency transfer equipment not suitable to be used	- Provision of temporary emergency transfer equipment - Function test on temporary emergency transfer equipment
	Oil spill	- Ensure temporary emergency equipment is installed in the transfer equipment system and is functioning properly - Ensure oil boom availability on site - Emergency shutdown system	- Emergency equipment not functioning - Oil boom damaged - System failure	- Ensure temporary emergency equipment carries out function test before installation - Ensure periodic inspections of oil boom - Manual emergency response plan (ICS)
	Environmental damage due to oil spill	- Ensure emergency equipment ready to use and functioning properly - Ensure oil spills do not reach the beach by using existing emergency equipment	- Emergency equipment not functioning - Oil spill reach the beach or has impact on living creatures	- Ensure emergency equipment carries out function test before installation - Manual emergency respond plan
	Corrosion occurs on the product transfer equipment	- Periodic inspection and maintenance - Immediately replace materials that have indications of corrosion found during inspections	- No periodic and regular inspections and maintenance - Material is not immediately replaced when an anomaly is found during inspection	- Ensure inspection and maintenance run according to schedule and reference - Ensure that anomalies found during inspections

8. Conclusion

Based on the analysis, it can be concluded that:

1. There are five tasks in the product transfer process and the most potential hazard failure of loading-unloading crude oil operations at FSO Y is inadequate quality of transfer equipment components.
2. Bow-Tie Analysis shows the threats of Inadequate quality of the transfer equipment components, namely corrosion, inadequate material quality, eroded material, service life, and material degradation. (material degradation). Meanwhile the Consequences shows five consequences: property damage, operational activities delayed, oil spill, environmental damage due to oil spill, and corrosion occurs on the product transfer equipment.

References

[1] H. Meng, L. Kloul, and A. Rauzy, "Production availability analysis of Floating Production Storage and Offloading (FPSO) systems," *Applied Ocean Research*, vol. 74, pp. 117–126, May 2018, doi: 10.1016/j.apor.2018.02.026.

[2] J. Yu, H. Ding, Y. Yu, S. Wu, Q. Zeng, and W. Ma, "A novel risk analysis approach for FPSO single point mooring system using Bayesian Network and interval type-2 fuzzy sets," *Ocean Engineering*, vol. 266, Dec. 2022, doi: 10.1016/j.oceaneng.2022.113144.

[3] M. H. Haghighi and M. Ashrafi, "A new qualitative and quantitative analytical approach for risk management in energy project time-cost trade-off problem under interval type-2 fuzzy uncertainty: A case study in the gas industry," *Energy Reports*, vol. 8, pp. 12668–12685, Nov. 2022, doi: 10.1016/j.egyr.2022.09.064.

[4] A. Gokce Cicek Ceyhun, "THE IMPACT OF SHIPPING ACCIDENTS ON MARINE ENVIRONMENT: A STUDY OF TURKISH SEAS," 2014.

[5] L. Averill, B. Durkin, M. Chu, U. Ougradar, and A. Reeves, "Deepwater Horizon disaster." [Online]. Available: www.deepwaterhorizon.co.uk

[6] J. E. Vinnem, "FPSO Cidade de São Mateus gas explosion – Lessons learned," *Saf Sci*, vol. 101, pp. 295–304, Jan. 2018, doi: 10.1016/j.ssci.2017.09.021.

[7] M. Abimbola, F. Khan, and N. Khakzad, "Dynamic safety risk analysis of offshore drilling," *J Loss Prev Process Ind*, vol. 30, no. 1, 2014, doi: 10.1016/j.jlp.2014.05.002.

[8] M. Kaptan, "Risk assessment of ship anchorage handling operations using the fuzzy bow-tie method," *Ocean Engineering*, vol. 236, 2021, doi: 10.1016/j.oceaneng.2021.109500.

[9] R. Mahapatra, N. Kushwaha, S. K. Singh, R. Gandhi, and P. Vishwavidyalaya, "Hazards Identifications and Risk Assessment in Port Berth Construction-A Case Study." [Online]. Available: www.ijert.org

[10] C. Sakar, M. Buber, B. Koseoglu, and A. C. Toz, "Risk analysis for confined space accidents onboard ship using fuzzy bow-tie methodology," *Ocean Engineering*, vol. 263, Nov. 2022, doi: 10.1016/j.oceaneng.2022.112386.

[11] Y. Shao, H. K. Kang, Y. H. Lee, G. Królczyk, P. Gardoni, and Z. X. Li, "A preliminary risk assessment on development the fuel gas supply system of a small LNG fueled fishing ship," *Ocean Engineering*, vol. 258, Aug. 2022, doi: 10.1016/j.oceaneng.2022.111645.

[12] S. Slatnick *et al.*, "Bow-ties use for high-consequence marine risks of offshore structures," *Process Safety and Environmental Protection*, vol. 165, pp. 396–407, Sep. 2022, doi: 10.1016/j.psep.2022.07.026.

[13] A. Windiargo and D. M. Rosyid, "Risk Analysis of Decommissioning Process: Case Studies of Lima-Compresor Platform," *International Journal of Offshore and Coastal Engineering*, vol. 4, no. 1, pp. 27–34, 2020.

[14] S. A. M. Youssef and J. K. Paik, "Hazard identification and scenario selection of ship grounding accidents," *Ocean Engineering*, vol. 153, pp. 242–255, Apr. 2018, doi: 10.1016/j.oceaneng.2018.01.110.

[15] Silvianita, M. F. Khamidi, I. Rochani, and D. M. Chamelia, "Hazard and Operability Analysis (HAZOP) of Mobile Mooring System," *Procedia Earth and Planetary Science*, vol. 14, 2015, doi: 10.1016/j.proeps.2015.07.103.

[16] "Risk management-Principles and guidelines Management du risque-Principes et lignes directrices," 2009. [Online]. Available: www.iso.org

Analysis of occupational accident risk in breakwater construction and floating piers

Silvianita[1], A P Atmaja[1*], N Shafiq[2]

[1] Department of Ocean Engineering, Institut Teknologi Sepuluh Nopember, Surabaya, Indonesia
[2] Department of Civil Engineering, Universiti Teknologi PETRONAS, Tronoh, Malaysia
*Corresponding author: silvianita@oe.its.ac.id

Abstract. In every implementation of the project, work accidents can occur which result in injury or other losses. One of the projects that has the potential for work accidents is the floating dock construction work. The purpose of writing this paper is to find out the dominant hazard variable in floating piers and breakwater construction work, to know the causes and effects of the dominant hazard variable, and to know the right dominant risk control risk control in floating wharf construction work. This research begins with hazard identification using several sources and also direct discussions with resource persons who understand this work. Then a questionnaire was distributed to obtain the likelihood and severity values of the existing hazard variables. From the results of distributing the questionnaires, tests were carried out using validity and reliability tests on each hazard variable that might occur. The results of the tests carried out obtained a valid hazard variable and were then analysed using HIRA which was assisted by a risk matrix. So that the results of the analysis are known to be the dominant hazard variable, namely workers falling and sinking into the sea. Furthermore, an analysis of the two variables is carried out to determine the causes and also the impact of risks that can occur using a bow-tie diagram to reduce the occurrence of dominant potential hazards.

1. Introduction

A port is an area of water that is protected against waves, which is equipped with sea terminal facilities including a pier where ships can moor for loading and unloading of goods, and warehouses where goods can be stored for a longer time, while waiting for delivery to the destination area [1]. The port is a gateway to enter a region or country and as a connecting infrastructure between regions or islands in that region or country. In the (physical) port construction work, there are several related works including preparatory work, north and south side breakwater work, floating wharf work, dredging work, development area work, construction of land facilities (terminal building), implementation of a safety management system construction and others. In the (physical) port construction work, there are several related works including preparatory work, north and south side breakwater work, floating wharf work, dredging work, talud work, development area work, construction of land facilities (terminal building), implementation of a safety management system construction and others. In the construction work of a floating wharf contains several sources of danger both from the environment, the use of heavy equipment, materials and so on. Inappropriate implementation of work methods can result in the risk of work accidents which can disrupt the course of project activities.

One way to minimize this is by conducting a work accident risk analysis. Risk analysis can be interpreted as a procedure to recognize a threat and vulnerability, then analyse it to identify an accident, and find out how the impacts can be eliminated or reduced.

This paper discussed the risk assessment in construction project work, especially in the construction of floating docks is required to deal with various risks that may occur. There are several methods to conduct risk assessment as described in Table 1.

The method used in this study combines Hazard Identification and Risk Assessment (HIRA) and Bowtie Analysis. Hazard Identification and Risk Assessment (HIRA) is used to identify potential job hazards by defining the characteristics of the hazards that may arise and assessing the risks that arise through risk assessment using a risk assessment matrix to obtain the most dominant risks. Meanwhile, Bowtie Analysis is used to analyze the impacts and causes that can be generated through initiating events or top events so that they can provide preventive recommendations and mitigating controls that will be used.

Table 1. Risk Assessment Methods

No References	Objectives	Methods
[2]	to develop methodology for investigation of Risk Based Maintenance (MIRBA)	FTA, AHP
[3]	to apply the HAZOP methodology in process and safety operations in the oil production industry	HAZOP analysis; hazard identification
[4]	to change old-fashioned HAZOP industrial practice and improve safety performance in oil and gas industry	HAZOP analysis
[5]	to understand the limit values and the safety range for every parameter with the aid of graphs and tables and so they achieved a quantitative HAZOP to resume the data they have found.	HAZOP, Risk Assessment
[6]	is to evaluate the safety of crude oil tankers using the methodology of Formal Safety Assessment (FSA). And provides a general approach to safety assessment and discusses the evolution of risk acceptance in recent decades.	Formal Safety Assessment, HAZOP, HAZID
[7]	to promote safety and knowledge transfer on the LNG fuel chain to interested stakeholders, such as port operators, maritime instructors, port authorities, and stakeholders in planning, licensing, and emergency management practices related to LNG handling.	HAZOP, HAZID
[8]	as a technique for examining the systematic and efficient application of HSE engineering by exploiting the HAZOP systematic risk analysis technique and a quantitative risk derivation method, which is an advantage of the Safety Integrity Level	HAZOP
[9]	provide an overview of the Hazard and Operability (HAZOP) and Hazard Identification (HAZID) studies as risk management techniques. The paper discusses the application of HAZOP and HAZID studies in identifying and analyzing potential hazards in various industries, including the chemical process industry, oil and gas industry, and other process industries.	HAZOP, HAZID
[10]	to propose qualitative and quantitative approaches for identifying and prioritizing different hazards associated with decommissioning of offshore wind farms.	HAZID, FTA, ETA, Risk matrix

2. Hazards Identification

2.1. Study of literature

This study collects all the literature needed to support this topic. The studies required are as follows: studies related to risk analysis, risk management and work safety, regarding the process or work method of floating wharf construction.

In the maritime industry, there have been many studies discussing risk assessment such as fire analysis in the offloading process [11], anchor handling [12], natural gas pipelines [13], container terminal activities [14]. In addition, in the field of construction, there are also many topics discussed related to risk assessment, such as research [15] which discusses the use of HIRA in construction projects, [16]which combines the use of HIRA and HAZOP in risk assessment, [17] which discusses the risk analysis model for offshore engineering project to determine the risks associated with a particular activity and the justification factors for proposed improvements.

2.2. Data collection

The data used in this study is the result of observations and discussions with workers who are directly involved in projects related to the work of the Sanu Port Facility Development project, Bali. The data obtained is in the form of detailed data on activities, work methods and several HSE documents related to this research. Hazard identification carried out in research is carried out by conducting discussions with project work implementers or supervisors. After the data is in accordance with what is in the field and also approved by the project executor, then the author compiles a questionnaire to be distributed to several workers related to the project work. Table 2 shows the hazards identification in floating pier construction work based on the respondents.

Table 2. Hazards Identification in Floating Pier Construction Work

ID	Activity	Potential Hazard	Possible Hazard	Hazard Effect
1A	Steel Pipe Fitting	Extreme Weather	High waves (tide), Rain, Storm, Lightning Strike	Workers fall into the sea
1B			Exposed to a lifting maneuver	Serious Injury
1C			Crane failure due to fuel and hydraulics	Polluted environment
1D		Transfer using a crane	The sling broke so that the crane overturned because it was overloaded	Damage to materials and tools
1E			Heavy equipment operator fatigue	Operator fell ill
1F		Unstable Pontoon	Unsafe pontoon due to waves, leakage, mooring failure	Workers fall and sink into the sea
1G				The heavy equipment fell into the sea
2A	Formwork, ironing and casting work on trestle	Use of sharp tools during formwork installation	Manual handling error	Members of the body are exposed to bar bending/bar cutting
2B				Scratched iron material
2C				The worker's limbs were cut off
2D		Use of heavy equipment at the time of casting (concrete mixer, concrete pump truck	Vehicle malfunction (loss of control of heavy equipment), Operator fatigue	Worker hit by heavy equipment
2E				Exposure to dust due to heavy equipment traffic
2F				Exposed concrete material
3A	Pile cutting	Extreme Weather	High waves (tide), Rain,	Workers fall into the

IOP Conf. Series: Earth and Environmental Science 1298 (2024) 012032 doi:10.1088/1755-1315/1298/1/012032

ID	Activity	Potential Hazard	Possible Hazard	Hazard Effect
			Storm, Lightning Strike	sea
3B		Pile cutting	Use of sharp tools	Scratched iron material
3C				The worker's limbs were cut off
3D		Electrical short circuit	Use of equipment that uses a source of electricity (generator)	The worker was electrocuted due to an electric short circuit
3E				There was a fire due to an electric short circuit
4A	Piles welding	Extreme Weather	High waves (tide), Rain, Storm, Lightning Strike	Workers fall into the sea
4B		Extreme Weather	Unsafe pontoon because of the waves	Workers fall and sink into the sea
4C		Electrical short circuit	Use of equipment that uses a source of electricity (generator)	The worker was electrocuted due to an electric short circuit
4D				Exposed to welding sparks
5A	Installation of floating dock accessories	Extreme Weather	High waves (tide), Rain, Storm, Lightning Strike	Workers fall into the sea
5B		Dragged by the ocean currents	Tidal Wave	Workers are swept away by the tide
5C		Installation of floating dock modules (HDPE) for floating dock bases	Installation is not in accordance with the position, less balanced because of the waves	Scratches during installation of HDPE
5D				Pinched during installation of HDPE
5E				Fall into the sea
5F		Installation of galvanized hollow frames, pile guides, fenders and WPC on floating piers	The equipment uses a power source	Exposed to welding sparks
5G			Improper installation position	Stuck in time mounting WPC with frames
5H			Unstable conditions	Fall into the sea

Table 3 shows the hazards identification in breakwater construction work based on the respondents.

Table 3. Hazards Identification in Breakwater Construction Work

ID	Activity	Potential Hazard	Possible Hazard	Hazard Effect
1A	Geotextile / Geocomposite laying	Extreme Weather	High waves (tide), Rain, Storm, Lightning Strike	Workers fall into the sea
1B		Pontoon Failure	Unsafe pontoon due to waves, leakage, mooring failure	Falling into the sea while laying geotextile/geocomposite
2A	Installation of Core Stone	Extreme Weather	High waves (tide), Rain, Storm, Lightning Strike	Workers fall into the sea
2B		Material transfer using excavators/dump trucks	Material transfer using excavators/dump trucks	Worker hit by heavy equipment
2C				Hit by heaps of landslides
2D				Exposure to dust due to heavy equipment traffic

ID	Activity	Potential Hazard	Possible Hazard	Hazard Effect
2E			Uneven ground	Tripped while walking on it
2F				Vehicle overturned / mired
3A	Installation of Secondary Layer (River Stone 500-1000 kg)	Extreme Weather	High waves (tide), Rain, Storm, Lightning Strike	Workers fall into the sea
3B		Material transfer using excavators/dump trucks	Vehicle malfunction (loss of control of machine)	Worker hit by heavy equipment
3C				Hit by heaps of landslides
3D				Exposure to dust due to heavy equipment traffic
3E			The base is less solid	Vehicle overturned / mired
4A	Installation of 1.3 ton and 3.6 ton BPPT Locks	Flatbed mobilization to transport BPPT Lock	Driver operator error	Drive over speed
4B			The base is less solid	Vehicle overturned / mired
4C				Exposure to dust due to heavy equipment traffic
4D		Electrical short circuit	Use of equipment that uses a source of electricity (generator)	The worker was electrocuted due to an electric short circuit
4E				There was a fire due to an electric short circuit
4C		Transfer of material using a crawler crane	The base is less solid	The crane overturned
4D				The worker is crushed by a tool/material
4E			Slings broke	the crane overturned
4F				Damage to materials and tools
4G			Heavy equipment operator fatigue	Sick operators
5A	Installation of Cyclope Concrete Cubes	Use of crawler cranes when erecting steel pipes	The base is less solid	Crane overturned/stumbled
5B			Heavy equipment operator fatigue	Sick operators

2.3. Risk analysis using HIRA (Hazard Identification and Risk Assessment) method

Risk analysis begins with processing data from questionnaires that have previously been distributed to predetermined respondents. The results of the questionnaire are used to measure the likelihood and severity of each activity variable that has been determined. Furthermore, the likelihood and severity of each hazard variable will be tested using validity and reliability tests to determine the accuracy of the instrument in measuring what we want to measure in research and also to determine the consistency of research instruments from time to time. Invalid and reliable variables will be eliminated later. The results of the multiplication between the likelihood and severity of the valid variables will later be plotted into the risk matrix to obtain the most dominant hazard variable risk results.

3. Result and Discussion

3.1. Likelihood and severity

After distributing the questionnaires to several respondents, the likelihood and severity of each hazard variable will be known. The following is the likelihood and severity of each hazard variable in floating wharf construction work as seen in Table 4.

Table 4. Likelihood and Severity Rating in Floating Wire Construction Work

ID	*Likelihood* Rating				
	1	2	3	4	5
1A	0	1	2	2	1
1B	1	1	2	1	1
1C	2	0	3	0	1
1D	1	1	2	2	0
1E	1	2	2	1	0
1F	1	0	2	2	1
1G	2	0	0	4	0
2A	1	1	1	2	1
2B	1	1	1	3	0
2C	1	2	3	0	0
2D	3	1	1	1	0
2E	0	0	4	2	0
2F	1	1	0	3	1
3A	2	1	1	2	0
3B	0	1	2	3	0
3C	1	3	0	2	0
3D	1	2	1	2	0
3E	1	1	2	1	1
4A	1	1	0	2	2
4B	1	0	1	2	2
4C	1	2	0	2	1
4D	1	1	1	3	0
5A	2	1	1	1	1
5B	1	1	1	3	0
5C	1	2	1	1	1
5D	1	0	2	2	1
5E	1	1	0	3	1
5F	0	1	2	3	0
5G	0	2	1	3	0
5H	1	1	0	4	0

ID	*Severity* Rating				
	1	2	3	4	5
1A	0	1	2	2	1
1B	1	2	1	1	1
1C	1	2	1	1	1
1D	0	1	4	1	0
1E	1	2	2	1	0
1F	0	1	2	2	1
1G	0	1	2	2	1
2A	2	2	0	2	0
2B	4	1	0	1	0
2C	0	3	2	1	0
2D	1	2	2	0	1
2E	0	0	4	2	0
2F	2	2	1	1	0
3A	0	2	3	0	1
3B	4	2	0	0	0
3C	0	4	1	1	0
3D	0	4	1	1	0
3E	2	1	2	1	0
4A	1	1	2	1	1
4B	0	1	2	3	0
4C	1	3	0	1	1
4D	2	2	0	2	0
5A	2	1	0	2	1
5B	0	3	2	0	1
5C	5	1	0	0	0
5D	5	0	0	1	0
5E	0	1	2	2	1
5F	2	2	1	1	0
5G	4	1	0	0	1
5H	0	1	2	2	1

3.2. Validity and reliability check

From the results of the likelihood and severity that have been obtained, then a test is carried out to find out the accuracy and consistency of each of the existing variables. To test the validity using Bivariate Pearson correlation (Pearson Moment Product) and Corrected Item-Total Correlation.

$$r_{ix} = \frac{n\,\Sigma ix - (\Sigma i)(\Sigma x)}{\sqrt{\sqrt{[n\,\Sigma i^2 - (\Sigma i)^2][n\,\Sigma i^2 - (\Sigma x)^2]}}} \quad (1)$$

Whereas for reliability testing using the Cronbach Alpha test to state whether the variable is consistent when tested repeatedly.

$$r = \left[\frac{k}{(k-1)}\right]\left[1 - \frac{\Sigma \sigma_b{}^2}{\sigma_t{}^2}\right] \quad (2)$$

In testing the validity of the question is considered valid if the calculated R value is greater than R table (N = 6, R table = 0.707 with a significance of 5%). The following is the result of testing the validity of the likelihood and severity of each variable as seen in Table 5.

Table 5. Validity Test of Hazard Variables in Floating Pier Construction Work

ID	r count	r table	description
1A	0.41	0.81	false
1B	0.84	0.81	valid
1C	0.94	0.81	valid
1D	0.95	0.81	valid
1E	0.55	0.81	false
1F	0.81	0.81	valid
1G	0.87	0.81	valid
2A	0.81	0.81	valid
2B	0.39	0.81	false
2C	0.60	0.81	false
2D	0.63	0.81	false
2E	0.70	0.81	false
2F	0.564	0.81	false
3A	0.77	0.81	valid
3B	0.03	0.81	false
3C	0.86	0.81	valid
3D	0.94	0.81	valid
3E	0.87	0.81	valid
4A	0.85	0.81	valid
4B	0.89	0.81	valid
4C	0.96	0.81	valid
4D	0.84	0.81	valid
5A	0.89	0.81	valid
5B	0.64	0.81	false
5C	0.12	0.81	false
5D	0.29	0.81	false
5E	0.95	0.81	valid
5F	0.92	0.81	valid
5G	0.28	0.81	false
5H	0.88	0.81	valid

ID	r count	r table	description
1A	0.46	0.81	false
1B	0.88	0.81	valid
1C	0.84	0.81	valid
1D	0.86	0.81	valid
1E	0.06	0.81	false
1F	0.91	0.81	valid
1G	0.86	0.81	valid
2A	0.88	0.81	valid
2B	0.52	0.81	false
2C	0.24	0.81	false
2D	0.51	0.81	false
2E	0.20	0.81	false
2F	0.45	0.81	false
3A	0.63	0.81	false
3B	0.20	0.81	false
3C	0.25	0.81	false
3D	0.83	0.81	valid
3E	0.82	0.81	valid
4A	0.87	0.81	valid
4B	0.85	0.81	valid
4C	0.96	0.81	valid
4D	0.87	0.81	valid
5A	0.89	0.81	valid
5B	0.46	0.81	false
5C	0.65	0.81	false
5D	0.65	0.81	false
5E	0.85	0.81	valid
5F	0.90	0.81	valid
5G	0.66	0.81	false
5H	0.85	0.81	valid

IOP Conf. Series: Earth and Environmental Science 1298 (2024) 012032 doi:10.1088/1755-1315/1298/1/012032

The variables of likelihood and severity which invalid are eliminated and not included in the next stages. The next data processing is risk assessment with a risk matrix.

Table 6. Reliability Tests on Hazard Variables in Floating Pier Construction Work

Reliability Check			
Cronbach's Alpha	0.967	N of Item's	30

From table 6 above it is found that the Cronbach's Alpha coefficient on the floating pier work item is 0.967 (ri > 0.7) which means that the test items used are reliable/can be used repeatedly.

3.3. Hazard identification and risk assessment (HIRA)

From the results of the validity and reliability tests that have been carried out, valid and reliable hazard variables are then used for further analysis by multiplying the likelihood index and severity index to later be plotted in the risk matrix to determine the dominant hazard variable.

To calculate the likelihood and severity index the author uses the equation below:

$$I = \frac{\sum_{i=1}^{N} ai \times xi}{5 \times N} \times 100\% \quad (3)$$

With, $ai = Value\ Konstanta\ (1\ s/d\ 5)$

$xi = Respondent\ Probability$

$N = Total\ Number\ of\ Respondents$

The results of calculating the likelihood index and severity index can be seen in Table 7.

Table 7. Likelihood Index and Severity Index

ID	Likelihood Index	Likelihood	Rating	Severity Index	Severity	Rating
1B	*70%*	*Very Effective*	*4*	*57%*	*Moderately Effective*	*3*
1C	*53%*	*Moderately Effective*	*3*	*57%*	*Moderately Effective*	*3*
1D	*57%*	*Moderately Effective*	*3*	*57%*	*Moderately Effective*	*3*
1G	*67%*	*Very Effective*	*4*	*70%*	*Very Effective*	*4*
1H	*60%*	*Moderately Effective*	*3*	*70%*	*Very Effective*	*4*
2B	*63%*	*Very Effective*	*4*	*47%*	*Moderately Effective*	*3*
3D	*53%*	*Moderately Effective*	*3*	*50%*	*Moderately Effective*	*3*
3E	*60%*	*Moderately Effective*	*3*	*47%*	*Moderately Effective*	*3*
4A	*70%*	*Very Effective*	*4*	*63%*	*Very Effective*	*4*
4B	*73%*	*Very Effective*	*4*	*67%*	*Very Effective*	*4*
4C	*60%*	*Moderately Effective*	*3*	*53%*	*Moderately Effective*	*3*
4D	*60%*	*Moderately Effective*	*3*	*47%*	*Moderately Effective*	*3*
5A	*53%*	*Moderately Effective*	*3*	*57%*	*Moderately Effective*	*3*
5E	*67%*	*Very Effective*	*4*	*70%*	*Very Effective*	*4*
5F	*67%*	*Very Effective*	*4*	*43%*	*Moderately Effective*	*3*
5H	*63%*	*Very Effective*	*4*	*70%*	*Very Effective*	*4*

Risk assessment using a risk matrix is carried out by multiplying the value of the likelihood and severity index which have previously been calculated. The hazard variables contained in the hazard identification are plotted as a whole into the risk matrix to find out which hazards are acceptable or unacceptable.

Table 8. Assessment with The Risk Matrix on Floating Works

		Severity				
		1	2	3	4	5
Likelihood	5					
	4			1B, 2A, 5F	1F, 4A, 4B, 5E, 5H	
	3			1C, 1D, 3D, 3E, 4C, 4D, 5A	1G	
	2					
	1					

Based on the risk matrix shown in Table 8, there are 5 risks that must be eliminated immediately because they are in the red area namely IF, 4A, 4B, 5E, and 5H codes. Based on the risk matrix, the five risks are unacceptable, so risk control must be carried out because they can affect the floating dock construction work. These dominant risks will be further analyzed using a bowtie diagram to determine the causes, impacts and mitigation of these risks. Bowtie diagram of the dominant variable can be seen in Figure 1.

3.4. Risk assessment in breakwater construction work

In carrying out a risk assessment on breakwater construction work, the analytical steps carried out are the same as when carrying out a risk assessment of floating pier construction work. Starting with hazard identification, calculating likelihood and severity, conducting validity and reliability tests, calculating likelihood index and severity index, and later obtaining the most dominant risk in breakwater construction work as seen in Table 9.

Table 9. Assessment with The Risk Matrix on Breakwater Work

		Severity				
		1	**2**	**3**	**4**	**5**
Likelihood	**5**					
	4					2B
	3			4K	2C, 3B, 6A	
	2			5C, 5D	4B, 4E, 4F, 5A, 5E	3E, 4I, 6C, 6D
	1					

Based on the risk matrix that has been made, there is one dominant hazard variable that occurs in breakwater construction work, which is variable 2B (worker being hit by heavy equipment). The dominant risk variables will be analyzed further using a bowtie diagram to find out the causes, impacts, and mitigation to prevent the risk variables from arising again. The bowtie diagram for this risk variable can be seen in Figure 2.

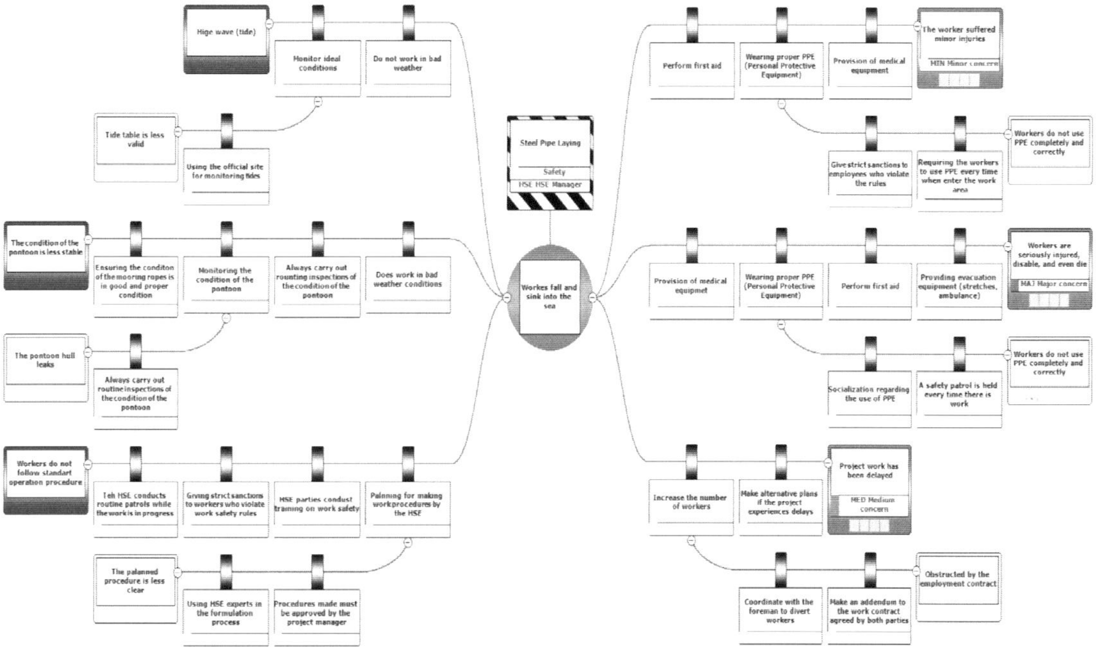

Figure 1. Bowtie diagram on the risk of falling and sinking into the sea

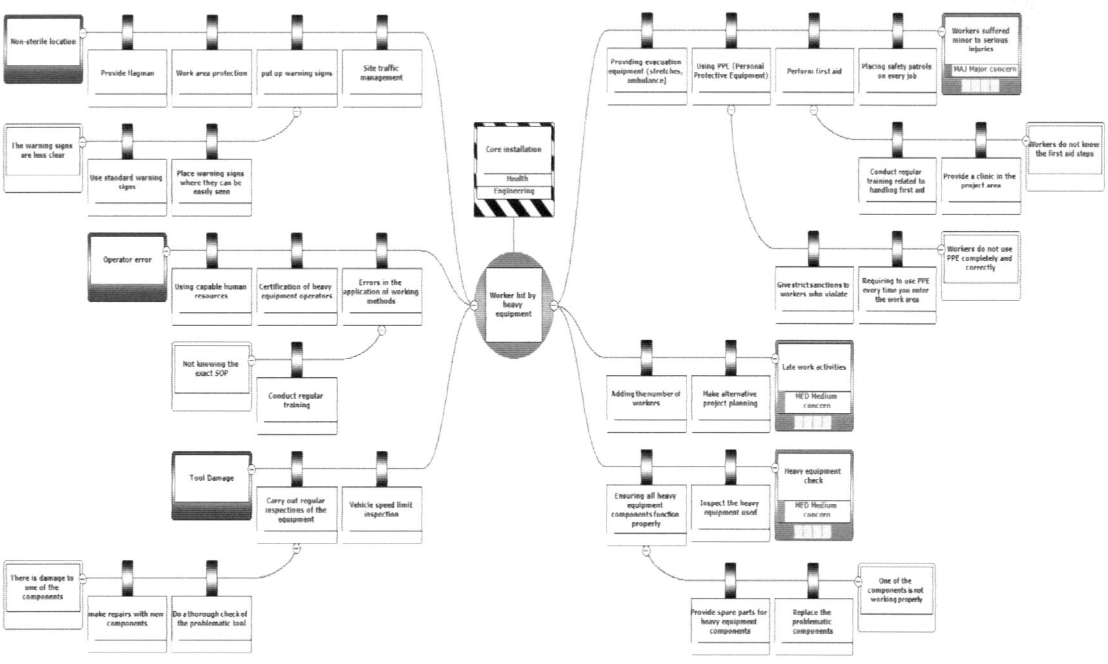

Figure 2. Bowtie diagram on the risk of workers being hit by heavy equipment

4. Conclusion

Based on the results that has been carried out in this study it can be concluded that there are 5 activities in the floating pier construction work and the extreme hazards is workers falling into the sea, that can be seen in Figure 1. Whereas in the breakwater construction work there is one dominant variable, namely workers who are hit by heavy equipment which can be seen in Figure 2.

References

[1] K. Miller and T. Hyodo, "Assessment of port efficiency within Latin America,"*Journal of Shipping and Trade*, vol. 7, no. 1, Dec. 2022, doi: 10.1186/s41072-021-00102-5.

[2] . S., Mohd. Faris Khamidi, . S., and D. Marina Chamelia, "Methodology for Investigation of Risk Based Maintenance (MIRBA) for Mobile Mooring System," *Res J Appl Sci*, vol. 13, no. 1, 2019, doi: 10.36478/rjasci.2018.26.33.

[3] A. de J. Penelas and J. C. M. Pires, "Hazop analysis in terms of safety operations processes for oil production units: A case study," *Applied Sciences (Switzerland)*, vol. 11, no. 21, 2021, doi: 10.3390/app112110210.

[4] J. Wu and M. Lind, "Management of System Complexity in HAZOP for the Oil &Gas Industry," 2018. doi: 10.1016/j.ifacol.2018.06.379.

[5] D. Arena, F. Criscione, and N. Trapani, "Risk assessment in a chemical plant with a CPN-HAZOP Tool," 2018. doi: 10.1016/j.ifacol.2018.08.487.

[6] P. Vidmar and M. Perkovič, "Safety assessment of crude oil tankers," *Saf Sci*, vol. 105, pp. 178–191, Jun. 2018, doi: 10.1016/j.ssci.2018.02.009.

[7] O. Aneziris, M. Gerbec, I. Koromila, Z. Nivolianitou, F. Pilo, and E. Salzano, "Safety guidelines and a training framework for LNG storage and bunkering at ports," *Saf Sci*, vol. 138, Jun. 2021, doi: 10.1016/j.ssci.2021.105212.

[8] N. Anwar Siddiqui, A. Nandan, A. Srivastava, R. Manohar, N. A. Siddiqui, and M. Sharma, "Risk Management Techniques HAZOP & HAZID Study." [Online]. Available: www.ohsfejournal.com

[9] J. Y. Choi and S. H. Byeon, "Hazop methodology based on the health, safety, and environment engineering," *Int J Environ Res Public Health*, vol. 17, no. 9, May 2020, doi: 10.3390/ijerph17093236.

[10] M. Shafiee and T. Adedipe, "Offshore wind decommissioning: an assessment of the risk of operations," *International Journal of Sustainable Energy*, vol. 41, no. 8, pp. 1057–1083, 2022, doi: 10.1080/14786451.2021.2024830.

[11] J. A. Suardin, A. Jeff McPhate, A. Sipkema, M. Childs, and M. S. Mannan, "Fire and explosion assessment on oil and gas floating production storage offloading (FPSO): An effective screening and comparison tool," *Process Safety and Environmental Protection*, vol. 87, no. 3, 2009, doi: 10.1016/j.psep.2008.12.002.

[12] M. Kaptan, "Risk assessment of ship anchorage handling operations using the fuzzy bow-tie method," *Ocean Engineering*, vol. 236, 2021, doi: 10.1016/j.oceaneng.2021.109500.

[13] M. Weber, "Some safety aspects on the design of sparger systems for the oxidation of organic liquids," *Process Safety Progress*, vol. 25, no. 4, 2006, doi: 10.1002/prs.10143.

[14] Sunaryo and M. A. Hamka, "Safety risks assessment on container terminal using hazard identification and risk assessment and fault tree analysis methods," in *Procedia Engineering*, 2017. doi: 10.1016/j.proeng.2017.08.150.

[15] N. R. Saisandhiya, "Hazard Identification and Risk Assessment in Petrochemical Industry," *Int J Res Appl Sci Eng Technol*, vol. 8, no. 9, 2020, doi: 10.22214/ijraset.2020.31583.

[16] B. Suhardi, P. W. Laksono, V. E. A. Ayu, J. Mohd.Rohani, and T. S. Ching, "Analysis of the potential Hazard Identification and Risk Assessment (HIRA) and Hazard Operability Study (HAZOP): Case study," *International Journal of Engineering and Technology(UAE)*, vol. 7, no. 3, 2018, doi: 10.14419/ijet.v7i3.24.17290.

[17] W. Li, "Risk analysis model of offshore engineering project management based on fuzzy membership function," *J Coast Res*, vol. 98, no. sp1, 2019, doi: 10.2112/SI98-023.1.

IOP Conf. Series: Earth and Environmental Science 1298 (2024) 012033 doi:10.1088/1755-1315/1298/1/012033

Formulation and selection of blue economy operational business strategy using Montecarlo Simulation and Multi Criteria Decision Making

N Siswanto[1*], M Rayhan[1], SI Rosdaniah[2]

[1]Department of Industrial and Systems Engineering, Institut Teknologi Sepuluh Nopember, Surabaya, East Java, Indonesia
[2]Interdisciplinary School of Management and Technology, Institut Teknologi Sepuluh Nopember, Surabaya, East Java, Indonesia

[*]Corresponding author: siswanto@ie.its.ac.id

Abstract. The Cirebon Government has provided guidelines for the establishment of mangrove ecotourism aimed at reducing the level of mangrove damage in Cirebon, which has reached 75%. One of the newly established mangrove ecotourism sites in January 2023 is the "Kasih Sayang Mangrove Ecotourism Area." This tourism destination focuses solely on the natural beauty of the mangrove ecosystem, generating revenue from entrance tickets and parking fees. Therefore, a business operational strategy is needed to increase income and the number of visitors. The strategy design is based on the Internal Factor Evaluation (IFE) Matrix, External Factor Evaluation (EFE) Matrix, Internal-External (IE) Matrix, and the General Strategy of Blue Economy, resulting in three alternative strategies: Mangrove Planting Scheme for Tourists, Paddling Attractions for Tourists, and Utilization of Mangrove Trees as Souvenirs. The selected alternative strategy, using the Monte Carlo Simulation and Multi-Criteria Decision-Making approach, is the Paddling Attractions for Tourists, with an average Net Present Value of Rp 92,210,01, an investment of Rp 544,298,000, an additional 5 local workers, and an estimated 14,598 kg of waste.
Keywords: Blue Economy, IE Matrix, Kasih Sayang Ecotourism Area, Montecarlo Simulation, Strategy

1. Introduction

The mangrove forest in Indonesia is the world's largest mangrove area, covering approximately 20% of the global mangrove ecosystem, with an area of about 3.36 million hectares [1]. Despite having the largest proportion of mangroves in the world, the mangrove area in Indonesia, especially along the Java Coast, has experienced a significant reduction of 74% from 173,000 hectares to 45,000 hectares in 2012 [2]. One of the regions that has suffered a considerable level of damage is Cirebon Regency, where six districts have experienced a 75% reduction in mangrove coverage [3]. The Cirebon Government is making efforts to conserve mangrove forests through ecotourism [4]. One ecotourism site established in early 2018 is the "Kasih Sayang Mangrove Ecotourism Area". However, the tourism site was closed again after 21 days of operation due to the lack of public facilities and insufficient funds for their construction.

Kasih Sayang Mangrove Ecotourism Area was officially inaugurated on December 29, 2022, by the Head of BBWS Cimanuk Cisanggarung as a tourist destination and opened to the public on January 1, 2023. According to information from the management chairman and Kuwu Mundu of Cirebon Regency, Kasih Sayang Mangrove Ecotourism Area is a new tourist spot, so there is a need for input regarding the business lines that can be implemented by the mangrove ecotourism. These business lines aim to generate additional income for the tourism site, which can be used for road track funding, supporting operational expenses, and addressing the fact that 77.83% of the 110 respondents have not visited Kasih Sayang Mangrove Ecotourism Area yet. Furthermore, the issue faced in Cirebon Regency is the relatively high unemployment rate, averaging at 11.52 percent, which is above the national average unemployment rate of 5.82% in 2022 [5].

This study will focus on designing an operational business strategy with the aim to clarify the planning process by specifically identifying business management. The strategy design will be approached with the concept of the blue economy, which takes into consideration economic, social, and environmental aspects. The research results will prioritize alternative strategies based on their NPV, investment amount, additional local workforce, and waste estimation.

2. Material and methods

Stages in the design and selection of a blue economy operational business strategy can be determined as below.

2.1. Strategic Formulation

Strategic formulation is the process of using existing information and data to determine business direction to achieve the company's business goals. The strategic formulation steps details as below.

2.1.1. Establish Vision and Mission

Creating vision is carried out to determine the management's goals regarding what the organization want to achieve in the long term. The creation of the vision and mission will be referred to elements of the Blue Economy which are based on the direction of national development [7][8].

2.1.2. Creating IFE Matrix and EFE Matrix

Based on previous research, observation results, interview results with manager based on strategic management criteria and blue economy indicators [6] [8].

2.1.3. Calculation IFE Matrix and EFE Matrix Values

The assessment for each component from manager Kasih Sayang Mangrove Ecotourism Area with the EFE Matrix rules is 4 = very good, 3 = above average, 2 = average, 1 = below average and the IFE Matrix is 4 = very strong 3 = strong, 2 = weak and 1 = very weak [6]. The weight value in each matrix uses the FUCOM (Full Consistency Method) method using a mathematical model [9].

$$\left|\frac{w_k}{w_{(k+1)}} - \varphi_{k/(k+1)}\right| \leq \chi, \forall j \tag{1}$$

$$\left|\frac{w_k}{w_{(k+2)}} - \varphi_{k/(k+1)} \ x \ \varphi_{(k+1)/(k+2)}\right| \leq \chi, \forall j \tag{2}$$

$$\sum_{j=1}^{n} w_j = 1, \forall j \tag{3}$$

$$w_j \geq 0, \forall j \tag{4}$$

Strategy determination is based on visi and mission Kasih Sayang Mangrove Ecotourism Area, IFE Matrix, EFE Matrix, IE Matriks, blue economy indicators, blue economy general strategy, dan blue economy strategy previous research [8][10][11].

2.2. Calculation of Strategic Economic Benefits

2.2.1. Cost Calculation

The cost calculation is adjusted to each strategy alternative with the components of raw material costs, labor costs, electricity costs, and marketing costs.

2.2.2. Revenue Calculation

Revenue from services or goods in each alternative strategy is obtained by the target market multiplied by the selling price of services or goods.

2.2.3. Feasibility Study

The variables used as a feasibility study are NPV and IRR with a Monte Carlo Simulation approach.

2.3. Determination of the Best Alternative Strategy

Determination of the criteria for selecting the best alternative strategy will be based on Blue Economy indicators which pay attention to 3 aspects, namely economic, social and environmental as well as ecotourism assessment indicators which are based on the Decree of the Minister of Tourism and Creative Economy Number 9 of 2021 which was adopted from the Sustainable Tourism Council. Determination of the best alternative strategy using the Topsis Method with the following steps.

1) Create a Normalized Decision Matrix

$$r_{ij} = \frac{x_{ij}}{\sqrt{\sum_{i=1}^{m} x^2 ij}} \quad (5)$$

r = Normalized matrix
m = Alternative
x_{ij} = Alternative i and criteria j

2) Create a Weighted Normalized Decision Matrix

$$y_{ij} = w_i \times r_{ij} \quad (6)$$

y_{ij} = Weighted normalized matrix of alternative i and criteria j
w_i = Alternative weights for i
r_{ij} = Alternative normalized matrix i dan criteria j

3) Determine Positive Ideal Solution Matrix & Negative Ideal Solution Matrix

$$A^+ = (y_1^+, y_2^+, \ldots, y_n^+) \quad (7)$$
$$A^- = (y_1^+, y_2^+, \ldots, y_n^+) \quad (8)$$

A^+ = Positive ideal solution
A^- = Negative ideal solution

4) Determine the Distance Between the Values of Each Alternative with the Positive Ideal Solution Matrix & Negative Ideal Solution Matrix

$$D_i^+ = \sqrt{\sum_{j=1}^{n} (y_1^+ - y_{ij})^2} \quad (9)$$

$$D_i^- = \sqrt{\sum_{j=1}^{n} (y_{ij} - y_1^-)^2} \quad (10)$$

D_i^+ = Positive ideal distance solution
D_i^- = Negative ideal distance solution

5) Determine Preference Value for Each Alternative

$$V_i = \frac{D_i^-}{D_i^- + D_i^+} \quad (11)$$

V_i = Assess alternative preferences i

6) Perform Preference Order Ranking

3. Result

3.1. Strategic Formulation

Formulation of strategies for the Kasih Sayang Mangrove Zone Tourism as stated below.

3.1.1. Vision and Mission Kasih Sayang Mangrove Ecotourism Area

The tourism vision of the compassionate mangrove zone is "The Compassionate Mangrove Zone as the Leading and Cheapest Tourism Destination Based on Community Education" with the mission of being committed to providing satisfaction to customers by maintaining quality of service, the use of technology in tourism operations, the existence of community participation in seeding and planting mangrove trees, and building partnerships with academics, government, and companies.

3.1.2. IFE Matrix and EFE Matrix Kasih Sayang Mangrove Ecotourism Area

Components, weight assessments and ratings of the IFE Matrix and EFE Matrix Compassion for Mangrove Ecotourism Area are listed in table and table 2.

Table 1. IFE Matrix Kasih Sayang Mangrove Ecotourism Area

Key Internal Factors	Weight	Rating	Weighted Score
Strength			
Being a relatively cheap mangrove tour in Cirebon	0.160	4	0.64
Appearance of 4 tourist icons for tourist trademarks	0.081	3	0.243
Decent public facilities	0.037	3	0.111
Has strategic location on the northern coast route	0.085	4	0.34
Involvement of local communities in tourism management	0.090	4	0.36
Profit sharing system on parking fees	0.040	3	0.12
Weakness			
The use of technology in operations is still use inactively	0.075	1	0.075
Padding facilities for tourists are not exist	0.055	1	0.055
Business line for planting mangroves is not made yet	0.130	1	0.13
Appearance of tourist souvenir center	0.094	1	0.094
Road access is inadequate	0.043	1	0.043
Lack of promotion for tourist attractions	0.060	1	0.06
Unstructured vision, mission, and organization on formal context	0.053	2	0.106
Total			2.38

Table 2. EFE Matrix Kasih Sayang Mangrove Ecotourism Area

Key Internal Factors	Weight	Rating	Weighted Score
Opportunities			
Already have clarity on the status of land and tourism ownership	0.056	2	0.112
Financial assistance from the government for tourism operations	0.046	1	0.046
There is an increase in siltation of mud which has the potential as a business opportunity for mangrove trees	0.120	2	0.24
Ample land is available for development	0.160	3	0.48
Support from the community regarding the business development of tourist attractions	0.097	3	0.291
Presence of similar mangrove tourism in Cirebon is rare	0.130	3	0.39
The increae of West Java population who own or control cell phones as of 2021 by 4.6%	0.045	3	0.135
Threat			
Level of competition from similar mangrove tourism in Cirebon	0.080	2	0.16
Lack of awareness of visitors in protecting the tourist environment	0.060	2	0.12
Damage to mangroves as a tourist icon caused by natural conditions	0.200	3	0.6
Total			2.57

3.1.3. IFE Matrix and EFE Matrix Kasih Sayang Mangrove Ecotourism Area

The results obtained from the IFE Matrix & EFE Matrix are in the V quadrant with the recommended strategy being product development and market penetration. Based on these results, three alternative strategies were formed, such as mangrove planting business scheme for tourists (alternative strategy 1), paddling rides for tourists (alternative strategy 2) and the use of mangrove trees as tourist souvenirs (alternative strategy 3). The tourist souvenir chosen by 110 respondents for the Kasih Sayang Mangrove Tourism was mangrove syrup with the result being 52% mangrove syrup, 28% dodol pidada and 20% did not buy tourist souvenirs.

3.2. Strategic Formulation

3.2.1. Master Plan dan Forecasting Demand Masing – Masing Alternatif

Table 3. Strategy Alternative Master Plan

Year	2024	2025	2026	2027	2028
Population	1,727,941	1,752,864	1,777,787	1,802,710	1,827,633
Target Market of Alternative Strategy Visit 1			4.47%		
Target Market of Alternative Strategy 1			8%		
Target Market of Alternative Strategy Visit 2			4.65%		
Target Market of Alternative Strategy 2			18.5%		
Target Market of Alternative Strategy Visit 3			4.34%		
Target Market of Alternative Strategy 3			18%		

3.2.2. IRR (Internal Rate of Return) of Each Alternative Strategy

Based on the target market and demand forecasting for each alternative, the results of the IRR (Internal Rate of Return) for each alternative strategy are stated below.

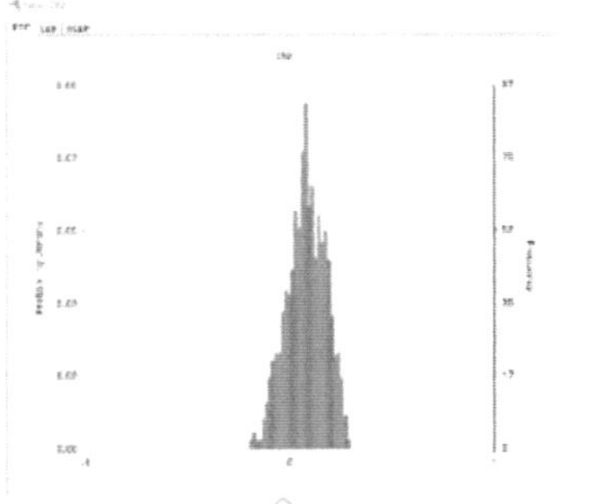

Figure 1. IRR Alternative Strategy 1

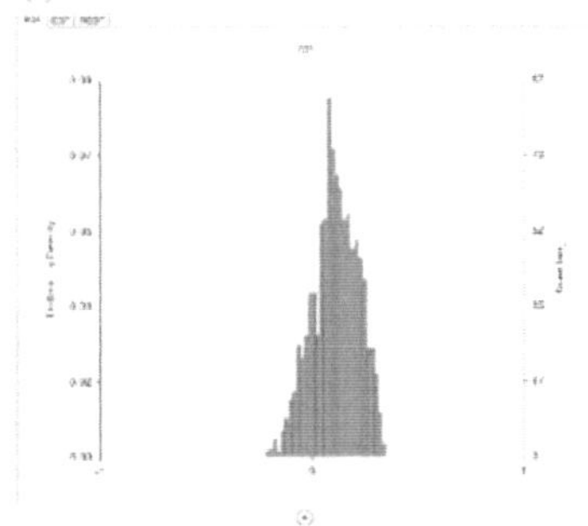

Figure 2. IRR Alternative Strategy 2

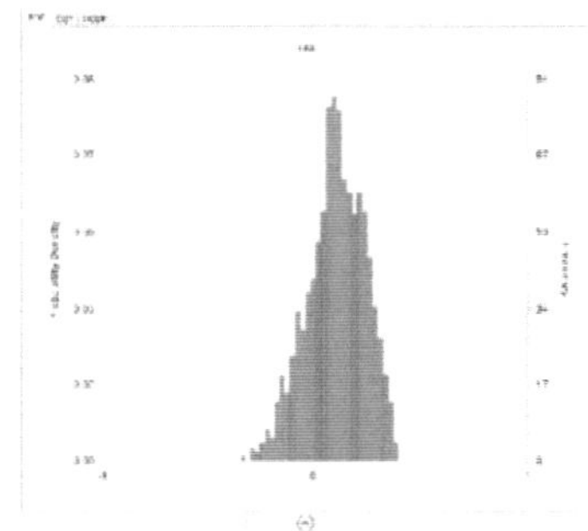

Figure 3. IRR Alternative Strategy 3

Based on figures 1, 2, and 3 with 1,000 simulation iterations, the average IRR for strategy alternatives 1, 2, 3 respectively is 7.8%; 11%; 10.8%.

3.2.3. NPV (Net Present Value) of Each Alternative Strategy

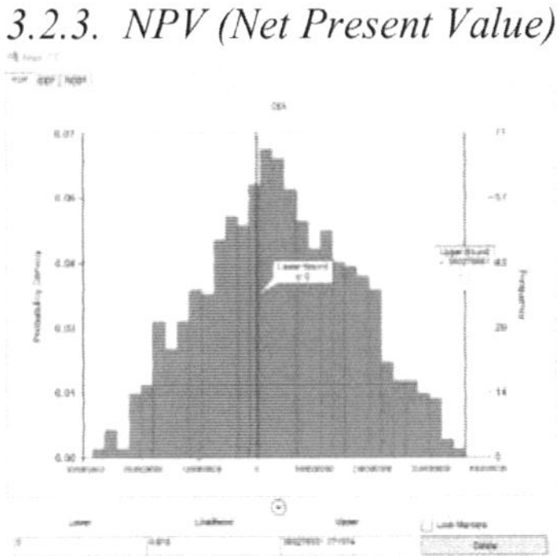

Figure 4. NPV Alternative Strategy 1

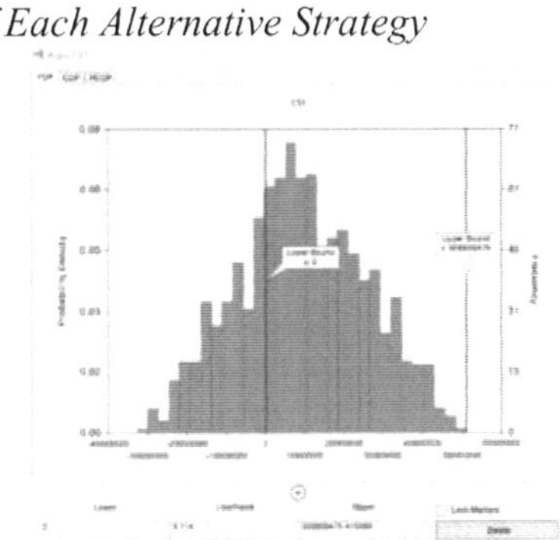

Figure 5. NPV Alternative Strategy 2

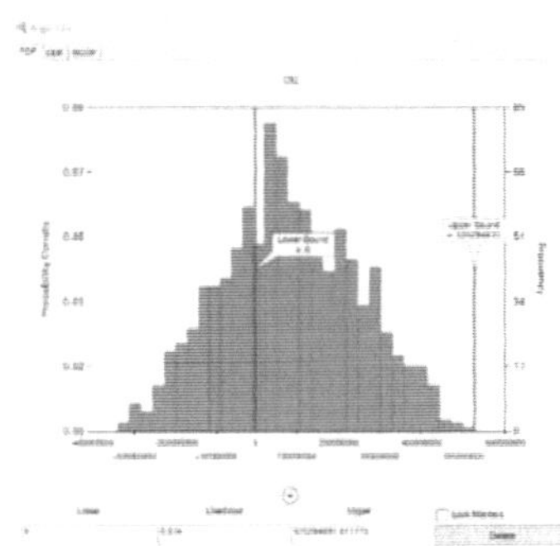

Figure 6. NPV Alternative Strategy 3

Based on figures 1, 2, and 3 with 1,000 simulation iterations, the average NPV of alternative strategies 1, 2, 3 respectively is IDR 38,765,646; IDR 92,210,018; IDR 77,539,201

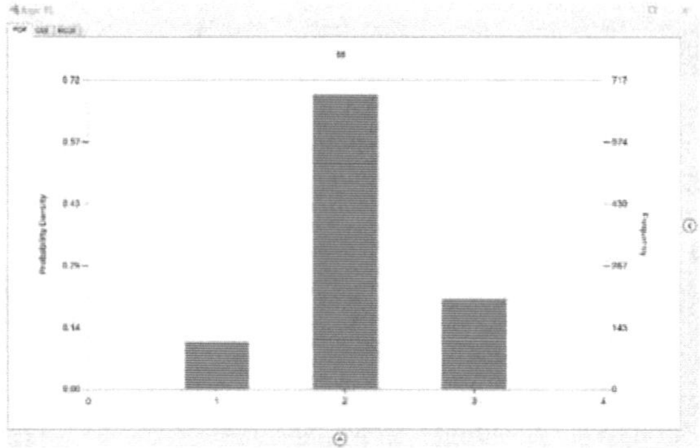

Figure 7. Winning Percentage Alternative Strategi

3.3. Selection of Alternative Strategies Using Multi Criteria Decision-Making

3.3.1. Alternative Strategy Criteria

Criteria for alternative strategies are listed in table 4.

Table 4. Alternative Strategy Criteria

Blue Economy Aspect	Assessment Criteria	Type of Criteria	Characteristics
Economic	Net Present Value	Benefit	higher is better
	Investment Amount	Cost	lower is better
Social	Number of Field Workers	Benefit	higher is better
Environment	Garbage Waste	Cost	lower is better

3.3.2. Weighting Criteria for K-Decision

Weighting method for criteria uses Full Consistency Method with respondents from 1 tourism manager and 5 tourists. The weighting results for the criteria are listed in table 5.

Table 5. Mean Criteria for K-Decision

Criteria	K1	K2	K3	K4	K5	K6	Mean
C1	0.13	0.27	0.29	0.2	0.12	0.24	0.21
C2	0.3	0.43	0.27	0.28	0.16	0.2	0.27
C3	0.35	0.13	0.23	0.22	0.24	0.17	0.22
C4	0.22	0.18	0.22	0.3	0.48	0.4	0.30
Total	1	1	1	1	1	1	1

3.3.3. Normalized Matrix

Normalized matrix for selecting alternative operational business strategies is shown in table 6.

Table 6. Alternative Strategy with Normalized Matrix

Alternative Strategy	K1	K2	K3	K4
Alternative Strategy 1	38,765,646	496,358,000	4	13,229
Alternative Strategy 2	92,210,018	544,298,000	5	14,598
Alternative Strategy 3	77,539,201	502,252,490	6	18,673
R1	0.306	0.557	0.422	0.487
R2	0.729	0.610	0.527	0.538
R3	0.613	0.563	0.738	0.688

3.3.4. Weighted Normalized Matrix

Weighted Normalized Matrix is shown in table 7.

Table 7. Weighted Normalized Matrix Alternatif Strategi

V	NPV	Investment Amount	Number of Field Workers	Garbage Waste
V1	0.06	0.15	0.09	0.14
V2	0.15	0.16	0.11	0.16
V3	0.12	0.15	0.16	0.2

3.3.5. Distance Measures Positive and Negative

Distance measures positive are listed in table 8 and distance measures negative are shown in table 9.

Table 8. Distance Measures Positive

Di	NPV	Investment Amount	Number of Field Workers	Garbage Waste	Di^+
$D1^+$	0.000	0.008	0.005	0.005	0.000
$D2^+$	0.000	0.000	0.002	0.002	0.000
$D3^+$	0.003	0.001	0.000	0.000	0.004

Table 9. Distance Measures N

Di	NPV	Investment Amount	Number of Field Workers	Garbage Waste	Di^+
$D1^-$	0.002	0.000	0.000	0.000	0.004
$D2^-$	0.003	0.008	0.000	0.001	0.002
$D3^-$	0.000	0.004	0.000	0.005	0.000

3.3.6. Relative Closeness to Ideal Solution

The results of alternative strategies ranking are shown in table 10.

Table 10. Ci-Value for Each Alternative Strategy

Alternatif Strategi	Ci
Alternatif Strategi 1	0.35
Alternatif Strategi 2	0.66
Alternatif Strategi 3	0.59

4. Discussion

The initial stage for strategy formulation is to create a tourism vision and mission with the vision created fulfilling SMART rules (Specific, Measurable, Achievable, Relevant, and Time Bound). This vision has answered specific aspects because it has answered things that need to be done. The reasons for this step are important is who involved to achieve these goals and the place of business activities; in the measurable aspect such as the measurement is clear as well as excellence can be measured from the number of tourists and cheapest in terms of entrance tickets; on the measurable achievable aspect of the existence of human resources who are experts in mangrove tourism. The relevant aspect is providing education to the community regarding the benefits of mangroves, mangrove seeding and planting, able to create jobs for the surrounding community; in the time bound aspect such as the community education aspect can schedule the opening of mangrove planting with the community according to the date you want to set.

The value of the IFE Matrix and EFE Matrix with the weighting scale and rating by the tourism management of the Compassionate Mangrove Zone are 2.38 and 2.57 respectively, so that the existence of the Kasih Sayang Mangrove Ecotourism Area is in quadrant 5 with strategies that can be carried out are Market Penetration and Product Development. An alternative recommendation for the Kasih Sayang Mangrove Ecotourism Area is a mangrove planting business scheme for tourists (alternative strategy 1), paddling vehicle for tourists (alternative strategy 2), and utilization of mangrove trees as tourist souvenirs (alternative strategy 3).

The average IRR value for each strategy alternative is above the average inflation rate for the past 10 years with 5.5%, so it can be concluded that strategy alternatives 1, 2, and 3 are feasible to implement. The average NPV of strategy alternatives 1, 2, and 3 respectively is IDR 38,765,646; IDR 92,210,018; IDR 77,539,201 and the opportunity when the NPV is more than 0 alternative strategies 1, 2, and 3 in a row is 61.8%; 71.4%; 67.4%. Results of the simulation carried out are alternative strategy 1 has a winning percentage rate of 10.8%. Strategy alternative 2 has a winning percentage rate for being selected of 68.3% and alternative strategy 3 has a winning percentage rate for being selected of 20.9%. Based on these results, the alternative strategy 2 is the chosen alternative strategy using only the NPV factor in its determination.

Impact of alternative strategies based on investment amount, the number of local workers, and waste for alternative strategy 1 has investment amount of IDR 496,358,000; additional number of local workers by 4; the average garbage waste is 13.229 kg. Furthermore, the alternative strategy 2 has investment amount of IDR 544,298,000; additional number of local workers by 5; the average garbage waste is 14,598 kg. Also, the alternative strategy 3 has investment amount of IDR 502,252,490; additional number of local workers by 7; the average garbage waste is 18.763 kg. Based on the input obtained and processing using the Multi Criteria Decision Making method, the TOPSIS method obtained the sequence are alternative strategy 2, alternative strategy 3, and alternative strategy 1. Based on two approaches, using Monte Carlo simulation and Multi Criteria Decision Making, both results are same.

5. Conclusion

The alternative tourism strategy for the Kasih Sayang Mangrove Ecotourism Area is the mangrove planting business scheme for tourists, paddling rides for tourists, and utilization of mangrove trees as tourist souvenirs. The selected alternative strategy using the Monte Carlo Simulation approach and Multi Criteria Decision Making are paddling rides for tourists with an average NPV of IDR 92,210,018; investment amount of IDR 544,298,000; the additional number of local workers is 5 and the estimated amount of garbage waste is 14,598 kg.

Acknowledgement:

This work was supported by the Directorate of Research, Technology, and Community Service; Ministry of Education, Culture, Research, and Technology; the Republic of Indonesia under Grant Number 1259/PKS/ITS/2023.

References

[1] Mudiyarso. D.. Arifanti. V.B.. Mulia. F.. & Windardi. A.C. (2020). Retrieved in April 2023 from http://pojokiklim.menlhk.go.id/read/ekosistem-mangrove-dan-aksi-iklim.

[2] Ilman. M.. Dargusch. P.. Dart. P.. & Onrizal. (2016). A Historical Analysis of Drivers of Loss and Degradation of Indonesia's Mangroves. *Land Use Policy*.448-450.

[3] Raharjo. P.. Setiady. D.. Zallesa. S.. & Putri. E. (2015). *Identifikasi Kerusakan Pesisir Akibat Konversi Hutan Bakau (mangrove) Menjadi Lahan Tambak di Kawasan Pesisir Kabupaten Cirebon* [Identification of Coastal Damage Due to Conversion of Mangrove Forests into Fishponds in the Coastal Area of Cirebon Regency]. *Jurnal Geologi Kelautan*. 9-24.

[4] Amelia. F. (2021). *https://kabarcirebon.pikiran-rakyat.com*. Retrieved in April 2023 from https://kabarcirebon.pikiran-rakyat.com/saba-desa/pr-2936055526/konservasi-hutan-mangrove

[5] Rakyat Cirebon. (2022). *https://rakyatcirebon.disway.id*. Retrieved in April 2023 from https://rakyatcirebon.disway.id/read/636502/angka-pengangguran-di-kabupaten-cirebon-tembus-1152-persen

[6] David. F. R.. & David. F. R. (2017). *Strategic Management concepts and cases A Competitive Advantage Approach* (Sixtennth ed.). England: Pearson Education Limited.

[7] Dewan Kelautan Nasional. (2012). *Kebijakan Ekonomi Kelautan dengan Model Ekonomi Biru*. [Marine Economic Policy with the Blue Economy Model]. Jakarta: Ministry of Maritime Affairs and Fisheries.

[8] Akhir. K.. Araujo. B. D.. & Farand. F. S. (2021). Designing and Mapping the Blue Economy Company Index (BECdex) to the Sustainable Development Goals (SDGs) for Maritime Companies in the Coastal States. *International Journal of Innovative Science and Research Technology*. *VI*(9). 417 - 426.

[9] Fazlollahtabar, H., Smailbašić, A., & Stević, Ž. (2019). FUCOM method in group decision-

making: Selection of forklift in a warehouse. Decision Making: *Applications in Management and Engineering*, 2(1), 49-65.

[10] PANN Maritim. (2022). Retrieved from pannmaritim.com: https://pannmaritim.com/strategi-ekonomi-biru-indonesia/

[11] Bidayani. E.. Soemarno. Harahab. N.. & Rudianto. (2016). Model Components of Mangrove Resource Management Based on Blue Economy Concept. *Economic and Social of Fisheries and Marine*. 1-15.

Risk Management in the LPG Refrigerated Green Terminal Project for Sustainable Development Goals

D Satrio[1,*], M A Anjelina[1], D M Rosyid[1], H Ikhwani[1], W N Fadilah[2], M A R Fauzi[3] and A Ridlwan[3]

[1] Institut Teknologi Sepuluh Nopember, Surabaya 60111, Indonesia
[2] Politeknik Elektronika Negeri Surabaya, Surabaya 60111, Indonesia
[3] Institut Teknologi Sumatera, Lampung 35365, Indonesia

*E-mail address: dendy.satrio@its.ac.id

Abstract. The LPG Refrigerated Green Terminal Project for Sustainable Development Goals seeks to enhance national stock resilience and reduce reliance on leased refrigerated LPG floating storage by a State-Owned Oil and Gas Company. Various challenges were encountered throughout the project, including resident demonstrations causing schedule disruptions, delays in project design due to client approval processes, and disparities in land data during tendering and implementation phases, among others. This research aims to identify the most critical risk by assessing the number of priority risks and risk categories. Effective risk management is pivotal in achieving project objectives. The methodology employed encompasses Fault Mode and Effect Analysis (FMEA), a 3D Risk Matrix, and a Pareto Diagram, supplemented by Analytical Hierarchy Process (AHP) software for risk weight determination. The study reveals a minimum Risk Priority Number (RPN) of 6, signifying low-risk severity, infrequent risk occurrence, and the company's ability to detect these risks. Conversely, the highest RPN is recorded at 36, indicating significant risk severity, frequent risk occurrence, and challenges in risk detection by the company.

1. Introduction

The LPG Refrigerated Green Terminal Project for Sustainable Development Goals encountered various challenges, for instance, disruptions to the schedule caused by protests from local residents, delays in project design due to client approval document holdups, and discrepancies in land data between the tender phase and project execution, among other issues.

Within the dynamic process of construction projects, various risks can impact productivity, performance, quality, and cost parameters [1]. Typically, these risks fall upon the project owner unless transferred or assumed by another party. Consequently, risk management is imperative for organizations as it plays a pivotal role in achieving project objectives [2]. A highly effective approach for assessing the reliability of products or processes is Failure Mode and Effect Analysis (FMEA), which serves as a valuable tool for identifying and prioritizing significant risks [3,4]. Thus, proactive measures are essential for pre-emptively addressing issues before they materialize.

Risks can be handled in a way that is reduced, transferred to other responsible activities or parties, absorbed, and avoided by implementing quality control practices and procedures [1]. ISO 31000 is a generic risk management standard. This standard has a broader and conceptual perspective [2]. This standard aims to guide industries to manage different types of risk from different industries with different organizational sizes, types, complexity, structures, activities, and locations [3]. This standard can be implemented in several fields, such as in power plants [4–10]. Research on the oil and gas sector still needs to be developed even though the ocean renewable energy sector is being researched a lot [11–15].

This research was conducted to identify, analyse, and evaluate risks to be given treatment to reduce the impact of risks regarding ISO 31000:2018. The risk obtained will be assessed using the FMEA method to compute the Risk Priority Number (RPN) and the risk matrix to categorize the existing risk. Subsequently, the most critical risk will be chosen through the Pareto chart approach. In order to mitigate potential consequence, preventive measure will be applied to the identified risks.

2. Material and Methods

2.1. The LPG Refrigerated Green Terminal Project for Sustainable Development Goals

As mentioned earlier, the project's objectives are enhancing national stock resilience and reducing dependencies on leasing the Refrigerated LPG floating storage in Remen Village, Jenu District, Tuban Regency, East Java. This project has a scope of work covering civil with the scope of fencing work, Double Wall Tank (DWT) tank foundation work, and spherical tank foundation work. The scope of mechanics includes DWT tank erection work, accessories, spherical tank work, and accessories. Also, the scope of testing and commissioning includes hydro test work, NDT work, mechanical test work, and Nitrogen purging work with a contractual duration of 30 months from October 2019 to March 2022.

This study obtained data from the October 2020 – July 2021 risk register. From October 2020 to July 2021, the tasks executed in this project encompassed the assembly and welding of both the external and internal shells, the construction of the roof, the elevation of the roof's air components, and the installation of the base for Double Wall Tanks T-201 and T-301. Additionally, the project involved the erection, inspection, and testing activities associated with Spherical Tanks V-101 and V-102.

Figure 1. Propane Double Wall Tank T-201.

Figure 2. Butane Double Wall Tank T-301.

Figure 3. Spherical Tank V-101 and V-102.

2.2. ISO 31000:2018

This study cites ISO 31000:2018 as it is a versatile standard applicable across industries. ISO 31000 is designed not to standardize risk management practices among companies but to offer supportive standards for implementing risk management that align with company objectives. The risk management process involves methodically applying policies, procedures, and practices to various activities, including communication, consultation, contextual setting, risk assessment, management, review, documentation, monitoring, and reporting [19,20]. This process is illustrated in Figure 2.

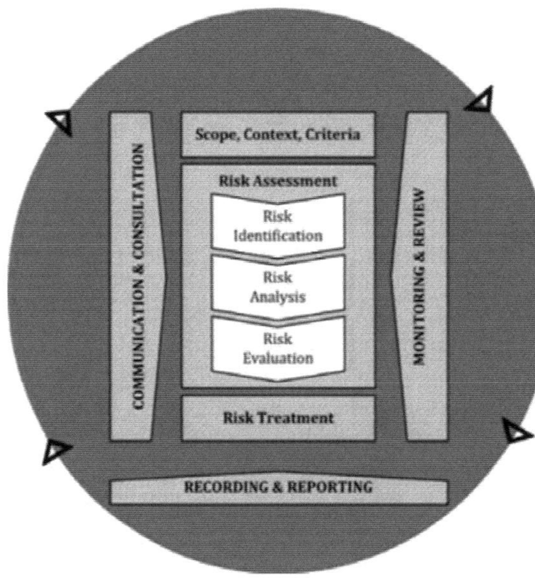

Figure 4. Risk Management Process.

In Figure 4, it is known that the process of performing risk management has several stages, namely:

1. Communication and consultation
2. Scope, context, and criteria
3. Risk assessment

   The risk evaluation needs to be conducted in a methodical and cooperative manner, based upon the insights and perspectives of stakeholders.

   - Risk identification

     Hazard distinguishing proof points to discover, recognize, and depict dangers that might help or avoid the organization from accomplishing its destinations. Significant, exact, and up-to-date data is basic in distinguishing dangers. Organizations can utilize different methods to recognize instabilities influencing one or more goals.

   - Risk analysis

     The reason for the chance investigation is to get the nature of the hazard and its characteristics. Hazard examination includes considering vulnerability, sources of chance, results, probability, occasions, scenarios, controls, and their viability. The detail and complexity of the chance investigation can vary depending on the reason for the examination, the accessibility and unwavering quality of the data, and the accessible assets. Expository strategies can be carried out qualitatively, quantitatively or a combination of those, depending on the circumstances and the objectives.

   - Risk evaluation

     Risk assessment aims to facilitate decision-making by comparing the outcomes of risk analysis to predefined risk criteria, identifying areas requiring further action.

4. Risk treatment

   Chance treatment points to choose and actualize alternatives for overseeing the chance. Chance treatment includes an iterative prepare from defining and selecting chance treatment choices, arranging, and executing hazard treatment, evaluating the adequacy of given chance treatment, and choosing whether the remaining chance is satisfactory.

5. Monitoring and review
   The reason of observing and survey is to guarantee and progress the quality and viability of handle plan, execution, and comes about. Continuous observing and intermittent survey of the hazard administration prepare, and it comes about ought to be a arranged portion of the chance administration handle, with clearly characterized obligations. Observing and audit ought to be carried out at all prepare stages, counting arranging, gathering, and dissecting data, recording comes about, and giving criticism. The comes about of observing and survey ought to be consolidated into all administration exercises, estimations, and announcing of the organization's execution.
6. Recording and reporting
   The hazard administration handle and it comes about ought to be recorded and detailed through fitting components. Recording and detailing point to communicate hazard administration exercises and comes about over the organization, give data for choice making, progress chance administration exercises, and help partner intuitive.

2.3. Fault Mode and Effect Analysis (FMEA)

FMEA was developed by the aerospace industry in the 1960s and is known as a design methodology with proven reliability and strict safety requirements. Ford implemented FMEA for production, design, safety and regulatory compliance in the automotive industry by the late 1970s. As a result, FMEA is further adopted by many industrial practices, such as aerospace, military, automobile, electrical, mechanical, and semiconductor [16].

The primary aim of FMEA is to proactively prevent the potential failure of a new design, process, or system in meeting specified requirements within defined conditions, including objectives and constraints. Additionally, FMEA aims to enhance design and testing methodologies to eliminate failures, ultimately leading to the development of highly competitive products [3]. In the case of Process FMEA, it identifies and mitigates failures, their effects, and associated risks within a process or product [21]. FMEA involves two distinct stages, which are:

1. Evaluating the risk of failure
   There are three determining factors to evaluate the risk and consequences of the failure:
   - Severity : The extent of consequences resulting from a failure
   - Occurrence : The likelihood or frequency of failure
   - Detection : The probability of identifying a failure before the consequences are realized
2. Assessing the Risk Priority Number (RPN)
   Perform the calculation of the Risk Priority Number using the following formula:

$$RPN = Severity \times Occurence \times Detection \quad (1)$$

2.4. 3D risk matrix

In a two-dimensional risk analysis, the risk is assigned an indicative level for Likelihood, Cost of Impact or Time of Impact. The higher the product or letter combination, the more attention is paid to the risk; thus, mitigation measures are designed. However, the risk is more complex in the real world, and two-dimensional risk analysis can only be used as a steppingstone.

The 3D risk matrix makes it possible to consider risk in the space formed between the coordinates, which is pyramidal in shape. This pyramid is formed from the edges of the cube (coordinates) and the inner surface between the edges, allowing the use of decimal numbers for the magnitude of Likelihood, Cost of Impact, or Time of Impact [17].

2.5. Pareto chart

As stated by Vilfredo Pareto in 1895, the Pareto principle states, 'In any set of elements controlled, small fraction chosen, in terms of the number of elements, always accounts for the large fraction in terms of effect'. This principle proposes that approximately 80% of the total impact of errors in any industry scenario stems from a small number of error types, referred to as 'vital few'. While the remaining 20% of impacts come from other types of errors, known as 'many trivial'. This concept was later popularized as "Pareto Analysis" [18].

Pareto analysis is instrumental in the initial phase of Six Sigma analysis, helping identify the most critical faults among the 'vital few.' These 'vital few' errors are investigated to determine their root causes, enabling proactive measures to prevent or mitigate their impact. Pareto analysis is pivotal in guiding quality problem-solving efforts and identifying crucial success factors for Six Sigma implementation. It is essential to emphasize that the utility of Pareto analysis results hinges on the data's source being a stable process.

The stages in conducting a risk evaluation using the Pareto Chart are as follows [3]:

1. Determine the level of importance between risks. The level of risk importance is carried out by comparing one risk to another.
2. Then, the assessment results of the importance of this risk are entered into the AHP software to produce output in the form of risk weights. This software is a program that is used as a tool to assist decision-making by AHP method. By structuring a hierarchy of various considerations to develop weights or priorities, The AHP method is effective to solve complex problems [19].
3. Next, the risk value is calculated by multiplying the Risk Priority Number with the previous risk weight.
4. Then, the risk value is divided by the total risk value to obtain the percentage of each risk.
5. Calculating accumulated risk is carried out by adding the percentage of the risk with the previously accumulated risk.
6. According to Pareto Chart theory, 20% of the total risk will be selected, which will provide 80% of the impact so that the risk can then be selected for risk treatment.

2.6. Risk treatment

Risk avoidance or prevention can be employed as an internal control for the identified risk. Risk avoidance is commonly enforced when the risk is deemed crucial enough. It is done by proactively avoiding actions that will lead to the identified risk. Evaluation and monitoring of the application of the chosen risk management technique are critical to determining the effectiveness of the option [20].

3. Results and Discussion

3.1. Risk identification

In this stage, risk identification is carried out to identify, find, and describe risks in the LPG Refrigerated Green Terminal Project for Sustainable Development Goals. This risk identification process is carried out by the company listed in the risk register. The risk register used in this study is a risk register from October 2020 - July 2021. The data obtained from the risk register is a list of risks, causes, and impacts of risk, a severity, occurrence, and detection rating. The company determines severity, occurrence, and detection rating by referring to the risk criteria determined by the company.

3.2. Risk assessment

Risk analysis seeks to determine the nature and characteristics of the individual risk. In this stage, two methods will be used: FMEA and 3D Risk Matrix. FMEA is used to find the RPN obtained from probability, consequences, and detection. The calculation results of the RPN are shown in Table 1.

Table 1. Calculation Results using the FMEA method.

Risk Number	Work Area	Risk	Causes	Impact	S	O	D	RPN
1	Construction	Potential disadvantages of fencing work	The value of the budget plan on fence work only accommodates permanent fences	More than fence work cost	1	3	2	6
2	Procurement	Inefficiencies from tank work	The price of the heating system when the tender offer uses a vendor outside the scope of the S-Tank	Inefficiency in procuring heating system items on the tank foundation	4	3	3	36
3	Quality Control (QC)	Inefficiencies in hydrotest tanks	Rejection of the tank hydrotest method using seawater	Additional costs for hydrotest water procurement	4	3	2	24
4	Engineering	Delays in overall engineering design projects	1. Delay of owner document approval 2. Late submission of vendor documents 3. There is a design change (addition or subtraction) from the client 4. The client does not use a good documentation system 5. The client does not have a dedicated team 6. Incomplete information and documents from the client	1. Delay of an issuance P.O for equipment/material 2. Delay of delivery timing	4	2	3	24
5	Procurement	Loss of procurement costs	1. Increase in the price of materials/equipment (import & local) and transportation services 2. Implementation of multi-year contracts	Additional procurement costs (inefficiency)	4	2	3	24
6	Finance (Business)	Increasing plate material prices	Exchange rate gap	Addition of the cost of procurement of tank plate material	4	3	3	36
7	Construction	Mechanical completion delays due to the rainy season	The civil work's/welding work's timing overlaps during the rainy season	1. Delayed schedule 2. Indirect cost increase	4	3	3	36
8	Construction	Lack of construction equipment and experienced local labor	1. Subcontractor has no experience in Refrigerated Tank 2. Lacking finance	1. Misscommunication with third country laborers may cause schedule delays or poor work quality 2. Impact on EPC Contractors' schedule due to the limited construction equipment 3. Difficulties on material procurement may affect the timeline on the construction work	4	3	1	12

9	Construction	Hydrotest is blocked	1. The water flow for the hydrotest is not enough 2. Water quality does not meet the requirements	Late hydrotest progress	3	2	2	12
10	HSE	Work accident when climbing the vertical ladder	During maintenance, use a vertical ladder	Stop Working Action	4	1	2	8
11	Construction	Delay on tank schedule	1. There is a difference in land data during the tender phase with the project implementation (there is a cavity in the actual conditions in the field) 2. From historical data, projects under owner Marketing, Extension of Time submissions for additional time exclude project management cost claims	1. Over five months of indirect costs & equipment standby costs, claim Subcontractor 2. Delay schedule mechanical	4	2	2	16
12	Construction	Terminated project activity	There is a demonstration from residents regarding the activity of disposing of water from the hydrotest into the sea	There is a delay from the predetermined schedule	3	3	2	18
13	Construction	Request for Maintenance/Preservation Minutes of Work Handover 2	1. Reference from the other project commissioning regarding interfaces with other contractors' work that has already Minutes of Work Handover 2 2. Interface with other contractors (in the case of phase 2 project work) 3. Nitrogen for more than one year is feared there will be a reduction in volume or loss	1. Addition of indirect costs after the treatment period is over 2. There is an increase in the cost of nitrogen filling requests after Minutes of Work Handover 2	4	2	2	16

The FMEA will measure any risk associated with the failure of critical process inputs but will not indicate what level should be defined as an acceptable or assumed risk. Therefore, a risk matrix is used to determine the risk level from the occurrence frequency and consequences arising from the risk.

In this study, a 3D risk matrix is used because a 2-dimensional risk matrix is insufficient to describe the risk's actual complexity. There are three components in the 3D risk matrix, namely severity, occurrence, and detection. Each component is on a different axis, namely the x, y, and z axes.

The color of the risk category in the risk matrix is determined by calculating the first and third quartile of the obtained RPN. Data in the first quartile indicate that the risk is included in the low category, while the risk in the third quartile indicates that the risk is included in the high-risk category [21].

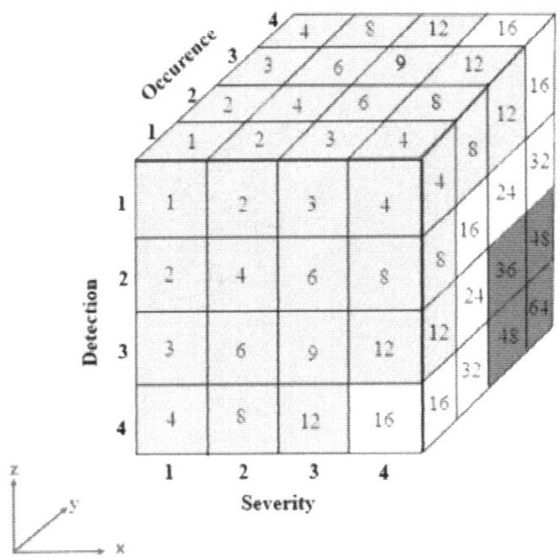

Figure 5. 3D Risk Matrix.

Based on Table 1 and Figure 3, the results of the analysis are as follows:

1. Risk with Low-Risk Category
   Through the calculation of the RPN and the Risk Matrix, four risks are in the low-risk category, which means that the risk is easy to detect and has a small impact. The probability of the risk occurring is small. The risk in the low category has an RPN value of 6 to 12. The risk with the smallest RPN is the risk of potential loss of fence work having an RPN of 6, then the risk of work accidents when climbing the vertical ladder with an RPN of 8, and the risk of lack of construction equipment and experienced local labor and hydrotest blocked by RPN 12.
2. Risk with Medium Risk Category
   From the results of the calculation of the RPN and the Risk Matrix, six risks are in the moderate risk category, which means that the risk is not easy to detect, has a moderate impact, and the probability of the risk occurring is moderate. The risk with this medium category has an RPN value of 16 to 24, namely the risk of delays in the tank schedule and requests for maintenance/storage of Minutes of Handover of Work 2 with RPN 16, the risk of terminating project activities with RPN 18, as well as the risk of hydrotest tank inefficiency, delays in all engineering project designs, and loss of procurement costs with RPN 24.
3. Risk with High-Risk Category
   Three risks are in the high-risk category, which means that the risk is difficult to detect or even cannot be detected, has a high impact, and the probability of the risk occurring is high. Risks in this category have a RPN of 36, the risks with the highest RPN include the risk of inefficiency from tank work, increase in plate material prices, and Mechanical completion delays due to the rainy season with an RPN of 36.

3.3. Risk evaluation

Due to the risk originating from various departments (areas) within the company, it is essential to do a risk weight assessment to determine the level of importance of each risk. The risk weight assessment is carried out by comparing the importance of one risk to another. The level of importance assigned to each risk is subjective to the authors. In determining the calculation of risk weights, the AHP software is used. Then the risk weight can be processed into a risk value.

Table 2. Pareto Chart Calculation.

Risk Number	Risk	Total Risk Value	Percentage of Risk	Risk Accumulation
6	Plate material price increase	6.2280	22.45%	22.45%
7	Mechanical completion delays due to the rainy season	6.2280	22.45%	44.89%
2	Inefficiencies of tank work	5.4720	19.72%	64.61%
3	Inefficiencies in the hydrotest tank	2.1840	7.87%	72.49%
4	Delays in overall engineering design projects	2.1840	7.87%	80.36%
5	Loss of procurement costs	2.1840	7.87%	88.23%
12	Terminated project activity	1.1160	4.02%	92.25%
11	Delay on tank schedule	0.6400	2.31%	94.56%
13	Request for Maintenance/Preservation Minutes of Work Handover 2	0.6400	2.31%	96.86%
8	Lack of construction equipment and experienced local labor	0.3120	1.12%	97.99%
9	Hydrotest is blocked	0.3120	1.12%	99.11%
10	Work accident when climbing the vertical ladder	0.1440	0.52%	99.63%
1	Potential disadvantages of fencing work	0.1020	0.37%	100.00%

Table 2 shows which risks need treatment. According to the Pareto Chart method, the risk will be selected with a total of 80% of the total accumulated risk. After sorting from the highest risk percentage to the lowest, then 20% of the total risk is selected because it is the risk that impacts 80% of the overall risk [18].

In Table 2, five risks are included in 80% of the total accumulated risk, namely the risk of increasing plate material prices, mechanical completion delays due to the rainy season, inefficiencies from tank work, inefficiencies in hydrotest tanks, and delays in the overall engineering design project. Then, risk number 5 was also chosen because it has the same risk value as risk values number 3 and 4. Meanwhile, other risks are not treated because they do not have a significant impact on project continuity. The calculation results in Table 2 can then be arranged into a Pareto diagram.

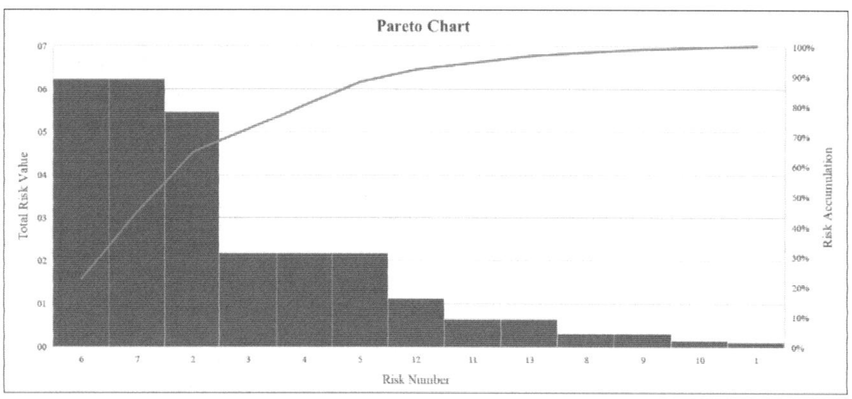

Figure 6. Pareto Chart.

Based on the diagram in Figure 4, the X-axis shows the number of risks. The total risk value located on the left Y axis shows the most influential risk value according to the predetermined weight in the form of a blue bar graph. The higher the blue bar, the higher the total risk value. While the accumulated risk is shown on the Y axis on the right, showing the cumulative value of risk, which is also represented through the Pareto line (yellow color).

3.4. Risk treatment

In this stage, the risks that have been previously selected are then treated as these risks. Risk treatment is given to minimize the impact that will arise from the risk. Risk treatment is determined by having discussions with company experts. The following is the treatment given for the selected risks:

Table 3. Risk Treatment.

Risk	Risk Treatment	Reason
Increasing plate material prices	Monitoring the movement of exchange rates and bank interest rates	This is done to determine if the rupiah's condition is weakening against foreign currencies, which are generally in the form of dollars. This will make the company spend more money than usual
Mechanical completion delays due to the rainy season	Night work permit on site	This is necessary to facilitate construction work to compensate for delayed work progress during the rainy season
	Maximizing pre-fabrication	Given the low productivity of work in the rainy season, it is necessary to manufacture factory construction parts to be directly installed and combine the core parts
	Document any work that is hampered by rain as evidence for an EOT (Extension of Time) claim to the owner	This is done to avoid the consequences of delays due to the occurrence of several disturbances due to external factors such as unpredictable natural conditions
Inefficiencies from tank work	Propose another vendor for the procurement of the tank foundation heating system	This is done as an alternative to avoid inefficiency in the cost of procuring a heating system
Inefficiencies in hydrotest tanks	Ask the owner/client to use seawater for hydrotest	Using seawater for hydrotest can save the costs needed to procure water used during the hydrotest. If the client agrees to use seawater, it can cut expenses.
Delays in overall engineering design projects	Document approval should be done as early as possible	The sooner the document is approved, the faster the EPC process will be carried out to avoid the risk of delays.
	Client/owner team placed together with contractor engineers	This placement is done so that the party representing the contractor and the owner/client can communicate effectively so that, if necessary, approvals and others can be carried out directly, effectively and efficiently.
	Accelerate the formation of the engineering team	With the formation of an engineering team, the earliest process in EPC, namely engineering, will be completed more quickly, which of course, will affect the start of the procurement process to construction
	Create records and Technical Query for proof of Change Order on design changes	If every change that appears in the field is recorded, submitted, and approved quickly, project delays can be avoided, and any changes can be directly executed in the project
	Adding engineering team manpower/manhours	With the addition of the number of workers and the work duration, the project's work process will be more likely to be completed on time or even before the deadline for completion
	Make notifications or reminders to the owner regularly	This needs to be done to prevent delays in document approval and others that can delay the implementation of project work
Loss of procurement costs	Review the contract again and submit a price increase claim to the owner	If there is an unavoidable price increase, the company can file a price increase claim to the owner to cover the additional costs incurred so that no losses occur
	Benchmarks of similar projects	By doing benchmarking, the company can find out the standard market prices that are commonly used in similar projects. This will be very useful to avoid buying at a price that exceeds the standard benchmark
	Make a Purchase Order before the plate price increases	This is done to avoid an increase in the price of materials/equipment from a predetermined price so that there is no inefficiency in procurement costs

4. Conclusion

The lowest RPN with a value of 6 is the risk of potential loss of fence work. In contrast, the highest RPN with a value of 36 is the risk of inefficiency from tank work, the increase in plate material prices, and the mechanical completion delays due to the rainy season. Six risks were identified as the 'vital few' through the Pareto Diagram method: the risk of increasing plate material prices, mechanical completion delays due to the rainy season, inefficiencies from tank work, inefficiencies in hydrotest tanks, delays in overall engineering design projects, and loss of procurement costs. These risks are then selected for treatment. The treatment given to the risk of rising plate material prices is to monitor the movement of exchange rates and bank interest rates. The treatment given for the risk of delay of mechanical completion due to the rainy season is a night work permit on site, maximizing pre-fabrication, and documenting delayed work for EOT (Extension of Time) claims. The treatment given for the risk of inefficiency from tank work is to propose another vendor for the procurement of the tank foundation heating system. The treatment given for the risk of inefficiency in the hydrotest tank is to propose using seawater for hydrotest. The treatment given for the risk of delay in the entire engineering design project is to approve documents as early as possible, place the client/owner team together with the main contractor engineer, accelerate the formation of the engineering team, create records and TQ for evidence, add manpower/manhours to the engineering team, and make a reminder to the owner. The treatment given for the risk of loss of procurement costs (material/equipment inefficiency) is reviewing contracts and submitting price increase claims, benchmarking similar projects, and conducting PO before the plate price increases.

Acknowledgments

The authors gratefully acknowledge financial support from the Institut Teknologi Sepuluh Nopember for this work, under the project scheme of the Publication Writing and IPR Incentive Program (PPHKI).

References

[1] Cervone H F 2006 Project Risk Management *OCLC Syst. Serv.* **22** 256–62
[2] Putra Z, Chan S and IHA M 2018 Desain Manajemen Risiko Berbasis ISO 31000 Pada PDAM Tirta Meulaboh *J. E-Kombis* **3** 52 – 71
[3] Dumara S 2017 *Analisis Risiko Berdasarkan Konsep Manajemen Risiko ISO 31000: 2009 pada Proyek Floating Storage Offloading (FSO)* (Institut Teknologi Sepuluh Nopember)
[4] Satrio D, Trisnayanthi N N A I and Pratilastiarso J 2021 Analysis of the Effects of Fuel Type Selection on the Performance and Fuel Consumption of a Steam Power Plant *Int. J. Adv. Sci. Eng. Inf. Technol.* **11** 2046–54
[5] Nyoman Ayu Indah Trisnayanthi N, Pratilastiarso J and Satrio D 2020 Effects of Fuel Replacement on the Performance and Fuel Consumption of Steam Power Plant *IES 2020 - International Electronics Symposium: The Role of Autonomous and Intelligent Systems for Human Life and Comfort* (Surabaya: IEEE) pp 67–73
[6] Kurniawati I, Pratilastiarso J and Satrio D 2020 Analysis of the Effect of Cooling Water Condenser to Power Plant Cycle Using Cycle-Tempo Software *IES 2020 - International Electronics Symposium: The Role of Autonomous and Intelligent Systems for Human Life and Comfort* (Surabaya: IEEE) pp 37–42
[7] Nikmah A, Amalia R and Satrio D 2022 Analysis of the Effect of Bend Angle Outlet Main Steam Line on the Steam Flow Characteristic *IOP Conference Series: Earth and Environmental Science* vol 972 (Surabaya: IOP Publishing Ltd) p 012064
[8] Simangunsong R G, Amalia R and Satrio D 2021 Numerical Analysis of a Shell and Tube Heat Exchanger Using Computational Fluid Dynamics Software *International Electronics Symposium 2021: Wireless Technologies and Intelligent Systems for Better Human Lives, IES 2021 - Proceedings* (Surabaya: IEEE) pp 630–5
[9] Ariansyah M N, Diana L and Satrio D 2022 Numerical Study the Effect of Gap Ratio on Flow Characteristics and Heat Transfer in Staggered Tube Banks *IOP Conference Series: Earth and*

Environmental Science vol 972 (Surabaya: IOP Publishing Ltd) p 012065
[10] Azizah A N, Amalia R and Satrio D 2022 Analysis of Combustion Characteristic in Tangentially Fired Pulverized-Coal Boiler with Tilting Burner Angle Variation using CFD *IOP Conference Series: Earth and Environmental Science* vol 972 (Surabaya: IOP Publishing Ltd) p 012067
[11] Satrio D, Suntoyo S, Albatinusa F, Junianto S, Musabikha S, Madi M and Setyawan F O 2023 Effect of Positioning a Circular Flow Disturbance in Front of the Darrieus Turbine *IOP Conference Series: Earth and Environmental Science* (Surabaya: IOP Publishing Ltd) pp 1–8
[12] Satrio D, Suntoyo S, Erwandi E, Albatinusa F, Tuswan T and Hakim M L 2023 The Benefit Using a Circular Flow Disturbance on the Darrieus Turbine Performance *Int. J. Eng. Appl.* **11** 55–63
[13] Junianto S, Fadilah W N, Adila A F, Tuswan T, Satrio D and Musabikha S 2022 State of the Art in Floating Tidal Current Power Plant Using Multi-Vertical-Axis-Turbines *IES 2022* (Surabaya: IEEE) pp 1–7
[14] Satrio D, Adityaputra K A, Suntoyo S, Silvianita S, Dhanistha W L, Gunawan T, Muharja M and Felayati F M 2023 The Influence of Deflector on the Performance of Cross-flow Savonius Turbine *Int. Rev. Model. Simulations* **16** 27–34
[15] Yahya N Y, Silvianita S and Satrio D 2023 A Review of Technology Development and Numerical Model Equation for Ocean Current Energy *IOP Conference Series: Earth and Environmental Science* (Surabaya: IOP Publishing Ltd) pp 1–13
[16] Sharma K D and Srivastava S 2018 *Failure Mode and Effect Analysis (FMEA) Implementation: A Literature Review Bottling of Biogas-A Renewable Approach View project*
[17] Antoniadis D and Thorpe A 2003 Evaluation of risks using 3D Risk Matrix *2nd Int. Conf. Struct. Constr. Eng. Syst. Vis. Strateg. Creat. Des.* 2179–84
[18] Sarkar A, Mukhopadhyay A R and Ghosh S K 2013 Issues in Pareto Analysis and Their Resolution *Total Qual. Manag. Bus. Excell.* **24** 641–51
[19] Saaty T L 2008 Decision making with the analytic hierarchy process *Int. J. Serv. Sci.* **1** 83–98
[20] Epstein A L and Harding G H 2019 *Risk management* vol 627 (Elsevier Inc.)
[21] Quantum TC Consulting 2017 Transform Your FMEA into a Risk Matrix for ISO 9001:2015

IOP Conf. Series: Earth and Environmental Science 1298 (2024) 012035 doi:10.1088/1755-1315/1298/1/012035

Morpho-dynamic Induced Rip Currents in Klayar Beach, Pacitan, East Java, Indonesia

Wahyudi[1,4], Vierda Khairene Tiffany[1], Yeyes Mulyadi[1], Haryo Dwito Armono[1], Kriyo Sambodho[1], Leo Eliasta Sembiring[2], Nguyen Trung Viet[3]

[1]Department of Ocean Engineering, Institut Teknologi Sepuluh Nopember, Kampus ITS Sukolilo, Surabaya 60111, Indonesia
[2]Directorate General of Water Resources, Ministry of Public Works, the Republic of Indonesia, Jl. Patttimura 20, Jakarta Selatan, Indonesia
[3]Department of Civil Engineering, Thuyloi University, 175 Tay Son, Dong Da, Hanoi, Vietnam

[4]wahyudictr@oe.its.ac.id

Abstract. Klayar Beach (KB), Pacitan, East Java, Indonesia is a rugged coast with a narrow sand beach, a relatively small embayment with rocky headlands, and a very exotic view as a coastal tourism destination. Despite its attractiveness, KB poses a hazard due to rip currents. From 2009 to 2022, KB visited more than 3 million people, and more than 30 people were reported dead due to drowning in rip currents. This paper presents the result of fieldwork to investigate morpho-dynamic driven rip current hazard of the study area. This study carried out bathymetric and topographic mapping and sediment sampling. The last 10 years' wave data was used for predicting hydrodynamic parameters. The study shows a good correlation between a variety of incoming significant wave heights and tide fluctuation vs. rip current velocity. The existence of headland plays a significant role in rip current formation in the study area. The headland influences either shadow or deflected rip currents formation for the incoming wave angle. The study also indicates that the type of rip current in the KB is structural boundary controlled, and the most important thing to be alerted for promoting beach safety in this area is that this rip current is predicted permanently to occur.

Keywords: Klayar Beach, headland, boundary-controlled rip, beach safety

1. Introduction

Klayar Beach (KB) is located in the southern part of Pacitan Regency, the southern coastal area of East Java Province, Indonesia (Fig. 1). As a coastal tourism destination for both domestic and foreign visitors, KB provides attractive views with mountainous and cliffed coast, wave cut terrace and sea stacks, and embayment beach with beautiful rocky headlands and spectacular incoming breaking waves. This feature is common to be found along the southern coast of Java. The last 10 years record of visitors to KB shows a significant increase (Fig. 2). Increasing tourism in KB is supposed to be a local economic driver for the local community as well as for coastal communities along the southern coast of Java. Coastal waters, particularly in the nearshore is an active zone that is very attractive to the tourist.

However, despite their attractiveness coastal water can be very dangerous to their visitors due to rip currents which are really inhospitable to them [1, 2]. Rip currents are among the most deadly coastal hazards for the swimmer in the coastal water [3, 4], particularly in a swell-dominated environment (as it is possessed by the southern Java coastal area), because the swell can result in more deadly rip current hazard than wind waves [5, 3].

Figure 1. Location map of the study area

Beach accidents caused by rip currents have been a major global hazard worldwide, but there is a lack of attention to rip currents compared to the other hazards [6], since rip currents never destroy buildings or infrastructures and never cause environmental damage and economic loss. In fact, however, it has been reported that an estimated 100 people drown in rip currents each year in the U.S., 21 people in Australia, and 53 people in Costa Rica [7]; and 71 % of all recorded incidents were due to rip current in the UK, and in many other countries (Brazil, Colombia, Israel, and New Zealand) have reported that rip current are major hazards [8]. [9] reported that more than 200 people had drowned in rip currents and 30 people were dead every year in Korea. Surfzone fatalities in China have been reported reached 660 from 2009 to 2019, even more than the 628 casualties of maritime disasters during the same period [10]. In Indonesia, fatalities due to rip currents on the Southern Coast of Java have been reported as many as 28 people dead each year in the last 5 years, only on the southern coast of Java [11]. Beach accident due to drowning in rip current in KB has been recorded as high as 30 people's death since 2009 (Fig. 2).

A rip current is defined as water masses that move contrary in direction to the incident wave which extend outside the surf zone normal to the beach [1, 2, 5, 6, 12, 13, 14], narrow, strong and concentrated seaward-directed flows [5, 6, 8, 14], associated with wave action in the breaker zone [12, 14, 15], its characteristics are depending on the incoming wave height variability [5, 6], and variability in the offshore bottom topography [12, 15, 16], it is much shallower than its wide and its lifetime is correspond to its development time [17]. This present study examines the existence of rip current and headland-induced morpho-dynamic processes in generating a rip current, and the implication for beach safety campaigns in KB, as well as on the southern coast of Java.

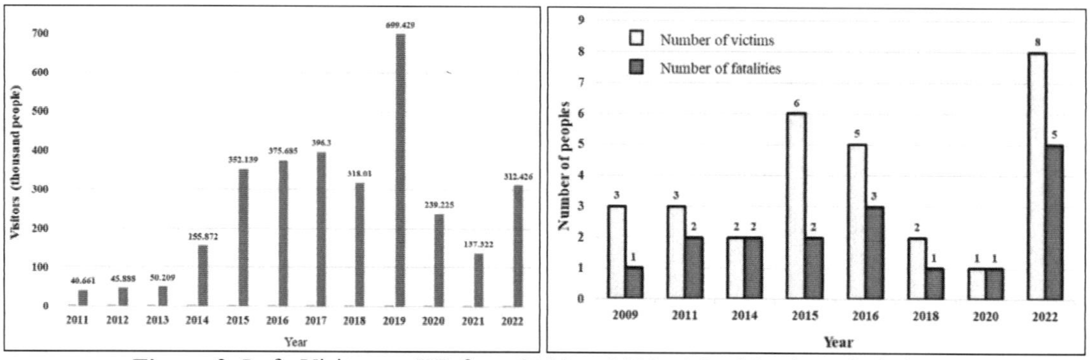

Figure 2. Left: Visitors to KB from 2009 to 2022; right: Number of victims [11]

2. Material and Methods

We carried out fieldwork to investigate the occurrence of rip currents in KB. We conducted shore topographic and seafloor bathymetric mapping, shore and sea bottom surface sediment samplings, and secondary waves data collecting. Before the field works, satellite imageries of the KB were collected to select locations where the rip current existed.

2.1. Site Selection

The KB is located in the southern part of Pacitan Regency, East Java Province, Indonesia with latitudes and longitudes boundaries as follows: 8°13'23"S; 110°56'46"E to 8°13'28"S; 110°56'43"E and 8°13'30"S; 110°56'54"E to 8°13'33"S; 110°56'30"E (Fig. 1). Coastal morphology of the study area shows highly exposed coast with wave dominated erosional coast which is characterized by the presence of sea stacks, wave-cut terraces, narrow sandy beach, natural structural headlands, and embayed beach. This is typical of the secondary coast which is dominated by wave actions (Fig. 1 and 3). This coastal feature is common in the southern coastal area of Java, Indonesia, and it is almost always crowded by beach visitors throughout, especially during the holiday season.

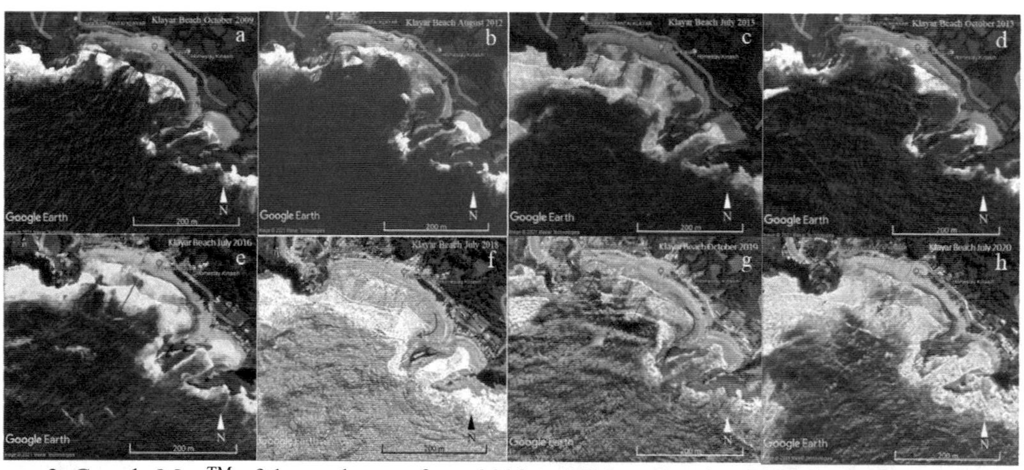

Figure 3. Google Map™ of the study area from 2009 to 2020. a. October 2009; b. August 2012; c. July 2013; d. October 2013; e. July 2016; f. July 2018; g. October 2019; and h. July 2020

This study used Google Earth™ map to investigate the existence of rip current in KB. Utilizing satellite imagery from Google Earth™ for rip currents identification has been done by previous researchers worldwide [18, 19, 20, 21, 22, 23]). KB was selected as the study area based on the result of the visual analyses of the eight satellite imageries of the Google Earth™ maps (Fig. 3). Eight satellite imageries from 2009 to 2022 show that there are rip currents persistently occur, particularly at the left-hand (east) embayment close to the headland.

2.2. Environmental Dataset

Shore zone and seabed topographic mappings were carried out using a single beam echosounder for bathymetrics (Fig. 4), and using Total Station for shore zone topographic. We collected the last 10 years' winds and waves data from secondary published data released by The Agency of Meteorological Climatological and Geophysical of The Republic of Indonesia (BMKG). We processed it to find significant wave height (H_s) and its peak periods (T_p) and to predict breaking wave heights (Tabel 1).

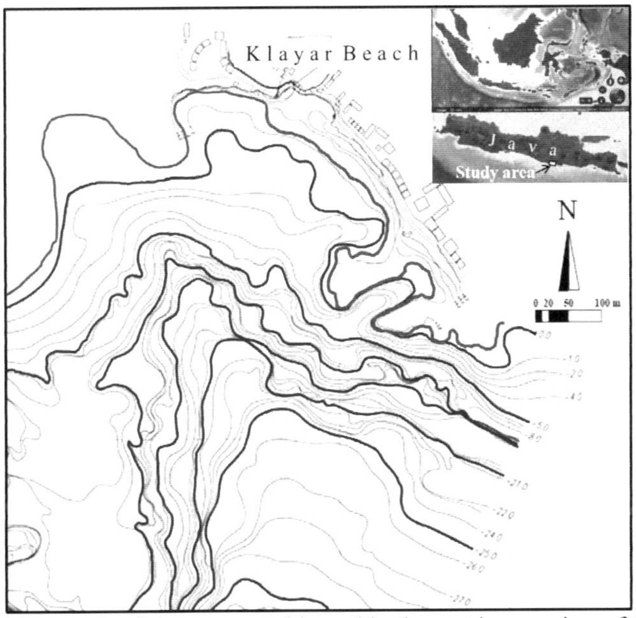

Figure 4. The result of the topographic and bathymetric mapping of the KB

Table 1. Wave characteristics of the study area extracted from the last 10 years of data

Wave Characteristics	Dominant wave direction (going to)				
	W1 N273°E	W2 N337°E	W3 N22°E	W4 N69°E	W5 N112°E
Peak period, Tp (s)	8	11	16	16	10
Significant wave height Hs (m)	3.87	4.60	6.34	6.04	5.23

Source: Meteorology Climatology and Geophysical Agency

3. Result and Discussion

3.1. Coastal Morphodynamic

Oceanographically, the area along the coast of the southern part of the Indonesian Archipelago especially the southern coast of Java faces directly the unlimited fetch of the Indian Ocean, consequently this area is characterized by swell swell-dominated environment with high to extremely high incoming wave height. Tectonically, Java Island is a part of the island arcs that has been being uplifted since at least ~2 ma [24], as a result of the subduction between the subducting Indian-Australian plate below overriding the Eurasian plate (Fig. 6). Because of this uplift, the southern coast of Java has high topography with hilly to mountainous morphology. Combined oceanographical and geological processes lead to the high energy of highly exposed coasts with wave-dominated erosional coasts.

IOP Conf. Series: Earth and Environmental Science 1298 (2024) 012035 doi:10.1088/1755-1315/1298/1/012035

Therefore, the southern coast of Java exhibits spectacular views characterized by high incoming waves height, cliffed coasts, small to large embayments with rugged rocky headlands, and beautiful sand beaches.

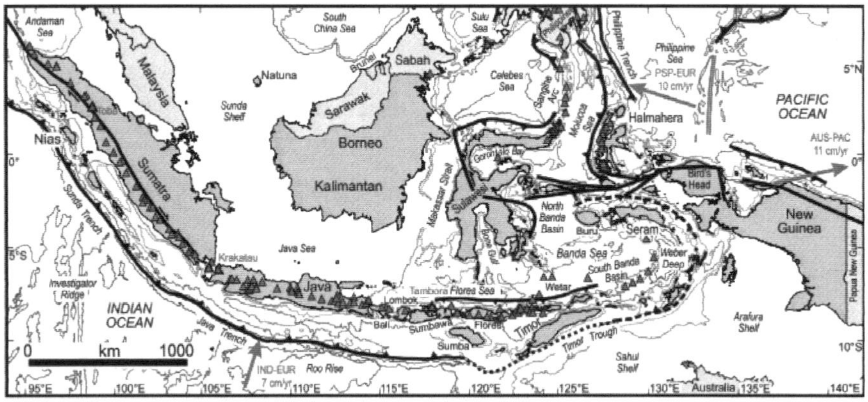

Figure 5. Tectonic setting of the Indonesian archipelago [24]

As a system, beach morpho-dynamic consists of elements of morphology, waves and flow, sediment transport and bed change, and as a process refers to interactions between morphology and hydrodynamic including wave, tide, and wave-driven currents involving sediment movement [25, 26, 27]). Interactions between beach morphology and hydrodynamic processes take place when waves reach the surf zone. As waves propagate from deeper to shallower water, the friction of the seabed starts to decrease wave celerity and deform wave characteristics, which are wave height becomes higher and wavelength becomes shorter, and eventually wave breaks when breaker index (γ_b) reaches 0.78. Breaking waves have been believed as a driving mechanism of rip current formation [2, 5, 12, 14]. Waves and tides are predominantly factors influencing coastal morphology and processes [25], and rip current is one of the phenomenons created by the interactions between waves and coastal morphology [8, 14, 25, 28], and often modulated by tides [2, 29]. [30] discovered that rip currents vary significantly in occurrence, spacing, and associated nearshore topography, depending on the predominant and the origin of wave-beach conditions.

Since morpho-dynamic processes can generate coastal hazards, one of those is rip current, therefore, in Australia, the ABSAMP (Australian Beach Safety and Management Program) used beach morphodynamics to develop a beach hazard rating system [31]. Previously, [30] combined sediment characteristics with breaking wave parameters to categorize beach types. He used the non-dimensional settling velocity (Ω) of [32] as follows:

$$\Omega = \frac{H_b}{TW_s} \tag{1}$$

where Ω is the non-dimensional settling velocity, H_b is the height of wave break, T wave period, and W_s settling velocity of the sediment.

Based on the Ω, [33] described beach types into three categories, as the followings: (1) reflective beach when $\Omega < 1$; (2) intermediate beach with $1 < \Omega < 6$; and (3) dissipative beach if $\Omega > 6$. In order to investigate coastal morpho-dynamic processes, usually researchers do not use only a single criterion with a single parameter, but they often use many parameters and combine more than one criterion [30, 33, 34, 35]. [33] developed a method for describing the circulation pattern in the surf zone of an embayment beach with a dimensionless scaling factor (δ'). He used δ' for predicting the influence of headland to the circulation pattern, which is obtained from:

$$\delta' = \frac{S_l^2}{100R_0H_b} \tag{2}$$

where S_l is the length of the embayed beach which is a sum of both lengths of the headland and beach (Fig. 7).

This method is now widely used to classify embayed beaches [35]. The surf zone circulation pattern can be described by the range of the δ'; normal beach circulation ($\delta' > 19$); transitional circulation ($8 < \delta' < 19$); and cellular circulation ($\delta' < 8$) [30]. Recently and more specifically, [35] developed a new and low-cost method for classifying swell-dominated embayed beaches utilizing morphometric parameters. They used this classification to determine morphological processes that help in effectively assessing the vulnerability of general beach embayments. To classify swell-dominated embayed beaches, [35] developed embayment morphometric parameter (γ_e) and utilized the dimensionless scaling factor (δ') of [30], and headland orientation (H_1 to H_6). The embayment morphometric parameter is defined as:

$$\gamma_e = \frac{a}{\sqrt{A_e}} \tag{3}$$

where a is an indentation of the embayment from the headland point to the back-beach limit, and A_e is the embayment area.

We followed the works of [30, 33, 35] to classify the beach type of the study area and to investigate the morpho-dynamic process and its influence in generating rip currents in KB. We summarized embayment geometry and its components, hydrodynamic characteristics and its interaction with the seafloor topography, and beach and surf zone sediments of the study area (Fig. 7), and used them to determine embayment morphometric parameter, dimensionless fall velocity, and headlands orientation (Table 2). Those parameters are then employed to investigate the dominant process that is playing an important role in generating rip currents in KB. According to the dimensions of the embayment parameter components (Table 2), KB has two inward-oriented headlands (Fig. 7) implying that shadow rip is more susceptible to occur than deflected one. The dimensionless falling velocity (Ω) ranges between 4-6, indicating that the study area is categorized as an intermediate beach, which rip currents are susceptible to occur. The Dimensionless embayment scaling factor (δ') of the KB falls to the range of 4.0 to 6.6 suggesting that the existence of the structures or headlands is very significant in influencing morphodynamic processes in the study area with cellular circulation pattern. The Embayment morphometric parameter (γ_e) of the study area is 0.72, indicating that the embayment of the KB consists of a relatively lower distance between headlands and greater coastal indentation.

Table 2. Beach embayment parameter and its components

Embayment/beach parameters of the Klayar Beach	
Alongshore beach length, X_d	400 m
Embayment length, S_l	600 m
Embayment indentation, a	145 m
Width between headland, R_o	280 m
Embayment area, A_e	40,300 m^2
West headland orientation to beach aspect, q_{hw}	inward
East headland orientation to beach aspect, θ_{he}	inward
Dimensionless falling velocity, Ω	4 – 6
Dimensionless embayment scaling factor, δ'	4.0-6.6
Embayment morphometric parameter, γ_e	0.72

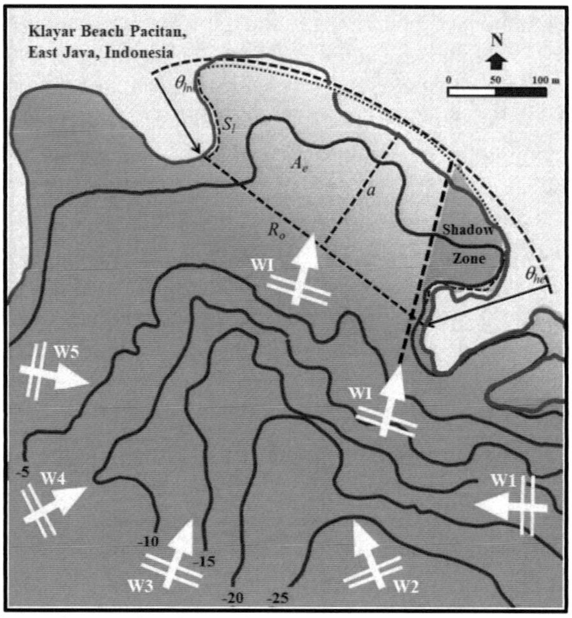

Figure 6. Morphometric and morpho-dynamic parameters of KB: area of embayment (A_e), the indentation of the embayed beach (a), width between headlands (R_o), the total length of embayment that is a sum of headlands and beach length (S_l), and orientation of headland direction to the beach (θ_{hw} and θ_{he}: west and east headland, respectively); dominant incident waves directions (W1 to W5), because of the valleyed seafloor, all direction of incident waves are refracted and eventually the prevailing wave direction entering to the nearshore zone tends relatively constant all the times (WI); east headland plays an important role in shadowing shadowed zone by blocking incident waves.

3.2. Type of the KB Rip Currents and Its Implication for Promoting Beach Safety

Classical taxonomy of any type of rip current comprises the rip feeder(s) which is initially part of a longshore current that is parallel to the coastline and becomes a feeder when its direction turns to the open sea and perpendicular to the coastline; a rip neck is the longest part of rip current with seaward direction flow and normal to the shoreline, and a rip head that formed outside of breaker zone. Even though they vary significantly in occurrence, spacing, and associated nearshore topography, rip currents do not occur on all-type beaches [32]. Rip currents could not occur on the reflective type beach swash oscillations perpendicular to the shore are dominant in this environment, and do not occur on fully dissipative beaches. Rip currents are a typical feature and essential element of the intermediate type of beach on the transition state from the intermediate to dissipative beach type. Based on the morphometric analyses, we found that all embayment parameter components of the study area suggest that rip currents are susceptible in the KB.

Coastal scientists and coastal engineers have identified and described rip currents to classify rip currents [32, 8, 36, 6] and [14]. [32, 8] classified rip currents into four types based on hydrodynamic and topographic controls. Those are accretionary and erosional rip currents and topographic and mega rips. Based on the generating mechanisms of rip currents, [36] described four categories of rip currents: (1) waves-structures interactions, (2) waves-waves interactions, (3) shear instabilities of strong alongshore currents, (4) hydrodynamic and morphologic instabilities. In addition, they showed that the rip current type that most frequently occurs is hydrodynamically generated by uneven seafloor topographic-induced wave refraction. To increase awareness and promote a better public understanding, [6] suggested using descriptive terminology. [6] classifies rip current types based on the description of environmental conditions including seafloor morphology and waves and sea conditions. He also describes the appearance and persistence of each rip current type, and its relative threats and proposes ways to avoid drowning in each rip current type. [6] proposed five categories of rip current type: (1) rips occur between bars, (2) cusped-shore type rips, (3) structurally generated rips, (4) flash rips, and (5) large-scale or mega rips. [14] considered interactions between wave breaking and morphology and tides influence induced flow behaviors to describe and classify the type of rip current. This type is categorized by the generation processes. [14] divide rip currents generating forces into four major mechanisms: (1) hydrodynamic-generated rip currents, (2) bathymetric-induced rip currents, (3) boundary-influenced rip currents, and (4) other generated-mechanism rip currents.

Referring to the above-mentioned four rip currents classifications, KB rip currents are categorized as topographic rips of [8]; classified as waves-structures interactions rip by [36]; identified as structurally-controlled rips of [6], and termed as boundary-controlled rips of [14]. Even though all classifications applied for KB rip currents are semantically different, they refer to the same driving mechanism, they mention the same processes, and they show the same appearances. Characteristics of sediment may be important as an indication to determine whether positive or negative feedback of the wave action in the study area. [37] also considered the availability of coarse sediment as an indicator of a wave-dominated coast, not only based on the absolute parameter of wave height. Sediment from KB consists of predominantly well to very well-sorted coarse sand [11] indicating that the study area is categorized as a wave-dominated coast [37]. In addition, sediment is only found at the shore zone, and during low water elevation, fringing reefs and bedrock are exposed. This is interpreted that KB is categorized as erosional rather than accretional beach. Therefore, according to [32], KB rip currents could have been classified as erosional rips. However, since the morphology of the embayment including the size of the embayment is relatively small, its geometry as shown by the sea bottom valley and direction of the east headland tends to refract and block the incident waves, respectively, then eventually, any direction of incident waves almost always perpendicular to the shoreline (Fig. 7) and are blocked by the east headland. Therefore, the above-mentioned classification of rip types for the study area is applicable. More specifically the existence of east headland which is playing an important role in blocking the incident waves that lead the down waves area as a shadow zone that is very susceptible for shadow rips to take place.

KB, especially at the east headland looks calm as protected from incident waves and breaking waves, also looks very attractive since very close to the cliffed headland. Despite that attractiveness, structural (headland) boundary shadow rip current in KB persistently occurs and could be very fast and dangerous to the bathers. Concerning the condition, it is suggested to formulate mitigation measures to lower the risk due to this deadly hazard. The first thing that can be done is to stake a red flag at the most dangerous location on this beach and make an alert announcement board there. These measures are necessary not only in the KB but also along the southern coast of Java Island

4. Conclusion

We have carried out fieldwork and analyzed the morphodynamics of the embayment beach concerning rip formation in the study area. Tectonically and oceanographically, the study area is categorized as a highly exposed swell-dominated coast that exhibits spectacular views characterized by high incoming wave height, cliffed coasts, small to large embayments with rugged rocky headlands, and beautiful sand beaches. Despite the attractiveness, the KB poses a deadly hazard due to rip current. The morphometric embayment parameters of the KB indicate that this area is susceptible to the rip current, this is consistence with the evidence from the satellite imageries. According to this, it is suggested to formulate mitigation measures to reduce the risk due to the deadly hazard.

Acknowledgments

This study was partly supported by the Research Fund of the ITS Research Center, Institut Teknologi Sepuluh Nopember (ITS) Surabaya Indonesia. The study would not have happened without the help of the captain and his crews of the survey boat during the field investigation. The tireless work of the research assistants and technicians for this study is highly appreciated. Thanks, are due to the anonym reviewer(s) for reviewing this manuscript of this paper.

References

[1] M. C. Haller and R. A. Dalrymple, Rip Current Dynamics and Nearshore Circulation. *Research Report No. CACR-99-05,* August, 1999. p. 144, Center for Applied Coastal Research Ocean Engineering Laboratory, Univ. Of Delaware, Newark, DE 19716, 1999

[2] J. H. MacMahan, Ed. B. Thornton, Ad. J. H. M. Reiners, Rip current review. *Coastal Engineering*, 53 (2006) 191-208. doi: 10.1016/j.coastaleng.2005.10.009, 2006

[3] M. J. Austin, T. M. Scott, P. E. Russell, and G. Masselink, Rip current prediction: development, validation, and evaluation of an operational tool. *Journal of Coastal Research*, 29 (2), 283-300. Coconut Creek (Florida), ISSN 0749-0208. Austin., 2012. DOI: 10.2112/JCOASTRES-D-12-00093.1, 2013

[4] L. E. Sembiring, *Rip Current Prediction System for Swimmer Safety toward Operational Forecasting Using A Process-Based Model and Nearshore Bathymetry from Video*. Ph.D. Dissertation Delft University of Technology. CRC Press/Balkema, PO Box 11320, 2301 EH Leiden, the Netherlands. ISBN 978-1-138-02940-8 (Taylor & Francis Group), 2015

[5] A. J. Bowen, Rip Currents 1. Theoretical Investigations, *Journal of Geophysical Research,* Vol. 74, No. 23, pp 5467-5478, 1969

[6] S. P. Leatherman, Rip Currents, in Finkl, C.W. (ed.), *Coastal Hazards*, Coastal Research Library 6, DOI 10.1007/978-94-007-5234-4_26, # Springer Science+Business Media Dordrecht 2013, 2013

[7] S. M. Trimble, Mapping Rip Currents at the Beach: Technology moves us closer to mapping and predicting deadly beach currents. *AGI Critical Issues Program*: www.americangeosciences.org/critical-issues, 2018

[8] Short, A.D., 2007. Australian Rip Systems – Friend or Foe? *Journal of Coastal Research*, SI 50, (Proceedings of the 9th International Coastal Symposium), 7–11, Gold Coast, Australia, ISSN 0749.0208.

[9] Shin, S.; Nam, J.; Son, S.; Kim, I.H., and Jung, T.-H., 2017. Field observation and numerical modelling of rip currents within a pocket beach. In: Lee, J.L.; Griffiths, T.; Lotan, A.; Suh, K.-S., and Lee, J. (eds.), The 2nd International Water Safety Symposium. *Journal of Coastal Research, Special Issue* No. 79, pp. 229-233. Coconut Creek (Florida), ISSN 0749-0208. DOI: 10.2112/SI79-047.1

[10] Zhang, Y., Huang, W., Liu, X., Chi Zhang, C., Xu, G., Wang, B., 2021. Rip current hazard at coastal recreational beaches in China. Ocean and Coastal Management, 210 (2021)105734, 1-8. https://doi.org/10.1016/j.ocecoaman.2021.105734.

[11] Wahyudi, Y. Mulyadi, H. D. Armono, K. Sambodho, Disaster Risk Reduction Due to Rip Currents Hazard in Southern Coast of Java. Research Report, Institut Tenologi Sepuluh Nopember, (ITS) Surabaya, 2022

[12] F. P. Shepard, K. O. Emery and E. C. LaFond, Rip Currents. A Process of Geological Importance, *J. Geology*, Vol. 49, p. 337-369, 1941, 1941

[13] R. S. Arthur, A Note on the Dynamics of Rip Currents, *J. of Geophysical Res*., Vol. 67, No. 7, pp 2777-279, 1962. doi:10.1029/JZ067i007p02777, 1962

[14] B. Castelle, T. Scott, R. W. Brander McCarroll R.J. Rip current types. Earth-Science Reviews 163 (2016) 1-21. http://creativecommons.org/licenses/by-nc-nd/4.0/, 2016

[15] F. P. Shepard, and D.L. Inman, Nearshore Circulation, Proc. First Conf. on Coastal Engineering, Council on Wave Research, Berkeley, Oct. 1950

[16] J. Yu, and D. N. Slinn, Effects of wave-current interaction on rip currents, *J. Geophys. Res*., 108(C3), 3088, doi:10.1029/2001JC001105, 2003.

[17] J. A. Smith, and J. L. Largier, Observations of nearshore circulation: Rip currents. *Journal of Geophysical Research*, Vol. 100 No. C6, pp. 10,967-10,975, June 1995.

[18] G. Dusek, and H. Seim, A probabilistic rip current forecast model Journal of Coastal Research, 29(4), 909–925. Coconut Creek (Florida), ISSN 0749-0208. DOI: 10.2112/JCOASTRES-D-12-00118.1, 2013

[19] M. C. Haller, D. Honegger, and P. A. Catalan,. Rip Current Observations via Marine Radar. *Journal of Waterway, Port, Coastal, and Ocean Engineering*, Vol. 140, No. 2, March 1, 2014. ©ASCE, ISSN 0733950X/2014/2-115–124. DOI: 10.1061/(ASCE)WW.1943-5460.0000229, 2014
[20] Li, Z., 2016. Rip current hazards in South China headland beaches. *Ocean & Coastal Management* 121 (2016) 23-32, http://dx.doi.org/10.1016/j.ocecoaman.2015.12.005, 2016
[21] G. Benassai, R. Montella, P. Aucelli, G. Budillon, M. De Stefano, D., Di Luccio, G. Di Paola, L. Mucerino, M. Sica, and M. Pennetta, Rip current evidence by hydrodynamic simulations, bathymetric surveys and UAV observation. *Nat. Hazards Earth Syst. Sci.*, 17, 1493–1503, 2017 https://doi.org/10.5194/nhess-17-1493-2017
[22] H.D. Kim, K.-H. Kim, Analysis of Rip Current Characteristics Using Dye Tracking Method. *Atmosphere* 2021, 12, 719. https://doi.org/10.3390/atmos12060719, 2021
[23] A. de Silva, I. Mori, G. Dusek, J. Davis, A. Pang, Automated rip current detection with region-based convolutional neural networks. *Coastal Engineering* 166 (2021) 103859. https://doi.org/10.1016/j.coastaleng.2021.103859, 2021
[24] R. Hall and I. Sevastjanova, Australian crust in Indonesia. *Australian Journal of Earth Sciences* (2012) 59, (827–844) ISSN 0812-0099 print/ISSN 1440-0952 online, 2012
[25] G. Masselink and A. D. Short, 'The effect of tide range on beach morphodynamics and morphology: A conceptual beach model. *Journal of Coastal Research*, 9, 785–800, 1993
[26] A. D. Short, Beach Morphodynamics in Australia 1970s–2010. *Geographical Research*, 50(2):141–153. doi: 10.1111/j.1745-5871.2012.00760.x, 2012
[27] B. Castelle and G. Masselink, Morpho-dynamics of wave-dominated beaches. *Cambridge Prisms: Coastal Futures*, 1, e1, 1-13, https://doi.org/10.1017/cft.2022.2, 2023
[28] A. D. Short and R. W. Brander, Regional Variations in Rip Density. *Journal of Coastal Research*, 15(3), 813-822. Royal Palm Beach (Florida), ISSN 0749-0208, 1999
[29] A. Withers, S. Maldonado, On the swimming strategies to escape a rip current: a mathematical approach. *Natural Hazards* (2021) 108:1449–1467. https://doi.org/10.1007/s11069-021-04740-7, 2021
[30] A. D. Short, Rip current type, spacing and persistence, Narrabeen Beach, Australia. *Mar. Geol.*, 65: 47-71, 198
[31] A. D. Short, and C. L. Hogan, Rip Currents and Beach Hazards: Their Impact on Public Safety and Implications for Coastal Management. *Journal of Coastal Research Special Issue*, No. 12: Coastal Hazards, pp. 197-209, 1994
[32] M. R. Gourlay, *Beach and Dune Erosion Report*. Delft Hydraulics Laboratory M935/M936, 1968
[33] A. D. Short, The role of wave height, period, slope, tide range, and embaymentisation in beach classifications: a review. *Revista Chilena de Historia Natural*, 69: 589-604, 1996
[34] B. Castelle, G. Coco, The morphodynamics of rip channels on embayed beaches. *Continental Shelf Research* 43 (2012) 10–23. http://dx.doi.org/10.1016/j.csr.2012.04.010, 2012
[35] T. E. Fellowes, A. Vila-Concejo, Shari L. Gallop, Morphometric classification of swell-dominated embayed beaches. Mar. Geol., 411, 78-87. https://doi.org/10.1016/j.margeo.2019.02.004, 2019
[36] R. A. Dalrymple, J. H. MacMahan, Ad J. H. M. Reniers and V. Nelko, Rip Currents. *Annu. Rev. Fluid Mech.* 2011.43:551-581, 2011
[37] R. A. Jr. Davis and M.O. Hayes, What is Wave-dominated Coast? *Marine Geology*, 60 (1984) 313-329, 1984

Exploring Thermal Barrier and Sedimentation Simulation for Enhanced Performance in Grati Combined Cycle Power Plant

Suntoyo[1*], I Cahya[2], M R Islam[3] and H Tanaka[4]

[1] Department of Ocean Engineering, Faculty of Marine Technology, Institut Teknologi Sepuluh Nopember, Surabaya, 60111, Indonesia
[2] Study Program of Ocean Science, Universitas Trunojoyo Madura, Jl. Telang PO. Box 2 Kamal, 69162, Indonesia
[3]Department of Civil Engineering, Faculty of Civil, Planning, and Geo Engineering, Institut Teknologi Sepuluh Nopember, Surabaya, 60111, Indonesia
[4] Institute of Liberal Arts and Sciences, Tohoku University, Sendai 980-8577, Japan
*Corresponding author: suntoyo@oe.its.ac.id

Abstract. PLTGU is a power generation facility that concurrently utilizes both steam and gas power plant technologies. It necessitates a cooling system to operate efficiently throughout its entire lifespan. If PLTGU Grati plans to increase its power capacity, there is a concern that the water discharged from the Water Outlet channel might not cool down sufficiently before re-entering the Water Intake channel. Additionally, sedimentation in the sea water uptake is causing siltation. Hence, this study focuses on the jetty extension project to address sedimentation and hot water spreading issues in PLTGU Grati. The numerical modeling analysis, conducted using the Delft3D software, indicates that in the case of power addition, Alternative Model 1 can reduce the highest temperature compared to the existing condition by 0.655°C. Similarly, Alternative Model 2 reduces the highest temperature by 0.090°C. Moreover, with power addition, sedimentation rate in Area 1 increases by 261.43 m^3/month in Alternative Model 1, while in Alternative Model 2, it decreases by 969.47 m^3/month compared to the existing condition. Considering the ability of Alternative Model 2 to effectively reduce the temperature in the inlet canals by 0.090°C, it provides the best solution to contain the spread of hot water in the PLTGU Grati area. Currently, PLTGU Grati employs a Cutter Suction Dredger (CSD) to periodically dredge the water inlet channel. Therefore, Alternative Model 2 is recommended as the optimal choice among the alternatives. Henceforth, detailed studies related to current patterns and sedimentation rates are presented comprehensively in this paper.

1. Introduction

PLTGU Grati Power Plant, as part of PT. Indonesia Power's portfolio, holds a pivotal role in advancing Indonesia's power generation capacity. Operating as a Gas and Steam Power Plant, PLTGU Grati employs cutting-edge technology to generate electricity. This plant efficiently converts gas and steam into electrical energy, catering to the escalating demands of a burgeoning population and thriving industrial sectors. The reliability and capacity of PLTGU Grati are instrumental in stabilizing and enhancing the national power grid. In December 2014, the Indonesian Ministry of Energy and Mineral Resources (ESDM) authorized the issuance of the Electricity Power Supply Business Plan (RUPTL) 2015-2024 [1]. In Java Island, among a total of 35 power plants with a cumulative capacity of 10.681 MW, 8 power plants, with a combined capacity of 4.905 MW, include both existing projects with completed procurement and new projects. Notably, the Gas & Steam Power Plant (PLTGU) Grati Add On Block 2, located in East Java Province, is among the new projects with an open procurement plan. This plant is planned to have a capacity of 150 MW.

With the increase in power capacity at PLTGU Grati or Grati Combined Cycle Power Plant (CCPP), it can lead to an increased demand for water volume required in the cooling system. Similarly, the volume of hot water discharged from the water outlet channel will also increase and spread into the surrounding waters. One way to increase the water volume for the cooling system is by extending the jetty in the water inlet channel. However, it is essential to consider the impact of sedimentation it may cause. Studies related to sedimentation, including wave movements and currents in coastal waters, have been conducted by e.g. [2, 3, 4, and 5]. Moreover, studies related to hydrodynamics, sedimentation, and morphological changes in power plant waters and coastal waters have been conducted by i.e., [6, 7, 8, 9, 10, 11 and 12] showing that coastal structures significantly affect morphological changes and sedimentation patterns. Experimental [13] and numerical simulations can both be employed to anticipate the pattern of fluid flow and the choice of the model of turbulence is cruicial for obtaining accurate predictions [14] including CFD applications [15].

The study of sedimentation rate and thermal circulation at the water inlet channel will reduce the efficiency of the power generation system at the Grati CCPP. Although fuel requirements are inversely proportional to efficiency, a decrease in inefficiency will result in a significant increase in fuel consumption and expenditure. Therefore, to prevent or reduce sedimentation and thermal circulation, a detailed and comprehensive study of the best jetty extension engineering to contain hot water discharge and prevent sediment entering into the inlet channel of water is crucial. In this study, numerical modeling is performed using Delft3D software with environmental data validation obtained through on-site surveys. It is expected that after the increase in power capacity at the PLTGU Grati, there will be no operational challenges that require a systematic and innovative approach to improve overall performance through the selection of the form and extension of the jetty structures in its water inlet channel.

2. The Location of Study Area

The study area is located at PLTGU Grati, situated in Wates Village, Lekok, Pasuruan District, East Java, along the Surabaya-Probolinggo Highway at kilometer 73. Geographically, the study area spans from 113° 00' 35.5" to 113° 02' 06.2" East Longitude and from 7° 39' 10.6" to 7° 39' 11.6" South Latitude (see figure 1).

Figure 1. Location of Study Area (PLTGU Grati)

IOP Conf. Series: Earth and Environmental Science 1298 (2024) 012036 doi:10.1088/1755-1315/1298/1/012036

3. Hydro-Oceanographic Data

The hydro-oceanographic data used in this study includes bathymetric data, tidal information, currents, temperature, wind, waves, river discharge, and sediment. Figure 2 presented the bathymetric data were obtained in November 2014, covering the approximately area 4397m × 2557m. Depths are referenced to the LWS (Lowest Water Spring). The mean depth within the inlet ranges from approximately -0.5 meters to -1.5 meters LWS. Tidal measurements were conducted over 15 days at one-hour intervals. The analysis of tidal, using the method of least squares, yielded a Formzhal value of 1.0725 (Figure 3), indicating that the tidal pattern in the PLTGU area is of the mixed diurnal double high and double low type (two high tides and two low tides occurring within a 24-hour period).

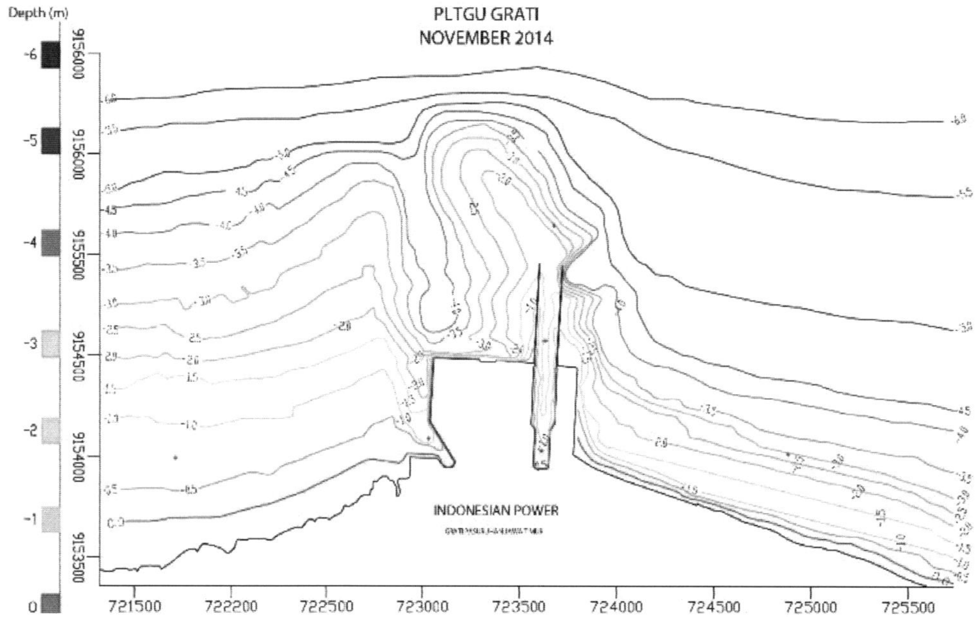

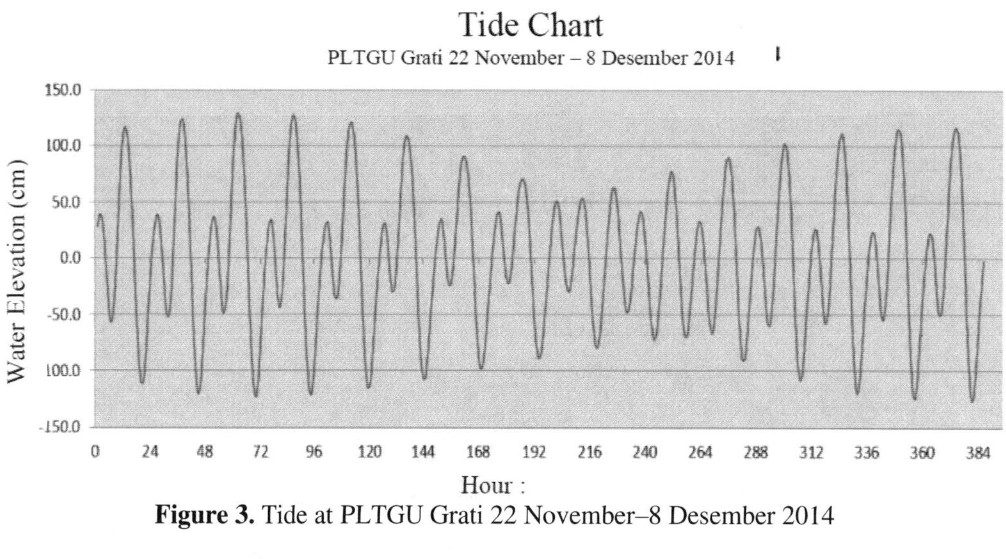

Figure 3. Tide at PLTGU Grati 22 November–8 Desember 2014

Wind and wave data for a period of 10 years (2004-2014) are also provided (figure 4), showing that the dominant wind directions are from the northwest, north, northeast, and east. The prevailing wave orientation originates from the east. The sea in the PLTGU Grati area is enclosed, with Madura Island to the north and Java Island to the south. This implies that the prevailing wave orientation from the eastern direction is quite plausible, as depicted in figure 5.

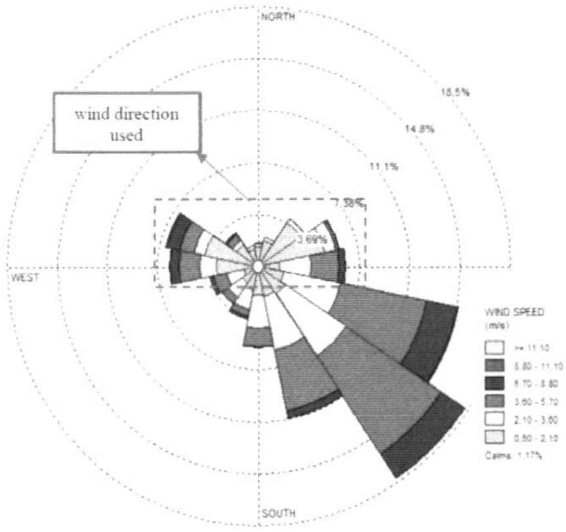

Figure 4. Wind Rose in PLTGU Grati Area

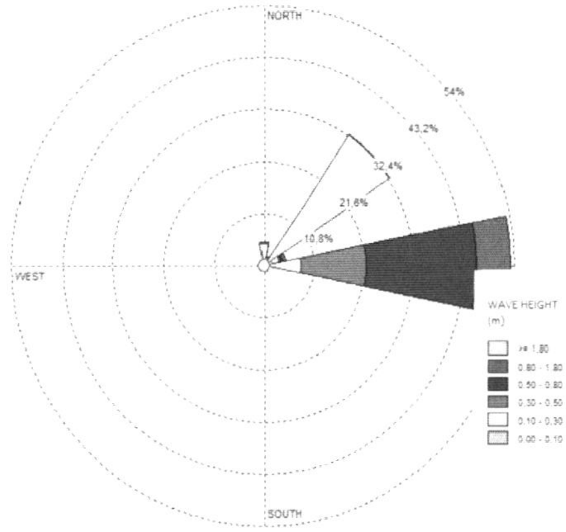

Figure 5. Wave Rose in PLTGU Grati Area

Figure 6. Data observation points

The collection of current, temperature, and sediment data was conducted at the 6 observation points, as shown in the figure 6 with the following coordinates; Point 1 (inlet canal - inner) : X=723615.2, Y=9154027.3; Point 2 (inlet canal - middle): X=723634.0, Y=9154573.1; Point 3 (north of inlet canal): X=723677.7, Y=9155145.8; Point 4 (eastern coast): X=724881.1, Y=9154013.6; Point 5 (outlet): X=723033.9, Y=9154090.1 and Point 6 (western coast): X=721710.9, Y=9153990.6

4. Delft3D Numerical Modelling

Delft3D is a widely used numerical modeling software package for simulating various physical and environmental processes in aquatic systems, including oceans, rivers, lakes, and coastal areas. It is developed by Deltares, a research institute in the Netherlands, and is particularly well-suited for hydrodynamic, sediment transport, and water quality modeling. When applied to the context of PLTGU Grati, which is interested in exploring thermal and sedimentation barriers, Delft3D can be used to perform numerical simulations to analyze and optimize these aspects. In this study, the modeling of heat distribution and sedimentation rate is conducted both for the initial jetty condition and for the modified layout and length of the jetty.

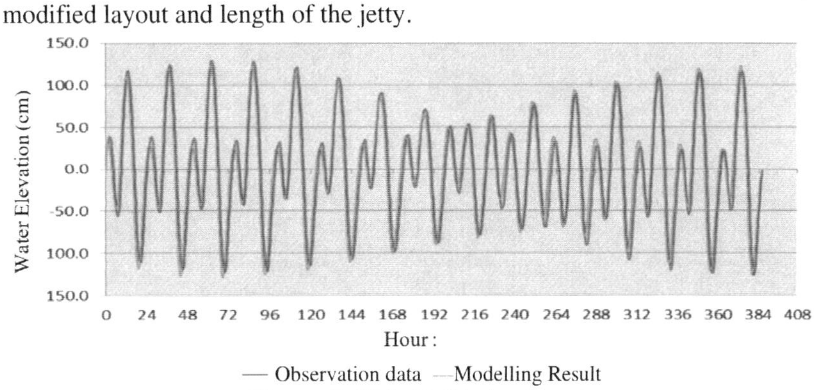

Figure 7. Tide Validation Result

IOP Conf. Series: Earth and Environmental Science 1298 (2024) 012036 doi:10.1088/1755-1315/1298/1/012036

5. Results and Discussion

5.1. Validation of the model with Tide Data

Simultaneously, the validation process at a given time step revealed an error of 8.97% when comparing observational data with model output data. Refer to figure 7 for the graphical representation of this comparison.

5.2. Results of the Existing condition model

The distribution of currents velocity, which is a combination of tidal currents and currents generated by wave breaking, is presented in this model. Figure 8 and figure 9 show the distribution of warm water exiting the outlet canal, which is highly dependent on the flow rate. It can be observed that an increase in flow rate (an increase in power) leads to a broader distribution of hot water flow. Meanwhile, the characteristics of current distribution show that during high tide, the dominant current direction is towards the coastline, whereas during low tide, the dominant current direction moves away from the coast, as shown in figure 10 and figure 11. During low tide, the dominant current direction moves away from the coast, allowing the distribution of warm water to reach the mouth of the inlet canal. A comparison of temperature differences at the inlet canal under each condition is presented in table 1. It is shown that the temperature difference between the existing condition and the condition with increased power (increased flow rate) is greater than 0.149°C for the lowest temperature and greater than 0.181°C for the highest temperature, this outcome aligns with the findings presented in [16].

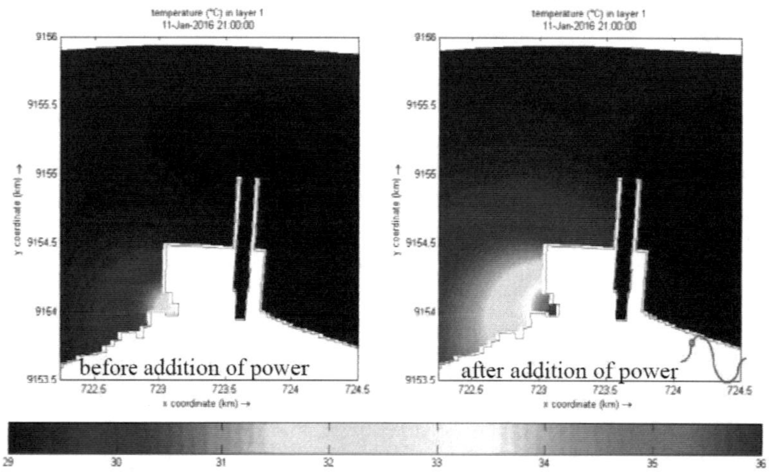

Figure 8. Water Temperature during High Tide in Existing Condition

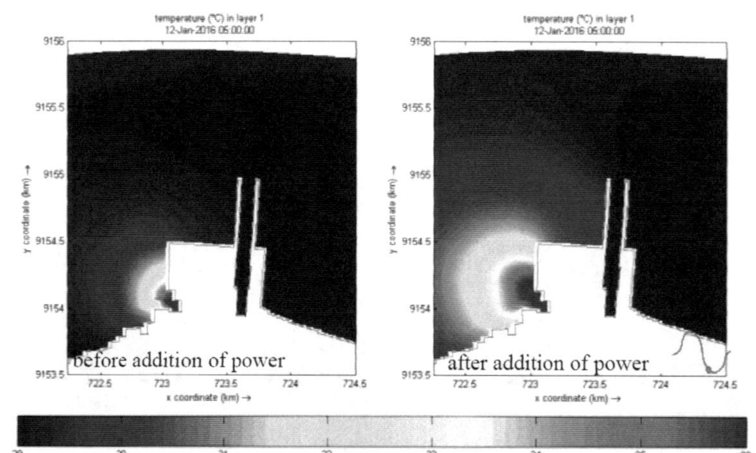

Figure 9. Water Temperature during Low Tide in Existing Condition

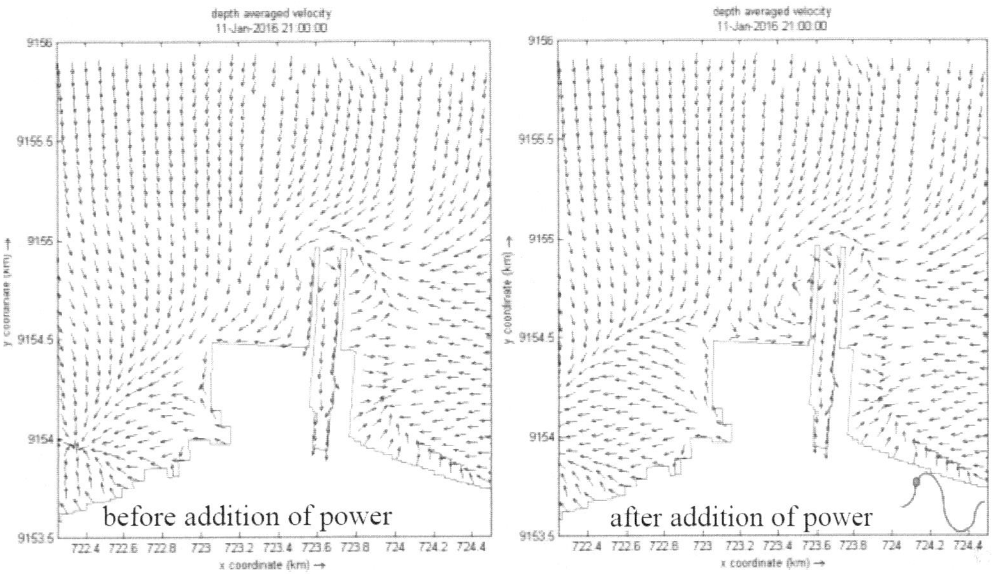

Figure 10. Current Direction during High Tide in Existing Condition

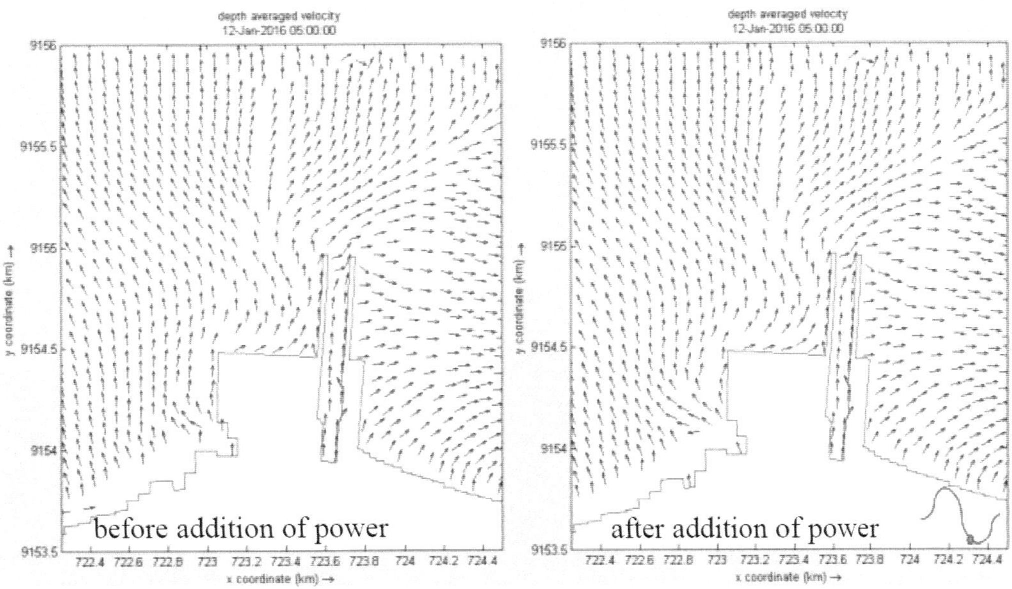

Figure 11. Current Direction during Low Tide in Existing Condition

Table 1. Comparison of Temperature in the Existing Condition Before and After Power Increase

Condition	Temperature in the Intake Channel	
	Lowest (°C)	Highest (°C)
Observation Data	27.869	28.934
Before addition of Power	28.369	28.580
After addition of Power	28.518	28.761

5.3. The proposed jetty extension

In this study, two forms of jetty extension for the PLTGU Grati's canal water intake are proposed. The first is Alternative 1, which involves an extension turning to the right at 51° degrees relative to the X-axis, with an additional length of 200 meters on the eastern side and 272 meters on the western side of the jetty, as shown in figure 12(a). The second is Alternative 2, which involves an extension turning to the left at 58° degrees relative to the X-axis, with a length of 200 meters on the western side and 272 meters on the eastern side of the jetty (figure 12(b)). Subsequently, both of these alternatives are simulated to assess the distribution of hot water, the patterns of current velocity, and the sedimentation rate, in order to evaluate which alternative better meets the needs of the PLTGU Grati Power Plant.

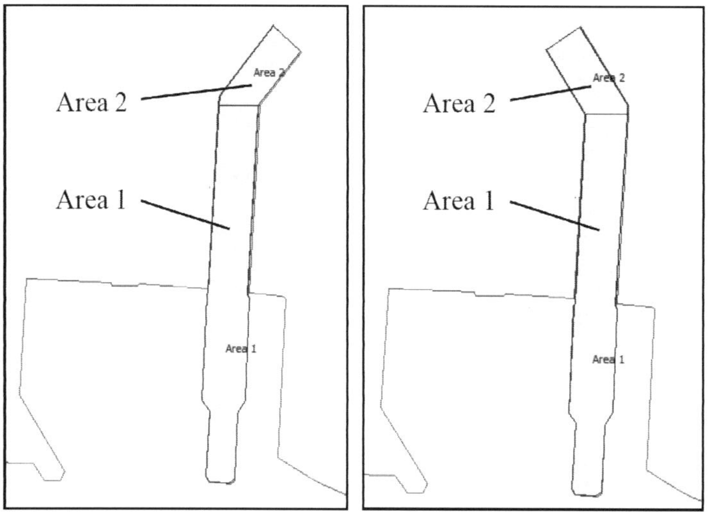

(a) (b)

Figure 12. The proposed jetty extension, Alternative 1 and Alternative 2

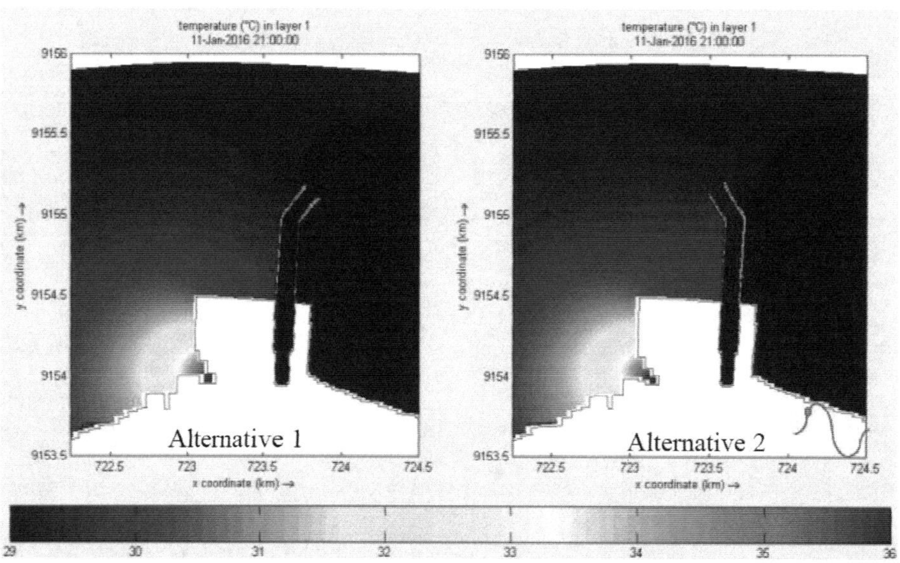

Figure 13. Distribution of hot water dispersion during Rising Tide Conditions of Power Addition Alternative Model

Figures 13 and 14 show the simulation results of hot water distribution for conditions before installation and during low tide for both Alternative 1 and Alternative 2. Low tide is a critical condition because the dominant currents flow towards the open sea. Both Alternative 1 and

Alternative 2 are quite effective in containing the spread of hot water compared to the average temperature simulation results for the existing jetty condition (28.640°C). This indicates that the hot water distribution in the simulation results for Alternative 1 and Alternative 2 is still below the existing condition, which is 27.792°C and 28.617°C, respectively. However, for sedimentation barriers, Alternative 2 shows better results than Alternative 1. The average sediment inflow rate into the canal water intake for Alternative 1 is 4207.08 m3/month, while for Alternative 2, it is 3751.10 m3/month. In contrast, the existing jetty condition with the addition of flow rate results in an average sedimentation rate of 5579.67 m3/month (as shown in figure 15 and figure 16). Based on the analysis of hot water distribution and sedimentation rate, it can be suggested that modifying the shape and length of the jetty for Alternative 2 can provide satisfactory results to support the operation of the Grati Combined Cycle Power Plant with the planned additional capacity. However, the selection of sediment transport formula and modifying the jetty extension are also very important to increase the accuracy of modeling results and the sedimentation barriers [17, 18, 19 and 20].

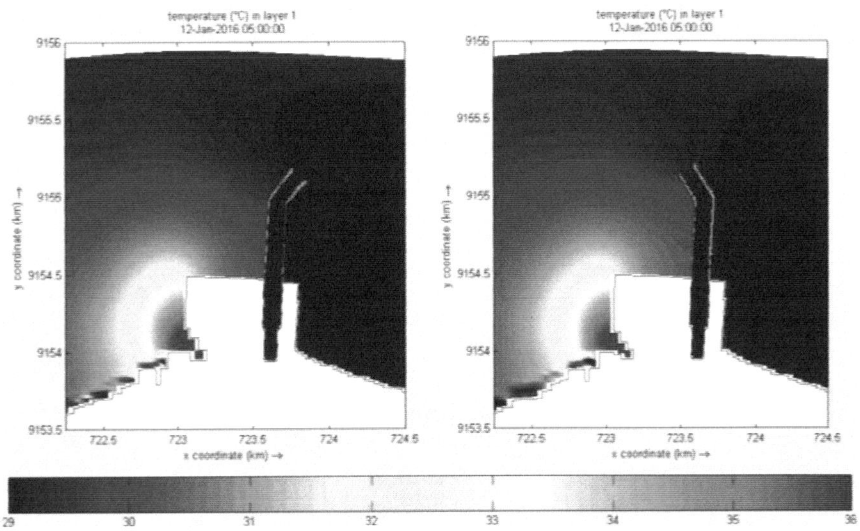

Figure 14. Distribution of hot water dispersion during Low Tide Conditions of Power Addition Alternative Model

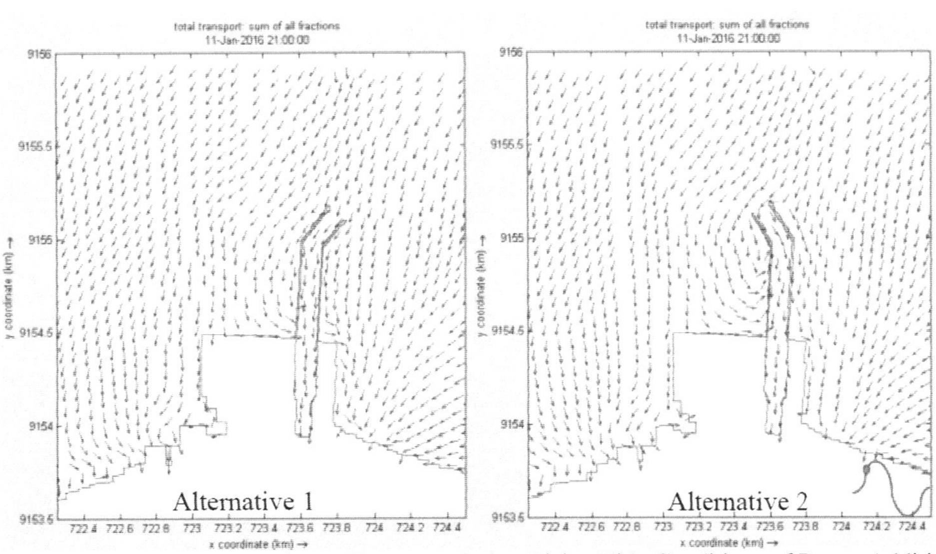

Figure 15. Direction of Total Sediment Transport during Rising Tide Conditions of Power Addition Alternative Model

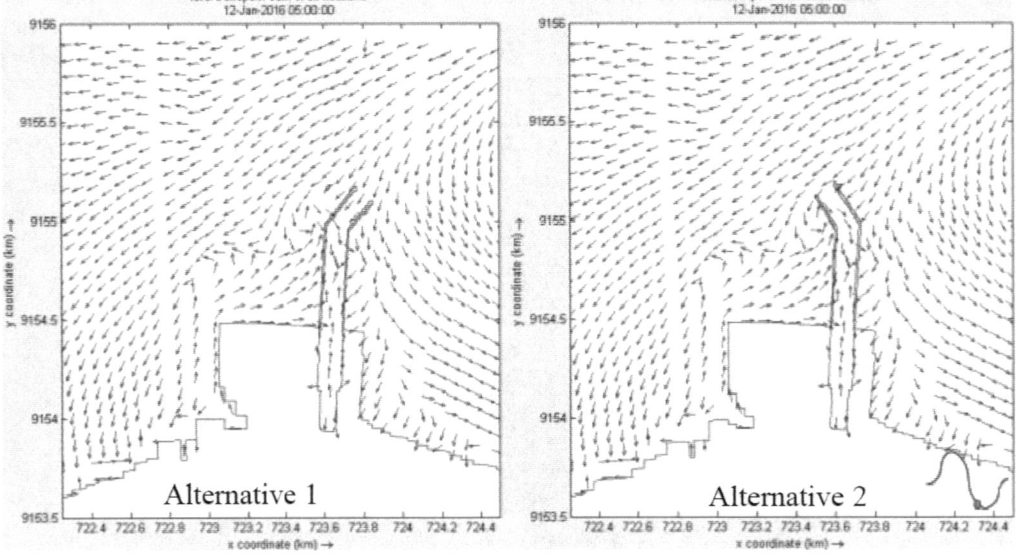

Figure 16. Direction of Total Sediment Transport during Low Tide Conditions of Power Addition Alternative Model

6. Conclusions

A study on sedimentation and thermal barriers related to the expansion of the Grati Combined Cycle Power Plant (PLTGU) has been conducted through alternative forms and extensions of the jetty. From the discussion above, it can be concluded that Alternative 2 is capable of reducing the average temperature significantly compared to the Existing condition, by approximately 0.023°C. As for sedimentation rates, Alternative 2 can reduce the sedimentation rate to 3751.10 m3/month, whereas Alternative 1 results in a sedimentation rate of 4207.08 m3/month. In contrast, the sedimentation rate for the existing jetty condition is significantly higher, at 5579.67 m3/month. It is evident that Alternative 2 can reduce the sedimentation rate by approximately 1828.57 m3/month compared to the existing jetty condition. Therefore, with the construction of the modified extension of jetty in Alternative 2, it is expected that this condition will increase the water intake capacity for cooling through the water intake canal, in line with the need for increased power and capacity at the Grati Combined Cycle Power Plant. Furthermore, dredging activities in the water intake canal are expected to decrease significantly, as Alternative 2 can reduce sedimentation rates by up to 32.77%.

7. References

[1] PT PLN (Persero), Rencana Usaha Penyediaan Tenaga Listrik (RUPTL) PT PLN (Persero) 2015-2024. Jakarta: PT PLN, 2014, pp. 27-29. Retrieved on 27 Juni 2016 from www.pln.co.id

[2] Camenen B and Larson M 2005 A general formula for non-cohesive bed load sediment transport. *Estuarine, Coastal and Shelf Science* 63 (1-2), pp. 249-260.

[3] Suntoyo, Tanaka H and Sana A 2008 Characteristics of turbulent boundary layers over a rough bed under saw-tooth waves and its application to sediment transport. *Coastal Engineering* 55(12), pp. 1102-1112.

[4] Suntoyo and Tanaka H 2009 Effect of bed roughness on turbulent boundary layer and net sediment transport under asymmetric waves. *Coastal Engineering* 56 (9), pp. 960-969.

[5] Suntoyo, Fattah A H, Fahmi M Y, Rachman T and Tanaka H 2016 Bottom shear stress and bed load sediment transport due to irregular wave motion. ARPN J. Eng. Appl. Sci. 11(2), pp. 825-829.
[6] Panda U S, Mohanty P K and Samal R N 2013 Impact of tidal inlet and its geomorphological changes on lagoon environment: A numerical model study. *Estuar. Coast. Shelf Sci.* 116, 29–40.
[7] Silva P D, Lisboa P V, Fernandes E H 2015. Changes on the fine sediment dynamics after the port of rio grande expansion. *Adv. Geosci.* 39, 123–127.
[8] Wijaya M M, Suntoyo and Damerianne H A 2016 Bottom shear stress and bed load sediment transport formula for modeling the morphological change in the canal water intake. *ARPN J. Eng. Appl. Sci.* 11(4), pp. 2723-2728.
[9] Fattah A H, Suntoyo, Damerianne H A and Wahyudi 2018 Hydrodynamic and Sediment Transport Modelling of Suralaya Coastal Area, Cilegon, Indonesia. *IOP Conf. Ser.: Earth Environ. Sci.* 135(1) 012024.
[10] Wahyudi N R, Suntoyo and Pratikto W A 2021 Hydrodynamic and Sediment Transport Simulation at the Port of the Electric Steam Power Plant Adipala and Serayu Estuary, Central Java Province, Indonesia. *IOP Conf. Ser.: Earth Environ. Sci.* 698 012006.
[11] Hidayati N, Suntoyo and Pratikto W A 2023 Bed shear stress and sediment transport equations for predicting morphological and shoreline changes: A review. *IOP Conf. Ser.: Earth Environ. Sci.* 1198 012036.
[12] Franzen M O, Silva P, Siegle E, Fernandes E H L 2021 Influence of long jetties on estuarine and coastal hydrodynamics in a microtidal estuary. *Regional Studies in Marine Science* 59, 102809.
[13] Suntoyo, Muslim T W, Wicaksana F T, Rahmawati S and Silvianita 2021 An Experimental Study on Hydraulic Model of Water Intake Canal at Steam and Gas Power Plants. *IOP Conf. Ser.: Earth Environ. Sci.* 698 012029
[14] Suntoyo 2020 Comparison of Turbulence Models in the Turbulent Wave Boundary Layer for Cnoidal Waves. *Int. J. on Engineering Applications (IREA)* 8(5), pp 202-214.
[15] Putri P D M and Suntoyo 2020 Tidal Effect on Sea Water Intake of Power Plant using CFD Model. *Proc. of the 6th Int. Sem. on Ocean and Coastal Engineering, Environmental and Natural Disaster Management* 1, pp 209-211. DOI: 10.5220/0009859102090211
[16] Cahya, I., Suntoyo, S., Lailiyah, I. and Nurdin, J.A. 2023. Jetty Extension Study Due to Addition of Power Capacity on Thermal Water Dispersion at PLTGU Grati Pasuruan. *Maritime Park.* (Jun. 2023), 53-60. DOI:https://doi.org/10.20956/mp.vi.27892.
[17] Hidayati, N, Suntoyo and Pratikto, WA 2023 Bed shear stress and sediment transport equations for predicting morphological and shoreline changes: A review, IOP Conference Series: Earth and Environmental Science, vol. 1198, no. 1, 012036. https://doi.org/10.1088/1755-1315/1198/1/012036
[18] Tanya M. Beck, Ping Wang 2019 Morphodynamics of barrier-inlet systems in the context of regional sediment management, with case studies from west-central Florida, USA, Ocean and Coastal Management (177), pp. 31-51, https://doi.org/10.1016/j.ocecoaman.2019.04.022.
[19] Wang, Y H, Tang, L Q, Wang, C H, Liu, C J and Dong, Z D 2014 Combined effects of channel dredging, land reclamation and long-range jetties upon the long-term evolution of channel-shoal system in Qinzhou bay, SW China, *Ocean Engineering* (91), pp. 340-349, https://doi.org/10.1016/j.oceaneng.2014.09.024.
[20] Angnuureng, B D, Adade, R, Chuku, E O, Dzantor, S, Brempong, E K, Mattah, P A D 2023 Effects of coastal protection structures in controlling erosion and livelihoods, Heliyon (9)(10), pp. e20633, https://doi.org/10.1016/j.heliyon.2023.e20633.

Acknowledgments
This work was partially suported by ITS's WCP-like program 2023. The authors would also like to acknowledge the PLTGU Grati for providing the information and the available data necesarry for conducting the research.

IOP Conf. Series: Earth and Environmental Science 1298 (2024) 012037 doi:10.1088/1755-1315/1298/1/012037

Pre-processing data and window function testing on wave spectrum analysis

S. Panalaran [1) 3)], Suntoyo[1)], Aries Sulisetyono [2)]

[1] Ocean Engineering Department, Institut Teknologi Sepuluh Nopember, Surabaya, Indonesia
[2] Naval Architecture Department, Institut Teknologi Sepuluh Nopember, Surabaya, Indonesia
[3] Regional and Infrastructure Technology Department, Institut Teknologi Sumatera, South Lampung, Indonesia

Corresponding author email: satriyo.panalaran@kl.itera.ac.id

Abstract. Pre-processing data in oceanographic data research is an important step in processing data into the required information. This phase often takes longer and more rigorous time to present the correct data. Data pre-processing involves the selection of data, and the modification of data so that it can be readable in a computational algorithm. The data selection process includes selection of sample numbers, correction and filling of data gaps.The data used in spectrum analysis is the recording of the water surface profile of the wave in the data resolution of 1 second. The data needs to be corrected by the process of detrending against the tidal height which become the still water level. The spectral analysis is processed using the pwelch function on MATLAB. The frequency spectrum curve of the measurement is obtained from the calculation of the power spectral density value. In order to obtain a stable power spectral density curve, the Hann window function is used with the number of window elements, nfft/16.

1. Introduction

The data read by measuring device is not always directly usable. Sometimes there is noise obtained during measurement. The noise may be caused by disturbance by activity around the recording device, or environmental anomalies, or the device's failure to function properly due to the power instability or due to limitation in the sensor used. Commonly known, there are several methods in measuring waves in the sea [1], among others: by using an accelerometer on a buoy, measuring electrical capacitance using wave staff, echosounder installed in reverse at the bottom of the water, the acoustic doppler current profiler (ADCP) for identify the orbital motion of water particles due to the wave movement, using pressure sensor [2], or using ultra-sonic sensors like the device developed by Adrianto [3].

In this article the author will share the process in the early stages of data processing performed for spectral analysis of wave measurement data carried out by Adrianto in 2016 in the West Jave Sea [3]. The wave recording device developed is the modified ultra sonic device (MUSD) which operates on the principle of sending ultra sonic acoustic waves from the transducer to the surface of the water. Then the height of the surface profile of water is measured based on the distance between the water surface and the transducer read from the travel time of the acoustic wave passes back to the receiver. The MUSD is designed to measure data at sampling frequency of 6 Hz, meaning that in 1 second, the device will record as many as 6 times in the same time interval. The objective is to capture the wave cycles in the high-frequency component, or more precisely the wave frequency components generated by the wind. This

IOP Conf. Series: Earth and Environmental Science 1298 (2024) 012037 doi:10.1088/1755-1315/1298/1/012037

component is important because in the analysis of random waves in a short time. The wave generated by the wind dominate every occurrence of waves event at sea. Note on Figure 1, the condition of the waves generated by the wind forming wave in a frequency component of around 0.05 Hz to 8 Hz.

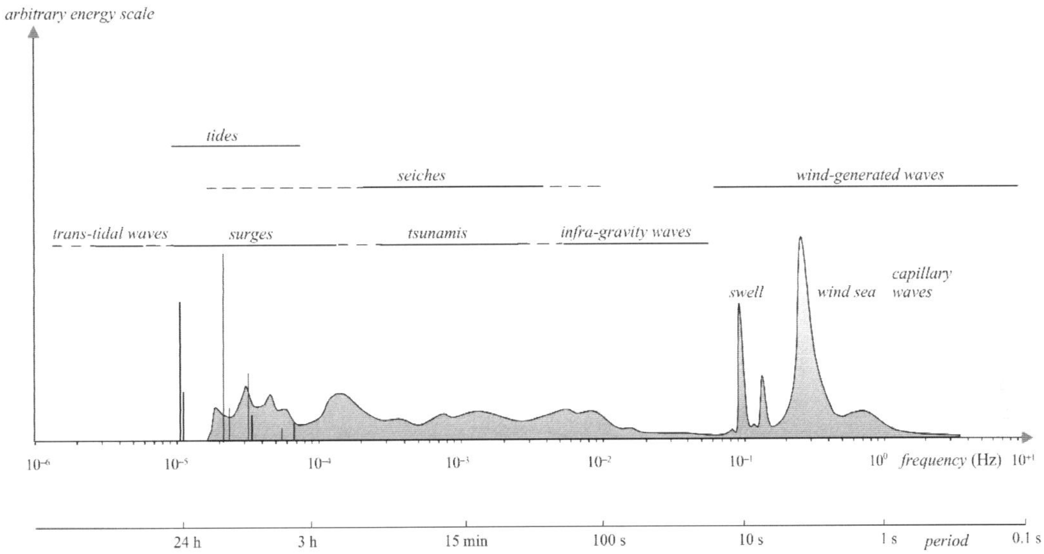

Figure 1. Wave based on classification of the constituent frequency components (after Munk, 1950 adopted from Holthuijsen [1])

In FFT analysis, the Nyquist frequency is known, which in order to identify a particular frequency component, requires a sample frequency that is greater or at least 2 times the highest frequency. For example, to examine a wave with a frequency of 1 cycle per second, then a sample frequency of at least 2 cycles per second is required (or 0.5 Hz). Thus, in order to review components of sea waves that have higher frequencies, a larger data recording interval is required. Recording carried out by Adrianto [3] with a sampling frequency of 6 Hz, would ideally be able to identify up to a frequency of 3 Hz. Where waves with a frequency of 3 Hz are capillary waves that appear to restore the equilibrium to surface water with the force of gravity. However, before the data is used, it is necessary to check whether the entire data can be used directly, or whether modification processes need to be carried out on the data such as smoothing, cutting, and filling.

2. Data Inspection and Editing

Data is surface recording of wave water with a resolution of 6 Hz. Where in 1 second consist of 6 data with the same time interval (1/6 seconds). However, most of data indicates a recording time gap with an irregular row, as shown in Figure 2. To maintain the reliability of data and prevent misinformation about the presence of other wave frequency components, only data with an interval of 1 second is used. Meanwhile, to the fill the gap between data in seconds to i and i+1, linear interpolation is performed on data time rows. Moreover, if the data is found to be more than 1 second apart, then the data at a time that does not have information will be positioned according to the still water level (which is zero elevation in measuring the variance of water level). The data is then modified in the water surface's high vector array and saved in the *.csv file.

Time	Date	Profile-1	Profile-2	Profile-3	Profile-4	Profile-6	Profile-6
0:00:00	22/08/2016	1.355	1.355	1.355	1.391	1.424	1.467
0:00:01	22/08/2016	1.669	1.747	1.824	1.886	1.962	1.993
0:00:03	22/08/2016	1.885	1.82	1.748	1.719	1.654	1.576
0:00:04	22/08/2016	1.464	1.471	1.471	1.477	1.537	1.537
0:00:05	22/08/2016	1.731	1.757	1.758	1.744	1.727	1.728
0:00:07	22/08/2016	1.679	1.679	1.665	1.656	1.633	1.633
0:00:08	22/08/2016	1.621	1.621	1.62	1.665	1.694	1.723
0:00:10	22/08/2016	1.829	1.857	1.858	1.856	1.854	1.843
0:00:11	22/08/2016	1.67	1.604	1.541	1.505	1.406	1.32
0:00:12	22/08/2016	1.231	1.231	1.373	1.397	1.434	1.554
0:00:14	22/08/2016	1.873	2.067	2.076	2.025	1.957	1.876
0:00:15	22/08/2016	1.606	1.431	1.431	1.431	1.432	1.51
0:00:16	22/08/2016	1.653	1.71	1.756	1.772	1.807	1.801

Figure 2. Loss of data from measurement

time Column1	date Column2	profile Column3	time difference Column4	time interpolation Column5	profile interpolation Column6
0:00:00	22/08/2016	1.355			
0:00:01	22/08/2016	1.669	0.0000115740741	FALSE	FALSE
0:00:03	22/08/2016	1.885	0.0000231481481	0:00:02	1.777
0:00:04	22/08/2016	1.464	0.0000115740741	FALSE	FALSE
0:00:05	22/08/2016	1.731	0.0000115740741	FALSE	FALSE
0:00:07	22/08/2016	1.679	0.0000231481481	0:00:06	1.705
0:00:08	22/08/2016	1.621	0.0000115740741	FALSE	FALSE

Figure 3. Data filling procedure with linear interpolation

3. Detrending Process to wave surface profil

It should be remembered that the analysis of the wave spectrum is based on the deviation of the water level at a time from the average water level. However, it is known that the water level always varies according to the cycle of the wavelength of tides. Assume if the sea turns calm, we will find that the height of the water surface is the sea level due to tidal factors alone. Thus, the middle water level of a pulse movement of the up and down waves more accurately indicates the position of the calm water level, rather than the average water level. This is especially true of a data set that has a duration of more than 4 hours (considering the shortest tidal component periods which are influenced by the sun's gravitational force in shallow waters [4]).

Figure 4 shows a plot of data measurement of water surface profiles at the Pabelokan wave measurement station [5] whic is overlaid with tidal water level data (i) published by the Geospatial Information Agency [6]; and (ii) Naotide tide predictions (credit: National Astronomical Observatory of Japan). The best tidal data should be provided by provided by the measurement staion. However, the availability of tidal data at Cilegon Sation is far from the location of wave measurements at Pabelokan Station, and is not in an area that geographically can represent each other. Therefore, water level recording data from the Cilegon tidal station does not represent the still water level at Pabelokan Station. From the results of tidal prediction using the Naotide program, it was found that the water level was exactly at the center line between the height of the rising and falling wave water surface profiles. This also adds confidence to the water surface profile recording data for wave analysis used in this study.

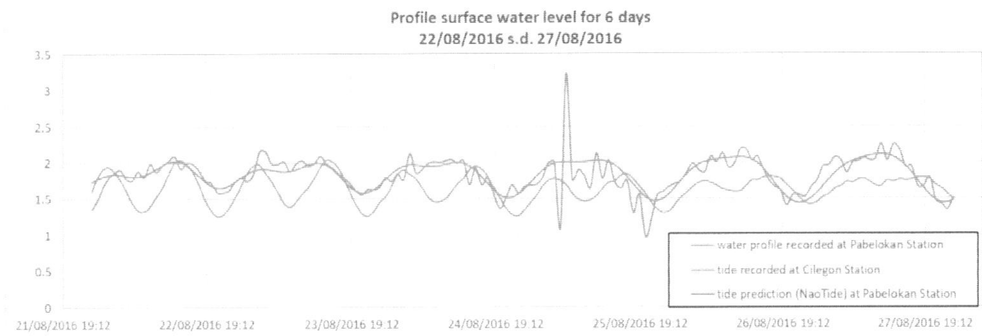

Figure 4. Plot data profile surface wave water and tidal conditions 1 hour interval over a duration of six days

Another use of aligning recorded water level data in wave and tidal measurements is to define water level heights in a time series where data is missing. This is illustrated in Figure 5 where there are no measurements from 16:00 to 17:00. In order to maintain the length of the data, then the information at that time must be filled with an altitude data. The most realistic things is to fill this information with data on the condition of still water level. This is done so as not to cause miss information from entering any data.

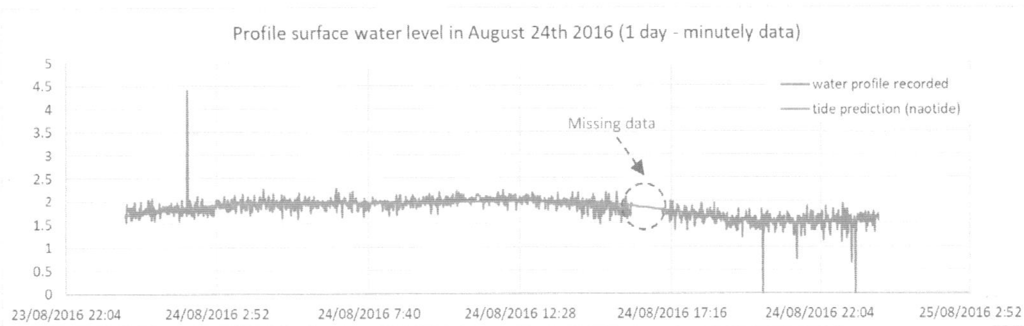

Figure 5. Plot data profile surface wave water and tidal conditions 1 minute interval over a duration of one days

The detrending process is performed against the data time series to eliminate the trend and periodic cycles contained in the data. The result of the detrending process is to calculate the deviation between the height of the water surface profile and the average value, or the value at the pulse reference height at rest. The result of the detrending process of wave water surface measurement data against tidal data are presented in Figure 6b. For short data (within a duration of less than 1 hours), this detrending process can be performed against the mean value by ignoring the error value that may occur even if the value is small.

4. Pwelch Function on Power Spectral Density

Wave spectrum analysis is based on the form of power spectral density (PSD) [2] using the Ocealyz Program [7]. Power spectral density is a measure of the power distribution of each frequency component of the wave composer, expressed in the unit "power per frequency" or "power per radians. Mathematically PSD is expressed in notation:

$$PSD = \frac{|X_n|^2}{f}\left(\frac{amplitudo^2}{Hz}\right) \quad (1)$$

Notation X_n is the amplitude value of the pulse of a wave in the frequency domain, so that the numerator term of the equation shows the power value (amplitude^2) of the wave per frequency bin. Computation process to convert waves from the time domain to the frequency domain is done using the

Fast Fourier Transform technique (FFT). One of the algorithms commonly used in calculating power spectral density is the Pwelch function [8].

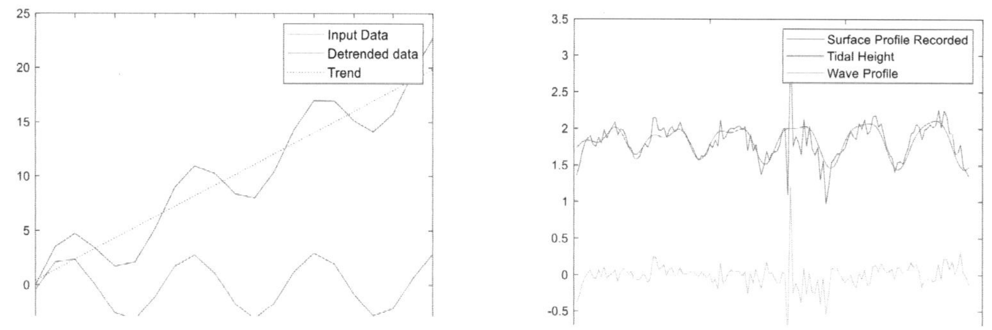

(a) (b)

Figure 6. a) Illustration of detrending on data with uptrend; b) Detrending data profile surface wave water against the tidal elevation

The Pwelch function uses the welch method to divide wave pulse into several overlapping segments. In order to reduce the loss of spectral information on the edge of the window, a data set overlapping between the window segments is performed. The process of transforming data into periodograms is done with discrete Fourier transform in the FFT algorithm. Each periodogram is then synthesized to obtain a PSD estimate. The advantage of using the Pwelch method is that it can reduce the spectral leakage that becomes readable on other frequency bins when analyzing signals in a limited duration. So the use of the Pwelch method can be useful to give better stability to the obtained PSD curve.

5. Windows Function on Power Spectral Density

Recording of the wave profile in a series of pulses in a series of pulses is carried out for a limited time. It is assumed that the wave pulses will have a repeating shape after a span of observation time. It is realized that this can cause wave discontinuity at the edge of each window segment. The result of this wave discontinuity can cause a leak of the energy contained in each frequency bin to another frequency bin, a phenomenon called a spectral leakage. The window function is applied to a time series signals as a consequence of using samples within a limited time. The window function has a role as a weighting function which will gradually weaken the signal amplitude to a value of 0 on the edge of the window. Thus, there is no significant frequency change due to sudden changes in the data.

There are many window functions that have been developed, each of which will provide different frequency response characteristics. In simple terms, the form of window function is classified into three main categories, namely: rectangular window functions, the sharp decreasing function called the tapering windows, and the cosine-shaped functions. The advantage of a rectangular window is that it can maintain the amplitude and wave energy, while the advantage of cosine and tapering windows is that they can shrink sidelobes in the frequency response to prevent spectral leakage. Meanwhile, the weakness of the cosine and tapering windows generally give smaller amplitude values to the spectrum and cause the loss of some detailed power spectrum information and frequency resolution.

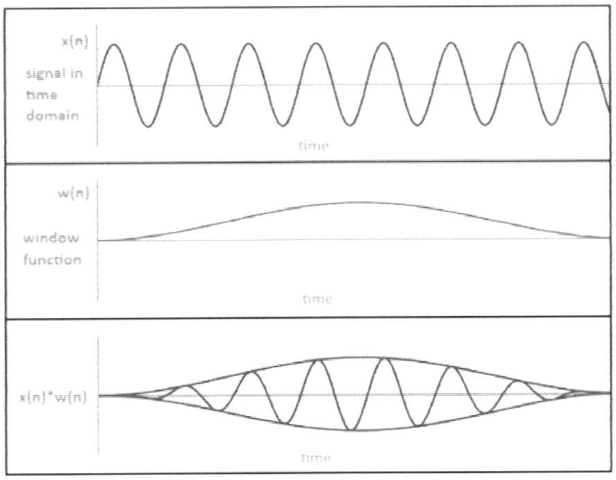

Figure 7. Illustrate the use of window functions

One of the basics for selecting a window function is based on the resolution and dynamic range produced by a window function. Resolution means the ability of the frequency response to show the difference between two adjacent frequencies. Dynamic range is the ability to detect weak frequency components in the presence of strong frequency components. A rectangular window a window that has the highest resolution and lowest dynamic range. This means that the rectangular window is good at distinguishing components of the same amplitude even if they are at very close frequencies, but bad at distinguishing components with different amplitudes even at distant frequency. In contrast to the rectangular window type, the cosine window type has the lowest frequency resolution but the highest dynamic range. An example of a taper type is a Bartlett window, while a cosine type has a shape that resemble a bell function like Gauss, Hamming, Hann, and so on.

In addition to the resolution and dynamic range, in selecting the window type it is also necessary to consider the stability of the spectrum estimate and its spectral leakage rate [9]. Stability is indicated in a condition to which spectrum estimates obtained from different segments can be aligned. Stability also implies smoothing on structures that are irrelevant to remove. Between these resolution and stability parameters is a contradictive consideration, where high stability is generally obtained from averaging a large number of periodograms, resulting in resolution reduction. The last consideration on the choice of window type is how much energy leakage can be minimized.

In spectrum analysis with power spectral density, the window function has an important role in obtaining PSD graphs. The Pwelch method divides the data into several segments for later averaging. Each segment is given a weighting function by applying a window function to avoid spectral leaks and obtain PSD information as expected. In this test, six window functions were applied to observe whether the selection of the window functions would have a significant influence on the spectral analysis of the wavefront profile. The six function windows tested are Rectangular, Gauss, Bartlett, Barthann, Hamming, and Hann windows. The frequency response given by the six window types is shown in Figure 8. The equation that forms the window weighting function are as follows [10] [11]:

a. Rectangular

$$w(n) = \begin{cases} 1 & 0 \le n \le N \\ 0 & n < 0, n > N \end{cases} \tag{2}$$

b. Gauss

$$w(n) = exp\left(-\frac{1}{2}\left(\alpha\frac{n}{N/2}\right)^2\right) = exp\left(\frac{-n^2}{2\,\sigma^2}\right) \quad ; -N/2 \le n \le N/2 \tag{3}$$

α inversely proportional to the standard deviation, σ ($\sigma = N/2\alpha$).

c. Barlett

$$w(n) = \begin{cases} \frac{2n}{N} & 0 \le n \le \frac{N}{2} \\ 2 - \frac{2n}{N} & \frac{N}{2} \le n \le N \end{cases} \tag{4}$$

d. Barthann

$$w(n) = 0.62 - 0.48\left|\left(\frac{n}{N} - 0.5\right)\right| + 0.38\,cos\left(2\pi\left(\frac{n}{N} - 0.5\right)\right) \quad 0 \le n \le N \tag{5}$$

e. Hamming

$$w(n) = 0.54 - 0.46\,cos\left(2\pi\,\frac{n}{N}\right) \quad 0 \le n \le N \tag{6}$$

f. Hann

$$w(n) = 0.5\left(1 - cos\left(2\pi\frac{n}{N}\right)\right) \quad 0 \le n \le N \tag{7}$$

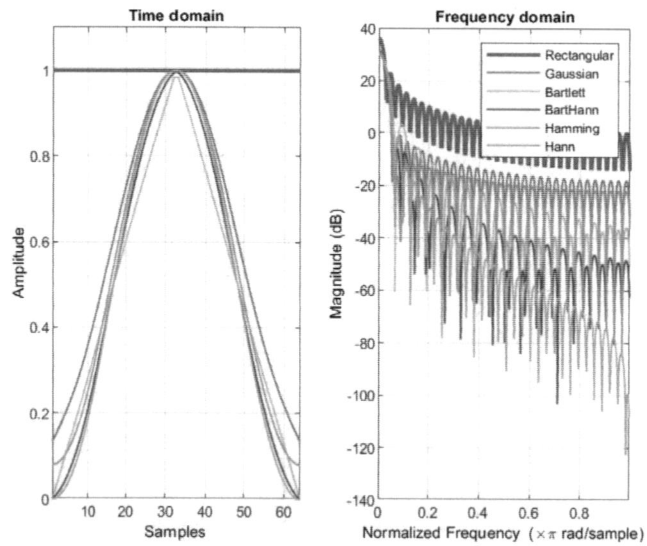

Figure 8. Frequency response for varying window function type

In general, the characteristics that can be seen from each type of window above are as follows [4]. In a rectangular window, the level of the first side lobes are 22% of the main lobes. Whereas in a Hann window the level of the side Lobes is much higher than that given by the rectangular window and results in a wider main Lobes. Meanwhile, the Hamming window produces a very low decrease in the level of the first side lobe (< 1% of the main lobes). However, after a very significant reduction in the level of the first side lobe, in the next side, the level of side lobes is not significantly reduced. On the Barlett window, it produces a narrower width of the main lobes, and a much smaller height of the side lobes than the rectangular window, which is about 5% of the heights of the main lobes. Referring to [9], to compare the effects of window types can be observed from the following characteristics: (i) normalized half main-lobe width, (ii) first side-lobe level, (iii) maximum side-lobe level, (iv) ratio of main-lobe energy to total energy, and (v) rate of fall-off side-lobe level.

Table 1. Comparison of window function properties [9]

Window	Normalized Half Main-Lobe Width	First Side Lobe Level (Db)	Maximum Side-Lobe Level (Db)	Ratio Of Main Lobe Energy To Total Energi	Rate Of Fall-Off Side Lobe Levels
Rectangular	0.5	-13.27	-13.27	0.902820	$1/\omega$
Gaussian/ Parabolic	0.708	-21.28	-21.28	0.992084	$1/\omega^2$
Bartlett/ Triangular	1	-26.53	-26.53	0.997057	$1/\omega^2$
Hamming	1	-44.05	-42.69	0.999632	$1/\omega^3$
Hann	1	-31.47	-31.47	0.999485	$1/\omega$

6. Window Function Settings Application in Power Spectral Density

The pwelch function used in calculating the power spectral density has default setting, namely: each segment applies the Hamming window weighting function; the amount of data in the FFT (nfft) process is 256 or the next power of 2 greater than the length of the segments; overlapping data between one segment and the next is 50% of the window length.

The application of the pwelch function by default to form a PSD curve from the wave surface profile data is shown in Figure 9. Both images show PSD curves for each day, namely August 23th and August 24th. The PSD curva on each day is then represented with a PSD average of 24 curves (24 hours) on that day, which is shown with a thick red line. It appears that high frequency resolution is unfavourable to form a smooth and well-defined PSD curve. In order to get a form of PSD curve to be the ideal formula, then a simple but firm curve shape is needed. Therefore, it is necessary to test to select the window type and set the number or element in the window averaging.

The testing of the PSD curve shape in varying window types and the number of elements in each window segment were performed on the result of PSD curves with the most dynamic values, namely the PSD curve on 23/08/206 at 21:00 and on 24/08/2016 by 22:00. The smooth and firm PSD curvature shape is expected to provide the highest reliability value that can represent spectrum estimates. Thus, to obtain a smooth PSD curve from the processing of the pwelch function, it is necessary to carry out the smoothing process which here uses the moving average technique. In Figure 10, it can be seen that the smaller the number of window elements used, the smoother the dynamic shape of PSD. This is because decreasing the number of window elements will cause an increase in the number of averaged windows, and will lead to a decline in the frequency resolution. This technique is done to achieve smoothness and statistical stability of spectrum estimates [9].

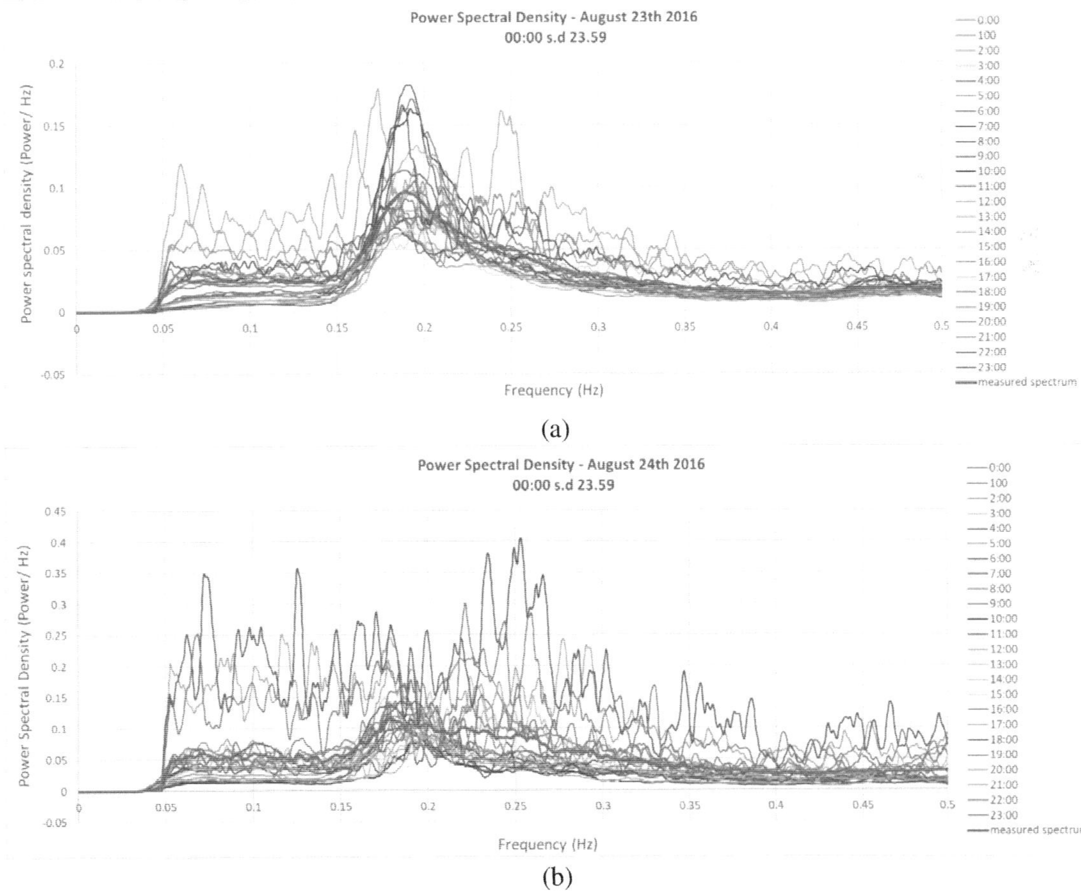

(a)

(b)

Figure 9. a) PSD curve for data on August 23th 2016; b) for data on August 24th 2016

IOP Conf. Series: Earth and Environmental Science 1298 (2024) 012037 doi:10.1088/1755-1315/1298/1/012037

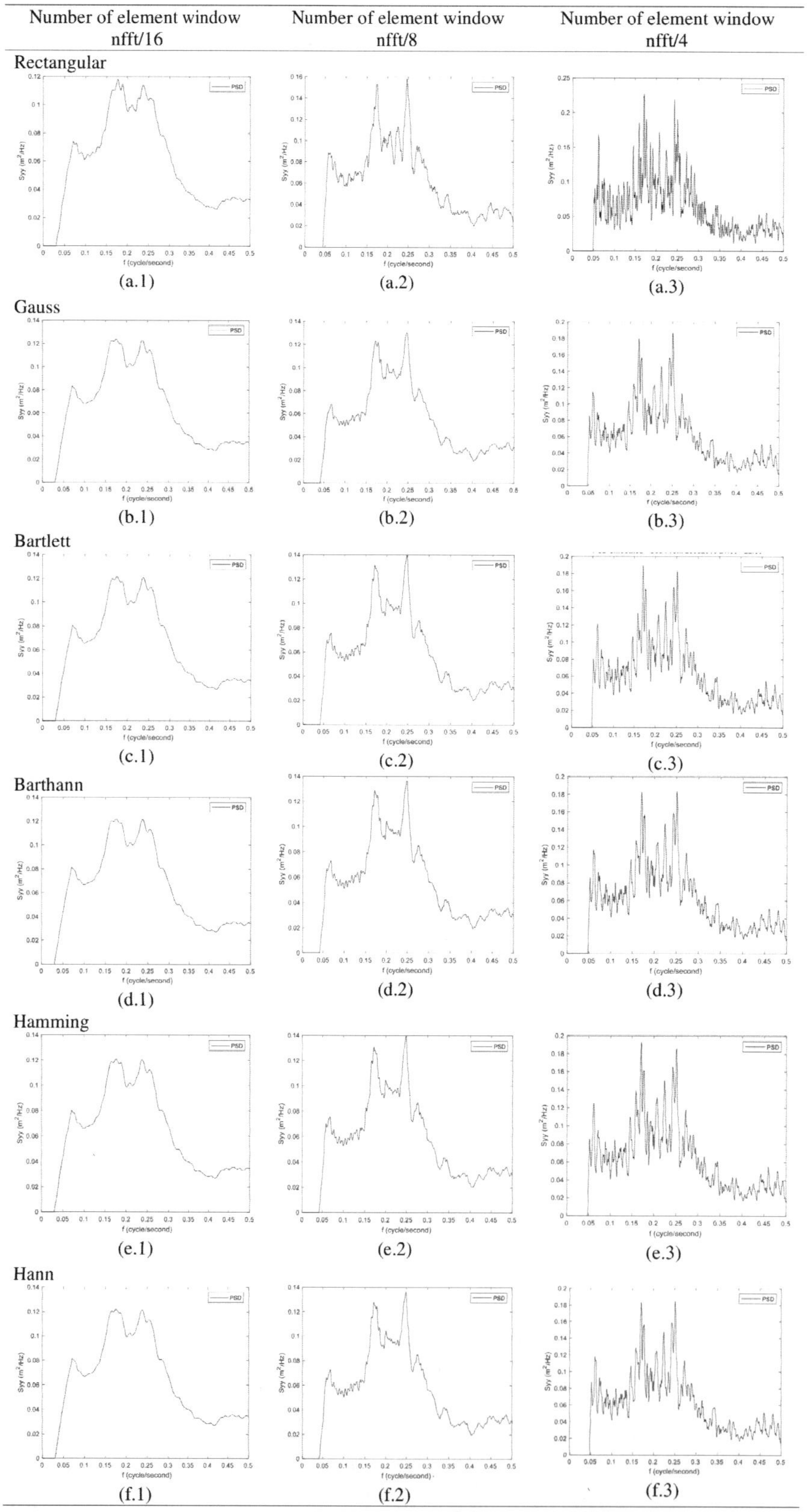

Figure 10. Test the variation of the window type and the number of window elements

The element window nfft/16 means that from a long set of data is divided into 16 windows, where the default of this pwelch function divides a set of data into 8 windows. By minimizing the window element, its frequency resolution value will decrease and can cause the height of the PSD in each frequencies bin to decreases. The setting of the value of a window element also cannot be immediately minimized, it is necessary to select a value of window element that can emphasize the shape of the PSD curve but also does not have a major consequence on the loss of amplitude power. Based on testing in Figure 10, it was finally determined that the number of window elements nfft/16 showed the most realistic shape.

In Figure 11 and Figure 12, comparisons were made of PSD results on data for August 23th 2016 (21:00) and August 24th 2016 (22:00). Each data was applied to the six types of spectrums that were tested, namely: Rectangular, Gauss, Bartlett, Barthann, Hamming, and Hann. The selection of the six types of window to be tested is random from the many types of window available. The only consideration is to give a representation of the general shape of the window type, for example, the triangular shape group represented by Bartlett, the power-cosine group of windows represented By Hann and Hamming, and the Parabolic group is represented by the Gaussian window.

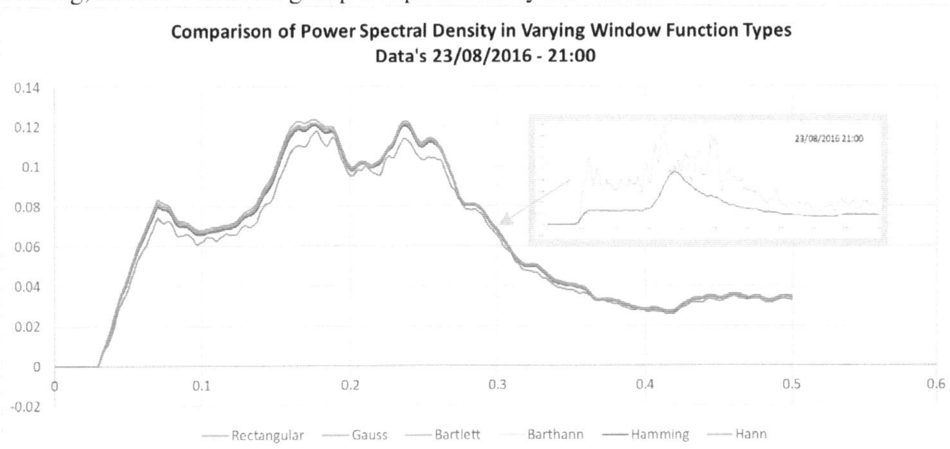

(a)

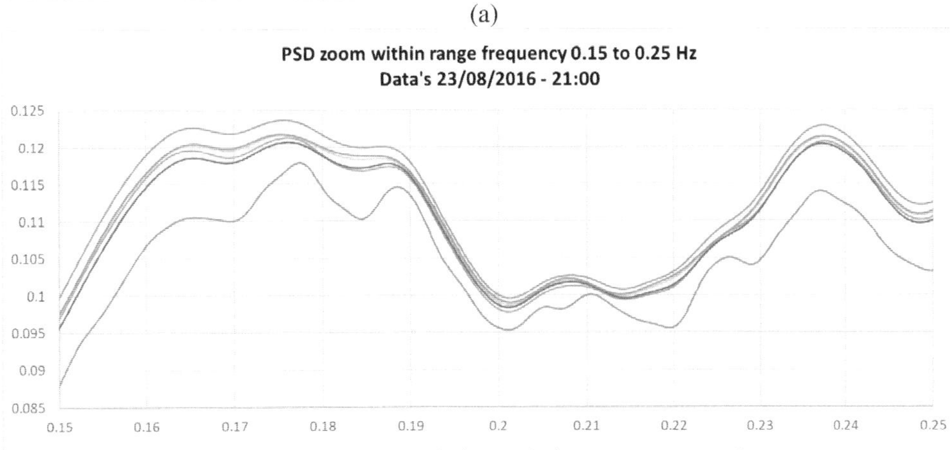

(b)

Figure 11. PSD with window type variation for data 23/08/2016 (21:00); nfft/16

From testing the six-window type, it was found that the rectangular type gave the most deviating results among the other types. This is normal because the rectangular window does not provide weighting to modify amplitude values for limited data. It can be seen that the frequency response provided by the rectangular type has the smallest power reduction. On the other five types give the same tendency to the results of the PSD curve. If these PSD curves are enlarged, as shown in Figure 11.b and Figure 12.b, then there seems to be a small difference that each window type produces. Nevertheless, among the five types, there is nothing significant in terms of reducing the power spectrum value. In spectral analysis applications to form frequency spectrum functions of random waves, the choice of window types can be slightly neglected. In essence, we need spectral shapes that has low resolution so that the spectrum curve can be more stable and more reliable in showing spectral estimates.

The selection of window type will generally consider two contradictory compensations, namely the quality of the frequency resolution or the received spectral leakage. The results of using the window function can be estimated based on the given characteristics. For the purposes of frequency resolution expected to be smaller, then can use the window function which has a narrow main lobe. On the other hand, window functions that have wider main lobes, it will produce high frequency resolution. Meanwhile, to obtain a low spectral leakage estimate, select the window functions with a low side lobes level and a high roll-off rate, whereas a window function with a high level of side lobes with a low roll-off rate results in a high level of spectral leakage [12]. To obtain good stability in the spectrum curve, a low frequency resolution is required, then select a window type with narrower main lobes width. So based on the window type applied in this study, then the Hamming, Hann, Barlett, and BartHann window types have the same main Lobes wide. So, these four window types can be taken into consideration. In line with this, in applications for spectrum analysis of ocean waves, the Hamming window or Hann window type is generally used for spectrum analysis of random waves [13], [14], [15].

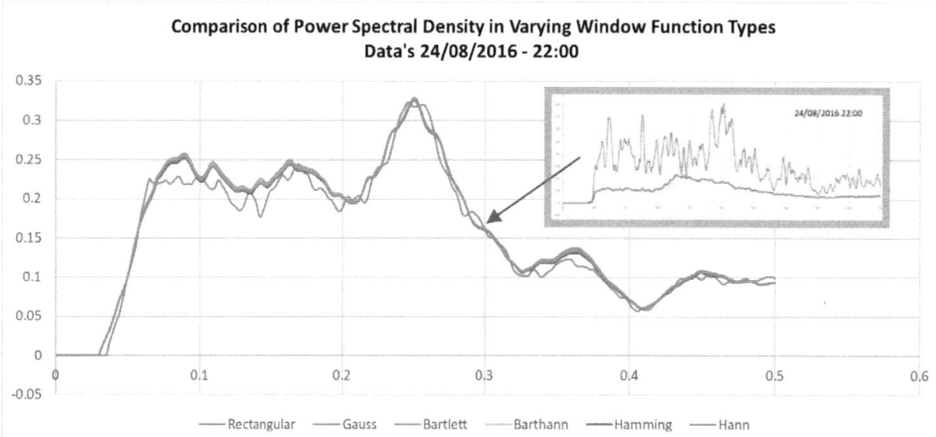

(a)

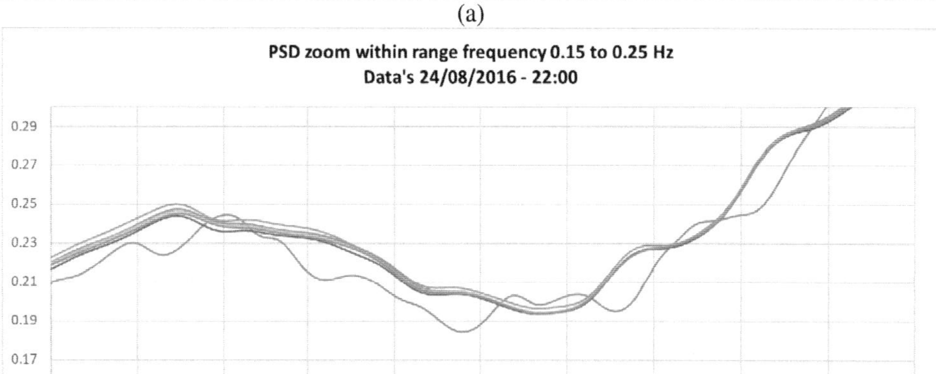

(b)

Figure 12. PSD with window type variation for data 24/08/2016 (24:00); nfft/16

The results of the implementation of the Hann window function on data dated August 23th 2016 and August 24th 2016 are presented in Figure 13. It appears that the PSD curve has become smoother and firmer to serve as a reference in providing estimated power spectrum values. Even though it appears in the data, there is a PSD curve that deviates significantly from the shape of its constituent, as shown in PSD 23/08/2016 (21:00); PSD 23/8/2016 (21:00); 24/08/2016 (21:00); 24/08/2017 (22:00); 24/8/2017 (23:00). On a curve which deviates substantially from the constituent, then it should be considered as an outlier value that does not represent PSD under normal conditions. So, it can be considered to be ignored in the next analysis process.

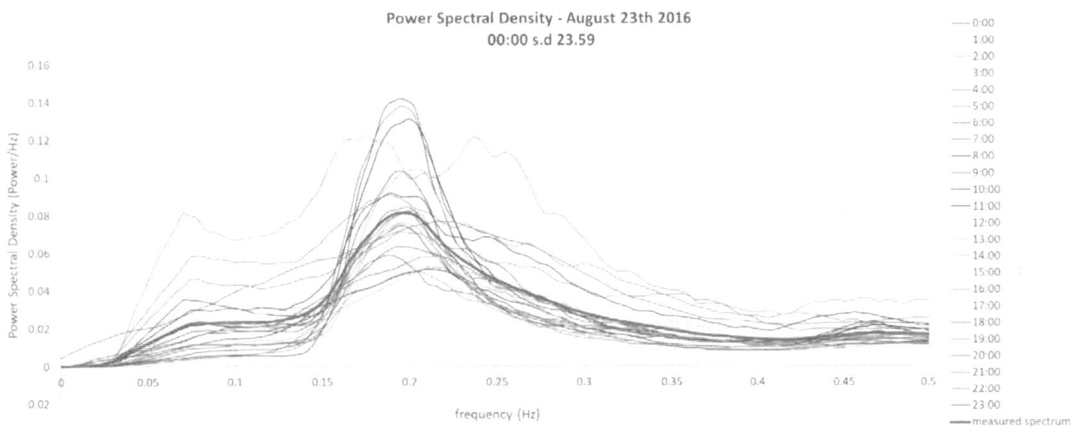

(a)

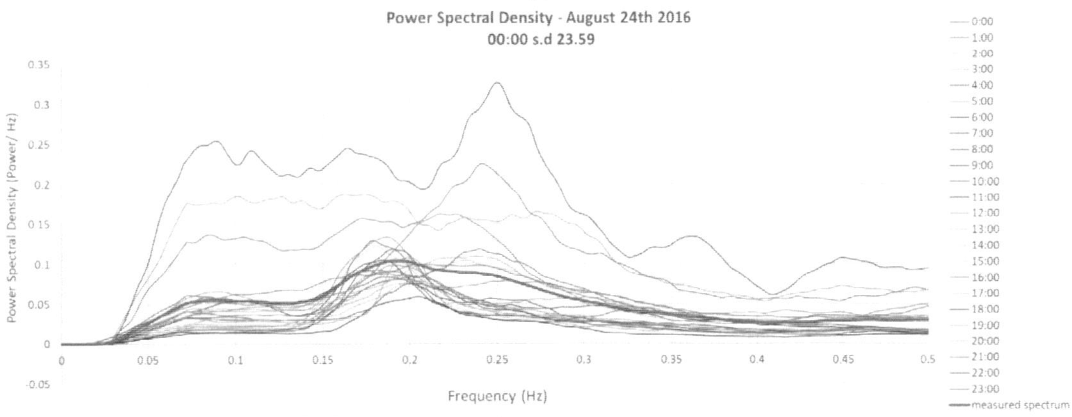

(b)

Figure 13. a) PSD curve using Hann window function, number of window elemets is nfft/16 for data on August 23th 2016; b) for data on August 24th 2016

7. Conclusion

The window function has an important role to play in preventing spectral leaks to other frequency bins. This function is applied as a consequence of wave data in limited measurement. In order to obtain a more stable PSD curve, it is necessary to estimate the value of a smooth and firm PSD Curve so that it does not produce a dynamic value in the frequency space. Nevertheless, the variation in the use of the window type does not result in significant differences between one and the other. The number of elements of each window is quite influential in producing PSD values according to the purposes of the research. In this test, the best PSD result is obtained from the Hann window function type with the number of window elements nfft/16, or the amount of data in fft processing divided by 16.

8. References

[1] L. H. Holthuijsen, *Waves in Oceanic and Coastal Waters*. New York: Cambridge University Press, 2007.

[2] A. Karimpour and Q. Chen, "Wind wave analysis in depth limited water using OCEANLYZ, A MATLAB toolbox," *Comput. Geosci.*, vol. 106, no. September 2017, pp. 181–189, 2017, doi: https://doi.org/10.1016/j.cageo.2017.06.010.

[3] A. Adrianto, E. B. Djatmiko, and Suntoyo, "The 6-Hz wave measurements in Western Java Sea and its preliminary characteristics analysis," *Adv. Environ. Sci. - Int. J. Bioflux Soc. (AES Bioflux)*, vol. 12, no. 1, pp. 66–82, 2020.

[4] C. Li, *Time Series Data Analysis in Oceanography. Applications using MATLAB*. Cambridge University Press, 2022.

[5] D. Adrianto, "Studi Karakteristik Gelombang Laut Jawa Bagian Barat," Institut Teknologi Sepuluh Nopember, 2020. [Online]. Available: https://repository.its.ac.id/id/eprint/82122

[6] "Data Pengukuran Pasang Surut BIG," *Badan Informasi Geopasial*, 2023. http://ina-sealevelmonitoring.big.go.id/ipasut/data/map (accessed Aug. 01, 2023).

[7] S. Panalaran and Suntoyo, "Study of Wave Parameters in the West side of Java Sea using Wave Spectral Analysis and Zero Up-Crossing Methods," *IOP Conf. Ser. Earth Environ. Sci.*, vol. 1198, no. 1, p. 012034, Jun. 2023, doi: 10.1088/1755-1315/1198/1/012034.

[8] P. Welch, "The use of fast Fourier transform for the estimation of power spectra: A method based on time averaging over short, modified periodograms," *IEEE Trans. Audio Electroacoust.*, vol. 15, no. 2, pp. 70–73, Jun. 1967, doi: 10.1109/TAU.1967.1161901.

[9] K. M. M. Prabhu, *Window Functions and Their Applications in Signal Processing*. CRC Press. Taylor & Francis Group, 2014.

[10] A. V. Oppenheim, R. W. Schafer, and J. R. Buck, *Discrete-Time Signal Processing*. Upper Saddle River, New Jersey: Prentice Hall, 1999.

[11] Y. H. Ha and J. A. Pearce, "A new window and comparison to standard windows," *IEEE Trans. Accoustics, Speech, Signal Process.*, vol. 37, no. 2, 1989.

[12] K. Gharaibeh, "Assessment of various window functions in spectral identification of passive intermodulation," *Electron.*, vol. 10, no. 9, 2021, doi: 10.3390/electronics10091034.

[13] S. Chakrabarti, *Hydrodynamics of Offshore Structures*. Springer-Verlag, 1987.

[14] S. Weichert, B. K. Smeltzer, and S. A. Ellingsen, "Biases from spectral leakage in remote sensing of near-surface currents," *IEEE Trans. Geosci. Remote Sens.*, pp. 1–16, 2023, doi: 10.1109/TGRS.2023.3321305.

[15] L. R. LeBlanc, F. H. Middleton, and S. D. Milligan, "Analysis and interpretation of wave spectral data," in *Offshore Technology Conference*, 1975.

AUTHOR INDEX

Institute of Physics
Dirac House, Temple Back
Bristol BS1 6BE UK

ISSN: 1755-1307
ISBN 978-1-7138-9123-9

9 781713 891239

CURRAN ASSOCIATES INC.
proceedings.com